Build Your Own Robot!

Build Your Own Robot!

Karl Lunt

A K Peters
Natick, Massachusetts

Editorial, Sales, and Customer Service Office

A K Peters, Ltd.
63 South Avenue
Natick, MA 01760

Library of Congress Cataloging-in-Publication Data

Lunt, Karl, 1952-
    Build your own robot! / Karl Lunt.
        p. cm.
    Includes bibliographical references and index.
    ISBN 1-56881-102-0 (alk. paper)
    1. Robots--Design and construction. I. Title.

TJ211.L86 1999
629.8'92--dc21   99-045830

Printed in the United States of America
04 03 02 01 00   10 9 8 7 6 5 4 3 2 1

To James M. Coffroth,
for his humor, his wisdom, and for
helping me start down the path.

Table of Contents

Foreword

Seeing the original Star Wars movie as a kid made a big impression on me. Shortly afterwards, I first touched a personal computer. The idea was obvious. Put the microcomputer into an appropriately shaped case, add some wheels and motors, and voila: R2D2.

It was not yet to be. During the next few years, many were bitten by this bug, and several companies started building robots for the home and hobbyist. One of the more famous robots from this era is the Heathkit HERO, which has now almost achieved the status of collector's item.

However, these robots were expensive, hard to program, and were a far cry from everyone's naive Star Wars expectations. Folks pretty much stopped making them in the early- to mid-'80s, and the Dark Age of hobbyist robotics began.

Happily, it was a short-lived hiatus. By the end of the '80s, microcontroller chips such as the Motorola 6811 combined reprogrammable memory and generous digital and analog I/O into a single package. Now a robot brain could be purchased for less than $20. Shrinking the electronics allowed shrinking the mechanics, making them cheaper as well, not to mention easier for someone to build without a machine shop. New software environments made programming these microcontrollers much easier than before.

I first met Karl in 1990 at a Seattle Robotics Society get-together, or "Robothon" as it is known to members. The SRS was one of the few groups to keep the hobbyist robot candle lit through the mid-'80s, and now Karl and the SRS were well-poised to help spread the word about the then emerging technology. The SRS members had amassed an impressive assortment of robots.

I also brought the latest technology I developed for the MIT Lego Robot Contest — a small 6811 board and the beginnings of the programming environment which would be later called Interactive C. The SRS Robothon always has at least one informal robot contest. That year's was a large maze with wooden sides. In a celebration of how easy robots were to build with the new technology, I built a new robot from legos the day of the contest.

I entered the robot's program on a laptop while listening to a lecture before the contest, and downloaded the code to the 'bot for a quick test just prior to its actual run. While this patchwork robot would have lost in a head-on collision with any of the sturdier robots present, it did have the distinction of being the fastest through the maze that day.

In the years that followed, the hobbyist robot community was still fairly diffuse. There were a few places in the country where the density of robotics enthusiasts was high enough to form clubs where ideas could be shared through personal contact. However, these were the exception rather than the rule, so most hobbyists worked in isolation.

This is why people like Karl are essential in order to allow hobbyists to progress as a group to tackle more powerful and compelling robotics projects. Articles such as Karl's have allowed readers to build on experiences and techniques they otherwise might not have known about, and not waste time and effort repeating mistakes.

Today, robotics has become quite a popular hobby. Witness the success of companies such as Mondotronics, and products such as the LEGO RCX robot controller. Making a simple wall-following robot in a day is no longer an unusual feat.

I think that today's success indicates that robotics has largely overcome the earlier barriers of cost, difficulty of assembly, and difficulty of programming, at least for the more simple tasks. The attention must now turn to the remaining ingredient missing from the Heathkit HERO and its brethren. That ingredient is intelligence.

Intelligence has long been the holy grail of robotics, and most would argue we are still far from achieving it. Even so, there are many signs of progress. Today's robot contests have progressed from the maze-solving contests of just a few years ago. Multiple robots now cooperate in teams to compete against other robot teams in sports like soccer.

Perhaps the most exciting new technology to enter the realm of hobbyist robotics is that of machine vision. Today the excitement is not that memory and I/O have all been combined onto a single chip, but that now single chips outperform the supercomputers of ten years ago. This, coupled with an explosion of CMOS and CCD videocameras, means that cheap robots can, for the first time, "see."

As an example, the new Sony AIBO robotic dog uses a camera to see its environment. While the AIBO is by no means "intelligent", it uses vision to sense and interact with the world in interesting ways, such as chasing a ball.

We're obviously not there yet. But we can only progress to the Star Wars-inspired future by not spending our time reinventing the present. So read these pages and use the ideas you find as stepping stones.

Then build upon them.

Randy Sargent
President, Newton Research Labs
October, 1999

Acknowledgments

This book came together over the span of several years. It started life as a series of 70 columns that ran each month for over five years, without interruption, in *Nuts & Volts* magazine. After I stopped the column, I spent a year and a half converting most of those columns into the form you see here. I assure you, those columns and this book could not have happened without the help, enthusiasm, and encouragement of many people. The least I can do, and it isn't nearly enough, is to thank those who have helped me so much over the life of this project.

Tops on my list are Larry Lemieux, editor of *Nuts & Volts*, and Robin Lemieux, the magazine's managing editor. When I first presented Larry with the idea of a few robot columns, I doubt either of us imagined how long our association would last and how much response the columns would generate. Larry and Robin gave me the freedom to write about whatever struck my fancy, turned my electronic scratchings into well-done columns, and always had the check out on time. In the writing game, that combination is hard to come by.

The crunch to bring this book together meant asking the *N&V* staff to wade through their archives and dig up material from two or three years ago that I needed to fill out some articles. Natalie Sigafus, who handles the classifieds for *Nuts & Volts,* showed great patience and dedication in helping me, even

though I know some of my pleas hit in the middle of her production schedule.

Randy Sargent, of Newton Research Labs, was instrumental in getting me off the dime and actually starting this book. It was over an Italian sub sandwich at Robo-Expo '98 that Randy convinced me that this book really was important and really needed to be done.

I have worked with Marvin Green, Dan Mauch, and Kevin Ross over the last several years on many robotic projects. They have proved best of friends, great inspiration, and a constant source of encouragement. Their names appear throughout this book, but I cannot convey how much fun these guys are to work with and how much they have helped me.

Through the years involved in this project, the Seattle Robotics Society membership has waxed and waned, members of long ago replaced by newer ones. But the group has always been supportive, always brought fresh ideas forward, and always shown the creative drive and enthusiasm that helps keep me going. I can only hope that what I've written here gives you even a glimpse of how good this club really is.

Klaus Peters, my publisher, has shown remarkable patience throughout this entire project. Klaus wanted me to do this book years ago, even though I doubted it showed any promise. He persisted, though, and when I finally decided to publish, I naturally asked him if he would handle the project. May his judgement prove as good as his persistence.

Kathryn Maier, my editor at A K Peters, has helped pull this whole project together. She and her staff have been very supportive and understanding of a novice book author. Writing magazine articles is nothing like writing book material, and Kathryn guided me past the rough spots.

And finally, I owe a special "Thank you!" to Ariel Jaffee, my production editor at A K Peters. I may have written the material and drawn the artwork, but that does not a book make. This volume looks better than anything I could have dreamed up, and its realization is largely due to Ariel's hard work and talent.

Introduction

The thought of building a robot fascinates people. Seeing the mechanical fruits of your labor roll, slither, stalk, or lurch across the living room floor has fired the imagination of tinkers of all ages. Whether your ideal machine mows the lawn, explores Mars, fetches beer, or just looks way cool, the feeling is, if you can imagine it, you can build it. Or at least, you can try to build it.

But first-time 'bot builders quickly hit one of many walls, and often call it quits. Unlike other high-tech hobbies, robot-building requires a workable tool set in a wide range of fields. You need mechanical tools for building frames and mounts, electronics gear for wiring circuitry, and software to write the code that makes everything work together. Few people, starting out, have a strong enough tool set in all three areas to pull off a first robot.

Even having a well-stocked workroom and a hurking PC isn't enough, because you also need the skill set to use all of these tools well. A strong frame loaded with top-notch electronics just gathers dust without robust software to drive it. The best robotics program written is worthless unless you can load it onto a working microcontroller with good mechanics surrounding it.

These seemingly insurmountable walls face anyone trying to build their first robot. Some people scale down their ambitions, opting for a simpler, though perhaps less satisfying, first project. Others charge ahead, sometimes creating a masterpiece but more often making a mess. All too many give up, postponing and eventually abandoning the dream of watching their own mechanical creation chase the family cat.

But the walls aren't insurmountable, only tall, and any task can be made simpler if you follow in the footsteps of others. It was to break down these walls, or at least break a trail around them, that I began writing a column on amateur robotics in *Nuts & Volts* magazine, back in October of 1992. Each month, I tried to provide one more foothold for those dreaming the dream. Topics included how to write motor control software, how to wire up a microcontroller, or how to make a super wheel mount. Scattered through the hard-core robotics info was the occasional discussion of famous or fascinating machines built by others, and sometimes I would include full instructions on a complete robotics project. Each column was different and, I hope, useful. I know they were lots of fun to write.

Yet even writing about robots can become wearying, and after nearly 70 columns, I decided to call it quits, to change direction. But the calls from readers asking for a collection of my columns, and for copies of older columns missed or lost, was incessant and, finally, decisive. So I present here a selection of my past *Amateur Robotics* columns.

These are my favorites, written with the beginner and intermediate builders in mind. Those of you who have never seen a microcontroller should be able to pick up a working knowledge without too much effort. If you have already built a couple of large electronics projects, you will find useful information specific to making a robot run. And those readers with a 'bot or two behind them already will find ideas for new robotics projects.

These columns represent tools, built from my experience, to make the hobby of amateur robotics more fun and more rewarding. Most of the tools herein are my own design, the fruits of my own hours. Others are collaborative efforts, the results of projects I completed with fellow robot hackers. Regardless of the source, think of each column as one more tool that you can bring to bear on a large and intricate problem, that of building your own robot.

Some of these columns show their age. Many appeared several years ago and deal with items no longer available. I doubt anyone will be able to find a Ready-Set-Go toy truck nowdays, and I'm sure all of the surplus bargains (and even some of the surplus outlets) have vanished by now. But the techniques I used for modifying or upgrading those items still have value, and you can learn a lot from the approaches I describe.

Other columns describe material that was novel at the time, but has since matured or even disappeared, replaced by newer and better. Yet the columns still contain useful information, and the recent history they provide helps illustrate how quickly this hobby is changing.

I tried to arrange these columns based on subject matter, but often an article covers multiple subjects. Thus, you might find a column that discusses infrared sensor technology and how to write a 68hc11 interrupt handler. To help you sort out what column handles which subjects, I've provided short descriptions in the table of contents. You can also use the index at the back of the book for more help. But I encourage you to view this mixture of subject matter as an inducement to browse, to read through each column repeatedly, sifting it for information and for ideas on your next robotic project.

This hobby is as much about people as it is about hardware. The fun I've had building robots over the years has been multiplied tenfold by the joy of working with the brightest, most capable group of hackers I've ever known. The membership of the Seattle Robotics Society served as springboard, catalyst, cheerleader, critic, and incubator for the ideas you see here, and I owe them all more thanks than I can express.

Finally, my wife, Linda, deserves both praise and apologies for putting up with the long hacking sessions, the too-short deadlines, and the frustrations that come with the hobby. I know she enjoyed the successes, the fun of watching me finish another machine, but she also had to put up with the stress when that machine didn't work, and her patience and support helped make the column and this book possible.

Keep on keeping on...

Karl Lunt
Bothell, WA

Part I
Getting Started

This section serves as your launching pad; it provides the basic information you need for building your first robot. The initial three columns in this section ran sequentially, and took readers through all phases of building a small robopet based on the 68hc11 microcontroller (MCU) and a pair of hobby R/C servo motors. These columns include a shopping list of parts and tools, sources for information, and instructions for modifying the R/C servo motors for use in a robot. The third column describes Huey, my first 68hc11-based robot. Though Huey only has 512 bytes of code space, the little beast's firmware exhibits some very robust behaviors. After you have Huey running around, you can use the supplied program as the start of your own experiments.

The last two columns in this section provide additional information on building techniques, frame materials, power systems, and other elements of hobby robotics design. Because these columns appeared fairly late in the life of the Amateur Robotics column, they use SBasic rather than assembly language for writing robotic firmware. SBasic is an ideal development tool for those starting out in software, and I think you'll find it easier to grasp some of the concepts involved when you can see them written in a friendlier language.

Inspiration and Implementation

This column, published in June of 1993, introduces the world to *Mobile Robots*, by Joseph Jones and Anita Flynn. This book was a breakthrough for the hobby robot builder, as it showcased two finished projects that any advanced experimenter could build. It also provided a strong toolset in all of the major elements of robot building, especially software, thanks to the use of Interactive-C. A second edition of the book came out in early 1998, and rates a "must-have" in any robotics library.

I also used this column to begin exploring the 68hc11 microcontroller (MCU). I stayed with the 68hc11, with minor side excursions, until the end of the column in 1998. There are more powerful MCUs around, but the 68hc11 offers a cheap, one-chip solution to most small robotics problems, and I still enjoy using it today.

This hobby suffers from a lack of concentrated, useful knowledge. Trying to build a working 'bot is difficult enough; having to round up enough information just to get started makes the task nearly impossible.

Happily, that may all have changed. The book *Mobile Robots*, by Joseph Jones and Anita Flynn, explains how to build two working robots. It uses these projects to describe building robots from the first stage of design to a working machine; as the authors put it, from "inspiration to implementation."

Tutebot (for tutorial robot) uses a couple of transistors and some relays, yet can demonstrate fairly sophisticated behavior. Rug Warrior uses a 68hc11 microcontroller with onboard RAM, and can serve as a platform for numerous robotics experiments.

Jones and Flynn are well-versed in designing and building robots. This book springs from work done by them and others in MIT's Mobile Robots group over the last six years. This group, part of the Artifical Intelligence (AI) Laboratory and headed by Rodney Brooks, has produced machines such as Genghis, a six-legged walking robot, and Goliath, then the world's smallest robot at just over one cubic inch in volume.

The book's first chapter introduces you to some of the history of robotics and details the philosophy behind Tutebot's design. It also touchs briefly on subsumption architecture, a method used to define a robot's behaviors; more on subsumption architecture later.

The introduction, and all other chapters, concludes with several paragraphs of references. The cited papers are generally written by MIT graduates and professors, and includes names such as Weiner, Minsky, and Brooks. The citation consists of only the author's name and a year of publication. Presumably, this is enough information to locate the cited document in the MIT library catalog.

Incidentally, you can order copies of most MIT master's theses, reports, or articles directly from the MIT Library. Check their web site of library services. The prices for these publications are reasonable; see the web site for pricing and ordering information. This site also allows you to search a database of MIT publications.

Chapter two of *Mobile Robots* fully describes the design and construction of Tutebot. The authors built this robot from several LEGO parts, including wheels, switches, gears, and motors. The "brain" is made of parts available at Radio Shack, wired up on a Radio Shack plug-in breadboard. The robot runs off four C-cell Alkaline batteries. Tutebot uses only two transistors and some relays for its computer, but you can alter its behavior using two potentiometers. The book describes how these two controls can turn Tutebot into a wall-following robot.

The authors assume that you have little or no experience in building electronics projects. They go into great detail on the functions of each component; they even provide drawings and schematic symbols of the parts. This chapter does an excellent job of introducing the beginner to the electronics needed to make a robot run.

The book's remaining chapters use Rug Warrior's design to describe areas of robotics such as motors, sensors, and computers. Chapter three covers the design of the 68hc11a0 computer used as Rug Warrior's brain. The schematic is simpler than the 80c51 computer used in my previous columns, but the finished hardware is far more powerful, owing to the 68hc11's on-chip resources.

The authors also use this chapter to introduce Interactive C (or IC), a development system written by Fred Martin and Randy Sargent of the MIT Media Lab. IC is copyrighted but free.

(Author's note: You can find IC at the Newton Research Labs' web site; see Appendix C. The current version is for sale, but you can download an (extremely old and unsupported) freeware version.)

Software used in *Mobile Robots* is presented in a mix of 68hc11 assembly language and C. You can download a free macro assembler for the 68hc11 from the Motorola FREEWARE web site; see Appendix C. Look in the /mcu11 directory for the pcbug342.exe file; it contains the asmhc11 assembler and a development package called PCBUG11 that we will use in upcoming *Nuts & Volts* articles.

Chapter three gives a concise description of the 68hc11's inner workings. The authors discuss elements such as the I/O ports, external bus, and analog-to-digital converter (A/D) system. You should also get the Motorola 68hc11 Reference Manual (M68HC11RM/AD) from the Motorola literature center, a Motorola field engineering office, or a local Motorola distributor. This chip is far too complex to adequately describe in just a few pages, though Jones and Flynn give it a good shot.

This chapter also includes some basics in 68hc11 assembly language programming, starting with an introduction to hexadecimal numbers. It covers the different 68hc11 addressing modes and registers, as well as how to use the chip's real-time clock, I/O ports, and interrupts. Their examples are good, but you need to hit the Reference Manual for all the details.

Chapter four describes designing and prototyping of electronics circuits. The authors discuss prototyping methods such as wire-wrap, Scotchflex, and Speedwire. The latter two systems use special tools for connecting pins together using 30-gauge wire-wrap wire. Jones and Flynn use Speedwire pins to fabricate connectors as needed, rather than buying off-the-shelf connectors such as the Molex KK-style parts I have used in my projects. Their method appears cheaper and more flexible; you may see more of this wiring system in upcoming columns.

Chapter five covers many different types of sensors and actuators. Rug Warrior uses bumper switches, a microphone, photoresistors for detecting ambient light levels, a pyroelectric sensor for detecting people, and IR emitter/detectors for finding objects and checking motor shaft rotation. For actuators, this little machine uses an L293 motor driver chip to control two reworked servo motors. It also carries a piezo buzzer for sound effects.

The authors provide excellent information on designing sensor systems and choosing the proper sensor and interface circuitry. Many of the robot's sensors are connected to the 68hc11's A/D channels, so the authors include sample code fragments (in C and assembler) showing how to use the A/D ports. I really appreciate the detail and range of sensors discussed. Jones and Flynn not only cover the Rug Warrior sensors mentioned above, but also IR modulation techniques, force sensors, bend sensors, microphones and amplifiers, piezoelectric film, gyros, compasses, tilt sensors, and sonar!

Chapter six covers methods of locomotion. The authors discuss the tradeoffs of different wheel alignments and describe how to find ready-made drive platforms at the local toy store. They also describe how to use two hobby servo motors to make a jointed leg for a walking robot such as Genghis.

This chapter also provides information on constructing your robot. Topics include wheel mounting and using materials such as sheet metal and acrylic. Surprisingly, the authors do not mention double-sided copperclad circuit board stock, a material I find very easy to machine and use.

The center section of the book includes pages of photographs featuring MIT robots, people, and events. The captions give a glimpse of the variety of possible robots and construction ideas.

Chapter seven is the largest of the book and discusses various types of motors and motor drivers. The authors cover topics such as internals of DC motors, electrical characteristics of motors, speed-torque curves, and how to read a motor data sheet.

The authors next discuss the Royal Titan Maxi servo motor, which they use as the driving motors in Rug Warrior. The authors advocate disconnecting the motor from its servo control printed circuit board (PCB) and using it as a simple DC gearhead motor.

This technique has many advantages. First, even though you are "throwing away" all the electronics, you still get a very compact DC gearhead motor for about $25. Second, the motor mounts easily to just about any frame. Finally, you can find motors and an assortment of attachments at nearly any hobby store.

Note, however, that this is only one way to use a hobby servo in a small robot. Jones and Flynn allude to another technique that leaves the electronics intact, but lets you control the motor as if it was still a servo motor. In upcoming articles, I will describe how to use a servo motor this way. The 68hc11 will directly drive up to four servos in this configuration.

The authors then describe various types of motor drivers, including integrated circuits such as the L293, used in the Rug Warrior. They also discuss using pulse-width modulation (PWM) as a means of motor speed control and give example C code for implementing PWM in the Rug Warrior. Next, they cover feedback control loops as implemented with Rug Warrior's shaft encoders. Again using C code, they show how to control both wheels of Rug Warrior so that the machine will compensate for rough terrain or uneven running surfaces.

Chapter eight provides an excellent tutorial on batteries and power systems. The discussion of energy density compares different types of batteries, such as NiCds and Lithium cells, to other power sources; the inclusion of uranium fission on the chart was an interesting touch.

The authors also cover recharging techniques, types of power regulators, DC-to-DC converters, and how to suppress electrical noise in the robot's power lines. This last topic is very important, as motors can introduce considerable noise into the system's wiring; the resulting glitches can look like software bugs or electrical errors, and can be very difficult to find.

Chapter nine covers robot programming. It first presents the traditional approach, in which the software starts with the sensors' input, fits it to the world as the robot understands it, uses the result to develop a planned motion or sequence, and executes that sequence. The authors touch on several drawbacks to this method of programming a robot, then propose an alternative.

Subsumption architecture views sensors as initiators of actions. An action, once triggered, may take priority over actions already in motion, effectively suppressing or subsuming those actions. For example, imagine a robot with the main action of "go forward." The robot will perform this function until its batteries run down or until another, higher-priority action is invoked.

A typical such action could be called "avoid," triggered by a closure of a bumper switch. When a bumper switch is closed, "avoid" subsumes "go forward" and becomes the robot's behavior. Eventually, the bumper switch is no longer closed, the "avoid" action is completed, and the robot returns to the "go forward" action.

This is a simple example of subsumption architecture, and could easily be implemented in either the traditional or subsumption approach. But if you build upon this set of behaviors, by adding other actions and other sensors, you quickly reach a level of complexity that strains the abilities of the traditional approach.

Jones and Flynn give some excellent examples of subsumption architectures, including the detailed design of Rug Warrior's behaviors. They conclude with several software examples, written in C, of subsumption architecture models. The examples are well designed and worth repeated study.

Chapter ten describes many unsolved problems in robotics design, essentially presenting the reader with ideas for upcoming projects. Topics include navigation, pattern and object recognition, learning, and cooperation among groups of robots.

The appendices include the full schematics for Rug Warrior, sources for parts of all kinds, trade magazines of interest, catalogs and technical references, the full source code (in C) for Rug Warrior's program, the resistor color code, and the ASCII chart.

The index is quite sparse and items that get considerable coverage in the text are oddly missing. For example, there are no entries for battery, NiCd, wire-wrap, 68hc11 or potentiometer. However, I was using an unedited review copy for this article; the finished book may have a more extensive index.

I would like to see a pinout diagram of the 68hc11's 52-pin PLCC socket (bottom view). Jones and Flynn assume you will be wiring a Rug Warrior using a point-to-point system of some kind. This wiring is a royal pain without a pinout diagram of the socket, and such diagrams are very difficult to find. In fact, I ended up making one of my own, and will publish it in the next *Nuts & Volts* column, entitled "Your First 68hc11 Microcontroller."

I was also disappointed that the authors did not provide the same level of detail in the wiring phase of Rug Warrior that they gave in the Tutebot section. Specifically, I think they should have provided numbered pinouts for all ICs used in this project, ideally in Appendix A.

You can deduce a large chip's pinout from their schematic diagrams, but they neglect to give pin numbers for smaller chips such as the LM386 and 74hc10, as well as modules such as the 8054 low-voltage inhibit and the shaft encoders. This lack will make construction by an experienced builder more frustrating than necessary, and will seriously hamper a beginner.

Despite the above shortcomings, Jones and Flynn deserve a great deal of support and appreciation for producing this book. They have brought together many of the topics in robotics design and explained them clearly, with excellent examples and good insights. Their writing is well crafted and easy to follow, and the code samples are well explained and commented. This book could easily serve as the core for an excellent class in robotics. In fact, the book was "beta-tested" by Bruce Seiger and his students at Wellesley High School.

[Author's update: The Second Edition of *Mobile Robots* is now available.] The book lists for $32

from the publisher, but Klaus Peters tells me he will give customers a 10% discount if the order includes the name of your robotics club. You can order a copy of *Mobile Robots* from A K Peters, Ltd.

Oddly enough, the Seattle Robotics Society has been moving steadily towards use of the 68hc11 in the last few months, and *Mobile Robots* provided the final impetus. I have already finished one small robot using the 68hc11 microcontroller (MCU); I will describe this robot next month, ("Your First 68hc11 Microcontroller.")

The 68hc11 offers a large array of features and on-chip resources. For example, it provides eight channels of 8-bit A/D; no need to add a MC145041 as I used with the 80c51 computer. It also sports 256 bytes of static RAM and 512 bytes of EEPROM in the 68hc11a1 variation, the chip my first project will use. Motorola makes the 'hc11 in a bewildering number of variations; other variants offer differing amounts of RAM, EPROM, EEPROM, A/D channels, and package types. Contact your Motorola distributor or field engineering office for details.

Besides the RAM and EEPROM, the 'hc11a1 also provides four PWM channels that run with no software overhead, an asynchronous serial port (the SCI), a synchronous serial port (the SPI), and at least 24 I/O pins, most of which are bidirectional. I say "at least" because the 68hc11 can be used in either single-chip mode or expanded mode. Single-chip mode (used in next month's project) means just that; the system uses only the RAM and EEPROM available on the 68hc11. In expanded mode, many of the I/O lines are used up in providing a 16-bit address/8-bit data bus for adding off-chip RAM and EPROM. Other plusses for this chip include its CMOS technology (low power consumption and high noise immunity), good interrupt support, and powerful programming model.

The programming model refers to the software resources available. This chip contains two 8-bit accumulators, two 16-bit index registers, and a well-designed set of instructions. You can write very compact assembly language code for this chip, getting a lot of performance from its tiny 512-byte memory. In fact, Marvin Green of the Portland Area Robotics

Society has built a line-following robot with this chip; the software only uses 85 bytes of EEPROM!

I will use point-to-point wiring for the first few projects with this chip; there is not yet a PCB available that does exactly what I want. Fortunately, you can wire up a working 68hc11 robot brain in just a few evenings.

You will need the following parts:

1	68hc11a1fn MCU
1	MAX232 RS-232 level-shifter IC
1	52-pin PLCC solder-tail socket
1	16-pin DIP solder-tail socket
1	8.0 MHz crystal
1	5.1M ohm, 1/4-watt resistor
4	4.7K ohm, 1/4-watt resistors
2	22 pf ceramic capacitors
2	10K ohm, 8-pin resistor SIPs or 14 10K ohm, 1/4-watt resistors
4	10 µf, 35 WVDC electrolytic capacitors, radial leads (Radio Shack 272-1025)
2	0.1 µf ceramic capacitors (Radio Shack 272-109)
1	2-pin 0.1-inch KK straight friction-lock header (Molex 22-23-2021)
4	3-pin 0.1-inch KK straight friction-lock headers (Molex 22-23-2031)
5	4-pin 0.1-inch KK straight friction-lock headers (Molex 22-23-2041)
1	2-pin 0.1-inch KK terminal housing (Molex 22-01-3027)
5	4-pin 0.1-inch KK terminal housings (Molex 22-01-3047)
	crimp terminals for KK terminal housing (Molex 08-50-0114)
1	6-pin 0.1-inch jumper block, arranged in 2x3 pins
2	0.1-inch shorting blocks or jumpers (Radio Shack 276-1512)
1	DB-25 solder-cup female connector (Radio Shack 276-1548)
1	DB-25 plastic connector hood (Radio Shack 276-1549)
1	experimenter's project board (Radio Shack 276-149)
1	4-cell AA battery holder (Radio Shack 270-391)

4   AA NiCd batteries
2   Futaba FP-S148 hobby servo motors
     30 gauge wirewrap wire

The protoboard called out above measures about 1.5 by 2.5 inches; this will be a small robot board. Note that the circuit uses no motor drivers of any kind. The entire robot electronics fits on this single board!

You will also need a copy of the *PCBUG11 User Manual*, mentioned in the text above. The quickest way to get a copy is to download the pcbug11.pdf file from the Motorola FREEWARE web site, then print it out using an Acrobat reader. Note that this file is about 450K bytes, so it will take a while to download. But it makes a great reference and is worth the effort.

You may have difficulty getting some of these parts. The 68hc11a1fn, in particular, may prove hard to find. B. G. Micro has been selling these chips for $5.95 each (a great price!), but their supplies have been sporadic. Best to place your order by phone so you can verify stock. Also ask B. G. Micros about prices on 52-pin PLCC sockets; they have had excellent deals recently. See Appendix A for contact information.

You can get the KK-style headers and receptacles from Active Electronics, DigiKey, or perhaps at some well-stocked local parts houses. I have built these connectors into all of my robots so far, and find them easy to use and reliable. Getting them is worth a little extra scrounging. See Appendix A for contact information.

Your First 68hc11 Microcontroller

This article, which appeared in *Nuts & Volts* in July of 1993, describes the first 68hc11 computer board built for the column. I built it using point-to-point wiring on an experimenter board; total construction time was just a few evenings. Even though there are now boatloads of wired and tested 68hc11 computer boards available, I still think building one or two on your own is the best way to get started in this hobby. You probably won't save any money over buying the board, but you will learn tons about what makes a microcontroller work, and even more about what makes them not work!

Previously I described a small 68hc11-based computer board that will serve as the core of future projects. The circuit is quite simple to build, and should go together in just a few evenings. You can find the parts list for this project in last month's column. Refer to the accompanying 68hc11 schematic (Figure 1).

Start by cutting several 2-inch pieces of 30-gauge wirewrap wire and stripping 1/4-inch of insulation from each end. You will use these wires in making point-to-point connections on the circuit board. I find that having a small bag of these wires already prepped makes the actual construction less tedious. You can also buy tubes or bags of wirewrap wire already precut and stripped; Active and Digi-key are just two suppliers to contact. These wires will be prestripped for wirewrap, so you will need to cut the exposed ends to length before using them for point-to-point wiring.

You will build the 68hc11 computer circuitry on a Radio Shack experimenter's board (276-149). The following paragraphs refer to holes in this board using the coordinates silk-screened on the board's component side.

Position the 8.0 MHz crystal so that its leads fit into holes B3 and D3. Similarly, place a 22 pf capacitor at B1 and C1, and another at D1 and E1. Finally, put the 5.1M ohm resistor at B2 and E2. Use the

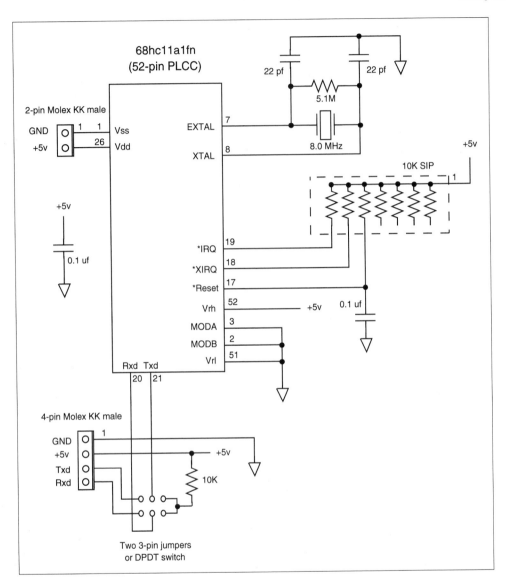

Figure 1. Basic schematic for the 68hc11 robot computer.

leads of each component to wire these parts together per the schematic. Leave the crystal's leads untrimmed; you will later connect these leads directly to the PLCC socket. You should not need to use any wirewrap wire for these connections. Work carefully and use good soldering technique.

Next, position the 52-pin PLCC socket so that pin 1 fits into hole G5. This should leave the socket placed firmly against the crystal, with only the A-row of pins exposed on the left side of the board. Carefully solder the crystal's leads to pins 7 and 8 on the socket and trim the wires. Make sure you don't

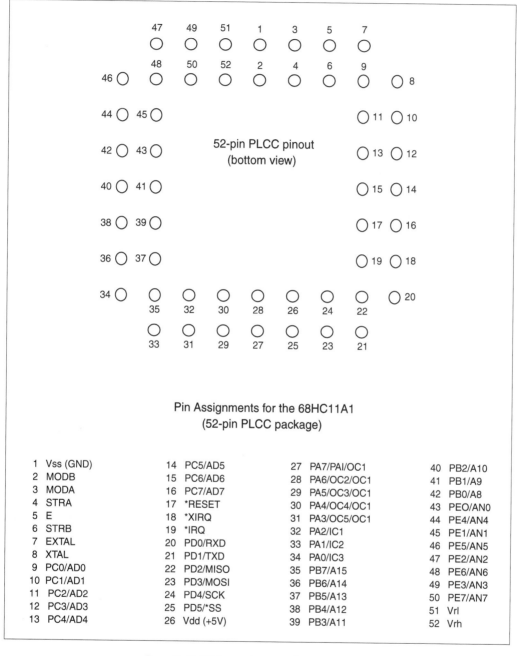

52-pin PLCC pinout
(bottom view)

Pin Assignments for the 68HC11A1
(52-pin PLCC package)

1	Vss (GND)	14	PC5/AD5	27	PA7/PAI/OC1	40	PB2/A10
2	MODB	15	PC6/AD6	28	PA6/OC2/OC1	41	PB1/A9
3	MODA	16	PC7/AD7	29	PA5/OC3/OC1	42	PB0/A8
4	STRA	17	*RESET	30	PA4/OC4/OC1	43	PEO/AN0
5	E	18	*XIRQ	31	PA3/OC5/OC1	44	PE4/AN4
6	STRB	19	*IRQ	32	PA2/IC1	45	PE1/AN1
7	EXTAL	20	PD0/RXD	33	PA1/IC2	46	PE5/AN5
8	XTAL	21	PD1/TXD	34	PA0/IC3	47	PE2/AN2
9	PC0/AD0	22	PD2/MISO	35	PB7/A15	48	PE6/AN6
10	PC1/AD1	23	PD3/MOSI	36	PB6/A14	49	PE3/AN3
11	PC2/AD2	24	PD4/SCK	37	PB5/A13	50	PE7/AN7
12	PC3/AD3	25	PD5/*SS	38	PB4/A12	51	Vrl
13	PC4/AD4	26	Vdd (+5V)	39	PB3/A11	52	Vrh

Figure 2. PLCC layout diagram and pin assignments.

accidently connect either crystal lead to any other pin on the PLCC socket. Refer to the accompanying PLCC socket layout diagram for pin locations (Figure 2). Note that this diagram shows the PLCC socket as viewed from the BOTTOM.

Install a 10K ohm SIP resistor pack by placing pin 1, marked with a painted dot, in hole A6. Wire two SIP resistors to pins 18 and 19 on the 68hc11 socket. You may use any pins of the SIP you like except pins 1 and 8. Pin 1 must be reserved; you will later wire it to +5V as a pullup.

Install a 0.1 μf capacitor in holes A14 and B14. Connect the capacitor lead at A14 to pin 8 of the SIP resistor pack. Wire this junction to pin 17 of the 68hc11 socket. The remaining capacitor lead will later be wired to ground.

Install a 4-pin Molex KK straight friction-lock header (male connector, Molex part number 22-23-2041) with pin 1 in hole O2 and pin 4 in hole L2. As with all my KK connectors, pin 1 is on the left as you view the connector from the front; the locking tab will be behind the pins in this orientation. This KK connector will carry the Rxd and Txd serial signals from the 68hc11.

Install a 6-pin jumper block (arranged as 2x3 pins) with the center pins of each row in holes H1 and H2. Note that you may replace this component with a miniature DPDT switch, if available. Wire the jumper block and KK connector per the schematic.

Install a 2-pin KK header (Molex part number 22-23-2021) with pin 1 in hole O6 and pin 2 in hole O5. Pin 1 will be ground and pin 2 will be +5V. Wire these two pins to all remaining ground and +5V pins. Remember to wire pins 2, 3, and 51 of the 68hc11 socket to ground. Also be very careful when wiring pin 52 of the 68hc11 socket. This wire connects to +5V; make sure you don't accidently short it to any of the grounded pins nearby.

This completes the first phase of construction. Use a digital voltmeter (DVM) to verify that the two pins of the power connector are not shorted together. Also verify the ground and +5V wiring by checking for continuity between the proper power connector pin and all other pins in that network.

After you have checked all connections and corrected any problems, you can install the 68hc11a1 MCU in its socket. To do this, first pick up the computer board and hold it so your skin is directly in contact with the ground pin on the power connector.

Pick up the 68hc11 MCU chip with your other hand and carefully orient it so the tiny dot on the chip's package aligns with the pin 1 marking on the PLCC socket's frame. Place the 68hc11 chip so that it rests flat on the socket's opening, sitting on the tiny silver leaves of the socket. Very carefully press the chip down into the socket. The chip must sit flat in the socket; if it is crooked or angled in any way, carefully remove it and retry.

The next phase of construction involves the MAX232 chip and the RS-232 connector called out in "Inspiration and Implementations's" parts list. This task requires you to build the MAX232 circuitry inside the plastic shell for the RS-232 connector. Use a solder-tail socket, insulated wire, and short leads on all components. Refer to the accompanying schematic (Figure 3).

Bring the four signals (GND, +5V, Rxd, and Txd) out from the RS-232 shell to the appropriate pins of a 4-pin KK terminal housing (female connector, Molex part number 22-01-3047). This connector will later plug into the matching 4-pin KK header described above. Be sure to verify signal wiring on both of these connectors as you wire the 4-pin receptacle.

If you feel that you just cannot build this circuitry into such a small space, you may choose to wire this section on a small piece of experimenter board, such as half of Radio Shack's 276-148. You will then have to find a suitable enclosure and cables. Regardless of your construction method, make sure you keep the leads between the computer board and the MAX232 circuitry no longer than 12 inches.

Note that you can replace the MAX232 and its four capacitors with a single MAX233 chip, available from some of the larger mail-order houses. This chip takes a different socket than the MAX232, and the pinout is obviously different. Consult the Maxim book for details on wiring in this chip.

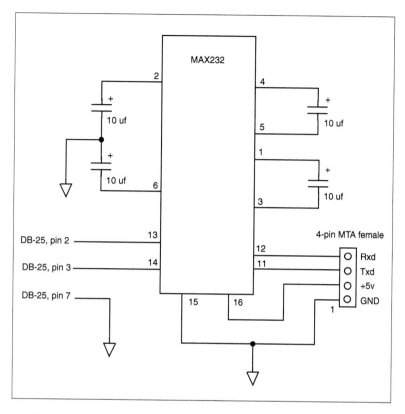

Figure 3. Schematic for RS-232 level-shifter. Build this circuit into a plastic DB-25 connector shell.

Testing Your Board

This computer board uses four AA-size NiCd batteries as its power supply. The Motorola spec claims a maximum operating voltage of 7.0 VDC for this chip, so you could conceivably run it from four alkaline cells. However, I will stick with the NiCds in my projects. Using the proper crimp-on terminal pins, connect the battery holder's black wire to pin one of a 2-pin KK terminal housing (Molex part number 22-01-3027). Similarly, connect the battery holder's red wire to pin two of this connector. Install four AA-size NiCd batteries in the battery holder; make sure you orient the batteries properly.

Next, configure the 68hc11 computer board for bootstrap mode. You do this by setting the two jump-

ers so that the Rxd and Txd pins are tied to the 4-pin KK connector, rather than to the pullup resistor; refer to the 68hc11 schematic (Figure 1). Note that both jumpers must be moved to the same side; the 68hc11 will not respond otherwise.

Plug the 4-pin serial connector from the MAX232 circuitry onto the computer board's 4-pin serial connector. Plug the DB-25 connector of the MAX232 circuitry onto the proper serial cable to your PC. Plug the battery's 2-pin connector onto the computer board's 2-pin power connector.

Install the pcbug11 software in a suitable directory on your PC. Start the program by entering the command:

pcgub11

The program will ask you a series of questions to determine your 68hc11 configuration and which COM port you are using. After completing the questions, jot down the command line that pcbug11 displays; you can then type the command directly the next time you want to run pcbug11. For example, my pcbug11 command line is:

pcbug11 -a port=2

because I am using a 68hc11a1 and my computer board is hooked to COM2.

pcbug11 will attempt to connect to the 68hc11 computer board and download a small program, called a "talker." If pcbug11 cannot complete this connection, it will provide an error message. In this case, you must troubleshoot your RS-232 connections.

This problem could be caused by a short or miswire in the MAX232 circuitry. It might also be caused by a crosswired RS-232 serial cable; if necessary, use a null-modem cable that swaps pins 2 and 3 of the RS-232 cable.

You can also try adding a small modification to the wiring on your DB-25 connector at the PC's end. Using a short length of bare wire, carefully connect pins 4, 5, and 6 together inside the connector's shell. This change effectively tells the PC that the equipment hooked to the serial port is always ready to receive data.

Another source of problems could be a miswired 68hc11 computer board. If you have an oscilloscope or logic probe handy, use it to verify that a 2 MHz signal appears at pin 5 of the 68hc11 socket. If this signal is missing, double-check your battery connections, crystal wiring, and 68hc11 installation.

After pcbug11 has established communications with the 68hc11 computer, you can begin exploring your new computer's resources. Refer to the *PCBUG11 User Manual* (Motorola M68PCBUG11/D2) for details; you can obtain a free copy of this manual by following the directions given in "Inspiration and Implementation."

PCBUG11 contains a terse but adequate set of help instructions. You can get help at any time by typing HELP at the command prompt and pressing Enter.

Exploring the 68hc11

Start by setting hexadecimal as pcbug11's default number base:

control base hex

Since you will use hex numbers almost exclusively, this command removes the need to add a '$' in front of all hexadecimal numbers (Motorola uses a '$' prefix to indicate hex numbers).

The 68hc11 contains 256 bytes of RAM, starting at address $0000 and running to $00ff. You can look at the contents of this RAM by entering:

md 0 ff

pcbug11 will display the requested addresses in hex.

You could try changing any of these addresses with the **mm** command; for example:

mm 10 a5

would write the byte $a5 into address $0010. However, pcbug11 uses quite a bit of the 68hc11's RAM, and writing values into that RAM will probably crash pcbug11. If that happens, simply turn off the 68hc11's power, reconnect the power, then enter the pcbug11 command:

restart

pcbug11 should then be able to reconnect to the 68hc11 computer.

The 68hc11 also contains a number of I/O registers, at addresses $1000 to $103f. These registers behave very much like memory addresses; you can read from them just as if they were memory. Some registers also control output lines or internal chip resources; you can alter these registers by simply writing to them as if they were memory addresses.

For example, the 8-bit bidirectional I/O port PORTC appears at address $1003; the eight matching I/O lines appear on the 68hc11 as pins 9 through 16.

The direction of each I/O line for PORTC is controlled by a corresponding bit in address $1007, which is PORTC's data direction register (DDRC). If a particular bit in DDRC is set to 0, the matching bit in PORTC is configured as an input line. If a bit in DDRC is set to 1, the matching PORTC bit is configured as an output line. The power-on default for the bits in DDRC is 0; this initializes all PORTC pins to be inputs.

You can change the configuration of any PORTC I/O line at any time by simply setting the proper bit in DDRC to the necessary state. To experiment with the PORTC I/O lines, connect a DVM between ground and pin 9 of the 68hc11's socket. You should read a very low voltage.

Now, use the command:

ms 1007 1

to configure bit 0 of PORTC as an output line. The DVM will likely show a value very close to 0 VDC. Now, anything you write to bit 0 of PORTC will appear on pin 9. If you enter:

ms 1003 1

pin 9 will output 5 VDC. If you enter:

ms 1003 0

pin 9 will read 0 VDC. You can control other bits of PORTC with similar changes to DDRC and PORTC.

68hc11 Assembly Language

The simplest way to start writing software for the 68hc11 is by using pcbug11's built-in assembler. You activate the assembler with a command such as:

asm <addr>

where **<addr>** is an address in the 68hc11's memory space where you want your assembly language program to start.

Since pcbug11 is using nearly all the 68hc11's available RAM, you have to write your assembly language programs into the 68hc11's EEPROM at addresses $b600 to $b7ff. Before you do that, however, you have to tell pcbug11 where the EEPROM is located. You do this with the command:

eeprom b600 b7ff

Now, when pcbug11 tries to modify addresses in this range, it will treat the memory locations as EEPROM and the data will be written properly.

For our first example, we will write a simple program to add two numbers. Start by invoking the assembler, beginning at address $b600:

asm b600

pcbug11 will respond by displaying the instruction currently at location $b600, which will likely be STX $FFFF. This doesn't mean someone assembled a STX $FFFF into your chip's EEPROM. This happens to be the instruction corresponding to $FF $FF $FF, which in turn are erased values in a block of EEPROM.

Now enter the assembly language line:

<sp>ldaa #55

(The **<sp>** means to enter a space before the ldaa; otherwise, pcbug11 will think you want to declare a label called "ldaa.")

Be sure you start each line you enter with CTRL-END, to erase the text initially displayed by pcbug11. If not, pcbug11 will tack the remainder of the original text onto whatever you enter and the resulting text will probably be an error.

If you entered the line properly, pcbug11 will display your entry as:

B600 86 55

which, in 68hc11 machine language, means "load accumulator A with the value $55." Note that since the default number base is hexadecimal, PCBUG11 assumed hex instead of a decimal 55.

When later executed, this instruction will put the value $55 in accumulator A, one of the two 68hc11 8-bit accumulators. Next, enter:

<sp>adda #2d
<sp>rts
<ESC>

This sequence completes the little program by adding the value $2d to the value already in accumulator A, then using the RTS instruction to return to the PCBUG11 monitor. The **ESC** key ends the assembly session and returns to PCBUG11's command prompt.

To execute the program, enter the commands:

call b600
rd

The **CALL** command starts the program by executing it as if it were a subroutine. The **RD** command forces pcbug11 to display the contents of its pseudo-registers, so you can see the changes your program caused in the 68hc11's registers. You should see the number $82 in register A.

This is only the most basic beginning in exploring the 68hc11 MCU. To get a better grasp on what this chip offers, consult the *Mobile Robots* book, described in "Inspiration and Implementation."

Another good source of 68hc11 material is *Single- and Multiple-Chip Microcomputer Interfacing*, by G. J. Lipovski (Prentice-Hall, Englewood Cliffs, NJ, ISBN 0-13-810557-X). I was able to get this book from the Motorola literature center; contact Motorola or your local Motorola Field Application office for details.

Be sure to get copies of the pcbug11 manual and the *68hc11 Reference Manual* (M68HC11RM/AD). Additionally, I have found the *68hc11 Programming Reference Guide* (MC68HC11A8RG/AD) to be very helpful. It is a shirt-pocket sized collec-

tion of information on register assignments, assembly language op-codes, and other important details.

Using Servos

My previous robots all used DC gearhead motors for moving about. These motors did the job; they moved the 'bot. But they were very large, heavy, and power-hungry. Since the 68hc11 computer board is tiny, light, and uses only about 10 mA of current, I decided to look for another type of motor.

Marvin Green of the Portland Area Robotics Society (PARTS) uses modified R/C hobby servo motors on most of his robots. I have built two robots using his techniques, and pass his ideas along to you.

In its unmodified form, an R/C hobby servo motor (or servo) will position a wheel or adapter at a selected point within an arc of about 90 degrees. The exact position of the wheel depends on the width of a positioning pulse that you (or your computer) supply.

A servo has three wires coming out of its case. On the Futaba FP-S148 servo, these wires are +5V (red), GND (black), and CONTROL (white). These wires terminate in a small, black connector that fits perfectly onto a 3-pin KK-style male connector.

The +5V and GND lines carry +5 VDC and ground, respectively. The CONTROL line carries a short pulse that the servo uses to position its output wheel. This pulse is usually anywhere from 0.7 to 1.7 msecs in duration. The shortest pulse causes the servo to rotate its wheel fully in one direction; the longest pulse will rotate the wheel fully in the opposite direction. A pulse of the proper duration (about 1.2 msecs in this example) will rotate the wheel until it is centered in its arc.

As mentioned above, the servo only rotates through about 90 degrees of arc. How can we use such a servo to drive a robot's wheel, which must rotate continuously in either direction as needed?

The servo's arc is defined by a mechanical stop or spur, molded into one of the gears inside the servo. If the spur is cut away or trimmed so it no longer functions, the servo's output shaft will rotate in a complete circle.

The servo also uses a small internal potentiometer, linked to its output shaft, to tell when the output shaft has reached the desired position. By   replacing this potentiometer with a suitable pair   of resistors, we can fool the servo into thinking it has never reached the target position. Thus, it will continue to rotate in the selected direction until told to stop.

You can easily modify a Futaba FP-S148 servo motor for robotics use. I chose this motor for several reasons. First, the modification is fairly simple; some motors (notably the low-end Hobbico servo) are too much trouble to modify. Second, this motor is cheap (about $15 mail order from Tower Hobbies, slightly more at your local hobby store) and easy to find. Finally, it packs about 42 oz-in of torque while only drawing about 8 mA of current. This in turn lets me build a tiny, working robot that only uses about 30 mA (total!) from a set of AA-size batteries.

Modify two Futaba FP-S148 servos; see Appendix B for instructions. Next month, we will complete the computer board's wiring and begin moving motors ("Allow Me to Introduce Huey.") We will also start on our first project using this new computer: Huey.

Allow Me to Introduce Huey

You never forget your first robot and Huey, described in this column from August, 1993, was my first 68hc11 robot. Huey, a type of robot called a robopet, still occupies a place of honor in my closet. Robots such as Huey make a great first project, since they are small, cheap to build, and can exhibit fairly complex behavior using simple software.

The software discussed in this column marked my first attempt to build templates and tools that other people, less skilled in program design, could use as a starting point. Studying the files mentioned in this article will give you a good foundation for 68hc11 assembly language and general program construction.

Last month, ("Your First 68hc11 Microcontroller"), I described a one-chip 68hc11 microcontroller board that will serve as the core for several upcoming projects. A few additional components and some software will turn this little board into a working Robopet.

We will now finish the 68hc11 computer board started in last month's article. Hole coordinates called out in the following paragraphs refer to the grid silk-screened onto the Radio Shack experimenter's board used as our construction base.

Install a 3-pin Molex KK straight friction-lock header (male connector, Molex part number 22-23-2031) on the 68hc11 board. Place pin 1 of this connector at M15 and pin 2 at N15. The locking tab at the back of this connector should lie towards the top (row 1) of the project board.

Install a second 3-pin KK male header on the 68hc11 board at locations M18 and N18. Orient this connector as described above. Wire pin 1 of both connectors to ground. Remember that pin 1 is on the left end of the connector, when viewed from the front (looking toward the pins, not the locking tab). Wire pin 2 of both connectors to +5 VDC.

Connect pin 3 of the KK header at M15 to pin 28 (OC2) on the 68hc11 socket. Connect pin 3 of the KK header at M18 to pin 29 (OC3) on the 68hc11 socket. Refer to the accompanying wiring diagram for connecting modified R/C servos (Figure 1). Also refer to

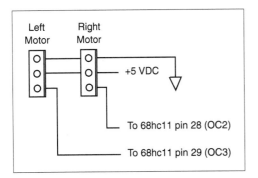

Figure 1. Wiring diagram for connecting modified R/C servos to the 68hc11 computer board.

the 52-pin PLCC pinout diagram in "Your First 68hc11 Microcontroller." Remember that this diagram shows the 68hc11's socket as viewed from the BOTTOM.

These two connectors will supply the signals for driving two hobby R/C servo motors, modified as described in Appendix B. The servos have 3-pin connectors that will plug directly onto the KK headers. Note, however, that the servo connectors do not have any keying built into them, so you could hook them up backwards. Remember that the ground (black) lead is always connected to pin 1, which is on the left end of the KK header in both cases.

Now install four 4-pin KK male headers (Molex part number 22-23-2041) on the 68hc11 computer board. Pin 1 of the first connector should lie at H16; pin 1 of the remaining connectors should lie at H19, H22, and H25. Pin 4 of these headers should lie at K16, K19, K22, and K25. As with the 3-pin connectors installed above, orient the locking tabs toward the top of the project board.

Wire pin 1 of these four connectors to ground. Wire pin 2 of all four connectors to +5 VDC. Finally, wire pin 4 of the first connector (at K16) to pin 43 (PE0) of the 68hc11's socket. Pin 4 of the remaining KK headers should tie to pins 45 (PE1), 47 (PE2), and 49 (PE3). Pay close attention to the layout diagram; it is easy to miswire these connections.

Install a 10K ohm SIP resistor pack with pin 1 at D14 and pin 8 at K14. Wire pin 1 to +5 VDC. Connect pin 43 (PE0) of the 68hc11's socket to any of the open leads on the SIP resistor pack. This acts as a pullup for the signal from the KK connector. Wire pins 45 (PE1), 47 (PE2), and 49 (PE3) to other open leads on the SIP resistor pack. Refer to the accompanying wiring diagram for connecting the A/D channels (Figure 2).

These four connectors serve as A/D inputs for the 68hc11. You can hook nearly any voltage source

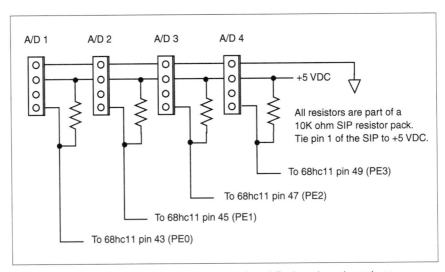

Figure 2. Wiring diagram for connecting the four A/D channels to the 68hc11.

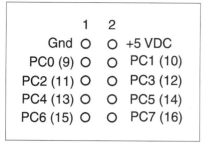

	1	2	
Gnd	O	O	+5 VDC
PC0 (9)	O	O	PC1 (10)
PC2 (11)	O	O	PC3 (12)
PC4 (13)	O	O	PC5 (14)
PC6 (15)	O	O	PC7 (16)

Figure 3. Pinout diagram of the 10-pin IDC connector used for the 68hc11's 8-bit I/O port. (Pin numbers for the 68hc11's socket appear in parentheses.)

or resistor divider network to them, provided the signal (applied to pin 4 of the connector) doesn't go above +5 VDC or below ground.

The last item we will install on this board is a 10-pin IDC male connector. This connector is arranged as two rows of five pins each; position pin 1 at A18 and pin 2 at A17. Hook pin 1 of this connector to ground and pin 2 to +5 VDC. The remaining pins are wired in order to pins 9 (PC0) through 16 (PC7) on the 68hc11's socket. Refer to the accompanying pinout diagram (Figure 3).

This connector provides eight lines of bi-directional I/O, along with +5 VDC and ground for powering external circuitry. You can wire lines on this connector to digital-type signals, such as switches or Hall-effect sensors. You can also directly drive actuators such as piezo beepers or low-current LEDs.

This completes the wiring for the 68hc11 computer board. As you can see, it is quite a compact board. It also draws very little current (total of about 10 mA). And this board provides plenty of I/O for your projects. It contains four A/D channels, at least 16 I/O lines (exact number is under software control), and two external interrupt lines.

The First Project

We will begin with a simple Robopet, similar to the machine described in a couple of previous columns. This machine will contain a bumper switch for de-

tecting contact with obstacles, as well as an infrared (IR) system for non-contact sensing of obstacles.

The accompanying photographs and sketches give some ideas for designing a working platform. I started with a piece of thin, double-sided blank printed circuit board (PCB) stock. You can use 1/16 inch-thick board, though thinner stock will also work.

Use common sense when working with tools, power or otherwise. Be very careful whenever you drill hardware, even plastic or PCB stock such as these robot parts. Always wear eye protection and clamp the part securely in position before you begin machining anything.

Cut a piece of PCB stock 4 by 3 inches, then drill four 1/8th-inch holes in a rectangular pattern toward one end, using the computer's project board as a template. You will later use these holes, with proper hardware, to mount the project board to the frame.

Put a 1/2-inch long 4-40 screw in each hole, with the screw's head against the underside of the PCB frame. On the top side of the frame, install a 4-40 split-ring lockwasher and 1-inch long threaded spacer on each screw end. Tighten down all four spacers. This will serve as the mounting platform for the circuit boards.

Next, position the two servo motors back-to-back, with their output shafts in line. Aligned this way, the motors' frames form a block almost exactly four inches wide. Position these motors so the two shafts lie about 3/4th-inch from the end of the PCB where you just drilled the project board's mounting holes.

The photos show motors mounted using a second, smaller piece of PCB as a mounting plate. The sketches, however, show a mounting system using nylon wire-ties (Figures 4 and 5). I much prefer the wire-tie method; it is easier to build, cheaper, lighter, and more easily adjustable.

Carefully mark two holes for each motor, so that the holes lie about 1/8th-inch beyond the motor's frame at the midline. Use the sketches for a guide (Figures 4 and 5). Drill the marked holes using a 1/8th-inch drill bit. If necessary, shim between the servo motors and the PCB frame with a square of corrugated cardboard, cut to size. This will prevent dam-

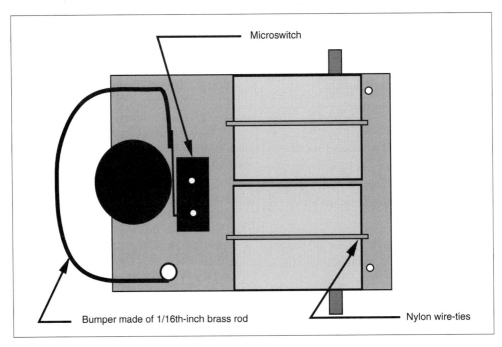

Figure 4. View of frame's underside, showing placement of servo motors, bump switch, and front skid.

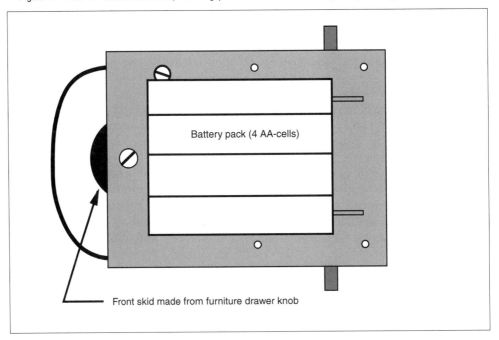

Figure 5. View of frame's top side, showing battery pack and bumper. The circuit boards and spacers are not shown.

age to the servo motors' housing when you tighten down the nylon wire-ties.

Using double-sided tape or masking tape formed into a large loop, sticky side out, tape each servo motor into position on the underside of the PCB frame. Make sure the motors' shafts extend beyond the PCB frame!

Run a nylon wire-tie through the two holes for each motor, then cinch each servo motor firmly in place and trim the wire-ties. I couldn't find wire-ties that were thin enough and also long enough, so I threaded one wire-tie into a second, making a long one out of two short ones.

The bumper switch consists of a small microswitch, a short length of 1/16th-inch brass rod, some heatshrink tubing, and a set of 4-40 hardware. Start by drilling two 1/8th-inch holes as indicated in the sketches, then mounting the microswitch to the underside of the frame using the 4-40 screws, washers, and nuts.

Drill another 1/8th-inch hole approximately as shown, then use 4-40 hardware to bolt one end of the brass rod to the underside of the PCB frame. Shape the rod into the figure shown, leaving the free end resting against the microswitch's lever. You may need to trim some length off of the brass rod to make the end meet properly with the switch's lever. Slip a small length of heatshrink tubing over both the brass rod and the microswitch's lever, then heat the tubing so it joins the two units together firmly.

Cut a 6-inch length of two-conductor wire and strip the insulation from all ends. At one end, solder one wire to pin 1 on a 4-pin KK female connector. Solder the other wire to pin 4 of the same connector. You will later plug this connector into A/D 2 (hooked to pin 45 of the 68hc11).

Solder the other two ends of the wire to the COMMON and NO pins of the microswitch; polarity is not important. Now, closing the switch will cause a short between ground and the signal pin on channel A/D 2. The software will in turn see this as a reading of 0, compared to the reading of 255 it will normally see if the switch isn't closed.

You will need a furniture drawer knob to use as a front skid for the robot. These usually cost about

$1 at the local hardware store, and are available in a variety of striking robo-colors; I favor fire-engine red myself.

Mark a hole at the front end of the PCB frame, then drill a hole large enough for the hardware that came with your furniture drawer knob. Fasten the drawer knob to the underside of the PCB frame by running the drawer knob's screw through the PCB, one or more spacers, and finally into the body of the knob. The exact distance between the PCB frame and the top of the knob will depend on the size of wheels you attach to your Robopet and the type of drawer knob you use.

Next, cut a piece of corrugated cardboard to match the shape of the battery holder's bottom. This cardboard will act as a spacer and mounting surface when installing the battery holder. Position the cardboard so that it lies between the two spacers that will hold the front edge of the circuit boards. Fasten the cardboard in place with double-sided tape. Then fasten the battery holder onto the cardboard square, again using double-sided tape.

Note: Although I have found the tape to be an adequate adhesive, you might want to drill a couple of holes as necessary and fasten the battery holder down with some thin, insulated copper wire. Run the wire through two holes in the battery holder and the PCB frame, then twist the ends together tightly.

Building the Wheels

The next phase of construction involves modifying the wheels for mounting to the servo motors. Begin with a suitably sized pair of Dave Brown Lite-Flite wheels, available from most large hobby stores. Exact size isn't critical; the larger the diameter, the faster your robot will move. The servo motors generate about 42 oz-in of torque each, so your robot will have plenty of power to move even 4-inch wheels easily. I have used 2-1/2 inch wheels and 3-1/4 inch wheels with no problems.

Refer to the accompanying photograph for details (Figure 6). The servo motors each come with a small, round disc (called a control horn) that is nor-

Figure 6. Closeup pf the wheel, with control horn attached. The tip of
the mounting screw is just visible inside the bore of the control horn.

mally mounted to the servo's shaft. The horn has many small holes in its flat surface, used to hook control lines and wires.

Place the horn's flat surface against the wheel's hub, so the bore of the wheel lines up with the horn's bore. Select two holes on the flat surface of the horn and mark through them onto the wheel's hub. Drill the two selected holes in the horn with a 5/64th-inch drill bit. Next, pry the wheel hubs apart; this might take a little effort. Work slowly and carefully, so you don't damage the hub or injure yourself.

Place the control horn's mounting screw in the horn's hub, so the screw head sits in the small cavity designed for it and the screw's length lies inside the coupling for the servo motor's shaft. Carefully align the flat side of the horn with the wheel's hub so the horn is centered on the outside surface of the hub.

Note: You may need to use a small file or Exacto knife to trim some of the plastic rim from the horn's flat surface, to get a more stable fit when you mate the horn to the hub. It also might help to countersink the hole in the wheel's hub, to allow space for the screwhead.

With the two units clamped together, make sure you can move the screw's tip back and forth inside the coupling. The screw should move easily inside the mounting. You should also be able to insert a small Phillips screwdriver through the hub of the wheel and turn the screw easily.

Now drive a pair of 1/2-inch, 2-56 sheet-metal screws (available at most hobby stores) through the two enlarged holes in the control horn and into the body of the wheel's hub. Turn these screws just enough to get them properly aligned and started; do not run them all the way down yet. Now reassemble the wheel by firmly squeezing the two interlocking hubs together.

Finally, run the two sheet-metal screws all the way into the wheel to secure the control horn to the wheel's hub. Verify that you can still freely move the screw's tip back and forth inside the control horn's coupling.

Perform this same modification on the second wheel. You now have two wheels that simply screw onto the shafts of your robot's servo motors. By modifying several pairs of wheels this way, you can

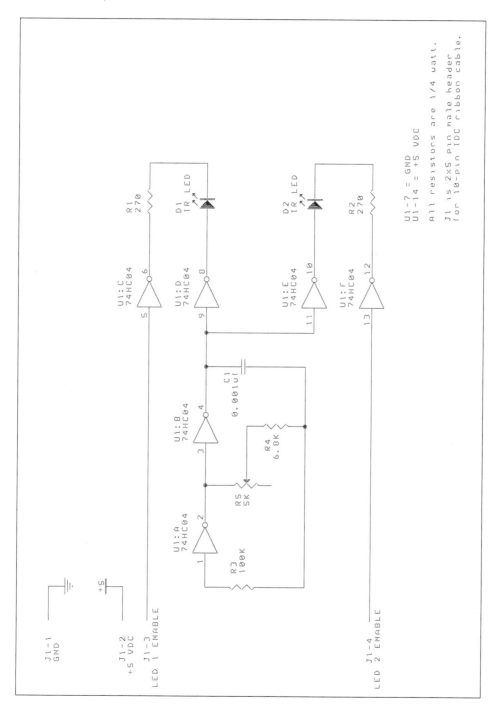

Figure 7. Schematic of IR LED driver.

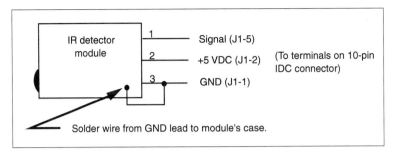

Figure 8. Details for wiring the IR detector module.

Object Detector

The accompanying schematic (Figure 7) serves as a source for two 40 kHz IR LED drivers. This circuit, combined with a single Radio Shack IR detector module (276-137), is all you need to build a reliable object detection system for your Robopet. The IR LED driver circuit appears in the book *Mobile Robots*, reviewed in "Inspiration and Implementation." Check that column for details on ordering your copy of the book.

Essentially, this clever circuit lets you turn two different 40 kHz IR LED drivers on or off using just two I/O lines from the 68hc11. If you bring U1-5 high by driving the 68hc11's PC0 high, D1 will flash at a 40 kHz rate. If you drop PC0 low, D1 goes dark. D2 behaves similarly, based on the states of PC1.

All that remains is to wire the signal output of an IR detector module to PC2 (pin 5 of the 10-pin IDC connector). Refer to the accompanying details for wiring the IR detector module (Figure 8). When the detector sees the 40 kHz signal from either IR LED, it will generate a low on its signal line. The 68hc11 can detect this condition simply by monitoring the state of input line PC2.

Begin by wiring the circuit as shown onto a Radio Shack experimenter's board identical to the board used to construct your 68hc11 circuitry. Use a 2x5-pin IDC male header for wiring J1. Solder about five inches of three-conductor cable to the terminals on the IR detec-

tor module. Connect the IR detector module and both IR LEDs directly to the appropriate points on the circuit board, rather than use connectors.

Finally, mount the IR LEDs directly to the circuit board by drilling 1/8th-inch holes at either end of the circuit board's front edge. Bolt the two IR LEDs in place so each points about 30 degrees away from the robot's centerline. Refer to the photographs for details (Figures 9 and 10).

Note that I used surplus, commercial IR LED emitter/detector pairs for my IR LEDs. I did this for two reasons. First, the mounting system is very compact and cheap. Secondly, and more importantly, the IR LED is already thoroughly shielded against IR leakage.

In fact, I tried unsuccessfully to use regular IR LEDs for this project and finally gave up in frustration; I couldn't get reliable readings because of IR leakage from the LEDs. Save yourself the grief and find some surplus IR LEDs, already encased.

Mount the IR detector module on a short length of 1/16th-inch brass rod. Bend the rod so it arches over the top of the robot, holding the IR module up above the circuit boards. You can mount the other end of the rod by soldering it directly to the surface of the PCB frame. Use a heavy-duty (100 watts or more) solder gun for this part; a 20-watt soldering pencil won't cut it.

The Last Bit

Only the software remains. The file huey.z00 (that's Z-zero-zero) contains the source code for this robot;

Figure 9. Close-up showing mounting of circuit boards and battery holder.

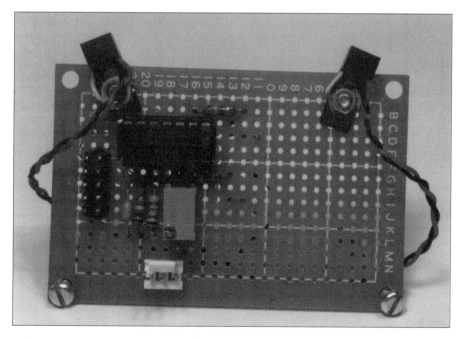

Figure 10. View of the IR object detector circuit board. Note the mounting of the two IR LEDs.

you can download the code from my web page. Use pkunzip to uncompress all the files. You will find the assembler source code (huey.asc), the object file (huey.s19), and a special boilerplate file (robot.blr).

huey.asc contains the full assembly language source for a Huey Robopet robot. Huey's code uses subsumption architecture to guide it around a room, avoiding any obstacles detected by the IR system or by the bumper switch. The code warrants study if you want to try designing your own robotics software. It is heavily commented and quite small, since the finished program has to fit in the 68hc11's 512 bytes of EEPROM.

huey.S19 holds the object file created by assembling huey.asc with the Motorola FREEWARE assembler asmhc11. This assembler runs on any PC clone and is available free for downloading. To get this assembler, and the required pcbug11 programming tools, download the file pcbug342.exe from the Motorola FREEWARE web page.

You "burn" the .S19 file into the 68hc11's EEPROM area with pcbug11's LOADS command. Full details of this operation are in the pcbug11 manual, though you shouldn't have any difficulty figuring it out from the on-line help text. See last month's column for details on hooking the 68hc11 computer to your PC and on using pcbug11 to talk to the 68hc11. (See "Your First 68hc11 Microcontroller.")

robot.blr is a nearly complete template for writing robotics software for these little 68hc11 computer boards. The file contains all the boring overhead text such as register equates, RAM and ROM assignments, and startup code.

It also contains working code for processing real-time interrupts and for controlling servo motors. The extensive comments will guide you in setting up your own working robo-code, starting from a copy of the robot.blr file. All you have to do is design your software, add it in at the indicated places in the file, then assemble, debug, and run it.

Figure 11. A small Robopet robot. The IR sensor sits at the top on a thin, brass rod. The red drawer knob acts as a front skid.

Figure 12. Rear view showing brass rod mounted to robot's frame, for holding the IR sensor.

Figure 13. Closeup of the 68hc11 computer circuit board. This layout is similar to that described in the article.

Future projects will use robot.blr as a starting point. If there is sufficient interest, I may devote an entire column to designing software for one of these small robots, using the robot.blr file.

One element of this software deserves special mention. The IR object detection system is based on the technique described in *Mobile Robots* by Joe Jones and Anita Flynn. Jones and Flynn used two IR emitters and one IR detector to "see" objects on the right, the left, and straight ahead. The technique depends on the placement of the IR LEDs (each aimed about 30 degrees off of straight ahead) and the IR sensor, which is pointed straight ahead.

The 68hc11 first activates the right LED, then checks the IR sensor to see if it detects a reflected signal. The 68hc11 then performs the same function with the left LED. If the IR sensor sees a signal when one LED is on but not the other, the computer turns away from that side. If the IR sensor sees a signal from both sides, the 68hc11 assumes there is an object dead ahead, and takes evasive action.

Jones and Flynn tried to avoid false signals by checking the IR sensor while the IR LED was on and while it was off. A signal wasn't a real signal unless the IR sensor saw something with the LED lit and didn't see anything with the LED off.

I didn't get satisfactory noise immunity with that scheme, so I used a variation. The 68hc11 turns the LED on and samples the IR sensor many times; the IR sensor has to see an object every sample time before the code assumes there might be something out there.

This technique has produced very stable operation, relatively free of false sensor readings. I say "relatively," because it still gets fooled sometimes if the IR sensor looks directly at strong incandescent light or sunlight. You can dabble with the parameters in the source file if you would like to try improving the system's performance.

One word of caution; this IR sensor scheme can see objects up to two feet away, depending on the object's color and texture. You may have to adjust the brass rod that holds the IR sensor, to prevent false echoes from the floor. If your robot spins around in a circle when you power it up, try tilting the IR sensor so it points slightly above horizontal.

The Basics of Hobby Robotics

Although this was one of the last Amateur Robotics columns I wrote, it was probably the one article of most help to beginners. Published in the April 1997 issue of Nuts & Volts, this column takes you through the basic concepts behind building your first robopet. In just a few pages, I cover frame materials and construction, fasteners, tools, motors, batteries, and other basics. This one column contains enough information to get you started properly in frame construction, and will carry you through your first several robots.

I've long favored hobby R/C servo motors for small robots, but you have to make some simple modifications to a hobby servo before you can use it in a robot. I present here a technique that appeared in an earlier Amateur Robotics column, but the information was valuable enough to warrant another printing. It's quick and easy to do, and you end up with a rugged, small gearhead motor in a sturdy, easy to mount case.

If you're looking to build your first 'bot, start out with this column. The rest of the chapters in this book will be much easier to follow with this one behind you. Even if you've built a 'bot or two already, the information in this column might get you started thinking along a whole new line, and that's always a Good Thing.

I receive several requests each month for help getting started in robot-building. The plea usually takes the form of "... my (son or daughter) wants to build a robot for (Science Fair, Scouts, fun); how do I get started?" Often, the parent has seen a copy of my column or found my name on the Internet. And some of the requests for help get pretty desperate, citing deadlines a week or so away.

I figure this would be a good time to condense some of the information from my previous 54 *Nuts & Volts* columns. After all, I've covered a lot of ground in the last 4+ years, and a review of the tools and techniques available in this hobby should be of interest both to beginners and to long-time readers.

How Do I Get Started?

People who've never built a 'bot before usually approach the task from one of two directions. One group simply jumps in and starts. Grab some DC motors (these old windshield wiper motors look strong enough), a hurking power source (Bob doesn't need this motorcycle battery anymore), and some type of base material (I'll take a sheet of that 3/4-inch plywood). Saw, hammer, glue, and bolt for a while, then step back and see what you've got.

Don't get me wrong; I'm all for this kind of experimentation. Of course, you might well end up with a 100-pound juggernaut that (hopefully) won't move too fast or too far, and you will probably learn a lot in the meantime. But you're just as likely to create a large wood and metal sculpture, and the frustration of not getting your first R2D2 to move at all might stop you altogether. If this happens, you will miss out on a fascinating hobby and a lot of fun.

The other group of beginners takes the time to sit down and think out the whole effort. They are quickly paralyzed by the sheer complexity of the task; the questions seem endless. Where do I find motors? How do I make the robot back up? Should I use AA batteries? How will I control the thing?

Some who start down this path do build a machine. Of course, the motors aren't held onto the base all that well, it only does right-hand turns, and the spare bedroom is filling up with used alkalines, but it does roll around under its own power. Others in this group end up like the old Star Trek trick of shutting down a rogue computer by giving it an infinite problem to solve; they get so mired down in the complexity that they finally give up in frustration.

The solution for both groups lies with the single most important resource in robot-building: information. The more sources of information you can tap, the greater your field of options and the more tools you can bring to bear on your project. And in the last four years, the amount of useful information for the novice robot-builder has exploded, thanks largely to the Internet.

When I started this column, Internet access was pretty much limited to colleges and universities, and to a few enlightened companies that provided their employees with 'net access in the form of email and news. In those days, there was no Web, and few individuals had built even one 'bot, let alone a herd of them.

How things have changed in such a short time! Today, Internet access is cheap and easily available, and the Web plays a major role in helping hobbyists pool their information. In just one afternoon of Web-surfing, you can visit dozens of sites filled with plans, information, pictures, and tools for helping you build your 'bot. If you are starting out in this hobby, the best help you can give yourself is to sign up with a reliable and supportive Internet service provider (ISP), hook up to the 'net, and get familiar with this new medium.

And since the Internet is so complex, it only makes sense that you need a tool to help you use the tool. This means a search engine, and my favorite is AltaVista. Once you've reached the AltaVista Web page, you simply type a keyword or phrase in the search box, click the Submit button, and the search engine will examine millions of entries in its database, looking for web pags that contain your keywords. Spend some time getting used to the Web and AltaVista (or whatever search engine you choose); it will be time well spent, and the information you dig up will save you plenty of frustrating trial and error later.

My web page can serve as a good starting point. Just aim your Web browser at my web page to see a collection of tools and information I've set up for robot builders. From my page, you can reach the Seattle Robotics Society page, which in turn will take you to pages for other clubs and individuals interested in robotics.

You'll also want to visit and/or subscribe to some of the Internet list servers. A list server is a cross between a bulletin board system and email. Each subscriber receives a copy of any email any other subscriber sends to the list. So if someone finds a cool surplus store selling motors, she can send a single email to the list server, and all 200 subscribers get a copy of the note. I'm subscribed to the 68hc11 list server and find it an invaluable source of information. Refer to the Motorola entry in Appendix C for information on subscribing.

More traditional media also provide a wealth of valuable information. Tops on my list of books for robot builders is *Mobile Robots*, by Joe Jones and Anita Flynn (ISBN 1-56881-011-3). This superb book, published by A. K. Peters, Ltd., covers many of the design elements needed to build small robots in general. It also contains detailed plans for building two small robot platforms for experimentation.

To find any weird or unusual technical books, check out Powell's Books and try their search system. Ideally, visit Portland, Oregon, and check them out in person. You gotta love Powell's City of Books, a bookstore that covers an entire city block. And a couple of blocks away, you'll find Powell's Technical Books, a normal-sized store filled with nothing but technical books of all types.

When you get to the low-level parts of your design, such as the electronic circuitry, be sure to contact the various manufacturers. In the last few years, more and more electronics firms have made it easier for the hobbyists to get up-to-date technical information. For example, National Semiconductor and Maxim, among others, let you order free samples directly from their web pages. And many companies, including Motorola and Atmel, provide top-quality technical info in the form of .pdf files. You can view these files with a free Adobe Acrobat viewer, available on most web pages, and print them out on your laser printer. You end up with a full-size, high-resolution technical summary, many including timing graphs, schematics, and application notes.

And remember to visit all the technical and surplus stores on the web. One of my favorites is Mondotronics, a great source for everything from kits for small robots to parts and books. Also check out Marvin Green's web site and take a look at his small BOTBoard circuit boards for building 68hc11 and PIC computers. And if all of the above isn't enough incentive to hit the Internet, I'll add one more item. You can send email to me directly at the address provided on my web page.

But getting onto the Internet remains the single greatest step you can take to ensure your successful start at robot building. I consider it so important that I use it as a benchmark for how serious a person really is in pursuing this hobby. If someone asks me for help in building robots, and then tells me they aren't now and don't intend to be hooked to the 'net, I assume they aren't all that interested in amateur robotics.

Getting Physical

Right. Now you have all the information you need, or at least you know where to find it. Time to get down to the tangible stuff. You can break down just about any robot design into four parts. The sensors collect information about the robot's internal and external environments. The actuators provide the physical interaction needed with the external environment. The controller, usually a microcontroller (MCU), coordinates the inflow of sensor data with the robot's program, and then controls the actuators to create the proper response. Finally, the power source provides the power, usually electricity, to drive the whole arrangement.

Whenever you want to start on a robot project, always begin by deciding what you want the robot to do. Whether you're aiming for the Holy Grail of robotics (a vacuum-cleaner robot), or something more modest like a robo-pet, take the time to spell out all of the robot's top-level goals. From here, you can determine the behaviors it will need and the information it must collect.

This phase can actually be the most fun, and it can shed light on how tough a design problem you've set for yourself. Start with something pretty outlandish, such as fetching a beer from the kitchen to the living room sofa. Then begin breaking this task down into its component parts. The robot must find the kitchen (don't laugh, that can be non-trivial, at least in my house), then find the refrigerator, then get the door open, find the beer, grab the beer, stash it somewhere safe, close the door, find its way out of the kitchen, find the living room (see above), find the sofa, move to but not into the sofa, and finally announce that the beer is served.

But a beer-fetcher is beyond the scope of a beginner article; sorry about that. Instead, I'll content myself to taking you through the design and construction phases of a small robot patterned after Arnold. Arnold is a robopet that wanders around my living room. It has fairly reliable object detection, in the form of bumper switches and an infrared (IR) reflector system, so it almost never gets trapped in a corner. It uses two small hobby servo motors, modified to spin continuously, for its drive system; power is provided by four AA alkaline batteries. Finally, its brain, built around a small 68hc11 computer board,

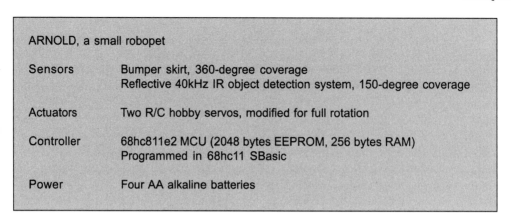

ARNOLD, a small robopet

Sensors	Bumper skirt, 360-degree coverage
	Reflective 40kHz IR object detection system, 150-degree coverage
Actuators	Two R/C hobby servos, modified for full rotation
Controller	68hc811e2 MCU (2048 bytes EEPROM, 256 bytes RAM)
	Programmed in 68hc11 SBasic
Power	Four AA alkaline batteries

Figure 1. A description of Arnold, a small robopet.

executes a program written in a Basic dialect to coordinate all its functions. Refer to the accompanying table of Arnold's design (Figure 1).

My original Arnold robot used a plastic frame designed and sold by Marvin Green. Dubbed the BBOT Frame, this circular plastic platform sports a clear plastic dome and a one-inch high bumper skirt that hangs around the frame's mid-point. Marvin may well still be selling the BBOT frames; if so, I highly recommend it as a simple and elegant starting platform. Check his web page for details.

But I'll assume that you don't have access to a BBOT frame, or prefer to roll your own, so I'll guide you through the basic elements of frame design and construction. I'm not going to give you step-by-step instructions, and you won't end up with a finished frame using only this article, but you'll understand the major problems of simple frame design and a couple of ways to attack them.

Many beginners start off by designing a frame that is too heavy or too large, or both. All of the components in Arnold, except the frame, weigh less than a pound. Using a three-pound frame to carry 12 ounces of electronics and batteries is gross overkill.

Besides being inelegant, an over-sized frame can prevent your robot from even working at all. Remember that your motors must provide motive power to carry the robot's full weight. Extra frame weight means larger motors and heavier wheels and drive system. This in turn means you need a beefier battery to get the same amount of running time, which adds still more weight. This vicious spiral, if left unchecked, can lead to a 100-pound robohog that gets 12 minutes of running time from a fully charged car battery.

Before you start your first robot design, resolve to build the lightest frame you can. If you build a light frame and experimentation shows it's too light, fine. You can always use the light platform as a template for making a sturdier second frame. But building the first version too heavy can prove very frustrating, and has ended more than one robot project.

There are many light frame materials available. One of the lightest is brass rod, available from the local hobby or hardware store. I use pliers to bend the rod to the desired shape, heavy nippers or horseshoe nail clippers to cut it to size, and a heavy-duty soldering gun and rosin-core solder to weld the pieces together. If you need to buy a soldering gun for this material, pick up a Weller gun rated at 100 watts or better. Note that you will NOT use this gun for soldering delicate electronics; it is only for welding brass and copper material.

Speaking of copper material, the blank copper printed circuit board (PCB) stock, available from many mail order houses, makes a terrific frame material also. Very light for its weight, you can cut the thinner stock with sheet metal shears or a hacksaw, drill holes in it easily, and weld brass rod directly to it. Go with the thinner material, say 0.062 inches thick, for your first robot.

There are other advantages to using copper-clad and brass rod as frame material. Many hardware stores carry brass hardware, such as nuts and screws. You can weld these onto the PCB material, then bolt other hardware to the frame using these threaded fasteners. And many surplus shops carry brass spacers, which weld easily to the brass rod and copper PCB.

I also use foamcore for making robot frames. This is a laminate of paper and rigid foam, available from many art supply stores. You can cut it easily with anything from a razor saw to a sharp knife. It weighs next to nothing, yet provides amazing strength. You can draw on it with a pencil, poke holes into it with an awl or a nail, and hot-glue odds and ends to it easily.

The big brother to foamcore is Gator-board. Also available at industrial art supply stores, Gator-board looks just like foamcore and weighs about the same, but it sports a tough plastic skin rather than paper. You will need a coping saw or a jigsaw to do a good job of cutting Gator-board, but you will like the end results. The material wears very well, is light but very strong, and hot-glue sticks to it well.

Your first robot shouldn't require anything more substantial than Gator-board or copper-clad as a frame material. If you need a heftier base material for a later robot, check out Sintra plastic. This is an opaque foam-based plastic, available in sheets of 1/4" thickness in various colors. It cuts with a saw, jigsaw, or even a tile knife, drills easily, and takes hot-glue well. It offers excellent strength for its weight, and you can usually find good-sized pieces in the scrap bin of your local plastics shop.

After choosing your frame material, you need to look into fasteners. My fastener of choice for small robots is hot-glue. You can pick up a suitable hot-glue gun from nearly any hardware store, and most stores sell a variety of glue sticks for different surfaces.

While you're out, pick up a roll of double-sided foam tape. You can find this invaluable tool at most hobby stores, and a few bucks worth will last you for several robots. To use the tape, simply cut a piece to length with an Exacto knife, peel the tape of the roll, stick it to one surface, peel the paper off the other side of the tape, and stick the second surface to it. I use foam tape for mounting everything from battery holders to servo motors. It holds light objects very well, though you need to ensure both surfaces are clean and free of grease.

My next fastener of choice is nylon cable ties. You can find cable and wire ties at any Radio Shack and most large hardware stores. Get a variety of lengths, including 4-inch and 6-inch. A bag of forty ties will last you through several 'bots, and you can sometimes get the ties in wild colors, which can add a little pizzazz to your design.

In some cases, I can't avoid using threaded hardware of some type. If my frame design uses copper-clad, I'll go to brass spacers and bolts if necessary. In nearly all other instances, I'll turn to nylon or fiber spacers. You can hot-glue these spacers in place, and the threaded spacers make it easy to mount circuit boards to nearly any surface. Often, you can find nylon spacers and bolts at the larger hardware or hobby stores, and sometimes the mail-order houses will stock a few.

As you can see from the above list, all of my frame materials and all of my fasteners have one element in common: they are lightweight. It doesn't take many pieces of aluminum or steel hardware to boost the weight of your robot to an unacceptable level. Stay with the lighter materials and you'll build a lighter, faster, longer-running robot.

Next up is batteries. Small, light robots don't need a lot of juice, and most of my small machines run just fine on four AA alkaline batteries. I don't use NiCd batteries for a few reasons. First, a NiCd cell only puts out 1.25 VDC, not the 1.5 VDC available from alkaline cells. This difference may seem small, but it means that four NiCds only yield 5 VDC when fully charged, compared to the 6 VDC available from fresh alkalines. This one-volt delta translates into longer running time for both the motors and the electronics.

And NiCd cells can exhibit what many people term a "memory effect," wherein the cells seem to go flat after shorter and shorter periods of use. The newer Renewal alkaline batteries, which I prefer, can be recharged up to 25 times and don't show this irritating behavior. Note that you MUST use a Renewal charger to recharge Renewal batteries; don't try to use a NiCd or other charger on these cells. Also, do not run a Renewal battery all the way down; this will effectively kill it. I've used the same Renewals for over three years now, and only had to throw away about eight. Four of those fatalities occurred because I accidentally drained a set down to zero.

To hold the batteries in your robot, you need a battery holder. Radio Shack sells a good assortment of holders made of nylon or similar light material. You'll likely only use the four-cell AA holders for your first robots, but it wouldn't hurt to pick up some two-cell holders and even some of the long, skinny four-cell units. These varieties give you greater flexibility in designing your robot frame, since you can build around the different holder shapes.

Let's move on to motors. For beginners, nothing beats the Futaba S148 hobby R/C servo motors. These go for about $16 each from Tower Hobbies and using them in your robot offers some real advantages. First off, they are compact, rugged, and fitted into a sturdy, cool-looking case. Next, the hobby world carries plenty of gadgets and doodads already designed for use with hobby servos, so attaching devices to your servo motors will be cheap and easy to do. Finally, these motors offer 42 oz/inches of torque in a small, 6 VDC package. The rotational speed isn't all that great, but you can't beat the convenience and flexibility.

Unfortunately, hobby servos are designed to go back and forth, since they normally control flaps and other such airplane things. For robots, they need to go 'round and 'round. The conversion from back and forth to 'round and 'round isn't tough to do, but you need to work carefully. I've included instructions for one type of conversion in Appendix B. Other methods exist, such as grinding down the shaft of the servo potentiometer, but these require more care than the technique I've outlined here.

After you've modified two motors, you are ready to cut the platform material to shape. For your first robot, start with a square base roughly four inches on a side. Cut a piece of foamcore or Gator-board to size and smooth the edges if necessary. Place the motors at one end, aligning them so the shafts protrude far enough beyond the edges of the base to give good clearance for the servo control horn. When you are happy with the motors' alignment, cut two

---

WARNING!

The BOTBoard was designed to use a servo other than the Futaba S148 used here. The power wiring for the servo motor is reversed from that needed by the Futaba. If you plug a Futaba S148 into the BOTBoard without changing the servo's wiring connector, you will burn up the servo!

The servo wiring connector is originally in white-red-black order. To modify the wiring connector, simply reverse the power (red) and ground (black) leads in the connector shell. You can do this by CAREFULLY prying up the black plastic finger that holds the red lead in place, then pulling the red lead out of the shell. Do the same for the black lead. Now swap the leads and push each back into the connector shell. Properly done, the connector wires should now be in white-black-red order.

pieces of foam tape for each motor, apply the tape to one surface of each motor, then press the motors back onto the foamcore in the previous positions.

Next, flip the platform over so the motors are on the base's underside, then position the battery holder on the top of the base, slightly forward of center. Fasten the holder to the base's surface using foam tape.

Now you need to add some wheels to the motors. Most hobby stores carry suitable wheels; I've always been happy with Dave Brown's LiteFlite wheels. These are a foam wheel, available in several different diameters. Choose a pair of wheels 2.5 inches or so in diameter for this first machine.

To mount a wheel onto a servo control horn, remove the control horn from the servo, then put the mounting screw back into the larger circular control horn that came with your Futaba servo kit. Trim a couple of pieces of foam tape to about 1/4 inch wide, then stick them onto the outer surface of the control horn. These pieces of tape act as a shim, leaving enough gap between the control horn's surface and the wheel so later you can tighten or loosen the mounting screw.

Working carefully, align a wheel against a control horn and press the wheel into place. Make sure the wheel is centered exactly on the control horn, with the wheel's bore exactly over the center of the mounting screw. If your first try doesn't match up perfectly, pry the wheel off the control horn, replace the tape if necessary, and give it another shot. When properly mounted, you should be able to slip a small Phillip's screwdriver through the wheel's bore and tighten the mounting screw onto the servo's output shaft. If the screwdriver reaches the screw but the

screw won't turn easily, you may have to remove the wheel and slightly countersink the wheel's hub around the bore next to the control horn. Reassemble the wheel and control horn.

After you've mounted the wheel to the control horn, run a bead of hot glue around the edge of the control horn where it meets the hub of the wheel. Perform the same operation for the second wheel, then mount the two wheels onto the robot's servo motors.

You have several options for the robot's front end. I usually stick a small caster onto the front underside of the frame, using foam tape to hold it in place. If you use such a caster, you will probably need to shim it out from the platform, so the platform sits level. You can also make a skid out of a large, colorful drawer knob. You'll find these at nearly any large hardware store, and the knob kit, which usually sells for less than a buck, includes a long mounting screw. To mount such a skid, just find or cut a piece of plastic tubing or spacer to the proper length, poke a hole in the platform at the proper location, and use the screw to bolt the whole assembly in place. Alternatively, you can mount the skid onto a separate piece of foamcore, about one inch square, and stick this foamcore to the underside of the platform where needed.

Now you have most of the base finished. Take a moment to check its weight. This base is suitable for many robopet projects, but weighs very little. Next month, I'll take you through the electronics involved. You can use nearly any small computer board to drive this machine. I usually stick one of Marvin Green's BOTBoards on my robopets; check his web page for distributors, or stop by Mondotronics.

An Intro to 68hc11 Firmware

This column from the May 1997 issue of *Nuts & Volts* continues where the previous article left off, getting you started with your first robopet project. These two columns give you enough practical instructions that you should be able to get a platform built and start writing your own SBasic robotics programs.

I start by listing some of the more common tools you will need, including a free software utility called pcbug11. Next, I show you how to use pcbug11 to explore the innards of the 68hc811e2 chip in your BOTBoard. This section includes how to modify the device's on-chip EEPROM and how to use pcbug11's built-in assembler.

From here, I move to the notion of higher-level languages, specifically my SBasic compiler. With three different programs, I cover topics such as controlling I/O port lines, using the **print** command for debugging, and setting up hobby servo motors. The last program, in particular, shows you how to get servo motors set up and working; you'll be surprised at how much you can do in just a dozen lines of SBasic code.

Previously, I took you through the design of a simple robopet frame, suitable for a beginner's robot–see "The Basics of Hobby Robotics." This month, I'll finish the robot by taking you through the steps needed to develop software and install it in your microcontroller (MCU) board.

Begin by building up one of Marvin Green's BOTBoards, a small 68hc11-based MCU board. This printed circuit board (PCB) measures just 3 x 2 inches, but holds a 68hc11 MCU, serial communications connector, four servo motor ports, and a small prototyping area. The BOTBoard is available several places, including Mondotronics. Be sure to get a 68hc811e2 chip for your BOTBoard. This device contains 2K bytes of EEPROM, enough room for some fairly strong robot code. The rest of this article assumes your board contains such a chip.

Carefully wire up the BOTBoard, following the instructions included. If you've never soldered any electronics before, hold off long enough to make a quick trip to the local Radio Shack and pick up Forrest Mims' excellent book on getting started in electronics. You will need information on how to read resistor codes, how to recognize different devices, and how to tell the polarity of components such as diodes and capacitors.

While you're out, pick up any tools you might need to round out your set. Ideally, you should have a 20- to 40-watt soldering iron with a grounded tip. You can get a fancy, temperature-controlled unit from Weller, the WCC-100, from most large electronics supply houses. This 60-watt unit has a dial for setting the tip temperature and comes with a wire cage for holding the iron safe from accidental contact. The rest of your tools list should include high-quality wire cutters (called "dikes"), needle-nose pliers, and a set each of Phillips- and standard-blade jeweler's screwdrivers.

Another real handy tool to have is a logic probe. This tool looks like a fat pencil with a thin cable coming out of the top. You hook these leads to ground and +5 VDC somewhere in your circuit. Now if you touch the probe's tip to a point on your circuit board, one of several LEDs will light to indicate the state of that point. Typically, a cheap probe will show logic high, logic low, changes between high and low, and a crude indication of frequency of change. Spend the ten bucks or so to get a basic logic probe; you'll be amazed at how often you rely on it.

Back at your workbench, finish wiring up the BOTBoard. Add a two-pin male Molex KK-style connector (available from DigiKey through the mail) for hooking up your four AA alkaline batteries, as indicated in the instructions. You will also need to add the matching female connector to the end of the battery leads on your four-cell AA battery holder. Be sure to get the polarity on both connectors correct; double-check the alignment of the red (positive) wire on the female connector with the corresponding pin of the male connector.

Also, add the mod in the BOTBoard instructions for setting the mode of the 68hc11 on reset. This mod involves cutting a trace on the underside of the PCB and adding a couple of wires and a resistor. Work carefully when making the mod, and verify your work with a magnifying glass if necessary. While you're at it, read over the theory of the mod so you understand what is happening in each of the two cases the mod supports.

At this point, you will likely have many questions on how the 68hc11 works. I can only refer you to the *Mobile Robots* book mentioned in previous columns (see "Inspiration and Implementation"), and to the assorted 68hc11 manuals available free from your local Motorola representative. The 68hc11 is far too complex to cover here, though I might devote a column or two to it another time. I realize that this can leave you hanging, since you may not know how to correct a problem that arises if you leave out a subtle but critical step. All I can say is you will have to do some digging on your own, use the Internet and Web resources, read the books, and ask questions. This phase is the hardest part of the hobby for most people, but (to me) is also the most rewarding. Making that slab of silicon wake up and do your bidding takes a lot of work, but the results can be way cool.

After you have wired up your BOTBoard, position it on the upper surface of your four-inch frame. Also place the battery holder where you think it fits the best. Play around with the placement of both pieces until you get an arrangement that you like. Remember to check for wire lengths and how you will run the cables from one element to another. The servo cables aren't normally a problem, given their length, but you want to place the battery holder so you don't need to lengthen its wires to reach the proper terminals on the BOTBoard.

Mark the final positions of the battery holder and the BOTBoard. Then use two pieces of double-sided foam tape to stick the battery holder in place. Add a dab of hot glue to two opposite corners of the BOTBoard to hold it in place.

If you haven't done so already, you need to make or buy a serial cable to connect between the BOTBoard and your PC. The BOTBoard docs include a schematic and parts list for building such a cable. If you can, you should build the MAX232 or MAX233 circuitry into a plastic DB-25 or DB-9 shell. This yields a compact, portable tool that you can use on several different robots as you get them built. Plus, you don't waste valuable board space and wiring to support the same circuitry on each robot; just plug in the serial cable and go.

(Author's note: As this book goes to press, Kevin Ross is selling a tiny circuit board for this function.

His PCB fits inside a DB9 connector shell; check his web site for details.)

Now is the time to install pcbug11 on your PC. You can pick up this valuable 68hc11 development tool from several places. Many vendors who sell BOTBoards also sell or give away a floppy disc containing pcbug11 and other 68hc11 tools. You can also find pcbug11 on the Internet, by browsing Motorola's FREEWARE web site. Note that the file you want is named pcbug342.zip.

Move the pcbug342.zip file into a suitable directory, such as c:\68hc11, and unzip it. You will now have a set of files including pcbug11.exe. pcbug11 uses a serial connection to your PC's comm port to exchange information with a tiny program, called a talker, that resides in the 68hc11. pcbug11 automatically downloads the proper talker program following invocation, provided you use the correct command line options and reset your 68hc11 so it comes up in special bootstrap mode.

Now you're ready to try starting up your BOTBoard. Connect the serial cable between the BOTBoard and the PC; I'll assume here that you are using COM1, though pcbug11 also works from COM2. Strap the BOTBoard for special bootstrap mode, then connect the battery pack and press the reset switch. The 68hc11 should now be sitting in a communications loop, waiting for instructions from the PC.

Now execute pcbug11 from a DOS prompt. Assuming you are using a 68hc811e2 in your BOTBoard, use the command line:

pcbug11 -a   if you're using COM1 or
pcbug11 -a port=2   if you're using COM2

If everything works properly, you should see a bright blue screen with a command prompt at the bottom. If so, you have cleared the first big hurdle in amateur robotics; you have a working MCU board.

If you got a blue screen with an error splat on it, start pressing the ESC or Enter key until you see the command prompt, then enter **quit y**. This kicks you completely out of pcbug11. Remove power from your BOTBoard, double check the wiring of your serial

cable, and try the above sequence again. The most common cause for this failure is a miswired serial cable; typically, the wires to pins 2 and 3 of the PC's connector are swapped.

If you are running Windows 95 and pcbug11 errors out to a black screen with a message about hardware failure, you may have to reboot your machine to DOS mode and try again. This happens on quite a few machines, and I haven't yet heard a good explanation for why it happens or how to get around it. My laptop, running Win95, cannot run pcbug11, though my main 486 can.

(Author's note: In some cases, you can fix this problem with a patched version of pcbug11, available from my web site. The file you want is called pcbug11a.exe; see my site for details on using this upgraded version of pcbug11.)

After you get a working connection to the BOTBoard, you can begin doing some serious playing. Enter the commands:

control base hex
md 1000 103f

The **control** command tells pcbug11 to assume any number you type is in hexadecimal, rather than decimal. The **md** command displays the 64 memory locations, starting at address $1000. If you check your 68hc11 reference manual (the "pink book") you will see that these addresses are actually the 68hc11's I/O ports. The 68hc11 places all of its I/O ports in addresses in the memory map, and you can use any machine language instruction that references memory to reference I/O ports as well.

Take a few moments to examine some of these addresses and compare the values in them to the technical information in the pink book. Note, for example, the contents of address $1004. In the manual, this is called PORTB and it is an output-only port. This means that whatever bit pattern your software writes to $1004 automatically appears directly on the port B output pins on the 68hc11.

The 68hc11 comes out of reset with PORTB holding a 0; thus, all output lines for port B should show a logic 0. You can verify this by touching your logic probe to some of the PORTB output pins and noting their logic level. Next, enter the pcbug11 command:

ms 1004 ff

This **ms** (memory set) command writes all 1s to PORTB, which in turn brings all port B output lines to logic 1. Again, verify this with your logic probe.

Now you're ready for your first 68hc11 program. Rather than jump right in to a major assembler program, we'll start with the pcbug11 in-line assembler. First, we need to choose a location in memory to hold our program. The 68hc811e2 contains 256 bytes of RAM from address $0000 to $00ff. pcbug11 uses all of this memory for its talker, so it is off-limits for your testing. Note that if you try to modify any of this RAM, pcbug11 will immediately crash.

The '811e2 also has 2K bytes of EEPROM from address $f800 to $ffff. This memory differs from RAM in two important ways. First, it is non-volatile; if you write something to EEPROM, it stays there forever until you erase or overwrite it. Second, you must use a special sequence to modify EEPROM; you just don't write to it as if it were RAM and expect the data to stay.

Fortunately, pcbug11 can help you in dealing with EEPROM. Enter the commands:

eeprom f800 ffff
ms 1035 10

The **eeprom** command tells pcbug11 to treat addresses from $f800 to $ffff as EEPROM rather than as RAM. Now any time you change an address in this range, pcbug11 automatically uses the proper ritual for changing EEPROM. The **ms** command changes a set of block-protection bits in the 68hc811e2. If you don't change the four low bits of $1035 to 0s, you won't be able to change EEPROM at all. (Actually, this isn't exactly true. Check the E-series pink book for full details on the functions of these four low bits.)

Now that you have set up pcbug11 so it will properly modify addresses in the EEPROM area, you

can use pcbug11's in-line assembler to write your first 68hc11 program. Enter the command:

asm ff80

This starts the in-line assembler, which will disassemble the contents of $ff80 and present it to you as a line of assembly-language source. You can then enter a new source line to overwrite the original. For example, assume pcbug11 gives you back the following response:

FF80 FFFFFF   > STX > $FFFF

This tells you that addresses $ff80 to $ff82 contain $ffffff, which disassembles to a STX instruction. Finally, pcbug11's cursor sits at the command prompt just before the STX, ready for your input. Anything you type now will overwrite the original text, and form your assembler source line. Enter the text:

ldaa $1004

Note that you MUST include a leading space before the **ldaa**, or pcbug11 will treat your **ldaa** as a label and the assembler will fail. Note also that you must overwrite all original text with either spaces or your own assembler source; any leftover original text will likely cause an assembler error. If you got it right, pcbug11 will open the next address for you, and display something like:

FF83 FFFFFF   > STX > $FFFF

Continue with the rest of the program:

        eora #$ff
        staa $1004
        jmp $ff80

After you've entered the **jmp** instruction, press the ESC key to leave the in-line assembler. Verify that your program is correct with the following disassemble command:

dasm ff80 ff8b

You should see your new four-line program echoed back to you correctly. Now you are ready to run it. Enter the command:

g ff80

pcbug11 won't appear to change much, but your program actually is running on the 68hc11. You can verify this by putting your logic probe on any port B output line. The probe will show a very high frequency signal on all port B output lines, with a duty cycle of 50%. You can stop your program by entering the command:

s

The program you entered creates a very tight loop that reads the value of PORTB, inverts all the bits, stores the new value back to PORTB, and starts over again. This causes all bits of PORTB to change from zero to one and back again at a uniform and very fast rate. Take some time to go through the pink book to understand exactly how this routine works. Though it is only four lines long, it uses some of the most common elements of hobby robotics software.

This idea of "change an output line" lets your robot begin to control its nearby environment. The simplest output device that works on this principle is an LED. Pick up a bright LED (rated at 20 mcd or more) and a resistor somewhere between 180 and 330 ohms. Wire one lead of the resistor to the LED's cathode, which is usually the lead next to the flat spot on the LED body. Wire the other lead of the LED (the anode) to line 0 of port B (PB0). Finally, wire the other end of the resistor to ground.

Now if you force PB0 high, the LED lights. Force the line low and the LED goes dark. Translated into pcbug11, this becomes:

ms 1004 01   to turn LED on
ms 1004 00   to turn LED off

If you run the little program we entered above, the LED will blink on and off so fast it will light at half-brightness.

But you can't keep typing your programs into pcbug11's one-line assembler; besides being error-prone, you have no easy way to maintain or to change your programs. You really need to use a text-editor and a stronger software tool such as an assembler or compiler. Armed with these tools, you can quickly go from your ideas for a program to an S19 object file, which you can then load into your 68hc11 via pcbug11. The S19 object file contains only the information the 68hc11 needs; that is, what opcodes get burned where in memory and how the program starts up following reset.

The best way to learn how an MCU really works inside is by learning how to write assembly language software for the chip. To make assembly language programs work, you have to know how the registers operate, what type of opcodes the MCU supports, and how the stacks and addressing modes behave. Learning 68hc11 assembly, however, is beyond the scope of this article, though I will point you to some helpful tools to get you started. The Motorola FREEWARE site contains two assemblers available for downloading. Look for files named as11 and asmhc11. Their user interfaces are both fairly user-hostile, unfortunately, so you might have to dig a little and ask some questions on the 'net if you get stuck trying to use them.

Another option for you is to use a compiler. This software tool takes a source file written in a high-level language, such as C, and translates it into assembly language for you. The compiler usually comes with other tools, such as a matching assembler, that completes the compilation job and leaves behind the S19 object file. A good compiler can really ease the task of writing robot software. You can focus on the larger picture of what you want your robot to do, and not get bogged down in the details of what value is in what register when.

Perhaps the friendliest compiler to start with is my SBasic compiler, available on my web site. SBasic lets you write programs in a simple dialect of Basic,

```
'
'     blinky.bas          a program to blink an LED slowly
'

include    "regs11.lib"

declare    n

main:
do
        n = peekb(portb)          ' get current port B
        n = n xor $ff             ' reverse all bits
        pokeb portb, n            ' change port B

        for n = 0 to $7ffe        ' software delay
        next

loop

end
```

Figure 1. SBasic program to blink an LED slowly.

then compile the source file down to an assembly language file. You then use the included asmhc11 assembler to assemble that file into the final S19 object file. To help you with SBasic, the distribution file includes a 60+ page manual on setting up and using the compiler. The rest of this article will use example code written in SBasic.

Let's go back to our LED and the program that blinks the LED too fast. One way to slow down the LED is to stick a software delay, called a timing loop or a delay loop, into the instruction flow. I'll rewrite this program in SBasic and stick in the delay loop. See the accompanying listing for details (Figure 1).

Notice how the SBasic listing lets you look at the concepts behind the program, rather than the lowest-level details. For example, the program reads and writes **portb**, not the address $1004. In fact, the SBasic program even buries the connection between **portb**, which is how you want to refer to the I/O port, and the address $1004, which is how the assembler must refer to the I/O port. The connection between these two is hidden in the **include** statement at the beginning of the program. This statement causes the SBasic compiler to switch over and begin com-

piling the file regs11.lib, which contains many lines of bookkeeping information such as:

const portb = $1004

You don't need to add these **const** statements to each 68hc11 program you develop; just add the **include** statement and the compiler handles the rest.

I'll assume you have written the SBasic program into a file named blinky.bas. To compile this program, enter the command:

sbasic blinky /v0000 /s00ff /cf800 >blinky.asc

This command causes SBasic to compile the file blinky.bas, placing the SBasic variables at address $0000, the 68hc11 stack at $00ff, the code for this program at $f800, and the output assembler source file to file blinky.asc. The code and stack locations are specifically set up for the 68hc811e2; you might need to change these for other variants of the 68hc11.

Check the blinky.asc file to see if you got any errors from the compilation; there shouldn't be any.

While you're looking at the file, compare the assembly language source for each SBasic instruction with the little program we used in pcbug11 above. You should notice several similarities.

To create the final S19 object file from blinky.asc, you need to run the asmhc11 assembler. Use the command:

asmhc11 blinky.asc

to perform the assembly. asmhc11 will leave behind a file named blinky.s19, which is the object file for your program. Remember that I said asmhc11 wasn't very user-friendly. You can run into a problem with the above statement if you did all of your work on the D: drive and try to assemble your blinky.asc file. asmhc11 wants to find your .asc file on the C: drive, and you will not be able to convince it to look into your working directory if that directory is not on C:. The simplest solution is to include a full path for the .asc file, as in:

asmhc11 d:\karl\68hc11\blinky.asc

This is awkward, I know, but asmhc11 is pretty old and in those days one hard drive was usually enough. A better solution is to build up a custom batch file for automating both the compile and assemble operations; you're on your own here.

Now that you have an S19 file for your blinky program, you need to move it into the BOTBoard so you can actually see it run. This gets us back to pcbug11. Restrap your BOTBoard for special bootstrap mode, then press the reset switch. At the pcbug11 command prompt, enter the command:

restart

This command causes pcbug11 to reload its talker into the 68hc11 and re-establish communications. This technique comes in handy when your 68hc11 program runs away and you have to reset everything to try again. Note that pcbug11 will still remember any previous **control** or **eeprom** commands you might have entered.

Now you are ready to load your S19 file into memory. Enter the command:

loads blinky

pcbug11 will open blinky.s19 and burn the object file into the 68hc11's EEPROM at the proper addresses. When the operation is done, which will only take a second or so, you can then enter:

verf blinky

This causes pcbug11 to reopen blinky.s19, but only for comparing the object file with the current contents of the 68hc11's memory. If everything matches, your program successfully burned and you are ready to go. Note that you can specify full paths on both the **loads** and **verf** commands, if necessary.

To run your program, restrap the BOTBoard for single-chip mode, then press the reset switch. You should see the LED begin blinking immediately. The length of time that the LED is on or off gives you a crude measure of how fast the 68hc11 is running. The program turns the LED on or off, then sits in an empty loop incrementing and testing variable **n** until **n** exceeds $7ffe. At this point, the program reverses the state of the LED and restarts the empty loop. Thus, the LED is on or off for the amount of time the 68hc11 takes to count to 32,767. As you can see, the MCU gets through the loop pretty quickly.

The blinky program is a long stretch from being a working robot program, but that isn't its purpose. I used it to show you how to go from a concept for a program to the SBasic source code to the final object file to burning the file into a 68hc11 and running the program. Everything else you ever write in SBasic will follow pretty much the same pattern, but the programs will be larger. blinky makes a good practice program to get you started.

Now that you know how to program your 68hc811e2 using SBasic and pcbug11, we can move onto more sophisticated ideas. One technique you will need to master fairly quickly uses SBasic's **print**

```
'
'  prtest.bas        a program to print debug info
'

include    "regs11.lib"

declare    n
declare    j

main:
pokeb  baud, $30                  ' 9600 baud
pokeb  sccr2, $0c                 ' enable SCI xmtr & rcvr

print  "prtest.bas"               ' here we are

j = 0
do
     for n = 0 to $7ffe           ' wait a bit
     next

     print "Loop"; j; "is done!"
     j = j + 1                    ' count this loop
loop

end
```

Figure 2. SBasic program that prints debug info.

statements to send debugging information over the serial port back to the host PC. Such debug info can be invaluable in pinning down errors in your program. Again, pcbug11 can help you in this. We'll start with the prtest.bas program shown in Figure 2.

Like blinky.bas, this program starts with the **include** statement to preload all of the 68hc11 register equates. Next, the program declares two variables for later use. The code following statement **main** sets up the 68hc11's serial communications interface (SCI) for uses with SBasic's **print** statement. Your program must set up these two registers properly, or any subsequent **print** statements simply won't work. As shown, these two statements set the SCI so it talks to the host PC at 9600 baud; by default, it also uses eight data bits, one stop bit, and no parity bit.

The rest of the program consists of the same high-speed empty loop we used in blinky.bas, along with a debug **print** statement that prints the value of **j** at the end of each loop. Compile and assemble this

program using commands similar to those above, then use pcbug11 to burn the object file into your 68hc11.

When the program is ready to run, restrap your BOTBoard for single-chip mode but don't press reset yet. Instead, enter the pcbug11 command:

term

This puts pcbug11 into a special terminal mode, much like a modem comm program. By default, pcbug11's terminal mode matches the configuration we set up in prtest.bas, so we should see anything the 68hc11 sends us following reset. Finally, press the BOTBoard's reset switch. You should see a stream of statements moving up the screen. Each statement should contain a higher value of **j**. If you wait long enough, you will see **j** count up to 32,767, switch to negative numbers, then finally reach 0 and start over again.

This program shows the basics of sending debug info back to the host PC. You can spiff up this routine by using other features of SBasic. For example, using **printx** instead of **print** causes all numbers to appear in hexadecimal, which can come in handy when you want to know what is appearing on an input port or in an A/D register. You can also use the **inkey()** function to wait for a character from the host PC before proceeding. This lets you halt the program while you examine the debug info, then press a key to continue.

We are still a long way from having a robot running, but you should begin to see how you can use SBasic and the 68hc11's I/O ports to get things done. I'll finish up with a final program for moving servo motors, which is how we got started down this path previously.

The 68hc11 can provide up to four low-speed PWM signals for controlling, among other things, hobby servo motors. The BOTBoard supports this feature with its four three-pin servo control ports; refer to the BOTBoard docs for details. All your program must do is configure the four corresponding timer channels, TOC2 through TOC5, so a signal of the proper duration appears at each servo control pin.

This setup is fairly straightforward, once you understand how the 68hc11 timer subsystems work. If you're trying to build your first 'bot, however, you probably want to just get on with it and figure out the low-level stuff later. (I know I would, at any rate.) To that end, my distribution package contains an SBasic library file called servos.bas. You can include this file in your SBasic program just as you include the regs11.bas file. When you do, your SBasic program can then simply invoke the new subroutine **InstallServo** to set up a timer channel for controlling a servo motor. See the accompanying listing of myservo.bas (Figure 3).

This program ties one servo to timer channel TOC2 and another to TOC3. It then writes a 16-bit value of $f500 to TOC2 and a value of $f700 to TOC3. The program concludes with an **end** statement, which SBasic compiles into a single instruction that jumps to itself. This creates an endless loop that effectively locks up the 68hc11.

The two timer values create two different servo control pulses; each affects its corresponding servo motor differently. The value you write to a TOC register determines the servo's position if you have not modified the servo as described last month. If you did the mods, then the TOC register value deter-

```
'
'    myservo.bas        a program to test servo motors
'

include    "regs11.lib"
include    "servos.bas"

main:
gosub InstallServo,  toc2
gosub InstallServo,  toc3

toc2  =  $f500
toc3  =  $f700

end
```

Figure 3. SBasic program that controls servo motors.

mines both the servo's speed and direction of rotation. Thus, by doing nothing more than writing a value to a register, your software can change the robot's direction of travel and its speed.

I can simply "fall" into an **end** statement at the end of this SBasic program because the 68hc11 timer subsystem is so powerful that it generates the servo pulses without any software overhead at all. You don't need to write any interrupt code and your program doesn't have to monitor any input lines. Your code simply writes a value any time you want the servos to do something different, then goes back about its business.

To watch this program work, download it into your BOTBoard and connect a pair of servos to the proper servo ports. Strap your BOTBoard for single-chip mode and press reset. You should see your servos turn 'round and 'round, or swing to position, whichever is appropriate. Experiment by using different values for the two timer registers to see what effects they have on the servo motors. Generally, you will be using values in the range $f400 to $f800, but the exact performance of each servo will vary.

If the servos don't move at all, make sure you have added the servo power jumper as called out in the BOTBoard docs. Without this jumper in place, the servos will not receive any power. If the servos move erratically or start to move and the 68hc11 seems to lock up, you might need to use fresh batteries, or move up to four C-cells. This runaway happens because the servos momentarily suck so much juice from the batteries that the 68hc11's supply voltage drops below the necessary minimum. In extreme cases, you can pull the servo power jumper and hook a second set of batteries to the board solely for supplying voltage to the servos.

That does it for this month. This should be enough to get you over many of the initial hurdles in designing your first robot. From here on out, it's mostly getting used to the tools, figuring out what you want your machine to do, and then using the tools at hand properly to get the task done. If you have access to my previous columns, take a look at some earlier robots; you will see elements of what we've done here in many of them. Build on those ideas, add your own innovations, and get that 'bot going.

Part II
Software

Perhaps the most intimidating task in building a robot involves writing the software to control it. At least with hardware and electronics, you're dealing with physical elements, but software is pure imagination. The columns in this section provide several powerful tools to help you design and debug your robot's software.

I developed my 68hc11 tiny4th compiler so I could write tiny programs in (guess what?) the Forth programming language. I've always liked Forth, and it is a natural for getting the most functionality in a limited amount of code space. Even if you have tried Forth and decided you don't like it, give tiny4th a look-see.

My 68hc11 SBasic compiler lets you write your programs in a friendly, well-known language, yet the resulting executables run nearly as fast as programs written in assembly language. What's more, the compiler's output is 68hc11 assembly language source. This means you can skim through the listing file created by SB to see how the compiler translates each of your SB source lines into the equivalent 68hc11 assembly language. I've written microcontroller software for years, and I still consider SBasic

my tool of choice for getting the job done quickly.

The first two columns in this section describe compilers, but the third column gives you a complete 68hc11 assembly language project geared specifically for remotely operated robots. You can load 811bug into a suitable 68hc11 board, hook up a serial link of some kind to your robot, and actually reload your robot's entire operating program over that link. I have used 811bug to upgrade the software in BYRD, my backyard robot, even when BYRD was downstairs and I was sitting in my upstairs lab.

The V25 microcontroller can execute programs written in a subset of the PC's assembly language. In fact, the fourth column in this section describes a complete system for storing .COM files in a simple file structure, then burning that file into EPROM and running those programs on demand with a V25-based robot brain. This system, dubbed BOTBios, lets you use your PC compilers and development tools to create robot code, then move that code into your microcontroller. The finished project even includes a command-line interface that runs from your robot's serial port, so you can run and debug the various .COM files.

The last column in this section covers the lowest levels of the 68hc11. I describe the reset sequence, how interrupts work, and the mechanics of logical operators. You have to understand how functions such as AND, OR, and XOR work if you are ever to succeed in writing robot software, and this column should get you over that hurdle.

My Tiny Forth Compiler

In this column from December of 1993, I described my newly-created 68hc11 Forth system, which I called tiny4th, or t4. My inspiration for t4 came from one of the early Seattle Robotics Society (SRS) campouts. A few of us would sit around the campfire on a summer weekend, discussing all things robotic, and I always ended up with some dynamite robotics ideas. One of those ideas, a tokenizing compiler that could generate executables small enough to run in 2K bytes of code space, became the tiny4th system you see here.

This column also marked my first use of the Rayovac Renewal alkaline batteries. Though this column describes them as being test-marketed, they obviously did very well, and are now widely available and a mainstay in hobby robotics. Just remember, you read it here first!

Previously, I introduced you to the Forth programming language, (See "Quick and Easy 68hc11 Expansion"). This column describes my 68hc11 Forth compiler, tiny4th. It concludes with a discussion of a new rechargeable alkeline battery introduced by Rayovac Corporation.

Tokens and Forth

I have written most of my 68hc11 robotics software in assembly language. I would have preferred using a high-level language such as Forth, C, or even Basic, but I couldn't find a compiler that met enough of my needs.

One of my biggest requirements is that the compiler generate code tight enough to run meaningful robotics programs in just 2K of EEPROM. This would let me build robots using a 68hc811e2 in single-chip mode. This in turn means tinier robots, programmed in a high-level language.

Another major requirement is cost; namely, the compiler should be free. There are compilers available for less than $100, notably Dave Dunfield's Micro-C compiler, but my budget runs more toward the public-domain and shareware prices.

I couldn't find a compiler anywhere that met these and some other needs, so I decided to write my own. Forth compilers are very easy to write, and

Forth happens to be my language of choice, so picking a language was easy.

For a couple of compelling reasons, I further decided that my Forth compiler would create tokenized object code, rather than the more traditional threaded object code. Tokenizing creates smaller object files than threading, but the code runs about half as fast. I felt, however, that the final results would still be fast enough for robotics work.

By tokenizing, I mean translating the Forth source code into specific bytes that are a shorthand for the original source. For example, the tokenizer might change the following Forth code:

1 2 +

into the sequence:

0c 01 0c 02 14

Here, the token $0c means "push the next byte onto the stack." Similarly, the token $14 means "add the top two items on the stack, and leave the result on the stack."

This list of tokens, stored in the 68hc11's memory, is the Forth code that the 68hc11 will execute. By changing the Forth source code and recompiling, you can create any type of robot code you want.

All that remains is a small 68hc11 assembly language program to read through the list of tokens, executing them as directed. This program is known as the runtime executive or kernel.

Using such a system to write robot code is pretty straightforward. You start by writing your program in Forth, then compiling the source code using the tokenizing compiler. Next, you write both the resulting object file and the runtime executive into the 68hc11's EEPROM. Finally, you reset the 68hc11 and, if everything went well, the kernel begins executing your Forth program.

I call my compiler tiny4th. The whole system consists of the compiler, the runtime executive, and the Motorola FREEWARE 68hc11 assember,

asmhc11. The system runs on a PC, and requires PCBUG11 to move the object code into your 68hc11 EEPROM or RAM. The system creates a .S19 file containing the runtime executive and the tokenized object. This means you could also burn the working code into EPROM, if necessary.

The tiny4th system contains the following files:

tiny4th.exe
tiny4th.asc
tiny4th.tok
t4.bat

plus the files necessary to support the asmhc11 assembler and PCBUG11. I'll describe each of the above files in detail.

tiny4th.exe

This is the PC-based Forth compiler. It takes as input a source file containing your Forth code, and creates an output file of 68hc11 assembly language source. You invoke the tiny4th compiler by typing:

tiny4th infile <options>

where **infile** is the name of the file containing your Forth source code, and **<options>** is a list of parameters that you can use to change the compiler's behavior. tiny4th assumes that the source file contains a .t4 extension.

The compiler will open the named input file, compile the source, and write the output assembly language source to the screen. You can write this output to a text file by using redirection when you invoke the compiler:

tiny4th infile >outfile

where **outfile** is the name of a text file. For example, to compile the source file test.t4 and write the assembly language output to test.asc, you would use the invocation:

tiny4th test >test.asc

tiny4th accepts several options that control the placement of program elements in the 68hc11's address space. You can specify where you want tiny4th to place the kernel's stacks, the variable space, and where you want your executable code to reside.

Options are indicated by a slash, followed by a single letter, followed immediately by a parameter, if required. For example, the option:

/cf800

tells tiny4th to compile the resulting code starting at address $f800 in the 68hc11 address map. Note that there is no space between the option letter and its parameter (in this case, $f800). The following paragraphs describe each option in detail.

The /c option fixes the starting address of the resulting 68hc11 code. This will be the first address of your executable. Normally, tiny4th creates code that starts at $8000, but you can use the /c option to change that if necessary.

For example, you might use this option if your 68hc11 system only has 8K of EPROM, starting at $e000. In this case, you would invoke tiny4th with a /ce000 option. You must use a four-digit hexadecimal parameter with the /c option.

The /r option fixes the starting address of RAM in your 68hc11 system. tiny4th will locate its internal variables starting at this address. Normally, tiny4th starts its internal variables at $0000, the default address of the 256-byte RAM block in most 68hc11s.

You can use the /r option to move tiny4th's RAM block, if necessary. For example, your system might have an 8K block of RAM at $2000 that you would like to use for tiny4th's RAM area. In this case, you would invoke tiny4th with a /r2000 option. You must use a four-digit hexadecimal parameter with the /r option.

The /v option fixes the starting address of tiny4th's variable area. Normally, tiny4th sets the RAM address for your variables to be slightly above

the address of its internal variables (see /r above). This is usually adequate.

If, however, you really need to start your program variables somewhere else, you can use the /v option. For example, you would use a /v4000 option to force tiny4th to place your variables starting at $4000. You must use a four-digit hexadecimal parameter with the /v option.

The /s option fixes the top of the two stacks used by the runtime executive in executing your program. Normally, the kernel's return stack starts at $00ff and the kernel's data stack starts $40 bytes below that. This arrangement is usually adequate.

If, however, you really need to move the stacks to another address, you can use the /s option. For example, a /se0ff option would place the top of the return stack at $e0ff. Note that the data stack is fixed at $40 bytes below the top of the return stack, regardless of where you place the return stack. You must use a four-digit hexadecimal parameter with the /s option.

Finally, the /e option assigns the above addresses as needed to create code for a 68hc811e2 running in single-chip mode. Specifically, the RAM block starts at $0000, the variables start at $0008, the top of the stacks is at $00ff, and the code starts at $f800. There is no parameter with the /e option.

tiny4th.asc

This file contains the source code for the run-time executive, written in 68hc11 assembly language. This file, added to the output of the tiny4th compiler, serves as the assembly language source for your 68hc11 program.

This file is not very large, but it is still a complete kernel for the tokenized tiny4th system. I encourage you to look through the source file, study the comments, and understand how the run-time executive works.

If you like, you can rewrite the kernel to try and increase its speed or decrease its size. You have full control over the source code for the kernel, so have fun. Let me know what improvements you make.

tiny4th.tok

This file contains the tokens used by tiny4th. Specifically, it contains 68hc11 equates that assign a hexadecimal byte to a string representing a token. For example, the line:

tk_dup equ $00

assigns the hex byte $00 to be the token that stands for the tiny4th word DUP.

The tiny4th run-time executive handles over 32 tokens; all tokens it understands are defined in this file. You may change the hex values assigned to these tokens, if you feel the need. Note, however, that you must make the corresponding changes in the token table in tiny4th.asc. Also, you may not change the string associated with any token; that string is set by the tiny4th compiler. Token values may range from $00 to $7f. Of these, I am reserving tokens from $40 to $7f for future expansion.

t4.bat

This file is a simple batch file for compiling tiny4th programs. You use it by invoking its name, followed by the name of your source file. For example:

t4 test

This command compiles test.t4, checks for errors, and, if no errors occurred, automatically assembles the resulting test.asc file. Feel free to modify this file if you like.

Inside tiny4th

tiny4th offers many of the features of a full Forth compiler. For example, it allows you to declare **VARIABLE**s for holding 16-bit values and **CONSTANT**s for declaring fixed 16-bit values. It supports ":" and ";", so you can compile Forth words. Refer to Figure 1 for a partial list of tiny4th's vocabulary.

dup	N	(n -- n n)	
swap	N	(n1 n2 -- n2 n1)	
drop	N	(n --)	
rot	N	(n1 n2 n3 -- n2 n3 n1)	
over	N	(n1 n2 -- n1 n2 n1)	
+	N	(n1 n2 -- n1+n2)	
-	N	(n1 n2 -- n1-n2)	
*	N	(n1 n2 -- n1*n2)	
/	N	(n1 n2 -- nquot)	
mod	N	(n1 n2 -- nrem)	
/mod	N	(n1 n2 -- nrem nquot)	
@	C	(addr -- n)	get 16-bit n at addr.
c@	C	(addr -- c)	get 8-bit c at addr.
!	C	(n addr --)	put 16-bit n at addr.
c!	C	(c addr --)	put 8-bit c at addr.
and	C	(n1 n2 -- n1&n2)	
or	C	(n1 n2 -- n1|n2)	
xor	C	(n1 n2 -- n1^n2)	
<<	C	shift nos left by tos bits	
>>	C	shift nos right by tos bits	
not	C	(n -- ~n)	1s complement of n
negate	C	(n -- -n)	2s complement of n
=	C	(n1 n2 -- f)	test for equality
<	C	(n1 n2 -- f)	test n1 < n2
>	C	(n1 n2 -- f)	test n1 > n2
U<	C	(u1 u2 -- f)	unsigned <
U>	C	(u1 u2 -- f)	unsigned >
0=	C	(n -- f)	test for 0
true	C	(-- f)	puts true on stack
false	C	(-- f)	puts false on stack
constant	I	(n --)	create named constant
variable	I	(--)	create named variable
timer	I	(--)	create named timer
decimal	I	(--)	switch to decimal radix
hex	I	(--)	switch to hex radix

Figure 1. A partial list of tiny4th's vocabulary.

It recognizes a few special words, as well. For example, the word **INCLUDE** lets you include other tiny4th source files inside your source file. This lets you build up libraries of working tiny4th programs, then include them in other, larger programs.

The syntax of the **INCLUDE** word is:

include file.ext

where file.ext is the file name and extension of the file to be included. You may specify a full path, if necessary.

The word **TIMER** declares a 16-bit variable of a special class. Variables declared with the **TIMER** word automatically count down at the rate of 1 count per 4.2 milliseconds (msecs). When such a variable's value reaches 0, the value stays at 0. You can use **TIMER** variables as downcounters, to control the timing of behaviors in your 68hc11 robot code. The syntax of the **TIMER** word is:

timer foo

which declares a 16-bit downcounting timer named foo. You may have as many timers as you want, subject to RAM availability.

TIMER and **INCLUDE** are examples of tiny4th interpret-time words. This means that they are only legal when the tiny4th compiler is interpreting your source file, not when it is compiling your source file. The difference is subtle but important, and using tiny4th requires that you understand the difference between interpret-time and compile-time.

Simply put, tiny4th is compiling when it is processing source text between a : and a ; in your file. At all other times, tiny4th is interpreting your source text.

For example, suppose your source file contains the text:

2 3 +

What tiny4th does with this source text is very different, depending on whether tiny4th is compiling or interpreting it.

If tiny4th encounters this text within a colon definition (that is, between a : and ;), it compiles the text into a series of tokens and writes those tokens to the output file.

If, however, tiny4th finds this text while it is in interpret mode, it pushes a 2 and a 3 onto its own data stack, then adds the values and leaves a 5 on its stack. Note that tiny4th does not write anything to the output file through this whole operation.

tiny4th has over 50 words in its vocabulary at invocation. These words are divided into three major classes, depending on whether they are legal in compile-time, interpret-time, or both.

Words that are only legal during compile-time are known as compile-only words. This group contains most of tiny4th's vocabulary, including **BEGIN**, **REPEAT**, **IF**, **@**, **C!**, and **U<**. A word is usually in this group because it only makes sense if executed by a Forth compiler. tiny4th's compiler is not a real Forth compiler, so it does not actually execute any Forth words.

Words that are only legal during interpret-time usually create timers, variables, constants, and new definitions. For example, **VARIABLE**, **TIMER**, **INCLUDE**, and **ALLOT** are all interpret-only words.

Another group of words is legal at both interpret-time and compile-time. These words are usually related to stack operations, and their actual behavior changes, depending on whether tiny4th compiles them or interprets them. Words such as +, **DUP**, **AND**, and **XOR** are examples of this group of words, known to tiny4th as normal words. Note that numbers fall into this group, as well.

Starting tiny4th

Begin using tiny4th by installing it on your PC, preferably in its own directory, such as c:\tiny4th.. Next, create a source file using an ASCII text editor. Nearly any editor will work; just make sure the editor doesn't add any hidden formatting characters to the text file.

Refer to the accompanying listing (Figure 2). Enter the text as shown into a file named test.t4. This simple program, when run, will add 2 to 3, then drop the result off the stack. Not very useful, but it will do for now.

Next, compile test.t4 with the command:

t4 test

```
timer  bar              ( an example of declaring a timer )
variable  mark          ( an example of declaring a variable )

: foo           ( -- )  ( a simple colon definition )
   2  3  +      ( add 2 to 3 )
   drop         ( now get rid of result )
;
```

Figure 2. Your first tiny4th program.

t4 will compile the program, then invoke the asmhc11 assembler to assemble the program. You can look through the resulting test.lst file to see exactly what the final program looks like.

Several items about the resulting test.lst file are worth mentioning. Start by moving to the end of the list file where the various 68hc11 vectors are declared. Note that tiny4th has set up the reset vector at $fffe to point to address START. START is the address for the beginning of the tiny4th runtime executive; all tiny4th programs will have this address as the reset vector address.

Now move to the part of the code at address START. Notice that a few lines below START, the kernel loads the X register with the address of FIRSTOP. FIRSTOP is the first tiny4th word that will be executed on powerup.

You never actually assigned any tiny4th word in your program to be FIRSTOP. tiny4th always uses the last word created in your program as FIRSTOP. Since you only created FOO in your program, the address of FOO became FIRSTOP.

Now look for the section of code labeled "Variables and timers," near the end of your tiny4th code. This section shows how tiny4th handles timers and variables. Note that the timer you declared as BAR is known in the list file as TIM000, just as the variable MARK is labeled VAR000. tiny4th automatically translates your names into short unique names of its own. tiny4th provides your original name in a comment at the end of the appropriate line, so you can tell which variable is which.

You can try using pcbug11 to download the resulting .S19 file into a 68hc11 and running the program, but you probably won't like the results. First of all, this program doesn't generate any output, so you couldn't tell if it ran, anyway.

But more importantly, the program doesn't end properly. By this, I mean that after the kernel gets through adding 2 to 3 and dropping the result, it reaches the end of the program, marked by the **;** in the source file. So, what happens then?

Well, the kernel "runs off" the end of the program and starts executing data left over in RAM or EEPROM from whatever program was last loaded into the 68hc11. To prevent this from happening, you must write your last tiny4th word so it never exits. The second program sample (Figure 3) shows one way to do this.

Here, I have added the phrase **BEGIN-REPEAT** to create a short endless loop at the end of FOO. It doesn't do anything useful, but it does stop the kernel from running off the end of the program.

The third example (Figure 4) shows how to access 68hc11 I/O registers, and actually affect the outside world with your tiny4th programs. This program sets the 68hc11's asynchronus serial port (the SCI) to 9600 baud, then endlessly transmits the letter A out the serial port.

You can compile this program, then download it into your 68hc11's EEPROM or battery-backed RAM, using pcbug11. After downloading, turn off the 68hc11's power and execute a PC-based communications program, such as Procomm or Crosstalk.

```
timer  bar              ( an example of declaring a timer )
variable  mark          ( an example of declaring a variable )

: foo            ( -- )  ( a simple colon definition )
   2  3  +              ( add 2 to 3 )
   drop                 ( now get rid of result )
   begin  repeat        ( loop forever )
;
```

Figure 3. Your second tiny4th program.

```
hex              ( switch to hexadecimal )

102b  constant  baud          ( addr of baud rate register )
102f  constant  scdr          ( addr of SCI data register )
102e  constant  scsr          ( addr of SCI status register )
102d  constant  sccr2         ( addr of SCI cmd register 2 )

: init-serial     ( -- )    ( initialize 68hc11 serial port )
   30  baud  c!    ( set baud = 9600 )
   0c  sccr2  c!    ( turn on transmitter and receiver )
;

: emit              ( c -- )   ( output char to port )
   begin
      scsr  c@      ( get port status )
      80  and       ( leave only xmtr status bit )
   until            ( loop until bit is set )
   scdr  c!         ( write tos to serial port )
;

: main              ( -- )   ( test program )
   init-serial      ( make port ready )
   begin
      ascii  A      ( get the character 'A' )
      emit          ( write to port )
   repeat           ( do forever )
;
```

Figure 4. Accessing the 68hc11 I/O registers in tiny4th.

When you apply power to the 68hc11, you should see an endless stream of As on your terminal.

This program, though short, shows many interesting elements of Forth in general and tiny4th in particular. Note the liberal use of comments, which helps clarify what might be termed cryptic Forth lines. As you can see, I also indent each control structure, such as **BEGIN-UNTIL** and **BEGIN-REPEAT** loops. Also, I like to include a short stack diagram in the initial line of a colon definition. This reminds me how to set the stack up before I invoke a word. Finally, notice that the lines in my tiny4th program are short. Each line usually contains a single operation or concept, which eases following the flow in the program.

More Details

Programs written in tiny4th are smaller than an equivalent assembler program, with some restrictions. If you simply wrote a dedicated assembly language program to write As to the 68hc11 serial port, you could certainly do better than tiny4th can.

But once your assembly language program gets much beyond the size of the run-time kernel, tiny4th's compiled code will prove to be the smaller of the two. This is because each tiny4th instruction is usually just one or two bytes. This becomes very important when you begin invoking your own colon definitions from within other colon definitions.

Additionally, you can usually leave most of your data on the stack, instead of having to load from and store to variables. Since tiny4th's stack operators are also very short, instructions for accessing data don't require much space.

All of this means that you can do quite powerful robotics applications in very small amounts of ROM or EEPROM. In fact, the 2K of EEPROM in a 68hc811e2 seems immense when you see how much robotics code you can pack into it. For example, the executable file for the third sample program only uses 669 bytes!

The flip side of small, however, is usually slow. A tiny4th program runs much more slowly than equivalent assembly language program, even a poorly-written one. It takes about 28 microseconds

for tiny4th to leave one tokenized operation and start the next. Thus, the fastest that tiny4th can run is about 33,000 instructions per second. Include the actual time used by a typical tiny4th word, and you will probably get about 25,000 tiny4th instructions per second on an 8 MHz 68hc11 computer.

This seems quite slow, compared to the 500,000 assembly language instructions per second you usually get from a 68hc11, but don't be deceived. tiny4th's execution times are still fast enough to do meaningful work with small robots.

I'll include a few comments here for those of you already proficient in Forth and interested in what is going on under the hood. The tiny4th compiler is written in Mix PowerC as several large switch statements. I don't like this from an elegance standpoint, but it did get the compiler up and running quickly.

Since tiny4th is not itself a Forth compiler (more of a translator, actually), it cannot create words that it can subsequently execute. Thus, tiny4th does not support the traditional behavior of **CREATE** and other compiler-building words.

Instead, all of tiny4th's compilation is geared to writing assembly language source files. That is why, for example, the tiny4th word **CONSTANT** doesn't actually make an executable token. It is enough for **CONSTANT** to generate the proper EQU pseudo-op and let the assembler deal with it later. However, the tiny4th compiler does have a data stack, and can support limited stack manipulations in interpret mode. This can make some **CONSTANT** declarations easier.

Where Do You Get It?

You can get a copy of the current version of tiny4th, complete with the required asmhc11 assembler, from my web site. Note that tiny4th is an evolving system. While I expect the system you see a month or so from now will behave much like the tiny4th described in this article, I can't guarantee it. The documentation files included with each tiny4th will provide full details on its features and how to use it.

You need to read through Leo Brodie's *Starting Forth* book (published by Prentice-Hall, Inc.,

Englewood Cliffs, NJ, 07632. ISBN 0-13-843079-9) for good background information on the Forth language. Realize that tiny4th is not an interactive system, so you won't be able to try many of Leo's exercises, but you will get a feel for the language's power.

Have fun, and let me know what you think of this new robotics tool.

Check This Out!

Rayovac Corporation is test-marketing a rechargeable alkaline battery, targeted at consumer electronics. The Renewal battery comes in D, C, AA, and AAA sizes and should be available nationwide in time for the Christmas rush. Fortunately for me, Portland and Seattle are two of the test cities, so I have had a chance to try out these batteries already. They are, in a word, awesome.

The Renewal batteries offer from three to five times the current of their NiCd counterparts. According to the Rayovac literature, you will typically get 1.7 Ahrs from a Renewal AA cell, 4.0 Ahrs from a C cell, and 8.0 Ahrs from a D cell! While this isn't quite as good as the nonrechargeable alkalines, it beats the socks off of NiCds. And Renewal cells cost no more (and usually less) than the same size NiCds.

Because these are actually alkaline cells, they exhibit many of the good qualities of traditional alkalines. For example, they exhibit very low self-discharge rates; usually about 0.2% per month. Compare this to the 30% self-discharge rate for NiCds.

Also, the Renewal batteries have a sloping discharge curve. This means that your robot can sense when its battery supply is running low, and still have time to do something about it before everything shuts down.

These batteries must be recharged using a newly-developed technique. Rayovac sells two suitable chargers. I bought the larger model, capable of recharing up to eight batteries of various sizes from AAA to D. Cost of this table-top unit was about $30.

You can also buy a wall-mount unit that plugs directly into an outlet. It will recharge up to four of the smaller cells, either AA or AAA. Rayovac has priced this machine at about $15 each.

Note that you must use these batteries in a Renewal charging system, and you cannot use a Renewal charger to recharge any other type of batteries, including traditional alkalines and NiCds.

I like other aspects of this battery system as well. Rayovac claims that these cells have no cadmium added to them, and they only use about 0.025% mercury. Rayovac notes that they intend to eliminate all mercury in their Renewal batteries by 1994.

The company estimates that consumers will see about 25 recharge cycles on each Renewal cell, about on par with actual life from most NiCds.

Unfortunately, Rayovac does not yet offer a Renewal 9-volt battery. This seems to be related to their recharging technology, which recharges each Renewal cell individually. This in turn means that you cannot recharge a pack of Renewal cells; at least, not yet. Rayovac has received questions about pack-recharging, and I would hope it becomes available in the near future.

Still, I really like the Renewal cells. Add up the long shelf life, low cost per cell, and high energy density, and I think Rayovac has a winner. Anyone want to make me an offer on some well used NiCds?

===

In the Robo-news

PolyPlus Battery Company of Berkeley, California, has announced development of a lithium battery that offers exciting possibilities.

The cell uses an anode of lithium about 50 microns thick, a separating film of polyethylene oxide about 30 microns thick, and a composite cathode up to 100 microns thick. Not counting any supporting frame, this results in a battery less than 200 microns thick.

And what a battery. Based on work originally done at Lawrence Berkeley Laboratories in the 1980s, the lithium battery generates 3 volts per cell, is fully rechargeable, and has more than three times the theoretical energy density of a NiCd cell. It is even recyclable.

What's more, the battery is completely dry. Nearly all commercial battery technologies use some type of liquid in their construction. This largely dictates the shape of common batteries, as they all are really liquid containers.

But the lithium cell uses no liquid in its final form, so it can be made literally as thin as a sheet of thick paper. For example, you could make greeting cards that sing and talk; the battery would actually be one of the pages.

Apparently, the technology exhibits almost no self-discharge problems. Studies indicate that the cell would take nearly 25 years to self-discharge 30% of its energy, something that a NiCd battery will do in just a month.

The cell's properties make possible safer, lighter, more powerful electric cars. Automobiles using this technology would not need heavy frames to support the weight of lead-acid batteries, or protective walls to shield occupants from acid spills during a crash. Instead, you could simply build the car's lithium batteries into the flooring and paneling of the auto.

Note that much of this technology is still in the research labs. PolyPlus claims that all the materials and processes needed to make these batteries already exist; no additional inventions are required. However, actual production is not yet underway.

To find out more about this fascinating technology, contact:

May-Ying Chu
PolyPlus Battery Co.
809 Bancroft Way
Berkeley, CA 94710
(510) 841-7242

(Source: Design News/9-6-93/pgs 113-115)

A First Look at SBasic

This column marks a pivotal point in my *Amateur Robotics* column. In this article from the May 1995 issue of *Nuts & Volts*, I introduce the first version of my 68hc11 SBasic (SB) compiler. I wrote SB to fill a long-standing need: a tool that would let me quickly write fast, small 68hc11 programs. I had tried and discarded several options before settling on the Basic language, but no one had written a 68hc11 Basic compiler that I could use for free. So finally, in desperation, I sat down and wrote my own, building on ideas from Herbert Schildt's excellent book, *The Art of C*.

This first version of SB is almost embarassingly primitive, yet it portends great things. SBasic grew from this simple, 26-variable program into a powerful, highly-featured compiler. Even today, I write nearly all of my 68hc11 programs in SB, and the compact code that the compiler generates lets me get a lot of functionality in just the 2K bytes of space in a 68hc811e2.

This column also contains a review of the book *Sensors for Mobile Robots*, by Bart Everett. I liked the book, the technology it addresses, and the foreword by Rodney Brooks. It still rates shelf space on any robot builder's desk.

A K Peters, Ltd., publishers of *Mobile Robots*, by Joe Jones and Anita Flynn, are scheduled to release another new robotics book the end of June, 1995. If the prepublication review copy I looked at recently is any indication, I think Klaus and his staff have another winner on their hands.

The book, entitled *Sensors for Mobile Robots*, was written by H.R. (Bart) Everett, a retired Commander in the United States Navy and a former director of the Office of Robotics and Autonomous Systems at the Naval Sea Systems Command in Washington, D.C.

Sensors deals almost exclusively with the theory and application of sensor systems. It covers a bewildering variety of sensors, describing the effect sensed, how to install and effectively use the sensor, weaknesses and strengths of the sensor and the returned data, and how to factor the returned data into developing a more accurate picture of the robot's world.

Often, the sensor discussed is available in one or more commercial configurations. In these cases, Bart usually includes a discussion of the more popular units, and sometimes provides pictures or graphics of existing robot systems built with the commercial sensor units. This helps you gauge how easily such a system might fit into your plans.

I knew I was going to like this book before I even finished the foreword, written by Dr. Rodney

A. Brooks of the MIT AI Lab. Rod Brooks makes a strong case for non-visual robotics sensors, arguing that some robot builders tend to focus on those sensors most like the system we find the most vivid: our eyesight. Nature, however, has devoted millenia to developing non-visual sensors that are remarkably effective. Rod mentions the cockroach specifically.

Any of you who have lived in the Arizona desert and tried to demonstrate your evolutionary superiority over a cockroach probably found it a maddening experience. Simply trying to stomp a cockroach usually makes you look like a fool. You can try the quick stomp, but the roach usually skitters away long before your foot gets there. You can try the sneaky method, where you slowly lower your foot until it's so close to the bug you couldn't miss if you tried, but you usually miss. And I won't even try to describe trying to hit a moving cockroach with anything slower than a 9mm slug. (No, don't ask!)

According to Rod, it isn't the bug's eyes that make it so hard to hit; it's the 30,000 wind-sensitive hairs attached to the roach's legs. These hairs detect minute changes in wind currents, and the bug can respond in only 10 milliseconds. The roach can detect your approaching foot, then change speed and direction far faster than you can react. And we think our vision sensors make us such hot stuff!

Back to the Book

Chapter 1 consists of an introduction describing some of Bart's earliest robots. WALTER, a Science Fair project from the mid-1960s, stood five feet tall and used an AC power cord for its current source. Bart scrounged actuators from sewing machines, a movie projector, a kitchen mixer, and bicycles.

WALTER used vacuum-tubes for its electronics, had two microphones for ears, and a capacitive sensor on one of the two grippers. Bart writes that this sensor was specifically added to discourage pulling and prodding from curious onlookers:

> ... any stray finger that poked its way into
> the open claw would be met by a star-

tling and decidedly effective warning snip. The resounding thump of the actuating solenoid only served to accentuate the message.

I'll leave the story of WALTER's demise undiscussed, other than to say it contains warnings all of us would do well to heed.

Bart went on to create CRAWLER I in the late 1960s, this time for a junior-year Science Fair. CRAWLER I was battery operated, fully autonomous, with tank-like treads for locomotion, a rotating photocell scanner for tracking a homing beacon, and tactile sensors made of guitar strings.

Lacking BOT-Boards in those days, Bart "programmed" his robot using an ingenious system built on a circular punched-paper disc. This program disc measured about 12 inches in diameter, and contained groups of punched holes. The hole patterns encoded each of the 16 possible input sensor arrangements, and the corresponding actuator responses. By altering the holes controlling the actuator patterns, Bart could precode his machine's responses to each possible input combination.

CRAWLER I simply spun the disc until its scanning photocells saw a group of sensor holes matching the current configuration. It then stopped the disc and read the actuator response for this pattern from the output section of the disc, and changed its actuators as needed.

This technique, updated to today's technology, could serve as a good stepping stone toward fully microcontroller-driven robots. Start by replacing the paper disc with a 27c64 or other cheap EPROM. Tie all of the robot's sensor signals to selected address lines on the EPROM. Wire some or all of the EPROM's data lines to your actuators. Then burn the proper responses to each possible input combination into the EPROM.

Voila! A one-chip robot controller, suitable for Science Fair projects and class demonstrations. And, incidentally, capable of quite sophisticated behaviors, given proper programming.

For example, you could hook one of the EPROM's data lines back to an address line, creating a feedback loop. Or, you could add a slow 555 timer as an input sensor, which would give your robot a "heartbeat" or a timing reference. (Actually, I like this idea a lot. You might see more on this concept in a later column.)

Bart followed CRAWLER I with CRAWLER II, a tethered robot, powered by wall current, that was apparently built in his early years at Georgia Tech.

ROBART I was built during Bart's thesis work at the Naval Postgraduate School in 1982. ROBART I contained a huge assortment of sensors, as well as a SYM-1 6502-based microcontroller board. Designed as a security and patrol machine, ROBART I could detect and respond to fire, smoke, flooding, earthquake, and toxic gas. It could detect intruders using optical or ultrasonic motion detection.

Bart's robots became increasingly sophisticated, with larger numbers of sensors and actuators, over the years. No doubt it helped that he didn't have to pick up the tab for all of his machines' additions; he was able to evaluate sensor systems beyond the budget of many robot hobbyists.

You have to see the pictures of ROBART III, a security robot that sports a six-barreled Gatling-style tranquilizer dart gun. The text contains quite a bit of detail on the gun's construction and operation. Very scary!

Chapter 1 concludes, as do all of the other chapters, with a multi-page list of references for relevant papers and publications. For all the information contained in this book, and it contains a lot, the extensive and up-to-date citation list may be the book's greatest asset. Bart has compiled over 500 references for this book, many of them as recent as 1995. Since the references are broken down on a per-chapter basis, you can skip to a desired chapter, read Bart's information, then find additional information by skimming the chapter's reference list.

The power of this technique becomes immediately apparent when you start on Chapter 2, dealing with dead-reckoning navigation. Bart describes several types of odometry sensors, such as potentiometers, synchros and resolvers, various kinds of optical encoders, and Doppler and inertial navigation systems.

He then writes about mobility configurations. This refers to the arrangement of steering and driving wheels; assuming, of course, that your 'bot has wheels or treads. Different configurations accumulate dead-reckoning errors differently, and you can improve your robot's ability to correct for these errors if you understand the weaknesses of your chosen system.

The chapter concludes with over 40 references on steering mechanisms and dead-reckoning navigation. If you want to know a LOT more about, say, Ackerman steering, read Bart's writeup, then head for the library or Internet and check out his references.

By the way, many of these citations actually refer to parts of a larger, more diverse reference. For example, Bart cites papers in Chapter 2 that describe robotic wheelchairs, autonomous robotic cars, mining robots, and tele-operation of off-road vehicles.

Following chapters go into similar detail on various sensor systems, and on using data collected from those systems. Chapter 3 discusses tactile and proximity sensors, Chapter 4 goes into triangulation ranging, Chapter 5 covers time-of-flight (TOF) ranging, Chapter 6 deals with various ranging systems built on phase-shift measurement and frequency modulation, and Chapter 7 covers several other ranging techniques.

Each of these chapters discusses a variety of methods for acquiring data. This usually includes both active and passive systems. It also includes, where appropriate, sensors that respond to a wide range of energies, such as infrared (IR), acoustic (sonar), radio-frequency (RF), laser, and microwave.

Chapter 8 goes into great detail on sonar systems, while Chapter 9 handles electromagnetic energy systems, including optical and radar systems. Starting with Chapter 10, on collision avoidance, Bart begins looking at sensor systems designed to fulfill specific tasks. For example, he discusses navigation control strategies, or general techniques your robot can use to avoid collisions with local obstacles. Each discussion is accompanied by suitable equations and charts to simplify and explain the concepts. Where necessary, Bart recommends sensor systems that fit well with each strategy.

Chapter 11 covers guidepath following, where the robot senses and tracks some type of course-defining agent or material. Here, the sensors must accurately locate the agent, then maintain "lock" as the robot performs its tasks.

Chapter 12 deals with magnetic compasses. This type of sensor has a large number of commercial designs available, and his comments on the different systems make interesting reading. This chapter covers a lot of ground, and I strongly recommend you check Bart's material on compasses before trying to add one to your own machine.

Chapter 13 discusses issues related to gyroscopes. While this is not usually a sensor one associates with a hobby machine, there has been considerable traffic in comp.robotics lately on using gyros, and this material could answer a lot of questions.

Chapter 14 handles position locating systems built around RF components. Such systems could be as simple as a couple of homebuilt RF beacons, or as technologically advanced as the global positioning system (GPS) satellites. Bart discusses the workings of several commercial systems, including Loran and Navstar.

Chapter 15 provides similar treatment of various optical and sonar position locating systems. In some cases, these designs are commercially available, either as components or as total robotic systems. For example, IS Robotics' Genghis walking robots use a pair of sonar pingers, located at two points on the floor or wall, to locate themselves within their world.

Chapter 16 covers world referencing using fixed landmarks, such as walls, doorways, and ceilings. This type of referencing is usually used to cancel out any accumulated dead-reckoning errors, since the robot has reached a known landmark. Bart describes how ROBART II uses elements of this system to locate and confirm its recharging station, and also to zero out its navigation errors.

The final chapter deals with a variety of application-specific sensors. This covers the strange sensors that you might have to add to, say, a security robot that would not be needed in your general-purpose mobile robot. Here, Bart limits his discussion to two applications with which he is most familiar: physical security and automated inventory assessment. Some of the more colorful applications, such as hazardous material handling, are not explored. Still, he provides good coverage of a variety of passive IR (PIR), acoustic, vibration, pyroelectric, and other sensors.

The book concludes with a listing of cited sources, so you know who to call to order copies of any of the cited literature. I didn't see much missing from this book, certainly not in the area of sensor coverage. Anything you would normally need in a robot confined to a human-benign environment is discussed.

I did notice that the book lacks an index. This could have been simply because my copy is not a publication draft; I certainly hope that is the reason. Failing to provide an index to a book like this would be a serious, though certainly not fatal, omission.

Another item curiously missing was any means of checking on-board power levels. Given the wide variety of batteries available, and the number of different power-control and charging systems in use, I would have liked some discussion on how a robot can determine that it is hungry. Perhaps the large number of possibilities to cover precluded any such discussion.

You might have noticed that I did not mention vision systems. *Sensors* focuses on non-vision sensors, leaving discussion of vision-based systems to other writers. Bart does, from time to time, make minor references to aspects of robot vision, but it is cursory only, and certainly not in the depth he devotes to sonar, for example.

And you will be disappointed if you go into *Sensors for Mobile Robots* expecting to find implementation-specific software. This book covers sensors only (well, and some actuators); you get to supply the software for your system.

But this book more than makes amends for these minor faults. This is, simply, an astounding work, in both the range and depth of its coverage. From the most mundane to some fairly bizarre sensors, Bart Everett provides both immediately useful information on theory and application, and (perhaps more valuable) citations to definitive references. These references are often papers developed by the sensor's designer, or in-depth reports of real-world robots that used the sensor in part of its suite.

Get a copy of this book, and keep it on the shelf next to your copy of *Mobile Robots*. I guarantee you'll go back to both works often.

68hc11 SBasic

For years, I waited for someone to develop a Basic compiler for the 68hc11, but no one ever did. At least, I haven't run across it yet in all my years of scrounging. There's something about Basic that is, well, basic. As in simple, easy-to-write software. Face it, almost anyone with a PC has access to some form of Basic, and most such robotic experimenters have tried their hand at writing Basic programs.

True, you can find a 68hc11 Basic interpreter on the Motorola FREEWARE web site. In fact, those of you interested in staging the Basic language on your 68hc11 robot should take the time to download this program and play with it.

But I wanted a compiler, not an interpreter. I'm perfectly willing to trade a little extra code size for a lot more speed. Well, no one ever got around to writing the compiler I wanted, so I am writing my own.

I call it SBasic, for Simple Basic. I started with the source code for a simple Basic interpeter that I copied from an excellent book by Herbert Schildt, entitled *The Art of C* (published by McGraw-Hill, Berkeley, CA ISBN 0-07-881691-2). Anyone interested in the nuts and bolts of simple, functional C programs should pick up a copy of *The Art of C* and spend some time looking through it. Herbert provides full source listings and files (on supplied disc) for fairly hefty PC projects. You get a simple serial-based local-area network (LAN), some simple video and graphics programs, a mouse program, and a simple Basic interpreter.

Starting with Herbert's Basic interpreter, which he calls sbasic, I started adding and changing, redesigning the code as I needed it. The program, vastly changed, no longer executes a Basic source file, but instead turns the file into 68hc11 assembly language source. Note that this is a modest start at a compiler, hardly a commercial grade product. Think of it more as tiny4th for the Basic language.

In its current form, SBasic supports the following operators:

```
+ - * /        Math functions
mod            Modulus
and            Boolean AND
or             Boolean OR
xor            Boolean XOR
=              Assignment, test equals
<>  ><         Test not equals
<              Test less than
>              Test greater than
<*             Unsigned test less than
>*             Unsigned test greater
               than
peek()         Fetch 16-bit value
peekb()        Fetch 8-bit value
poke           Store 16-bit value
pokeb          Store 8-bit value
if-else-endif  Structured if
while-wend     Structured while
interrupt      Interrupt service
               routine (ISR)
return         Return from ISR or
               subroutine
const          Declare a 16-bit
               constant
```

This list is very preliminary, and only describes SBasic as of this writing (end of March, 1995). The features will grow and mutate as time permits.

SBasic currently supports only 26 variables, each 16 bits wide. This will change in the future; I intend to support named variables and arrays, but this small, fixed variable set is a good starting point. I also have the first cut at **FOR-NEXT-STEP** loops installed. As of now, SBasic doesn't handle downcounting (negative step values), but that should change before long.

I also intend to add **DO-LOOP** structures, along with **WHILE-** and **UNTIL-** modifiers. And I want to add a whole slew of goodies to make SBasic more suitable for embedded control and robotics. Some of these features already exist in the list above, and deserve a little discussion.

I have added the <* and >* test operators, to allow unsigned comparisons of 16-bit numbers. Without them, $8000 is less than $7fff; with them, the comparison works the way you expect for a 65K number range. I also added the **INTERRUPT** statement, used to mark the beginning of a 68hc11 interrupt service routine (ISR) written in SBasic. Since SBasic compiles down to fairly tight assembly language, it is perfectly reasonable to write your ISRs in SBasic. I will, however, add the ability to include segments of 68hc11 assembly source, for those instances when you absolutely have to write in assembly language.

So how good is the compiler's output? I have included a very simple test case to show what kind of output you can expect. Note that the compiler does no optimization. It rearranges SBasic operands, where possible, to take advantage of register usage, but it does not optimize from one SBasic source line to the next.

The sample SBasic source file (Figure 1) contains a small program that alternates the output lines of port B between 0 and $ff. After each change, the code sits in a tight **FOR-NEXT** loop, running a counter from 0 to 10,000. My test system is a BOT-Board with an LED hooked to one of the port B lines. This give me a visual indication of the port's state.

After writing a 0 to port B, the code also glances at the SCI status register, in case the host computer sent a character. If so, this code echoes the character back to the host. Note that two lines of code at the top of the source file initialize the SCI for 9600 baud, assuming a crystal of 8 MHz. The SBasic compiler translates this file, prog3.bas, into a corresponding assembler source file named prog3.asc. The prog3.asc file is properly formatted to assemble with the Motorola FREEWARE asmhc11 assembler.

Take a few minutes to examine the accompanying listing (Figure 2) of the output from the SBasic compiler. All lines that do not begin with an asterisk in column one will be treated by the assembler as either source code or assembler directives.

Most of the comment lines (those with an asterisk in column one) are the original SBasic source lines, copied into the output file by the SBasic compiler. This helps you follow the transformation between SBasic source and 68hc11 assembler source.

SBasic automatically reserves 52 bytes of RAM space for its 26 variables, each of which gets 2 bytes.

Any reference to a variable, such as X in source line 26, is converted into an offset added to the address VARBEG. VARBEG is assigned at the top of the output file, and is currently hard-coded to $00c0. Later versions of SBasic will allow full control over the placement of VARBEG, CODEBEG, and STKBEG. The latter two values set the start of the code and stack, respectively.

Check the code generated by source lines 42 and 43, an empty **FOR-NEXT** loop. This represents the minimum amount of code that a **FOR-NEXT** loop will create, and indicates the top speed you can reach in a simple counting loop. This empty loop consumes 28 cycles for each iteration, or 14 μsecs using an 8 MHz crystal. Since the example loops count to 10,000 on each pass, the LED will stay on (or off) for 10,000 times 14 μsecs, or .14 seconds.

Looking at this from another angle, your software could count through this empty **FOR-NEXT** loop more than 71,000 times per second. Obviously, you will have to replace a large portion of those empty loops with meaningful code. Still, the code generated by the SBasic compiler is far faster than tiny4th's interpreted code, and plenty fast enough to do useful robot tasks.

SBasic is one of those evolving software projects; they always seem to take longer than I expect when I start them. But I will continue to refine SBasic, and hope to have a strong version, ready for release, by the time you read this.

I want to make SBasic so simple to use that even total 68hc11 novices can sit down at a PC, write a few lines in a language they already understand, and watch their robots move. Initial feedback from those SRS members who have been kind enough to beta-test my early SBasic executables is encouraging.

Incidentally, I have designed SBasic so that all code generation is confined to a single, well-defined module. I already have plans to make a second code generator, to create SBasic output for the 68302. This will give me the tools I need to develop serious robot code for my hacked Practical Peripheral's ProClass external modem; see "Hacking a 68302 Modem Board" and "Hacking a 68302 Modem Board, Part 2" for details.

```
'
'  prog3.bas      test  program  to  demonstrate
'                 68hc11  SBasic
'

'
'  Declare  some  constants
'

const  portb  =  $1004
const  baud   =  $102b
const  sccr2  =  $102d
const  scsr   =  $102e
const  scdr   =  $102f

const  MAXN   =  10000
const  MINN   =  1000


'
'  Initialize  the  target  variable  (x)  to  the
'  default  maximum.  Initialize  the  serial  port.
'

x  =  MAXN
pokeb  baud, $30
pokeb  sccr2, $0c

'
'  Do  forever...
'

while  1  =  1

'
'  Initialize  a  counter,  turn  the  LED  on,  then
'  count  up  until  the  counter  hits  the  target.
'

      pokeb  portb,  $ff
      for  n=0  to  x
      next

'
'  Initialize  a  counter,  turn  the  LED  off,  then
'  count  up  until  the  counter  hits  the  target.
'

      pokeb  portb,  0
      for  n=0  to  x
      next


'
'  Test  for  a  character  from  the  serial  port,
'  and  echo  anything  received.
'

      if  $20  and  peekb(scsr)  =  $20
            c  =  peekb(scdr)
            pokeb  scdr,  c
      endif
wend

end
```

Figure 1. Sample SBasic source file.

```
*
*  SBasic  compiler  (Beta  0.3)  for  the  68HC11
*
varbeg          equ     $00c0
codebeg         equ     $b600
stkbeg          equ     $00ff

        org     varbeg          start  of  variables
        rmb     52              room  for  26  16-bit  vars


        org     codebeg         start  of  code
start
        lds     #stkbeg         top  of  stack


*  foo.bas(1):  `
*  foo.bas(2):  ` prog3.bas      test  program  to
*  foo.bas(3):  `                68hc11  SBasic
*  foo.bas(4):  `
*  foo.bas(5):
*  foo.bas(6):  `
*  foo.bas(7):  ` Declare  some  constants
*  foo.bas(8):  `
*  foo.bas(9):
*  foo.bas(10): const  portb  =  $1004
*  foo.bas(11): const  baud   =  $102b
*  foo.bas(12): const  sccr2  =  $102d
*  foo.bas(13): const  scsr   =  $102e
*  foo.bas(14): const  scdr   =  $102f
*  foo.bas(15):
*  foo.bas(16): const  MAXN  =  10000
*  foo.bas(17): const  MINN  =  1000
*  foo.bas(18):
*  foo.bas(19):
*  foo.bas(20):
*  foo.bas(21):  `
*  foo.bas(22):  ` Initialize  the  target
*  foo.bas(23):  ` default  maximum.  Initialize
*  foo.bas(24):  `
*  foo.bas(25):
*  foo.bas(26): x  =  MAXN
        ldd     #10000
        std     varbeg+46

*  foo.bas(27): pokeb  baud,  $30
        ldd     #48
        stab    4139

*  foo.bas(28): pokeb  sccr2,  $0c
*  foo.bas(29):
        ldd     #12
```

Figure 2. Output from the SBasic compiler.

```
              stab  4141

*  foo.bas(30):  `
*  foo.bas(31):  `  Do  forever...
*  foo.bas(32):  `
*  foo.bas(33):
*  foo.bas(34): while  1  =  1
*  foo.bas(35):
whl000
        ldd   #1
        cpd   #1
        beq   *+5
        jmp   whl001

*  foo.bas(36):  `
*  foo.bas(37):  `  Initialize  a  counter,  turn
*  foo.bas(38):  `  count  up  until  the  counter
*  foo.bas(39):  `
*  foo.bas(40):
*  foo.bas(41):      pokeb portb, $ff
        ldd   #255
        stab  4100

*  foo.bas(42):      for  n=0 to x
        ldd   #0
        std   varbeg+26
for000
        ldd   varbeg+26
        cpd   varbeg+46
        ble   *+5
        jmp   for001
*  foo.bas(43):      next
        ldd   #1
        addd  varbeg+26
        std   varbeg+26
        jmp   for000
for001

*  foo.bas(44):
*  foo.bas(45):  `
*  foo.bas(46):  `  Initialize  a  counter,  turn
*  foo.bas(47):  `  count  up  until  the  counter
*  foo.bas(48):  `
*  foo.bas(49):
*  foo.bas(50):      pokeb portb, 0
        ldd   #0
        stab  4100

*  foo.bas(51):      for  n=0 to x
        ldd   #0
        std   varbeg+26
for002
```

Figure 2 continued.

```
        ldd    varbeg+26
        cpd    varbeg+46
        ble    *+5
        jmp    for003

*  foo.bas(52):     next
        ldd    #1
        addd   varbeg+26
        std    varbeg+26
        jmp    for002
for003

*  foo.bas(53):
*  foo.bas(54):  `
*  foo.bas(55):  `  Test  for  a  character  from  the
*  foo.bas(56):  `  and  echo  anything  received.
*  foo.bas(57):  `
*  foo.bas(58):
*  foo.bas(59):     if  $20  and  peekb(scsr)  =  $20  clra
        ldab   4142
        anda   #0
        andb   #32
        cpd    #32
        beq    *+5
        jmp    if000

*  foo.bas(60):          c  =  peekb(scdr)
        clra
        ldab   4143
        std    varbeg+4

*  foo.bas(61):          pokeb  scdr,  c
        ldd    varbeg+4
        stab   4143

*  foo.bas(62):     endif
if000

*  foo.bas(63):  wend
        jmp    whl000
whl001

*  foo.bas(64):
*  foo.bas(65):  end
        bra    *

*  foo.bas(66):
*  foo.bas(67):
        org    $fffe        reset  vector
        fdb    start

        end
```

Figure 2 continued.

Remote Reloads with 811bug

This very long article appeared in the January 1996 issue of *Nuts & Volts*, and describes the firmware I used to make BYRD, my Back Yard Research Drone, sing and dance. Much of BYRD's power and flexibility stems from the 811bug firmware that I burned into the top 2K of code space in BYRD's 68hc811e2 MCU. In this article, I go under BYRD's hood to show you how the firmware works.

The main reason for this column's length lies with the extensive listings I included. I hesitated to take up so much magazine space with firmware listings, but decided to go ahead, because I felt the firmware design was educational enough to warrant inclusion. Writing software to make robots run is probably the most difficult part of robotics, and I think that well designed and presented software has a place in any robotics article.

Along with 811bug, this article also touches on the main program, specifically the code that generates the pulse-width modulation (PWM) signals, used to provide motor speed and direction. The code is straightforward, and similar code appears in several of my robot designs. Best of all, it is written in SBasic, which makes it much easier to understand than comparable code written in 68hc11 assembly language.

Last month, I discussed the mechanics and electronics of BYRD, my Back Yard Research Drone robot (See "Build BYRD, a Back Yard Research Drone"). As you'll recall, I built BYRD by mounting a motor base onto the underside of a Rubbermaid roll-around beer cooler. This month, I'll describe the software needed to make BYRD fly. The software consists of two major elements. One program, dubbed 811bug, resides in the 2K of EEPROM in the 68hc811e2 I use in BYRD's computer board. I started a discussion of 811bug last month; this time, I'll provide the listing and more depth.

The other program, which provides BYRD's smarts, is downloaded through a serial link into the 32K of battery-backed RAM on the computer board. This program, called BYRD04, supplies all the tele-operation functions I need to run BYRD over an RF link from my laptop computer.

811bug

I wrote 811bug in 68hc11 assembly language, though it could have easily been written in SBasic as well. I designed this software to solve a couple of problems specific to BYRD and its intended use. BYRD is big; way too big to conveniently cart up and down the stairs whenever I need to download a new set of software. Therefore, I wanted 811bug to let me download a new operating program at any time, even if BYRD is outside on the deck.

Since robot code is, by its very nature, always in the prototype stage, I needed 811bug to serve as an emergency parachute; if my main software crashed, 811bug must always be able to recover and at least shut off the motors. I also wanted 811bug to let me trigger execution of any downloaded software from my remote system. I didn't want to have to traipse out into the back 40 to change a switch and press a reset button just to run my software.

My 811bug program meets all of these requirements and gives me an excellent development tool. Refer to the accompanying listing for details (Figure 1, at end of article). 811bug resides in EEPROM from $f800 to $ffff. When the '811e2 is reset in expanded mode, it automatically fetches its reset vector from $fffe. This starts execution of 811bug by jumping to address $f800.

811bug's first task is to clear a latch on BYRD's computer board. This immediately shuts off both motors. If you modify 811bug for your own use, this is a good place to put those panic operations that you need to have done first. Next, 811bug checks the state of port E, bit 7. This is connected to pin PE7 on the '811e2, an input pin that normally serves as an A/D port. In this case, however, my code just checks the pin for a logical 0 or 1.

If it detects a logic 1, meaning that the pin has been pulled high, 811bug blindly jumps to address $8000, the beginning of battery-backed RAM. Since I have wired pin PE7 to a DIP switch, I can manually set up BYRD to perform this auto-jump or not.

If PE7 is pulled low, 811bug instead jumps to the address MAIN, which acts as the beginning of a large control loop. The first tasks performed here are to clear the motor latch once again, initialize the stack pointer, and then set up the Serial Communications Interface (SCI) for 9600 baud operation. The top of the main loop also rewrites the RAM jump vectors.

Recall that 811bug resides from $f800 to $ffff. This means that all of the interrupt vectors, which run from $ffd6 to $ffff, contain values burned into them when I download 811bug. If I later use 811bug to download a program into static RAM and start it

up, the vectors still point into 811bug; they cannot be modified by the download operation.

To get around this problem, I designed 811bug so that most interrupt vectors point to fixed locations in the '811e2's on-chip RAM. Any other program can modify these addresses, from $00b0 to $00ec, to jump to the desired interrupt service routines.

Since the vectors in static RAM should always point somewhere useful, I use the routine SETVECS to modify the RAM jump vectors so they point back to the start of 811bug. This means that even if a downloaded program fails to set up a vector and the corresponding interrupt occurs, control will still go somewhere safe, namely back to 811bug.

Note the inner workings of SETVECS. It steps through the low RAM area, writing three bytes for every interrupt vector. In each case, the first byte written is $7e, which is the 68hc11 JMP extended opcode. The next two bytes written form the 16-bit address of the start of 811bug. Any downloaded program that modifies the low RAM jump vectors must maintain this pattern of a JMP opcode followed by an address.

Jump ahead in the listing to the section at the end, beginning with the ORG $ffd6 statement. This section defines the contents of the EEPROM vector area that 811bug needs. Most of the vectors point to fixed locations in low RAM, as discussed above.

Two vectors, however, do not. I intentionally force the Illegal Opcode and Computer Operating Properly (COP) Failure vectors to always point to 811bug. These two hard-coded vectors serve as a last line of defense against runaway code. Since most crashes usually end up executing bogus instructions, the Illegal Opcode vector will eventually get taken. The Illegal Opcode interrupt has another valuable feature: it cannot be masked by software. Thus, even if the downloaded program had interrupts disabled when it ran away, the Illegal Opcode interrupt will still work.

Now back to the top part of the listing. After my code has set up the RAM jump vectors using SETVECS, it waits for an incoming character from the SCI. This character serves as an action selector.

If the character is an uppercase-X, 811bug automatically jumps to address $8000. This blind jump lets me remotely restart a downloaded program, without having to change a DIP switch or redo the download.

If the character is a lowercase-x, 811bug jumps to the address saved from the S9 record encountered during the most recent download operation. This comes in handy if I want to download multiple files. I just send the main program as the last file, then hit an 'x' and the main program starts up.

If the character is a lowercase-j, 811bug uses the next four characters as a hex address. It saves this address for later use with an 'x' command, described above. This lets me override the execution address in a download, and jump instead to another starting address.

Note that 811bug is very unforgiving when it handles the 'j' command. You must immediately follow the 'j' with four hex characters. You cannot put in a space, and you cannot use a backspace to overwrite an error. 811bug is intended to talk to some type of host-based program, though you can use a comm package such as Crosstalk in a pinch.

Finally, if the character is an uppercase-S, 811bug begins reading the characters of a Motorola S0, S1, or S9 object record. These records are generated by nearly any software that creates executables for Motorola processors, including SBasic and tiny4th.

811bug ignores S0 records, as these usually just contain header information of use only to humans. 811bug automatically parses any S1 records, storing the supplied data into the appropriate addresses. Finally, 811bug saves the execution address, contained in the final S9 record, for later use with the 'x' command.

The rest of the code in 811bug supports character and record parsing. This design reduces the downloading operation to very few steps. I simply turn on BYRD, start up a comm package on my laptop, and send the appropriate .S19 object file through the serial port. No monitor commands, no startup incantation; just send a file, then press 'x'.

As you can imagine, this really makes life easy if I need to download while BYRD is sitting outside in the rain. As the human in the team, I shouldn't have to get out of my comfy chair to go reset a fallen robot.

The Leash

Before I start on BYRD's code, I need to discuss a hardware addition I made recently. Large robots usually need a leash, since they are too large to carry; BYRD is no exception. I could have used a standard Atari-style pushbutton joystick, but that would mean using four input lines just to get steering control. Instead, I built a network of five resistors and five switches, to give me right, left, forward, reverse, and stop commands.

The voltage developed by this network assumes one of five different values, which appears on a single A/D port of the 68hc811. Logic in the program can then translate the voltage into a specific command.

The leash uses only three wires; power, ground, and signal. This in turn uses up minimal I/O resources, requires little power, and allows me to use a thin, flexible cable between robot and human. See the accompanying schematic for details (Figure 2).

BYRD's Code

The listing for BYRD04.BAS is too large to show here completely. I'll instead discuss key sections, including parts of the code where appropriate. The code at the label RTIISR acts as the ISR for the 68hc11 Real-Time Interrupt (RTI). When properly set up, the RTI provides a timer interrupt at a fixed rate of one interrupt every 4.1 msecs. BYRD04.BAS uses this periodic interrupt as a master clock, controlling features such as the motor PWM speed control. Refer to the accompanying listing of the RTI ISR code (Figure 3).

The first few lines of code simply test the variable WAIT, and decrement it if it hasn't already reached zero. WAIT serves as a typical downcounting timer variable, used to time events to 4.1 msec resolution.

The next section of code is a rather juicy bit; it provides PWM speed control of BYRD's DC gearhead motors. Unlike the servo motors I normally

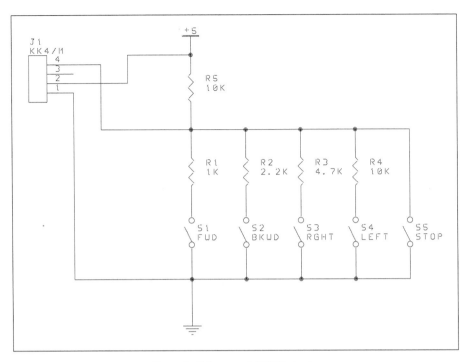

Figure 2. Schematic of BYRD's leash.

```
    interrupt                    ' interrupt routed through RAM

    rtiisr:
    if wait <> 0                 ' drop wait if not yet 0
       wait = wait - 1
    endif
    pokeb tflg2, %01000000       ' rearm the RTI subsystem

    mtr_mask = 0                 ' start with motors off
    rmtr_pwm = rroll(rmtr_pwm)   ' rotate right motor PWM
    if rmtr_pwm and 1 = 1        ' if got a 1, turn motor on
       mtr_mask = mtr_mask or RMTR_ON
    endif
    lmtr_pwm = rroll(lmtr_pwm)   ' now do left motor
    if lmtr_pwm and 1 = 1
       mtr_mask = mtr_mask or LMTR_ON
    endif
    mtr_mask = mtr_mask or mtr_dir      ' add direction info
    latch_ram = latch_ram and MTR_AND_MASK
    latch_ram = latch_ram or mtr_mask
    pokeb LATCH, latch_ram       ' update latch
    end
```

Figure 3. Listing of RTI ISR code.

use, a gearhead motor is either on or off, depending on the state of its control bit. Each motor has one control bit, available through the address LATCH.

To get speed control over a gearhead motor, you simply turn the motor on and off very quickly. The overall speed is roughly equivalent to the ratio of on-time to off-time, provided the frequency of the switching is fast enough.

Since I put the PWM code inside the RTI ISR routine, and since the RTI occurs once every 4.1 msecs, I end up with a PWM frequency of about 250 Hz. This is a little slow but functional, and it's free in that I was doing the RTI anyway.

The PWM code begins by setting the MTR_MASK variable to zero, which by default will turn both motors off. Next, it rolls the variable RMTR_PWM right one bit position. This roll operation, known in 68hc11 assembler parlance as a rotate, simply moves all the bits of RMTR_PWN one position to the right. The low bit, shifted out of position 0, is rotated into position 15. Thus, all the original bits in RMTR_PWN are preserved, they just change positions. After 16 successive rolls, RMTR_PWM will contain its original value.

The code then tests the value of the new bit in position 0 of RMTR_PWM, using a logical-AND operation. If the new bit is a one, the code ORs the proper bit for turning on the right motor into the variable MTR_MASK. If, however, the new bit is a zero, MTR_MASK isn't changed; the right motor will be turned off. The routine performs the same operation for LMTR_PWM, used to control the speed of the left motor.

Next, the code ORs the current direction bits for both motors, in variable MTR_DIR, into MTR_MASK. The direction information is updated by other sections of code; any direction changes don't occur until the next RTI, when this code actually sets up the change.

This code merges the new value in MTR_MASK with the last value written to the I/O latch, as preserved in variable LATCH_RAM. Since the I/O latch is write-only, the variable LATCH_RAM serves as its memory; it should always hold a copy of the last value written. Finally, this code uses a POKEB to update the physical latch with the value stored in LATCH_RAM. This step actually changes the electrical signals routed to the motor controller, causing the newest speed and direction information to occur.

Based on the above description, you can see how easily BYRD's software provides speed control. If code in the main section writes $ffff to RMTR_PWM, the right motor will be turned on at each RTI, and it will not be turned off. Thus, the motor will run at full speed. If code writes $0000 to RMTR_PWM, the right motor will be turned off at each RTI, which stops it.

To get speeds between full stop and full speed, just write values with varying numbers of logic 1s and logic 0s to RMTR_PWM or LMTR_PWM. For example, to get half speed, you could use values with eight logic 1s and eight logic 0s. But realize that some half-speed values give better performance than others. My code uses a value of $5555 or $aaaa to get half-speed; the alternating ones and zeros provide a smoother speed than a value of, say, $ff00.

I have included a table showing the various motor speed constants I use in BYRD (Figure 4.). Note that I don't use any speed below 50%. Lower speeds provide proportionately lower torque, and anything below 50% doesn't give enough torque to be useful.

The following paragraphs discuss the design of the main loop in BYRD's program. Refer to the accompanying listing of the program's main section (Figure 5). The large **DO-LOOP** following the label

```
const SPEED_100 = %1111111111111111
const SPEED_094 = %0111111111111111
const SPEED_087 = %0111111101111111
const SPEED_081 = %0111101111011111
const SPEED_075 = %0111011101110111
const SPEED_069 = %0110110110110111
const SPEED_062 = %0101101101011011
const SPEED_056 = %0101010110101011
const SPEED_050 = %0101010101010101
const SPEED_000 = %0000000000000000
```

Figure 4. Table of motor speed constants.

```
do
   key = inkey$() and $ff
   if key = 0
      gosub read_joystick
   endif
   select key
      case 'h'                           ' need help?
      case 'H'                           ' both cases
      gosub help
      endcase

      case '~'                           ' return to 811 monitor?
      interrupts off                     ' shut off all ints
      n = peek($fffe)                    ' find addr of monitor
      gosub n                            ' force return to 811bug
      endcase

      case 'd'                           ' show all a/d ports?
      pokeb adctl, $10                   ' convert chnls 0-3
      for n = 0 to 4                     ' silly wait loop
      next n
       print "Data (0-7): "; peekb(adr1); peekb(adr2);
       print peekb(adr3); peekb(adr4);
      pokeb adctl, $14                   ' convert chnls 4-7
      for n = 0 to 4                     ' silly wait loop
      next n
       print peekb(adr1); peekb(adr2); peekb(adr3); peekb(adr4)
      endcase

      case 13                            ' cr?
      if mwait = 0
         print "Working...";
      endif
      print
      endcase

      case '5'                           ' stop all motors
      case ' '                           ' emergency stop
      push RMTR_FWD + LMTR_FWD           ' first, release the relays
      gosub set_direction
      gosub all_stop                     ' now change the PWM pattern
      print "All stopped, sir!"
      endcase

      case '8'                           ' go forward
      push RMTR_FWD + LMTR_FWD
      gosub set_direction
      print "Forward, aye!"
      endcase

      case '2'                           ' go backward
      push RMTR_REV + LMTR_REV
```

Figure 5. Listing of main loop in BYRD's code.

```
      gosub set_direction
      print "Reverse, aye!"
      endcase

      case '6'                            ' turn hard right
      push RMTR_REV + LMTR_FWD
      gosub set_direction
      print "Hard a-starboard, aye!"
      endcase

      case '4'                            ' turn hard left
      push RMTR_FWD + LMTR_REV
      gosub set_direction
      print "Hard a-port, aye!"
      endcase

      case '+'                            ' increase speed
      if topspeed = '9'                   ' if already at top speed...
         print "Sorry, Captain, but she canna give no more!"
      else
         topspeed = topspeed + 1
         print "Speed now up to warp ";
         outch topspeed
         print
         gosub set_topspeed
      endif
      endcase

      case '-'                            ' decrease speed
      if topspeed = '1'                   ' if already at low speed...
         print "Sorry, Captain, but we are already at warp 1!"
      else
         topspeed = topspeed - 1
         print "Speed now down to warp ";
         outch topspeed
         print
         gosub set_topspeed
      endif
      endcase

      case 'v'                            ' turn camera on/off?
      case 'V'                            ' both cases work
      interrupts off                      ' protect the latch
      temp = latch_ram xor VIDEO_ON       ' switch the bit
      latch_ram = temp                    ' save in ram copy
      pokeb latch, temp                   ' now send to I/O
      interrupts on                       ' ready to go
      if temp and VIDEO_ON = 0
         print "Main viewer off, sir!"
      else
         print "Main viewer on, sir!"
      endif
```

Figure 5 continued.

```
        endcase

        case 'z'                          ' turn device #2 on/off?
        case 'Z'                          ' both cases work
        interrupts off                    ' protect the latch
        temp = latch_ram xor DEV2_ON      ' switch the bit
        latch_ram = temp                  ' save in ram copy
        pokeb latch, temp                 ' now send to I/O
        interrupts on                     ' ready to go
        if temp and DEV2_ON = 0
           print "Device 2 off, sir!"
        else
           print "Device 2 on, sir!"
        endif
        endcase

        case 'r'                          ' servo to right
        case 'R'                          ' do both cases
        temp = peek(PAN_TOC)              ' get current servo position
        if temp <* PAN_MAXR               ' if room to go...
           temp = temp + PAN_INC
        endif
        poke PAN_TOC, temp
        endcase

        case 'l'                          ' servo to left
        case 'L'                          ' do both cases
        temp = peek(PAN_TOC)              ' get current servo position
        if temp >* PAN_MAXL               ' if room to go...
           temp = temp - PAN_INC
        endif
        poke PAN_TOC, temp
        endcase

        case 'c'                          ' servo to center
        case 'C'                          ' do both cases
        poke PAN_TOC, $f500
        endcase

     endselect
     if key <> 0
        mwait = MORE_DELAY
     endif

loop
```

Figure 5 continued.

MAIN provides BYRD's functionality. What BYRD does at any time is controlled by one of two events, either a character arriving on the serial port (SCI) or a value appearing on A/D port 3, pin PE3.

The code first uses the **INKEY$()** function to test for an incoming character. If no character arrived, the code then executes the READ_JOYSTICK subroutine to see if a button on BYRD's leash has been pressed.

The READ_JOYSTICK routine samples the joystick port and tests the value read. It then converts this value into one of six characters, simulating the arrival of steering and motor control characters. Note that one character, null or $0, signifies that no button was pressed.

This routine also uses the variable LAST_KEY to record a key press. This means that the joystick buttons act as latching switches; press one, and the code remembers it as pressed until you press another. The character saved in variable KEY, regardless of its source, acts as a selector in a large **SELECT** structure. A value of $0, which signifies that no command was issued, is ignored. The program recognizes commands for stop, right turn, left turn, forward, reverse, and speed control. It also supports commands for controlling BYRD's power control port, one line of which activates the video camera and video RF transmitter.

The code inside the **SELECT** structure executes the desired command, then prints a whimsical confirmation back to the serial port. In the cold light of reality, the link between a motorized beer cooler and the starship Enterprise seems tenuous, but the messages always get a chuckle at the SRS meetings.

Note that some commands must modify the value in the I/O latch. Since the latch can be changed in the RTI code as well, the main code must protect the value in LATCH_RAM while using it. It does this by turning interrupts off, reading or writing the LATCH_RAM value, then turning the interrupts back on.

This concept holds anytime mainline code and interrupt code must access the same variable or port. The mainline code must ensure that the interrupt code can't muck with the variable or port at the same time the mainline code accesses it, or you'll end up with some very nasty bugs to shoot.

One command is of paramount importance. The tilde character ('~') forces a return to 811bug. The command works by using a PEEK() function to find the address of 811bug, then doing a GOSUB to that address. This lets me abandon the current version of BYRD's software and prepare to load another via the RF modem. Without the tilde command, I'd have to go find BYRD and press the reset switch to set up the next download.

This completes my discussion of BYRD. You will probably see BYRD mentioned again in future columns; I'm having a lot of fun with my big remote 'bot. Think I'll go add a strobe and siren ...

```
*
*  811bug     EEPROM-based S19 loader for the 68hc811e2 MCU
*
*  This program accepts S1-S9 records from the SCI and stores
*  them at the proper addresses in memory. It also accepts a
*  minimum number of commands from the serial port, to allow the
*  user to execute the downloaded commands from a comm program.
*



*
*  Declare a few I/O register addresses
*

iobase        equ     $1000                    i/o regs start here
porte         equ     $0a
baud          equ     $2b
sccr2         equ     $2d
scdr          equ     $2f
scsr          equ     $2e

latch         equ     $4000                    special output port



*
*  Define the stack address and a few vital RAM locations.
*
*  This program configures the low-area RAM ($00 - $ff) as follows:
*
*  $ff        +----+       <- Top of low-area RAM
*             |    |        811bug variables reside here.
*  ramvecs    +----+       <- Top of RAM vector area
*  + $3c      |    |        These addrs are targets of the 811bug
*             |    |        vectors; see fdbs at $ffc0 below.
*             |    |
*  ramvecs    +----+       <- Bottom of RAM vector area
*  stkbeg     +----+       <- Top of 811bug stack area
*             |    |        The stack grows downward from here.
*             |    |        Reserve at least $10 bytes for stack.
*             |    |
*  $9f        +----+       User program may use rest of low RAM.
*             |    |
*             |    |
*  $00        +----+       <- Bottom of low-area RAM
*


stkbeg equ    $af                   top of stack

ramvecs equ   stkbeg+1              jump vectors ($3c bytes)
```

Figure 1. Listing of 811bug program.

```
vars            equ     ramvecs+$3c         addr of variables
knt             equ     vars+0              byte variable
addr            equ     vars+1              word variable
chk             equ     vars+3              byte variable
xqtaddr         equ     vars+4              word variable
flag9           equ     vars+6              byte variable



*
*  Start of 811bug code.
*
*  The '811e2 should be configured to have EEPROM from $f800
*  to $ffff. This puts the vector addresses in EEPROM, so
*  811bug is always present.
*

     org    $f800               start of code

*
*  Following powerup or reset, 811bug checks the state of port
*  E, bit 7. If that bit is high, control passes to the main
*  loop in 811bug. If that bit is low, control automatically
*  jumps to address $8000.
*
*  The jump to $8000 is blind; no check is made to ensure that
*  code actually exists there. This leaves plenty of time for
*  the code at $8000 to alter the time-sensitive I/O registers,
*  if needed.
*

start
     clra                       special for my 'hc11 board
     staa   latch                   must clear output port!

     ldaa   porte+iobase    read digital port
     bpl    main            if bit 7 is high...
xxqt
     jmp    $8000           force start of user's prog


*
*  This is the main entry point to 811bug. Each time control
*  reaches this point, the code reset the stack pointer and
*  the SCI, then reloads the low-RAM vectors. It also loads
*  the 811bug variable xqtaddr with the address found in the
*  reset vector at $fffe. This effectively makes 811bug
*  the default program for use with the 'x' command.
*
*  This code then prints out a simple hello message.
*
```

Figure 1 continued.

```
main
      clra                            special for my 'hc11 board
      staa    latch                        must clear output port always!

      ldx     #iobase         point to i/o
      lds     #stkbeg         always reset stack
      ldaa    #$30            set baud = 9600
      staa    baud,x
      ldaa    #$0c            turn on sci
      staa    sccr2,x
      jsr     setvecs         setup vectors
      ldd     $fffe           get start of 811bug
      std     xqtaddr         use as default S9 addr
      ldy     #hello          point to string
      jsr     prints          print it

*
*  This is the main loop for accepting input from the host
*  and processing commands. Commands are determined by the
*  first letter following a CR. LFs are always ignored when
*  811bug is looking for a command.
*
*  Legal 811bug commands are:
*
*  S    marks the start of an S19 object record. The entire
*       line is treated as an S19 record, and is parsed and
*       loaded into memory.
*
*  x    transfers control to the last legal execution address,
*       as passed in the most recent S9 record, or as set by the
*       j command below.
*
*  X    transfers control to $8000, regardless of any S19
*       object records previously sent.
*
*  j    assigns a jump address for use with a subsequent 'x'
*       command as above. The j must be followed immediately
*       by a four-digit hex address, followed immediately by a
*       CR. The input cannot contain backspaces or other cursor
*       control characters, and it cannot contain embedded
*       spaces or tabs.
*

loop
      lds     #stkbeg         always reset stack
      ldx     #iobase         always point to i/o
      jsr     crlf            make it pretty
loop1
      jsr     getche          get first char
      cmpa    #$0a            LF?
      beq     loop1           always ignore
```

Figure 1 continued.

```
        cmpa    #'S'              load S19 record?
        beq     loads19           if so, do it
        cmpa    #'x'              execute prog?
        beq     xqt               branch if so
        cmpa    #'X'              execute at $8000?
        beq     xxqt              branch if so
        cmpa    #'j'              set jump addr?
        beq     jmp               branch if so
        cmpa    #$0d              empty line?
        beq     loop              if so, ignore it

*
* This code eats all incoming characters until a CR is detected.
* It then displays a '?' as an error indicator, then returns to
* the top of the loop for the next command.
*

eaterr
    bsr     eatline           error, waste the line
error
    ldaa    #'?'              get error flag
    jsr     outch             send it
    bra     loop              and leave


*
* eatline -- eat all characters, without echo, until a CR
*
* This routine absorbs all incoming characters until a CR is
* reached. This effectively ignores all text after an error
* has been detected.
*

eatline
    jsr     getch             grab a char (no echo)
    cmpa    #$0d              wait for CR
    beq     eatlx             loop until hit it
    jsr     outch             echo char
    bra     eatline           do some more
eatlx
    rts


*
* xqt -- process the 'x' command
*
* This routine jumps to the address stored in 811bug variable
* xqtaddr.
*
```

Figure 1 continued.

```
xqt
     ldy     xqtaddr              get execution addr
     jmp     0,y              and do it

*
*  jmp — process the 'j' command
*
*  This routine collects a four-digit hex address and a terminating
*  CR. It then stores the address into the 811bug variable
*  xqtaddr, for use with a later 'x' command.
*

jmp
     jsr     byte             get MSB
     staa    addr             save it
     jsr     byte             get LSB
     staa    addr+1            save it
     jsr     getch             get last char
     cmpa    #$0d             better be cr
     bne     eaterr            oops, that's bad
     ldd     addr             OK, get real addr
     std     xqtaddr           and save it
     bra     loop             do another one


*
*  loads19 — process an S19 record
*
*  This routine reads the rest of an S-record, processing the
*  characters as needed.
*
*  This routine supports the following S-records:
*
*  S0    header record; ignored by 811bug.
*
*  S1    16-bit data record; 811bug stores data into assigned
*        addresses.
*
*  S9    execution record; 811bug records the execution address
*        in 811bug variable xqtaddr for later use with the 'x'
*        command.
*
*  This routine automatically calculates and tests the checksum
*  for each record. Illegal records and any record with a bad
*  checksum are flagged to the user with an error indication.
*
*  NOTE: 811bug does not currently verify that data written to
*  a memory address actually got there.
*
```

Figure 1 continued.

```
loads19
     clr     flag9             ssume S1 record
     jsr     getche            rab next char
     cmpa    #'0'              S0 record?
     bne     loads1            branch if not
loads0
     jsr     eatline           waste whole line
     bra     loop              get next line
loads1
     cmpa    #'9'              S9 record?
     bne     loads2            no, try next one
     inc     flag9             show S9 record
     bra     loads4            continue processing
loads2
     cmpa    #'1'              S1 record?
     bne     eaterr            if not, error
loads4
     jsr     byte              get byte count
     staa    knt               save for now
     dec     knt               count this byte
     staa    chk               start checksum
     jsr     byte              get MSB of addr
     staa    addr              save addr
     adda    chk               add to checksum
     staa    chk               save it back
     dec     knt               count this byte
     jsr     byte              get LSB of addr
     staa    addr+1            save addr
     adda    chk               add to checksum
     staa    chk               save it back
     ldy     addr              get addr
     dec     knt               count this byte
loads3
     beq     loads5            if 0, all done
     jsr     byte              get data byte
     staa    0,y               write to addr
     iny                       point to next addr
     adda    chk               add to checksum
     staa    chk               save it back
     dec     knt               count this byte
     bra     loads3            go test knt
loads5
     jsr     byte              get checksum
     coma                      get one's comp
     cmpa    chk               does it match?
     bne     loadsf            branch if fail
     jsr     eatline           eat the rest
     tst     flag9             was this S9 record?
     beq     loadsx            branch if S1
     ldd     addr              get start address
     std     xqtaddr           save for execution
```

Figure 1 continued.

```
loadsx
      jmp    loop                do next command
loadsf
      jmp    eaterr              show bad record


*
* hexbin -- convert an ASCII character in AR to its binary value
*
* This routine expects an ASCII character in the A register. It
* converts the character to uppercase, then converts it to a
* binary value from 0 to $f.
*
* If this routine detects an illegal (non-hex) character, it
* automatically jumps to eaterr to handle the error condition.
*

hexbin
      jsr    ucase               make it uppercase
      cmpa   #'0'                test low range
      blt    hexnot              branch if fail
      cmpa   #'9'                test decimal
      ble    hexnmb              found hex number
      cmpa   #'A'                test low hex
      blt    hexnot              branch if fail
      cmpa   #'F'                test high hex
      bgt    hexnot              branch if fail
      adda   #9                   set up for conversion
hexnmb
      anda   #$0f                get binary value
      bra    hexx                time to leave
hexnot
      jmp    eaterr              fail, show error
hexx
      rts



*
* ucase -- convert an ASCII character in AR to uppercase
*

ucase
      cmpa   #'a'                check for low end
      blt    ucasex              leave if too low
      cmpa   #'z'                check for high end
      bgt    ucasex              leave if too high
      suba   #$20                make it uppercase
ucasex
      rts
```

Figure 1 continued.

```
*
* byte -- collect two ASCII hex characters and convert them to
* a byte
*

byte
     pshb                     save b
     jsr     getche           get a char
     bsr     hexbin           convert it
     asla                     move to MSB
     asla
     asla
     asla
     tab                      now save in b
     jsr     getche           get 2nd char
     bsr     hexbin           convert it
     aba                      make a byte of them
     pulb                     restore b
     rts



*
* getch -- get a character from the SCI; polling loop.
*

getch
     ldaa    scsr,x           get status
     anda    #$20             anything there?
     beq     getch            loop until data
     ldaa    scdr,x           get the data
     rts



*
* getche -- get a character, with echo, from the SCI;
* polling loop.
*

getche
     bsr     getch            get a char
     bra     outch            and echo it



*
* crlf -- send a CR/LF sequence to the SCI
*

crlf
     ldaa    #$0d             get cr
     bsr     outch            send it
```

Figure 1 continued.

```
        ldaa    #$0a               get lf
        bra     outch              send it, then exit


*
*  outch -- send a character to the SCI; polling loop.
*
outch
        tst     scsr,x             get status
        bpl     outch              loop until ready
        staa    scdr,x             send it
        rts


*
*  prints -- print a string to the SCI
*
*  This routine prints a text string to the SCI. The address
*  of the string is passed in the Y register, which is destroyed.
*  The string must be null-terminated.
*
prints
        ldaa    0,y                get next char in string
        beq     printsx            leave if done
        bsr     outch              send it
        iny                        bump pointer
        bra     prints             do another
printsx
        rts


*
*  setvecs -- set up the RAM vectors
*
*  This routine overwrites low RAM with jumps to the start of
*  811bug. This means that subsequent unexpected interrupts
*  should automatically vector to 811bug, restarting the monitor.
*  This isn't a foolproof way to save a runaway program, but it
*  is better than nothing.
*
setvecs
        pshx                       save xr
        ldx     #ramvecs           point to destination
setv1
        ldaa    #$7e               jump opcode
        staa    0,x                put in ram
        inx                        point to next cell
        ldd     #start             default vector addr
```

Figure 1 continued.

```
        std     0,x             put in ram
        inx                     bump pointer
        inx                     (twice for word)
        cpx     #ramvecs+$3c    reached the end?
        bne     setv1           branch if not
        pulx
        rts


hello
        fcb     $0a,$0d
        fcc     "811bug v1.0"
        fcb     0


        org     $ffd6                   start of 68hc11 vectors

        fdb     ramvecs+$00             sci
        fdb     ramvecs+$03             spi
        fdb     ramvecs+$06             paii
        fdb     ramvecs+$09             paovi
        fdb     ramvecs+$0c             toi
        fdb     ramvecs+$0f             oc5i
        fdb     ramvecs+$12             oc4i
        fdb     ramvecs+$15             oc3i
        fdb     ramvecs+$18             oc2i
        fdb     ramvecs+$1b             oc1i
        fdb     ramvecs+$1e             ic3i
        fdb     ramvecs+$21             ic2i
        fdb     ramvecs+$24             ic1i
        fdb     ramvecs+$27             rtii
        fdb     ramvecs+$2a             irq
        fdb     ramvecs+$2d             xirq
        fdb     ramvecs+$30             swi
        fdb     start                   illegal op
        fdb     start                   cop failure
        fdb     ramvecs+$39             cop monitor failure

        fdb     start           reset

        end     start
```

Figure 1 continued.

The Ultimate
PC Robot Tool

This article, from the May 1996 issue of *Nuts & Volts*, was one of the more rewarding projects in the history of my column. Bill Bailey, a most bodacious robot hacker and (naturally) an SRS member, spent several days working with me to develop a system for using 80x86-class chips as robot brains. The result of our efforts, which we called the BOTBios system, lets you use low-cost PC development tools to write code for cheap PC-compatible microcontroller boards. The flexibility, power, and simplicity of BOTBios adds a whole new dimension to PC-based robots.

I first describe the general layout of the BOTBios, a small BIOS that you can program into an EPROM and insert into the address space of your microcontroller. BOTBios supports a limited number of BIOS interrupts, rewritten so they behave properly in a robotic environment. I also cover some of the inner workings of BOTBios, so you can write programs on your PC that, when run on your robot, will take advantage of BOTBios' features.

Next, I describe ROMMAKER (RM), a companion PC-based tool used to create a single EPROM image containing multiple PC .COM files for execution on your robot. You use RM to create what Bill and I called a ROM disc, a collection of programs that can be run either following reset or by entering commands over the serial port. If your hardware permits, you can have multiple ROM discs on a single robot, increasing your machine's capabilities.

I conclude by describing how Bill and I modified our New Micros' V25 microcontroller boards to use the ROM discs developed with our BOTBios system. I also offer suggestions for compilers that can create suitable .COM files, including a shareware BASIC compiler that we used during our testing phase.

This all started, naturally enough, at a Seattle Robotics Society meeting. The SRS members constantly explore different chips and boards, looking for ways to build better robots faster. At the January 1996 meeting, the group tech session turned to using PC tools as a method of speeding up a robot's design.

Dick Martin mentioned the Intel 386ex microcontroller (MCU). As you can surmise from its name, this chip contains an 80386 core, built of CMOS technology and able to run any 80386 software. Additionally, Intel grafted on gobs of I/O normally needed to build a PC. This includes dynamic RAM (DRAM) refresh circuitry, parallel ports, serial ports, and timers.

The ability to work with a machine so much like a PC, yet containing valuable robotic staples on-chip, makes the 386ex a serious consideration for large robot systems. The major disadvantage, from a hobbyist standpoint, is the chip's packaging. Intel used a plastic quad flatpack (PQFP) design, with 132 pins on very fine pitch spacing. This layout forces the use of a fine pitch circuit board, most likely double-sided, just to hold the chip. Ambitious hobbyists can probably fabricate an adaptor board to hold the 386ex and bring the signals out on wirewrap or IDC-type connector pins. This would let you plug a larger adaptor board into your main board, so you could use the 386ex on a prototyping board.

Others at the SRS were thinking along similar lines. Mark Castelluccio brought in a small V25 MCU board he had purchased through the mail from Tern, Inc. The company offers several variations of V25 (and other) boards, plus a suite of add-ons specifically tailored for embedded control. You can get more info on Tern's web site.

The board Mark showed off contains the V25 chip and two 32-pin memory sockets. The V25 supports up to 1M of memory, so you could drop a 512K RAM chip in one socket, then stick a 512K EPROM in the other. One nice aspect of the V25 chip: it runs the 8088 instruction set directly. Thus, you can create working code for the V25 using just about any PC-based compiler.

The Start of BOTBios

All this talk about PC-compatible robot boards got to a couple of us software types. Bill Bailey, who does a lot of work with PC-based software, was talking about an idea he had for mutating PC software so it could run on an 8088-type chip, such as the V25. Our conversation stretched out, off and on, over

several days and took plenty of twists and turns before settling down to a concept Bill calls BOTBios. I'll describe the BOTBios system using a V25 chip, but remember that it works equally well with any 8088-compatible micro.

BOTBios resides in an EPROM of up to 64K in size, and is wired into the top portion of the V25's memory space. The BOTBios ROM contains code for handling power-on reset, plus support for selected V25 interrupt vectors. See the accompanying layout of the BOTBios ROM (Figure 1).

The layout shows the BOTBios ROM extending from some starting point up to linear address FFFFFh. I say "linear" because the V25 uses a segmented architecture similar to the PC's. Thus, it divides its 1MB address space into 16 64K blocks. You normally refer to a specific address by supplying both the 16-bit memory segment and the 16-bit

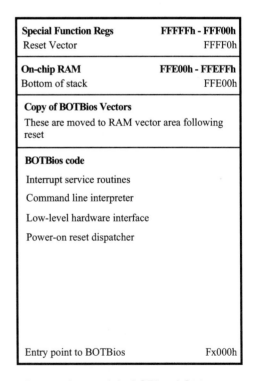

Figure 1. Layout of the BOTBios ROM image.

segment address. For example, you might refer to the address of the reset vector as segment FFFFh and address 0000h.

The V25 chip internally combines these two numbers to create a single, 20-bit linear address. It does this by multiplying the segment by 16, then adding it to the segment address. For the reset vector example, this becomes FFFF0h added to 0000h, or FFFF0h. Since I'm accustomed to thinking in linear address spaces, I'll generally refer to addresses in this article in their linear form.

The starting address of a BOTBios ROM varies, depending on the size of the EPROM defined when the code is assembled. Bill has so far assembled his BOTBios code to reside in an 8KB EPROM (27c64), so the BOTBios entry point for such an EPROM is at FE000h.

While the layout shows the EPROM extending all the way up to FFFFFh, the V25 actually overrides the top 512 bytes of this address space. The MCU reserves these bytes for its Special Function Registers (SFRs) and 256 bytes of on-chip RAM. With few exceptions, data written into the BOTBios EPROM at these addresses will not appear in the address space.

One obvious exception to the above is the 12 bytes of memory starting at the reset vector at FFFF0h. Following reset, the V25 sets its program counter (PC) to linear address FFFF0h and starts execution at that address. Bill's code assembles a far jump to the start of BOTBios at that address, so the V25 can reach code cleanly after reset.

Once the V25 completes its reset sequence and begins executing BOTBios, the fun really starts. Bill's code provides some very powerful services that meet the needs of robot designers, while remaining compatible with PC software tools such as Basic and C compilers.

The best example of this merging lies in the BOTBios interrupt service routines (ISRs). As the name implies, BOTBios provides a limited set of basic I/O service (BIOS) support, similar to that found in a PC BIOS ROM. Bill has, of course, modified these ISRs to fit into the robot world.

Most code generated with PC compilers relies on BIOS and DOS interrupt services to perform low-level I/O tasks. For example, the PRINT statement in most Basic programs eventually ends up executing a BIOS interrupt 10h, service 0eh (Write Character in Teletype Mode). When the Basic program invokes this interrupt on a PC, BIOS code writes the character to the video monitor so the character occupies the current logical output position on the screen.

Since most robots don't have video screens, Bill coded BOTBios so this interrupt and service combination sends the specified character out serial port 0 on the V25. Thus, a PC program that prints "Hello, world!" to the PC's monitor would, if run under BOTBios, send the same text out the serial port instead.

Similarly, Bill has written ISR code for interrupt 16h, service 00h (Read Next Keyboard Character) so it accepts input from the same V25 serial port. This means that a C program that uses the getch() function to read a character from the PC's keyboard will work just fine under BOTBios; it will simply read a character from the serial port instead.

I intentionally used two different languages, Basic and C, in my examples above to make a point. It doesn't matter what language you use to write your robot code, so long as the compiler uses the PC's BIOS routines in a well-behaved manner. Whether it's Pascal, Fortran, C, Forth, assembler, or COBOL, you can use it to write robot code and run that code in the BOTBios environment.

Having said that, I need to add a couple of caveats. First, your selected compiler must actually generate ROMmable code; specifically, it must be able to create .COM files. Not all "compilers" can do this. Some, like the recent Microsoft Basics, can only execute a program from within the Basic development environment.

Second, Bill's BOTBios must include support for the interrupt services your program needs. Obviously, Bill has not added disc services, though discussions are under way. Similarly, those video services that deal with graphics modes and screen manipulation aren't available in BOTBios. But if Bill has provided the needed support routines, and your compiler makes a .COM file of your program, you should be able to run your PC application under BOTBios.

One of the first options that BOTBios opens up is real-time debugging output from your robot code. Since your compiler does the nasty formatting work before it invokes the BIOS routines to send data to the screen, you can add all kinds of fancy debug statements to your code. You might hesitate to write a formatted output for eight different 16-bit values in assembler, but under BOTBios, you just make a call to the printf() function and you're done.

As the BOTBios layout shows, Bill has added several features beyond BIOS support. BOTBios also contains a command-line interpreter (CLI). The CLI works just like the DOS prompt on a PC, in that you can type commands at the prompt and see them executed.

Since Bill's BOTBios has already replaced the PC's keyboard and video display with the V25 serial port, it was a natural step to connect the BOTBios CLI to the serial port as well. You can hook a robot running BOTBios to the serial port on your PC and fire up a comm program such as ProComm. When you reset the robot, you will actually see a command-line prompt on your comm program.

So what kind of commands can you run from the BOTBios CLI? Recall that your compiler must be able to create .COM files. BOTBios automatically reads the command you enter and scans its ROM, looking for a .COM file of the same name. If it finds a match, BOTBios copies the file into on-board RAM and executes it. If the program later terminates cleanly, using interrupt 21h, service 00h (Terminate), control will return to BOTBios and it will prompt you for your next command.

I really like the power this feature offers. I can have a tethered robot capable of many different functions, waiting for my command. If I issue a command to send temperature data, it dutifully runs the GETTEMP command and streams the data back down to me over the serial line. If I then need to determine the robot's battery charge level, I just run that command.

Since the BOTBios CLI will run just about anything you can turn into a .COM file, you could even put small compilers on-board. I'm thinking specifically of a modified Forth compiler for the PC. Most Forths come with source code, and it would be a fairly simple task to strip out all of the disc I/O words. This would leave you with an interactive program development system running on your robot.

BOTBios also includes software interfaces to much of the V25 on-chip hardware. As noted, many of the standard PC BIOS routines have been rewritten to use one of the V25 serial ports instead. As of this writing, Bill is also looking into adding newer interrupts for accessing V25 subsystems such as the timers and analog ports. Much of this work is still ongoing, and I'll keep you posted as it evolves.

One low-level feature that is already available, however, is the Intel hex file loader. At the CLI command prompt, you can simply begin downloading an Intel hex file to your robot. The colon (:) at the start of each line marks it as an object record, and BOTBios parses the record out, stores the data into the appropriate area of RAM, and prompts for the next command. Thus, you can send a complete hex file anytime you see the command prompt, and you can send as many hex files as you like.

The built-in downloader makes program development a snap. Just create your .COM file, translate it into a hex file, and send it over the serial port to your robot. Issue the BOTBios built-in GO command, and your program runs. When your program ends, BOTBios regains control and you are ready to send the next file.

The last feature on the BOTBios layout diagram is the power-on reset dispatcher. This concept has no analog I know of in the PC world; it is primarily a robotics device, and a very powerful one at that.

Immediately after reset, BOTBios checks the low four bits of the V25's port 0 (P0.0 through P0.3). It uses the state of these four bits, usually wired to a set of dipswitches, to generate a hex number from 00h to 0fh. BOTBios then uses this number, called the dispatch value, to determine what to do next.

If the dispatch value is 00h, BOTBios automatically transfers control to the CLI. This is the normal method of startup, and assumes you have a serial terminal of some type hooked to the robot.

If the dispatch value is 0fh, BOTBios does a blind jump to address 00100h. This is the normal execution point of a .COM file, and it is also the starting point of any Intel hex file that you download into RAM. This feature lets you load a hex file into battery-backed RAM, then reset your robot so all V25 devices come up cleanly. Control will automatically start running your program, bypassing much of the BOTBios startup code. Note, however, that if your program terminates, BOTBios will again gain control and you will see the CLI prompt.

Finally, if the dispatch value is in the range from 01h to 0eh, BOTBios assumes you want to run a stored .COM file. It automatically searches through its ROM space, locating .COM files stored there when the ROM was programmed. It finds, loads, and runs the N'th .COM file, depending on the dispatch value. Thus, a dispatch value of 03h causes BOTBios to load and run the third .COM file it finds.

This feature will prove must useful on multi-purpose robots. If you design a robot that competes in, for example, the line-following and dead-reckoning events at a club competition, you could burn the necessary .COM files into the BOTBios ROM. At the first event, just set the dipswitches on port 0 to the proper value and hit reset; your robot runs the correct program. When the next event rolls around, change the dipswitches, hit reset again, and the same robot runs the program for that event.

As of this writing, Bill intends to distribute his BOTBios software by making it available on my web site. I don't know for sure what he expects to include in the distribution, but it may include full source code for the BOTBios routines, as well as sample implementations.

The ROMMAKER Program

I've repeatedly mentioned above the concept of burning .COM files into a ROM, and BOTBios' ability to search a ROM for named .COM files. The program that provides BOTBios with these functions is my contribution to the BOTBios system, and is named ROMMAKER.

ROMMAKER, or RM, is a PC-based program that creates an EPROM image from a collection of .COM files. RM places the .COM files in a sequential order, prepends a custom directory and header, and outputs the entire image as a binary file, ready for burning into an EPROM.

Bill and I refer to this resulting EPROM as a ROM disc. A ROM disc can reside anywhere in the V25 address space, though it makes most sense to place it at 20000h and above, to leave room for a minimum of 128K of system RAM. A brief tour of RM's output file will show how well the design meshes with BOTBios. Refer to the accompanying layout of a ROM disc (Figure 2).

Unlike a BOTBios ROM, a ROM disc basically "starts" at the lowest address, which is ROM address 0000h. All elements in a ROM disc are completely relocatable, and a ROM disc works, regardless of where in the address space you install it. Therefore, I'll refer to addresses inside a ROM disc as relative to the start of the EPROM, rather than relative to the V25 address space.

The first block of information in a ROM disc is the disc's header, used by BOTBios to find the ROM disc when it searches memory. The header takes up 16 bytes, and starts with four identifying bytes. The first two bytes, 05aa5h, mark the beginning of a ROM disc, while the next two bytes carry a specific ROM disc identifier code. For the current version of RM, these bytes are 0ff0h.

The next two bytes contain version information for the ROM disc. The RM program lets you supply a version number in the form x.y, and embeds these two characters as version information. BOTBios will display this version data when it locates a ROM disc. This lets you track what revisions you have put into a ROM disc while it is still in your robot.

The next six bytes contain a time stamp, giving the time and date at which RM began constructing the ROM disc file. As with the version information,

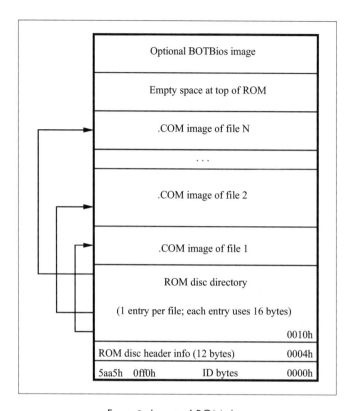

Figure 2. Layout of ROM disc.

BOTBios will display the time and date stamp to help you track the revisions of your ROM disc. The remaining bytes in the 16-byte header are reserved and will be written with 00h by RM.

Immediately after this header, RM places a directory describing the files contained in the ROM disc. Each directory entry contains the name of a .COM file, up to eight characters long, as well as the location of the initial stack pointer and where in the ROM the binary image for that file begins. The actual length of the directory depends solely on the number and size of the .COM files involved, and the size of the final ROM. If you have a large EPROM, say 512KB or so, you could probably fit a dozen or more small .COM files into your ROM disc.

I won't go into a complete discussion of RM; you can download the distribution files from my web site and read the included manual for that. I will,

however, walk you through a single example of using RM to build a ROM disc. This will give you a feel for the power in the BOTBios/ROMMAKER system.

Assume you have written three programs to run on your V25 robot. GETTEMP.COM returns some temperature readings, SONAR.COM takes a 360-degree sonar sweep and returns the data, and SETUP01.COM lets you download some information to the robot describing the tasks it must perform in the next 24 hours.

You know the amount of stack space each file requires, and you know that GETTEMP.COM can actually run out of ROM. This means that BOTBios won't have to copy the GETTEMP program into RAM before executing it; it will simply execute the ROM image directly.

Regardless of which languages you used to write these programs, the resulting executables exist as .COM files on your PC. Now you want to use RM to create a ROM disc image for burning into a 256KB EPROM. Just for grins, you also decide you want this ROM disc to contain a copy of the latest BOTBios binary image, which is in the file V25BIOS.BIN. Finally, you want this ROM disc to contain a version number of 2.1, so you can track the changes you made since the last revision.

You start by opening a text file (call it MYDISC.BLD), and entering the following lines:

```
*
* MYDISC.BLD     builds a ROM disc
*

ROMSIZE 256
VERSION 2.1

FILE  c:\v25\bin\gettemp.com
        gettemp 100 ROM
FILE  c:\v25\bin\sonar.com sonar 200
FILE  d:\foo\setup01.com setup 100
BIOS  d:\incoming\v25bios.bin
```

The above commands are pretty self-explanatory, and illustrate the simplicity of the RM command file. The ROMSIZE command declares the size of the final ROM image, in kilobytes. The VERSION command provides a three-character version identifier, of the format x.y. Each FILE command provides the path to the desired .COM file, the name for that file that will appear in the ROM disc, and the amount of stack space, in 16-byte paragraphs, needed by the program. If you include the optional keyword ROM at the end of a FILE command, that directory entry will be flagged as capable of running from ROM; see the FILE command for GETTEMP above.

Note the FILE entry for the SETUP01.COM file. This shows an instance when the PC version of the .COM file is named differently from the directory entry as it will appear in the ROM disc. This lets you change the revision of the underlying .COM file, while maintaining the same name across several ROM discs.

Finally, the optional BIOS command causes RM to append the file V25BIOS.BIN to the top of the

ROM image, located so it will execute properly on reset of the target V25 board.

To actually create your ROM disc, run the ROMMAKER program by entering:

```
ROMMAKER MYDISC
```

ROMMAKER will open the MYDISC.BLD file, execute the commands it finds, and leave behind a binary image file named MYDISC.BIN. This binary file will be exactly 256 KB long, and can be directly loaded into most EPROM programmers, such as the Needham PB-10.

The Hardware

During our development phase, Bill and I relied on the New Micros' NMIX-0025 single-board computer. I looked at a couple of different V25 boards, and settled on the New Micros' unit for a couple of reasons. First of all, I've done several projects, both hobby and professional, on New Micros hardware, and like their board designs a lot. Second, all of their development boards come with a very large prototyping area and full connector layout, making it easy to upgrade and add on to the board. Finally, their boards include three 32-pin memory sockets, giving you lots of options as to RAM/EPROM mix and sizing.

I quickly added the appropriate robot-type connectors, such as my standard four-pin male Molex KK-style serial connector, a two-pin KK-style power connector, a four-position dipswitch, power switch, and power-on LED. I also wired a bicolor LED to two of the V25 output lines, so my robot board could at least signal when it was alive.

I then rewired the jumper section so my V25 board supports a 256K static RAM from 00000h to 3ffffh in the first memory socket, a 32K EPROM from 80000h to 87fffh in the middle memory socket, and a 32K EPROM from 0f8000h to 0fffffh in the third memory socket. This arrangement lets me have a BOTBios and ROM disc combination in the top memory socket, as well as a secondary ROM disc in the middle socket, for experimentation. Since BOTBios can locate and use multiple ROM discs, this arrangement gives me ample flexibility for playing with new .COM files.

You can use the BOTBios/ROMMAKER system on other V25 boards, including the Tern board. In this case, you only have two memory sockets to play with, so you'll have to use RAM in one socket and put a BOTBios/ROM disc EPROM in the second. That's a little less flexible than the New Micros' board, but certainly workable.

Once we had our V25 boards modified, Bill and I sat down at our respective machines and began coding our assigned portions of this project. Bill did all of his work in MASM, the 80x86 PC-based assembler, while I wrote ROMMAKER in Microsoft C.

The coolest part, however, was developing the .COM files we intended to try out on our new V25 hardware. Rather than stick with MASM or MSC, I snooped around on the net, looking for a Basic compiler that could generate .COM files. I finally found a program called ASIC (for almost Basic) version 5.0. This shareware compiler makes VERY small .COM files that run fast, just what you need for those robot programs.

Bill and I took turns writing small .COM files, burning them into a disc ROM, installing them in our V25 boards, and testing the results. Initially, we used

ASIC to weed out the BIOS interrupts that Bill needed to support. Later, we used ASIC to create programs that directly manipulated the V25 I/O subsystems, using PEEKs and POKEs.

If you would rather use a more full featured Basic for your programs, see if you can scare up one of the older Microsoft Basic compilers. The older versions, up to about 5.0, actually created .COM files for standalone execution. This is ideal for use with the ROM disc system we've devised.

In Conclusion

This was a large software effort, but it unleashes power way out of proportion to the effort Bill and I expended. Whether you run BOTBios on a V25 or a 386ex board, you will find it to be a simple, powerful system that offers great flexibility in dealing with robot software design. The ROM discs open the door for adding PCMCIA cards to your robot, or hacking some of those surplus EPROM cards that are always appearing on the market.

My thanks to Bill Bailey for the opportunity to work with him on a project that took me into personally unexplored realms, the innards of the PC BIOS. And thanks, as always, to those in the SRS who once again aimed me in an interesting direction.

Inside the 68hc11

All three major elements of robotics design are difficult to some degree, but perhaps the biggest obstacle to success lies with software. People with no electronics or mechanical experience can solder a board or build a frame, even if it is only a sort of a board or a sort of a frame. But software has always posed a special challenge to those starting out, and I used this column from the June 1997 issue of *Nuts & Volts* to help beginners over that first hurdle.

Like the other two elements, software has its own special toolset, and I start by describing some of the most basic parts of that toolset. Boolean operations, essential for testing and changing the values in variables and I/O ports, get special attention. I receive plenty of email asking for help in understanding how such operations work, so I used this column to explain the techniques.

Next, I go into some detail on what takes place inside the innards of the 68hc11 as it comes out of reset. This transition state, when the chip goes from a cold slab of silicon to a high-speed robot brain, can seem like magic to someone starting out, but you have to understand how this change occurs and what your program must do to prepare for reset if you are to have any hope of writing working code.

Using the interrupts available on your microcontroller lends power to your code, so I spend some time discussing how to use a typical 68hc11 interrupt subsystem, the real-time interrupt, or RTI. This section includes sample code and a detailed explanation, so you can add this source to your own program.

For those who would like a little bit more information on the above topics, I include a review of an excellent little book by Jim Sibigtroth of Motorola, entitled *Understanding Small Microcontrollers*. Jim was one of the major forces behind Motorola's 68hc12 MCU family, and he's also a helluva writer. Novices and pros alike will find much of value in his book; recommended.

This column concludes with a short quiz on boolean operations. I've included the answers, in the form of snippets of SBasic code. If you can answer all of these questions correctly, you're well on your way to a successful robotics project. If you can't, read through this book, Jim's book, and any other material you can find until you have the boolean operators down cold. Their proper use in robotics programs is essential, and there is no time like the present to get proficient.

Previously, I finished the design of the beginner's robopet and started you off with writing software for your new machine. But designing software for a working robot often proves the greatest hurdle, and too many people manage to finish a robot only to be stopped cold when faced with writing a suitable program.

I have been writing software, in some form or another, for nearly three decades now. Over the years, I've picked up some general practices that serve me well, and I'll pass them along in hopes that they will help a few of you out there as you face the last great obstacle in getting your first robot running. The examples that follow use SBasic, since I think that offers a beginner the best chance of getting a machine up and running quickly, but the concepts embodied here apply equally well to all software designs.

Many people who tackle their first software project are intimidated because there is no tangible element to the task; you must solve a problem with no physical properties. You cannot weld, solder, bend, staple, or mutilate either the problem or the solution to that problem. This vagueness sometimes proves too daunting, and causes some people to give up, or worse, not even try.

I suggest that you view this intangible property of software as a plus, not a minus. If you can't bend or solder it, you also can't physically damage it. Your program won't take up any space on your desk, can't fall off the table and break, and won't catch fire if you plug the batteries in wrong. The worst you can do is write something that doesn't work, but we all do that.

What's more, you can carry your program, or more accurately, the design of your program, with you wherever you go. After all, it only really exists in your mind. You can work on it anytime you want, be it on the bus, sitting in the park, or in the shower. And you can change your program's design with virtually no effort; simply think up a few more ideas and graft them onto the program in your mind.

Even the testing of your program's design doesn't actually require a working program. As you gain experience in designing software, you will find yourself testing your designs in your head, running your imaginary program through its imaginary paces. You will learn to check for boundary conditions, those nasty inflection points where a program's design, given an overflow or a significant change in data, requires special attention.

The above concepts, carried to their logical conclusion, produce my guiding precept in software design. **Writing the code to any program is one of the very last steps in that program's design, not one of the first.**

Of course, if you are just starting out in software, that may not make a lot of sense. After all, if you don't write something, how can you ever get anything to run? You'll just have to trust me on this one. If you've only written a few programs, or maybe none at all, just go for it. Sit down with a text editor, put your robot on the table next to you (size and weight permitting) for inspiration, and flail away. But save that sentence above, perhaps in your robot notebook, and look at it again in a few months. The longer you stay with this hobby, the more reasonable that precept will become.

The Toolset

Just because your program is intangible doesn't mean you lack tools to solve a problem. Software, just like electronics and mechanics, has its own suite of tools and its own foundation of knowledge. Know your tools and understand the foundation that you start from, and you will have much more success in your software designs.

Every datum your robot must deal with ultimately is expressed as a combination of ones and zeros. Your program turns on a motor by writing a logic 1 (or logic 0, depending) to a motor control port. To sense the state of a bumper switch, your code must first read an input port. Even stimuli we normally think of as analog, such as light levels, must ultimately be reduced to ones and zeros. A brightly lit room, for example, could yield a value of $25 when your program reads an A/D port attached to a photocell.

I won't get into hexadecimal math here; I'll assume you already know that. For now, just remember that I write all hexadecimal numbers using Motorola's convention, which is a leading dollarsign. Thus, $100 is the hexadecimal value, or 256 in decimal. If you don't understand hex numbers, or if you need additional information on any of the basic concepts I discuss here, see my review below on Jim Sibigtroth's excellent book, *Understanding Small Microcontrollers*.

The Motorola 68hc11 microcontroller, or MCU, can address 64K bytes of memory. This spans an address range from $0000 to $ffff. Of these 64K possible addresses, some are given special meaning by the 68hc11's designers. For example, all of the 68hc11's I/O ports, used for contact with the external circuitry or for controlling the chip's internal characteristics, appear in the address range $1000 to $103f. The remaining discussions all assume that you are using the 68hc11 in single-chip mode, which frees up all of the chip's external I/O ports.

We can start exploring the 68hc11 as a robot controller by looking at one of these I/O ports. Port B (PORTB) appears as a single byte at address $1004. PORTB is an output-only port. Any pattern of ones and zeros that your code writes to PORTB appears immediately on the eight pins assigned as PB0 through PB7.

So simple tasks, such as writing a known value to all eight PORTB lines, are easy. Your SBasic code might look something like this:

pokeb portb, $ff

This works fine if all eight actuators you have wired to PORTB must change state at the same time, and they all change to a known state each time. But believe me, robot code doesn't work that way. More commonly, you will want to change one line of PORTB, say PB2, without altering any of the other lines. The following code writes a one to PB2 without disturbing the other PORTB lines. Assume that earlier SBasic statements have assigned the value $1004 to the constant PORTB:

```
n = peekb(portb)    ' get last known
                      state of PORTB
n = n or $04        ' set the third bit
pokeb portb, n      ' write the new
                      PORTB value back
```

This sequence of operations forms the core of nearly all interactions your code will have with the 68hc11 I/O ports. You will see variations of these lines in every program I discuss and in every program you design. If you have any trouble understanding what just took place, read the following discussion and Jim's book until you have this concept down cold.

The first line above reads a byte from I/O port PORTB. It doesn't read a word (16 bits) because I used the **PEEKB** function, which reads a single byte at the given address. If I read a word, using the **PEEK** function, I would get not only PORTB but the next byte (address $1005) as well. The value read from PORTB will contain the eight bits most recently written to that address. Thus, if my code most recently wrote the value $40 to PORTB, the first line above will read $40 from PORTB and store that value in variable N.

The second line above modifies the value in variable N by performing a logical OR between the value currently in N and the hex number $04. This OR operation guarantees that, no matter what value my code read from PORTB, the value in N will now hold a logic 1 in bit 2. For example, assume that N held $40 at the start of the second line's execution. An OR of $40 with $04 yields $44, which leaves bit 2 of N set to one.

The third line above writes the low byte of variable N back to I/O port B. It doesn't write the full 16 bits of N because I used the **POKEB** statement, which writes a single byte to a given address. If I had used a **POKE** statement, I would have modified not only PORTB but the next byte (address $1005) as well.

These three lines gets us what we want. Our code will write a logic 1 to the third output line of PORTB without changing the state of any other output line. The $04 used above is known as a mask. It represents a pattern of bits used to modify some variable or register. In this case, each bit position in the mask that contains a logic 1 causes the OR operation to set the corresponding bit in variable N.

You can use the other two major boolean operations to clear a selected bit or to toggle a selected bit. To toggle a bit, that is, to reverse the state of a given bit, use the exclusive-OR operation:

```
n = peekb(portb)    ' get last known
                      state of PORTB
n = n xor $04       ' toggle the third bit
pokeb portb, n      ' write the new
                      PORTB value back
```

This works just like the OR operation above, except that the mask in an exclusive-OR operation reverses the state of each corresponding bit in the other argument, which is variable N here. Thus, if variable N held $37 before executing the second line above, it would hold a $33 afterwards. The mask causes the third bit to toggle, changing from a one to a zero.

Clearing a bit using the AND operation is a little trickier. Here, the mask contains ones in all bits that you **don't** want to clear, and zeros in any bit that you **do** want to clear. The following example shows this:

```
n = peekb(portb)    ' get last known
                      state of PORTB
n = n and $fb       ' clear the third bit
pokeb portb, n      ' write the new
                      PORTB value back
```

Now the second line performs a logical AND operation between the value in N and the mask $fb. Since $fb contains ones in all bits except the third bit, all bits in N will remain unchanged except the third, which will be cleared to logic 0.

(Note: The last sentence above isn't exactly true. Since N is a 16-bit variable, the AND operation also clears all eight upper bits. But the **PEEKB** and **POKEB** operations only use the lower eight bits, so changes to the upper eight bits never appear on PORTB.)

The above three logical operators form a universal software tool. All viable languages for developing robotic or embedded control software must provide these three functions, though the syntax for using the functions might vary. What's more, you cannot develop working robotics code of any complexity without using these logical operators. Understanding how these operators work is crucial, and you should spend whatever time and effort you need on their study, until you are perfectly comfortable with their use.

To help you get up to speed, I've included a short quiz on logical operators (Figure 1), along with the solutions for each quiz question at the end of this article (Figure 3). The questions get progressively more difficult, and by the time you get through the whole quiz, you should have a good grasp of these tools.

The logical operators are a universal tool, in that you must use them regardless of the language you use or the type of MCU in the target hardware. Other tools are specific to the target hardware, and to write working robotic software, you must have a good understanding of the MCU involved. For this discussion, that means the 68hc11.

Using SBasic or any other high-level language can shield you from having to use assembly language for the underlying MCU, though I've long felt that knowing the assembly language of an embedded MCU makes you a more effective designer of embedded control software. However, no high-level language will remove the need to understand how the MCU's I/O registers and subsystems work. These registers and subsystems are the only way your code can sense or alter the outside world; the deeper your understanding of them, the more versatile and robust your robot design will be.

I'll start with what may be the biggest mystery to those beginning in robot design. The instant after you apply power, how does the MCU know where to start running your program? I'll answer this question for the 68hc11, but similar concepts hold for all MCUs used for embedded control.

Immediately after you apply power to the 68hc11 and its external circuitry, the MCU's clock begins to

Quiz!

Write an SBasic code fragment to perform each of the following tasks. Unless indicated otherwise, assume all addresses hold 8-bit bytes, not words. Similarly, treat all variables as holding 16-bit words. All logical operations must leave any unspecified bits of an address or variable unchanged.

1. Set bit 0 of address $1000.

2. Toggle bits 9 and 6 of address $1024. This address holds a 16-bit word.

3. Using whatever mask is in variable N, toggle the bits of address $1234.

4. Determine the state of bit 3 in address $1fff.

5. Toggle bit 5 of the value in variable N if bit 1 of $1004 is clear.

6. Clear bits 12 and 3 of the value in variable N if bit 11 of the value in variable K is set.

7. Execute the subroutine FOO if bit 4 of the value in variable N is clear.

8. Execute a tight loop until bit 0 of address $1005 is set. Use logical operators, not SBasic's **WAITUNTIL** statement.

Figure 1. A quiz on logical operators.

oscillate, which provides the "brainwave" to the MCU. This master clock, which runs at 8 MHz for most 68hc11 designs, dictates the speed of all internal operations such as memory accesses and timer functions.

After a fixed number of master clock pulses have elapsed, the MCU enters its reset state. All I/O registers are set to preassigned values, all external lines get set or cleared as the chip's designers intended, and the internal logic responsible for accessing memory is prepared with the first address to read.

You should refer to the Motorola 68hc11 reference manual, commonly called the "pink book" because of its color, to determine the reset state for all relevant ports and registers. When you first start out, you don't really need to know the reset state of a port or register unless your code depends on it at some point. But the reset state of some registers could subtly impact the behavior of your code, and

understanding these reset states becomes more important as your code becomes more complex and sophisticated.

At the end of the reset state, the MCU's internal logic prepares to read the first memory address. By design, the 68hc11 always reads a 16-bit value from address $fffe (known as the reset vector) immediately following reset. The reset logic finishes its job by taking whatever value it reads from this 16-bit address and copying it to the MCU's program counter (PC). With reset operations complete, the MCU now enters its normal run mode.

Run mode is an infinite loop that consists of fetching an instruction (opcode) from the byte addressed by the PC, executing that instruction, then fetching the next byte addressed by the PC. This sequence, with some variations, occurs for so long as the MCU power remains on and you don't hit the reset switch.

This reset sequence is a tool specific to the 68hc11, and you need to use it properly when designing your software. With this tool and a little more knowledge of your specific 68hc11 chip, you can ensure that your program begins running properly after reset. I'll continue this discussion by assuming your hardware uses the 68hc811e2 MCU, which has 2K of EEPROM from addresses $f800 to $ffff.

The '811e2 has on-chip memory that normally occupies the reset vector at $fffe. This eases our job of getting the MCU to start our program properly. Remember that the 68hc11 uses the 16-bit value stored at $fffe and $ffff as the address of the first instruction to execute after reset. If we can write the address of our program's starting point in those two bytes, then the reset sequence will jump to the beginning of our program.

SBasic automatically creates the proper set of assembly language instructions to set up this jump each time you compile a program. I suggest you write a small SBasic program, compile it, then use an editor to look at the resulting assembly language listing file, which usually has a .lst extension. Near the end of the program, you will see where SBasic has generated a Form Double Byte (FDB) instruction that is fixed (ORGed) at address $fffe. This instruction sets up the reset vector properly. All that remains is to move your SBasic object file into the '811e2's memory, using pcbug11 or a similar tool. Refer to the article entitled "An Intro to 68hc11 Firmware" for details on using pcbug11.

As you can see, it is the interplay between SBasic and pcbug11 that lets you write the proper address into the 68hc11's reset vector. If you prefer to write in assembly language, you have to insert your own FDB instruction at the proper location in order to set up your program's reset vector. You will then need to follow up with the same pcbug11 upload, to get your program into the proper area of memory.

You can do a lot of robotic software using nothing more than the reset vector and your SBasic program. But to appreciate the true power of the 68hc11, you need to take advantage of the 68hc11's interrupt capability. Interrupts form one of the most powerful tools available to any robotics designer, and using them properly will let you do some serious robo-magic. Note that almost all MCUs have support for interrupts of some kind, though the details vary based on the device.

The 68hc11 has a lot of interrupts available, but I'll focus on the real-time interrupt (RTI) for this article. I chose the RTI because of the power it grants your design when you use it properly. For example, you can use the RTI to create magic variables that automatically count down from some value until they reach zero, where they stay. These variables decrement at a fixed rate, letting you time various events or procedures with good accuracy.

The 68hc11 comes out of reset with the RTI subsystem disabled. Thus, your code can freely ignore all registers associated with the RTI if you choose. But if you want your code to take advantage of the RTI's power, your code needs to modify a few registers to "wake up" the RTI. See the accompanying RTI interrupt service routine (ISR) code sample (Figure 2).

The RTI subsystem consists of a timer chain inside the 68hc11, fed by clock pulses from the main timer. By default, the RTI timer chain is set to generate an interrupt once every 4.10 msecs, given a crystal of 8 MHz, which is fairly standard for 68hc11 systems. This means that once you enable the RTI interrupts, the 68hc11 will automatically generate a RTI at about 250 times per second.

The key here is that, following reset, your code must enable the RTI interupts. As I said earlier, the default condition out of reset leaves the RTI interrupts disabled. Enabling RTI interrupts lies with two I/O registers called TMSK2 and TFLG2, both one byte wide. TMSK2 lies at $1024, and bit 6 of this register controls whether the RTI subsystem can actually generate an interrupt. If this bit is cleared (the default following reset), the RTI subsystem will not generate an interrupt. This means your program will never see an RTI interrupt. Setting this bit, however, permits the RTI subsystem to generate interrupts.

Whether the interrupts ever actually occur, however, is controlled by the state of bit 6 in the TFLG2

```
include  "regs11.lib"


interrupt $fff0                     ' RTI
if wait <> 0                             ' if wait not yet 0...
     wait = wait - 1                ' drop it by one
endif
pokeb tflg2, %01000000              ' rearm RTI
end                                      ' back to main code


main:
pokeb tflg2, %01000000             ' clear RTI flag
pokeb tmsk2, %01000000             ' allow RTI interrupts
wait = 0                                ' start wait at 0
interrupts on                           ' enable all interrupts

do                                          ' start an endless loop
     if wait = 0                    ' if timeout happened...
               '
               ' do something here
               '
               wait = 250          ' set a 1-second timer
          endif
loop                                          ' loop forever
end
```

Figure 2. Setting up and using an RTI ISR in SBasic.

register. This bit is cleared by default after reset. So long as this bit is cleared, an RTI interrupt can take place (subject, of course, to TMSK2; see above). When an RTI interrupt does occur, bit 6 of TFLG2 is automatically set, locking out any subsequent RTI interrupts. To allow the next RTI interrupt to occur, your code must clear bit 6 of TFLG2 after it is set by the RTI interrupt. If your code does not clear this bit, your software will never see another RTI interrupt.

To clear bit 6 of TFLG2 after an RTI interrupt, your code must write a new value to TFLG2, with a logic 1 (NOT a logic 0!) in bit position 6. This might seem a little odd at first, but that's the way the 68hc11 designers set it up, so that's what your code has to do.

Even now, you're not quite done. Your code must complete the setup for RTI interrupts by finally enabling all interrupts in the 68hc11. You can do this with SBasic's **INTERRUPTS ON** statement, which enables system-wide interrupts. Note that this doesn't mean that your 68hc11 will suddenly begin firing off bunches of unknown interrupts. Remember that by default the 68hc11 comes out of reset with all interrupt subsystems turned off. Turning on the system-wide interrupts only allows those subsystems to work that your code has specifically enabled.

To review, your code turns on the RTI interrupts by using the following steps. First, it sets bit 6 of TMSK2, which enables the processing of RTI interrupts. Next, it clears bit 6 of TMSK2 by writing a one to that bit position. This resets the RTI interrupt flag, permitting the next RTI interrupt to happen. Finally, your code enables system-wide interrupts.

This completes the RTI setup, but how does your code actually process an RTI when it occurs? Recall the discussion of the reset vector above. When an RTI interrupt does occur, the 68hc11 momentarily suspends whatever it is doing, and saves all of its working registers in memory for later retrieval; this operation is known as a context save. Next, the 68hc11 moves the 16-bit value stored at address $fff0 into the PC. Finally, the 68hc11 restarts executing code, this time beginning with the instruction at the current address in the PC. As you can see, this technique of starting a program based on an address stored somewhere in memory exactly matches that used by the 68hc11 following reset. The address $fffe is called the reset vector because it holds an address to use following reset. Naturally, locations such as $fff0, which hold addresses of code to execute following an interrupt, are dubbed interrupt vectors.

After the 68hc11 has fetched the address in the interrupt vector, the chip begins running your code. This section of code is called the interrupt service routine, or ISR. After performing all the tasks necessary, your ISR can force the 68hc11 back to its original code by executing a ReTurn from Interrupt (RTI) instruction. In SBasic, this is simply an **END** statement at the end of an **INTERRUPT** routine.

I realize that this all sounds terribly complex, but the code involved is actually quite simple, especially when written in SBasic. After all of this setup and bit-clearing, your program is ready for the real magic to happen. Once every 4.1 msecs, the 68hc11 will stop whatever it was doing and execute your ISR. After it finishes your ISR, the 68hc11 will go back to what it was doing without missing a beat. Thus, your ISR executes automatically, with no support from your other code required.

Take a look at the accompanying RTI ISR code. The main code consists of a large empty **DO-LOOP** structure. The only code inside the loop is a test of the value in variable WAIT. Notice, however, that the mainline code never decrements WAIT; it only tests WAIT. All changes to WAIT occur in background; that is, within the RTI ISR code at the top of the file.

WAIT starts out at zero, so the first time through the **DO-LOOP**, the mainline code will notice that WAIT has reached zero and change it to 250. Since the RTI interrupt occurs once every 4.1 msecs, this value of 250 corresponds to slightly more than one second of delay.

As soon as the next RTI interrupt occurs, the ISR code sees that WAIT is no longer zero and begins decrementing the variable again. One second later, WAIT again hits zero and the cycle starts all over. Note that even if WAIT hits zero, the interrupts still continue. Thus, your code doesn't have to do anything special to restart the interrupts. As soon as something makes WAIT non-zero, the variable will begin decrementing automatically.

Take the time to look through the assembly language file that SBasic generates for this program. You will find the compiler's code very instructive in setting up and using interrupts.

This completes my series on getting your first robot going. These three articles have covered the electronics, mechanics, and programming of a simple robopet. Future articles will continue from here in various directions, but you should have more than enough info to start your own projects. Be sure to download the program samples from my web site; I think you will find many of them instructive.

Worth a Look

Jim Sibigtroth, one of the all-time good guys at Motorola, wrote a Motorola text book back in 1992 that only recently attracted my attention. *Understanding Small Microcontrollers* serves as an excellent introduction to the Motorola 68hc05 series of small embedded micrcontrollers. What's more, much of the information is directly useable on the larger 68hc11 and on microcontrollers in general.

Jim's book begins with good material on hexadecimal arithmetic, ASCII codes, and binary-coded decimal (BCD) numbers. This type of low-level information has become scarce in today's world of multi-megabyte compilers and visual-everything, but you absolutely have to know this material to have a prayer of pulling off a robot design.

Jim also explains how a small MCU's innards work, including how it uses the on-chip peripherals, memory, and interrupt vectors. He also touches on the various types of memory found on the MCU, describing how, for example, volatile memory such as RAM differs from non-volatile memory such as EEPROM.

Next, Jim covers what is called the programmer's model. This describes how an MCU's register set appears to the programmer. Here is where you find out how many and what kind of accumulators and index registers the chip contains. This model also covers related topics such as the available flags, stack pointers, and interrupts. Following this, he gives you a quick look at how the 68hc05's instruction set works, including the various addressing modes.

The next chapter covers how to write a simple program. Here Jim introduces you to the steps nec-essary to designing your own program. He starts with a description of the program, draws up a flow-chart of the solution, then turns the flowchart into actual code. The process is very instructive, and serves as a good model for any software design. Jim then goes through the various MCU opcodes and assembler directives used in the program. He even includes a very instructive section on various ways to code the same function and explains the advantages and disadvantages of each.

Jim concludes this excellent little book with a larger example program that covers the real-time interrupt (RTI). You can use this program as the starting point for many other 68hc05 projects that require RTIs or other interrupts. Finally, Jim throws in a detailed discussion of the many 68hc05 on-chip peripheral systems.

You can probably pick up a free copy of this book by contacting your local Motorola distributor or by calling the local Motorola field application office. Be sure to mention the literature number M68HC05TB/D.

Solutions to the quiz on logical operators

```
1.  pokeb $1000,  peekb($1000)  or  1

2.  poke $1024,  peek($1024)  xor  %1001000000

3.  pokeb $1234,  peekb($1234)  xor  n

4.  n = peekb($1fff)  and  %1000
    if n = 0
        print "0"
    else
        print "1"
    endif

5.  if peekb($1004)  and  2 = 0
        n = n xor $20
    endif

6.  if k and $800 <> 0
        n = n and %1110111111110111
    endif

7.  if n and %10000 = 0
        gosub foo
    endif

8.  do until peekb($1005)  and  1 = 1
    loop
```

Figure 3. Solutions to the quiz on logical operators.

Part III
Electronics

I devote the largest section of the book to electronics. The circuits in this section include subsystems, such as motor drivers and microcontrollers, as well as tools, such as battery chargers. Some of these projects are available ready-made, saving you the hassle of finding the parts and wiring the project yourself. In other cases, you can wire the circuit as needed for your specific robot design, customizing the project if necessary.

The first two columns describe a pair of low-priced 68hc11 microcontroller boards, ideal for use in small robots. Marvin Green's BOTBoard has become a staple item for builders of small robots, and the CGN 1101 has all the glue logic on-board for adding memory to your 68hc11 designs.

The next several articles include cheap and simple designs for small motor drivers, a gel-cell battery charger, and two switching power supplies. I particularly like the Junk-Box Switcher, a small power supply you can build from nearly any kind of parts. It is a great educational project, as well; spend some time with it if you have questions on how switching power supplies work.

The following two columns deal with robotic optics. One article includes a circuit for adding lots of IR LEDs to your robot, useful for a robot that must detect objects in several directions. The other article describes a simple optical array that you can mount underneath a robot frame. With this array and some software, you can get started on a robot for line-following contests.

Stepper motors serve an important function in various kinds of robotics, and the next article covers a practical controller board for steppers. It also discusses the two major types of controller electronics, and why you should choose a chopper drive system over the older and less efficient L/R designs.

The last two columns in this section not only cover two powerful new MCUs for hobby robotics, they were great fun to write. Marvin Green, Kevin Ross, and I spent some quality hacking time on two newly introduced 68hc12 variants from Motorola. In one case, we were some of the first people outside of Motorola to get our hands on these chips, and we had working systems up and running quickly.

Quick and Easy 68hc11 Expansion

My *Nuts & Volts* column from November 1993 showcased the CGN 1101 expanded-mode 68hc11 microcontroller board. This nifty little device contains all the logic needed for adding extra memory to a 68hc11 and brings the needed connections off the board as wirewrap pins. Just add your own 68hc11 chip and a few components, and you're good to go. Way easier than building up a full expanded-mode system by hand!

I've developed several Forth compilers over the years, including my favorite, tiny4th. This column prepared readers for my first Forth projects, by introducing them to the language. I describe many of Forth's key elements, such as the data stack, and give some simple examples. Later columns built upon this foundation, so if you want to try out this interesting robotics tool, start with this article.

The CGN 1001 68hc11 module, described in a recent column "The Rapid Deployment Maze," lets you build a single-chip computer for robot control. The single-chip aspect limits you to a maximum of 2K bytes of EEPROM for program storage, if you use a 68hc811e2 MCU. If you use the more common (and cheaper) 'hc11a1 version, you get only 512 bytes of EEPROM.

With careful design, you can write quite sophisticated programs in such a small space, but not everyone is fluent in assembly language or has the bucks for high-powered compilers. What do you do if you simply need more code space?

One answer lies with the CGN 1101 module. This board, like its smaller 1001 brother, contains a 52-pin PLCC socket and necessary support circuitry for running a 68hc11 in single-chip mode.

But the 1101 also contains all the support circuitry needed to run a 68hc11 in expanded mode. This mode gives the 68hc11 access to a full 64K address space. By wiring in your own sockets for addition RAM and EPROM, you can build a system as large as you like.

The 1101 board is roughly 2 inches by 3 inches and contains four surface-mount technology (SMT) chips that handle the digital signals needed for expanded mode. As with the 1001, you make electrical connections to the 1101 using the board's wirewrap posts and 30 AWG wire.

Adding an extra 32K of static RAM involves little more than wiring the proper connections between the 1101 and a 28-pin socket. The component side of the 1101 has silk-screened legends beside each pin to aid you in locating your connections. CGN also supplies a pre-printed paper template that slips down over the wirewrap pins; the template carries the pins' legends on it as well.

To use the CGN 1101 board as a robot controller, you need to add the customary servo motor, A/D, serial, and power connectors. Refer to my previous articles for part numbers, orientation, and wiring details.

The accompanying schematic shows how I wired my first 1101-based robot computer (Figure 1). Note that I hooked up my static RAM with an on-

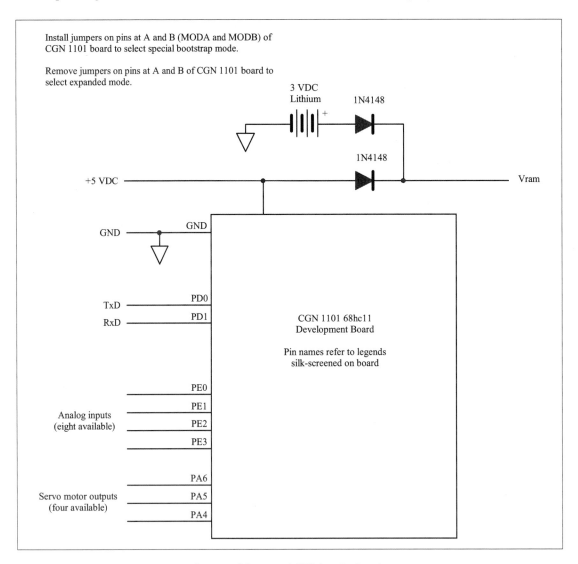

Figure 1. Schematic of CGN 1101 board.

board 3 VDC lithium battery. This battery-backed RAM lets me download software into my robot, then turn off power without losing the program.

A Closer Look

The CGN 1101 board packs all the boring glue logic into a single, finished module that you can treat as a "super-chip." I consider its price of $39 (less the 68hc11 chip) to be a bargain, as I have built too many of these expanded-mode systems by hand already.

I won't go into detail on the actual circuitry. You can consult the Motorola *M68HC11 Reference Manual* or related books for a full discussion on the circuitry for using expanded mode. You can find an even better discussion on designing a 68hc11 expanded-mode system in the book *Single- and Multiple-Chip Microcomputer Interfacing* by G. J. Lipovski. This excellent college-level text is published by Prentice-Hall and distributed by Motorola; it is literature number TB316. Contact your local Motorola office or the Motorola Literature Distribution Center for ordering information.

You can also use the ideas in the Lipovski book as a starting point for your own 68hc11 designs. The book describes all the timing and electrical considerations involved in doing an expanded-mode system. You can build on these ideas and create your own mixes of RAM, ROM, and I/O. A highly recommended book.

The CGN expanded-mode circuitry provides 16 address lines (A0-A15) and eight data lines (D0-D7), brought out on 24 wirewrap pins along the board's perimeter. You must connect all data lines and A0-A14 of the address lines to the appropriate pins of your 62256 32K static RAM socket.

But the RAM chip needs several additional signals for timing and chip-enable. The CGN board doesn't derive these signals for you, so I included a 74hc139 that handles the job. The 74hc139 is a dual 1-of-4 decoder/demultiplexer. This sounds pretty intimidating, but actually the chip performs a very simple function.

Each half of the 74hc139 uses two digital inputs to select one of four outputs. I'll describe the

half associated with pins 1 through 7; the other half works identically using pins 9 through 15. Refer to the 74hc139 circuitry in the accompany schematic (Figure 2).

Pin 1 is a select line. When it is high, all output lines (pins 4 through 7) are automatically high. When pin 1 is low, the signals on pins 2 and 3 are used as a two-bit binary number. The corresponding output line is brought low while the others remain high.

For example, suppose pin 2 is low and pin 3 is high when pin 1 goes low. Since pin 2 is the less significant bit of the two input lines, this means the input lines hold a binary 10, or decimal 2. The output lines are arranged so pin 4 is the least significant and pin 7 the most significant. Thus, pin 4 goes low when the input lines hold a binary 00, and pin 7 goes low when they hold a binary 11. The example has the input lines carrying a binary 10, so pin 6 will go low. All other output pins will remain high.

The 74hc139 circuitry must provide a reliable write-enable signal for the static RAM. By this, I mean the signal must tell the RAM chip when the data bus contains valid data anytime the 68hc11 performs a write instruction.

Now consider the wiring in the schematic. Pin 1, the select line, is tied to the R/W line from the CGN board. This signal is normally high, but goes low whenever the 68hc11 prepares to write data to the data bus. Using this as the select line into the 74hc139 means that all output lines will remain high so long as the R/W line is high, or for so long as the 68hc11 does not want to write data to its bus.

But suppose the 68hc11 wants to write data to the data bus. It will bring the R/W line low, put the data on the bus, wait a short time, then bring R/W high again. Unfortunately, the circuitry that wants to read the data written to the bus by the 68hc11 cannot guarantee exactly when the MCU will pull the R/W line high. This means that your static RAM chip, for instance, cannot rely solely on the state of R/W to decide when to read the states of the data bus lines.

That is because the 68hc11 R/W line is not a timing signal. However, the signal labeled E is a tim-

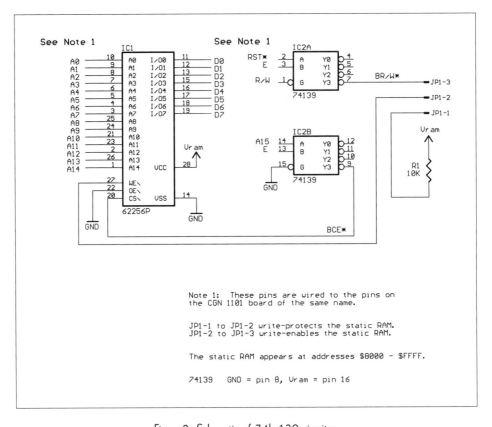

Figure 2. Schematic of 74hc139 circuitry.

ing signal and provides a very stable clock for deciding when the signals on the data bus are valid. E carries a 50% duty-cycle square wave that serves as the master clock for the 68hc11. During a write-cycle, the data on the bus are guaranteed stable 30 nanoseconds after E drops from high to low. This gives your RAM chip time to latch the data from the bus and save it.

The 74hc139 has E wired to pin 3 and RST* (or reset) wired to pin 2. Since RST* is normally high, only the states of E and R/W determine the values of the output pins. During a write-cycle, R/W is low and the two input lines are active. When E is high, output pin 7 will go low. When E is low, output pin 7 will go back high. Additionally, pin 5 will go low, but that is incidental.

The end result is that pin 7 provides a properly conditioned signal for controlling the write-enable input of a static RAM chip. Wiring it to the RAM chip's WR* line (pin 27) ensures that the RAM chip will read the data bus correctly.

The RST* Line

But why involve the RST* line? 68hc11 chips have a nasty problem involving memory chips and power supplies. The 68hc11 is sensitive to how quickly its RST* line rises to a logic 1. Specifically, the RST* line must go from a logic 0 to a logic 1 in less than two E cycles. E oscillates at a frequency equal to one-fourth of the crystal frequency. Thus, in this circuit, E must rise from a logic 0 to a logic 1 in less than 1 microsecond.

Note that I said "logic 1" and not +5 VDC. The 68hc11 RST* input sees any voltage above 80% of Vdd (the power supply voltage) as a logic 1. This means that if the power comes on slowly and RST* tracks it closely, the 68hc11 can come out of reset before the power supply has reached the proper working voltage.

If this happens, the 68hc11 begins fetching and executing random data, treating them just like a program. This is almost guaranteed to corrupt unprotected RAM. In fact, it can easily corrupt the 68hc11's on-chip EEPROM. The solution lies in properly conditioning the signal you apply to the RST* line of a 68hc11. CGN has added considerable circuitry in an effort to protect the 68hc11's RST* line.

This brings us back to the 74hc139 and the RST* line tied to one of its inputs. Since CGN generates the RST* signal, I tried using it to protect the static RAM from false writes. The theory is that RST* will not go high until power is stable and the 68hc11 can safely come out of reset.

Unfortunately, this idea just doesn't work reliably. I tested this by downloading a program into the battery-backed RAM, then repeatedly resetting the 68hc11 by turning its power off, then back on. About every third time the static RAM was corrupted and the 68hc11 "ran away," executing garbage instructions.

I noticed that the CGN 1101 uses a reset network made of several passive components, a 74hc132, and a transistor. Motorola recommends using the MC34064P low-voltage inhibit (LVI) IC as a way to condition the RST* signal. This chip comes in a three-pin TO-92 package, and is available at most Active Electronics stores for about $1 each.

Using this chip is simple. You wire pin 1 to the RST* input of the 68hc11, pulled to +5 VDC with a 10K ohm resistor. Connect pin 2 to +5 VDC and pin 3 to ground, and you're done. You should use this chip, or something similar, in any 68hc11 design you build. Motorola lists the Seiko S-8054HN as a replacement.

Oddly enough, CGN uses the Seiko chip in their single-chip 1001 module. I don't know why

they abandoned the simple, three-pin LVI chip for the more complex set of parts used on the 1101. I have only had one bad reset out of hundreds of cycles on the 1001, but the 1101 has proven much less reliable.

Use That Jumper

Since the data in the battery-backed static RAM were getting clobbered quite regularly on power cycles, I was forced to develop a different way of protecting the data during power-up. I don't like having to do it, but I resorted to using a three-pin jumper block. Refer to the schematic (Figure 1).

Connecting a jumper between pins 1 and 2 ties the RAM's WR* line high, preventing it from going low and causing a false write. Note that the pullup resistor for this pin is connected to Vram instead of +5 VDC. This means the WR* line is held high, even when the board's +5 VDC has been removed. You will normally use the write-protected mode when running your robot or just before you remove power from the board.

Plugging a jumper across pins 2 and 3 allows the 68hc11 to bring the WR* pin low as needed, using signal BR/W*. This is how the board should be configured during downloading. If you don't like the hassle of moving jumpers around, you can replace the jumper block JP1 with a small SPDT switch.

The Address Bus

The second half of the 74hc139 controls the chip select line of the static RAM. Again, the E signal is used to insure that the output line is properly synchronized to the 68hc11's address and data busses. In this case, I tied the select line (pin 15) to ground so this half of the 74hc139 is always enabled. This means one of its output lines will always be low.

The static RAM must see a low on its chip-enable line (pin 20) for a short time while the address bus is stabilizing. When the bus has settled, CE* must change from low to high. The contents of the address bus at this instant determine what address inside the RAM chip will be accessed.

Using a 32K RAM chip means the RAM can most easily appear in one of two places, either the high or low half of the 68hc11's address space. Since I want to store programs in the memory and later run them on powerup, I need the RAM to appear from $8000 to $FFFF. (68hc11s store the power-on reset vector at $FFFE, so there better be memory up there.)

I want the RAM enabled whenever E is high and the highest 32K of address space is selected, so I use E as the most significant input and A15 as the least significant. Using these signals as inputs gives a binary 11. The corresponding output, pin 9, will perform the desired CE* function.

Note that this half of the 74hc139 can also detect when addresses in the lower 32K of memory are accessed. In this case, E will be high and A15 will be low, giving a binary 10 on the inputs. This means that pin 10 will go low during reads or writes to addresses from $0000 to $7FFF.

Although I haven't used pin 10 for anything in this design, you could use it as a chip select for some type of I/O device in the lower half of the address space. This would make it easy to add, for example, additional parallel ports.

Generating chip selects with the 68hc11 timing signals is an old problem, and I have seen many techniques used. This idea comes from Lipovski's book and works very well when you are using 32K RAM chips.

Needing a Language

This and the previous several articles have developed methods of building the mechanics and electronics for simple robots. I have avoided software details intentionally, by developing the code for you and making it available on my web site. But this method is only a stopgap solution to a fundamental problem. If you want to experiment with robot behavior, you must develop your own robot code.

Until recently, I had to write my robot code in 68hc11 assembly language. I don't personally consider this a problem, since I have been writing such code for several years. But I assume many (or most) of you don't have the time to spend becoming proficient enough in 68hc11 assembly to write tight, well-tuned code for your little machines.

So I started looking for an alternative to assembly language. My earlier designs, using an 80c31, relied on a BASIC compiler distributed by Suncoast Technologies. Unfortunately, I couldn't find a 68hc11 BASIC compiler that was free and that built tight code. I am not a firm believer in BASIC, so I probably didn't look for one very hard anyway.

My language of choice for robots (or anything else) is Forth. Since Forth compilers are quite simple to write, I sat down and wrote one. It isn't ready to release yet, so I will spend the rest of this column getting you started with Forth.

Incidentally, one of my goals in writing this compiler was that it generate programs small enough to run in a single-chip 68hc11 system with only 512 bytes of EEPROM. It does this quite well, although you really need an expanded-mode system (or at least a 68hc811e2 chip) to do serious robot software.

Starting Forth

Forth is a programming system that gives you very low-level access to all of your machine's resources, yet lets you use high-level source code to do so. The smallest element of a Forth program is called a word. You string Forth words together to make other words. Eventually, the last word you make is your finished program.

When a Forth word executes, it gets any data it needs from Forth's data stack. When the word finishes, it leaves any results on the stack. Thus, the data stack serves as the data area for Forth words. Although Forth also supports variables, many Forth programs don't use any variables at all. Instead, they rely solely on data stored on the stack.

Suppose you wanted to declare PD5 as an output bit. To do so, you must write a $50 to address $1028, then write a $20 to address $1009. In Forth, this becomes:

hex
50 1028 c!
20 1009 c!

The first line tells the Forth compiler that any following numbers are in hexadecimal. A similar word, **DECIMAL**, switches the Forth compiler into decimal mode.

The second line pushes a $50 onto the data stack, then pushes a $1028 onto the data stack. Finally, **C!** actually writes the $50 to $1028. The third line performs the same function by writing $20 to $1009.

These last two lines show the behavior of the Forth word **C!**. It uses the top item on the stack as an address. It then stores the next item on the stack AS A BYTE to that address.

In most Forth texts, "top item on the stack" is usually shortened to TOS. Similarly, "next item on the stack" is usually written as NOS. Thus, **C!** writes NOS to TOS as a byte. A similar Forth word, **!**, writes NOS to TOS as a word (16-bit value).

Forth texts use a shorthand method of describing the stack's contents before and after a Forth word executes. This stack diagram helps clarify what a word expects to find on the stack when it starts, and what it will leave on the stack when it completes.

For example, the notation for **C!** is:

(c addr —)

C! (pronounced "c-store") expects an 8-bit value (shown as c) as NOS. It also expects a 16-bit address (shown as addr) as TOS. The TOS is always on the rightmost list of stack items in this notation. The two dashes "—" indicate that the word's execution occurs. Since no values follow the dashes, this word does not leave any values on the stack after it finishes.

The stack diagram for **!** ("store") is:

(n addr —)

It behaves just like **C!**, except it writes a 16-bit word n to a 16-bit address addr.

The reverse operation, fetching a byte or a word from memory, is done with **C@** ("c-fetch") and **@** ("fetch"). Their stack diagrams are:

C@ (addr — c)
@ (addr — n)

C@ uses the TOS as a 16-bit address. It fetches a byte from that address and leaves the byte as a 16-bit value in the TOS. **@** does the same, except it fetches a 16-bit word and leaves it in the TOS.

Note that both words leave behind a 16-bit value. This is because all items on Forth's data stack consist of one or more 16-bit values. (Strictly speaking, this isn't always true; but it is true for most of the 68hc11 code we will write.)

Forth supports simple 16-bit integer arithmetic with words such as:

+ (n1 n2 — n1+n2)
- (n1 n2 — n1-n2)
* (n1 n2 — n1*n2)
/ (n1 n2 — n1/n2)

If you use an RPN (reverse-Polish notation) calculator, you probably recognize the method used here. First you put the values on the stack, then you perform the function. Thus, you add 3 to 5 by:

3 5 +

leaving the result on the stack.

Note that ***** ("star") leaves a 16-bit integer. Thus,

1000 70 *

does not produce 70,000; instead, it produces 70,000 modulo 65,536, or 4,464.

Note also that **/** ("slash") does integer division and discards the remainder. Thus,

32 5 /

will leave the result 6 on the stack; the remainder 2 is gone.

Forth also supports logical operations:

OR (n1 n2 — n1|n2)
AND (n1 n2 — n1&n2)
XOR (n1 n2 — n1^n2)

and stack operations:

DROP (n1 —)
DUP (n1 — n1 n1)
SWAP (n1 n2 — n2 n1)
OVER (n1 n2 — n1 n2 n1)
ROT (n1 n2 n3 — n2 n3 n1)

The last word, **ROT**, is pronounced "rote" and is short for rotate. You will use stack operators quite often as you will need to rearrange the stack to reach desired items.

Forth also handles conditionals such as **IF** and **ELSE**:

IF (f —)
ELSE (f —)

Conditionals use the TOS as a 16-bit flag. If the flag is FALSE (or 0), one action is performed. If the flag is TRUE (not 0), another action is performed.

Note that Forth considers 0 to be FALSE and anything else to be TRUE. Note also that conditionals use the TOS as a flag and leave nothing when they finish.

Several Forth words handle comparisons of two values:

= (n1 n2 — f)
< (n1 n2 — f)
> (n1 n2 — f)
U< (n1 n2 — f)
U> (n1 n2 — f)
0= (n1 — f)

= leaves a TRUE if n1 = n2, else it leaves FALSE. < leaves a TRUE if n1 is less than n2, else it leaves FALSE. This comparison, and that of >,

are done with signed 16-bit numbers; that is, numbers that run from -32,768 to 32,767. If you need to do unsigned comparisons, use **U<** and **U>**. They treat the stack items as numbers that run from 0 to 65,535.

The Forth word **0=** ("zero-equals") compares the TOS to 0 and returns a flag. This word is normally used to invert the state of a flag. For example, if a logical operation leaves a 3 on the stack, **0=** will return a FALSE.

Forth offers some looping constructs such as **BEGIN, UNTIL,** and **REPEAT**:

BEGIN (—)
UNTIL (f —)
REPEAT (—)

You use these words to set up loops within your program. You start a loop with **BEGIN** and terminate the loop with either **UNTIL** or **REPEAT**. Note that **UNTIL** uses an item on the stack to decide what to do. If the flag is FALSE at **UNTIL**, control returns to the matching **BEGIN**. If the flag is TRUE, however, control continues to the next Forth word. **REPEAT** does not use an item on the stack; it always returns to the matching **BEGIN**.

A short example will show how to use these words:

```
hex
begin
    102e c@
    20 and
until
```

This fragment sets up a loop that reads the 68hc11's SCI status register. The loop continues until bit 5 (the RDRF) is set, indicating that a character has arrived over the serial line.

Finally, you use the Forth words **:** and **;** to define new Forth words. The word **:** marks the start of a new word; its name follows immediately. The word **;** marks the end of the word. For example:

```
hex
: wait-for-char
    begin
        102e c@
        20 and
    until0
;
: foo
    wait-for-char
    102f c@
;
```

The new Forth word, **WAIT-FOR-CHAR**, sits in a loop until a character is available in the SCI's data register. The word **FOO** calls **WAIT-FOR-CHAR** to wait until a character arrives, then reads the new character from the SCI's data register and leaves it on the stack. The above code fragment shows that even though we defined the word

WAIT-FOR-CHAR, it is used just like any other Forth word.

By far the best introductory text on the Forth language is Leo Brodie's *Starting Forth*, published by Prentice-Hall. It is available in most technical and university bookstores. In particular, read the book's foreword, written by Forth's inventor, Charles Moore. It is a beautiful summary of the power and elegance of the Forth language.

This article covers a lot of new ground regarding software. Don't worry about having to learn Forth to get your 'bots running. If you want to continue with assembly or C as your programming language, by all means do so. But if you aren't already fluent in a programming language, or if you are curious about Forth, you will likely find Forth easy to pick up. Future articles will provide extensive examples of Forth code, and I think you will find the language's features make it an excellent choice. I know I do.

Introducing the BOTBoard

This column from January 1994 marked a milestone of sorts in the amateur robotics field. It introduced Marvin Green's BOTBoard, a small 68hc11-based printed circuit board (PCB) that quickly became a mainstay in the hobby. Even now, there is no faster, cheaper way to get a small 68hc11 project built and running than using one of Marvin's BOTBoards. Many of the other columns in this book rely on the BOTBoard, and I still keep several lying around the lab for quick lashups.

Previous columns dealt with using infrared (IR) LEDs for object detection, but getting the LEDs aligned and shielded can prove difficult. This column answers many of the questions posed by readers who had a frustrating time with the devices.

I also used this column for a quick review of icc11, a recently-introduced C compiler by Richard Man of ImageCraft. The compiler, much enhanced, is still available and gets good marks from many people on the Internet.

I have had some feedback about problems dealing with IR sensors. Getting the emitters and detectors properly aligned can prove difficult, so I'll pass on a few tips and a tool.

I have not yet been able to get unmounted IR LEDs to work well as IR emitters. I refer here to the individual LEDs such as the Radio Shack 276-143 IR unit. The major problem with this type of component is the amount of radiation emitted in directions other than forward. This radiation can fool your IR sensor, regardless of its mounting location, into reporting an echo when it is really seeing direct IR.

You can get a feel for the problems involved by lighting a high-output red LED such as the Radio Shack 276-087. Simply connect the LED's cathode to the negative side of a 5 VDC supply, the anode to one end of a 180 ohm resistor, and the other end of the resistor to the supply's positive lead.

While lit, notice the amount of light emitted through the clear base and the side of the LED. Granted, the light emitted through the front lens is far brighter, but these light leaks are still quite powerful.

Masking this type of unwanted radiation from an IR LED has proven nearly impossible. I originally tried mounting the LED in various opaque tubes or holders, but that didn't work for two reasons. First, I cannot see the IR light being emitted, so I cannot see where the light leaks through my homemade mounting device. But more infuriating, many common lab materials are translucent (or worse) to IR. Just because you cannot see through a piece of plastic is no guarantee that it will stop any IR light.

The only method I have found for getting predictable IR patterns uses commercial IR emitter-detector pairs, usually available from surplus outlets. I am talking here about the tiny, prepackaged units intended for short distance detection of black-or-white markings. Many of these units are prefocused at 1/4 inch or so.

If you wire up just the emitter half of one of these units, you get a prepackaged, easy to mount IR source with a predictable IR pattern. Unfortunately, you can still run into problems in using these units in an object detector system. You need to remember that the close focal point means that this emitter is sending out a field of IR shaped roughly like a hemisphere. Anything that lies within that hemisphere is likely to generate an IR reflection. This includes parts of your robot frame, the floor, circuit boards, and nearly anything else.

I have had my best results by pointing the IR emitter forward and nearly straight up. This means the leading edge of the IR hemisphere just reaches the floor in front of the robot. You can also improve

object detection by strategic mounting of your IR detector modules. One good place to put the detector is on the underside of the robot's chassis, down near the floor. This placement uses the robot frame as an IR shield and helps prevent IR reflections from overhead objects. In fact, about the only signals the detector sees come from objects on the floor, directly in the robot's path. That, after all, is the purpose of the object detector system.

The IR Robo-Tool

I developed my IR Robo-Tool to help in all phases of robotics IR work. See the accompanying schematic for details on the first part of the circuit, the transmitter (Figure 1). The IR Robo-Tool transmitter consists of a 555 oscillator that provides an adjustable high-frequency chain of pulses suitable for driving LEDs. The circuit uses four AA-size batteries for power, so the tool is quite small and portable (Figure 2). As designed, you can hook up a variety of LEDs (both IR and visible) for experimentation.

The second part of the circuit lets you wire in a three-pin IR detector module, such as the Radio Shack 276-137. The output from the IR module lights a small red LED whenever the module detects sufficient IR energy of the proper frequency. Refer to

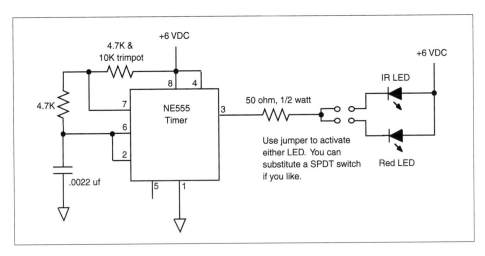

Figure 1. Schematic for IR Robo-Tool transmitter.

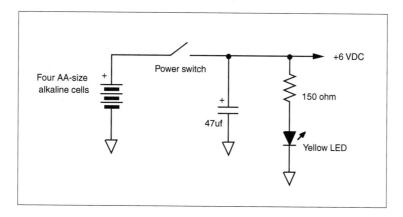

Figure 2. Schematic for IR Robo-Tool power supply.

the accompanying schematic of the IR Robo-Tool receiver (Figure 3).

Construction method is not critical. I built mine point-to-point on a Radio Shack 276-149 experimenter's board, and had lots of extra room left. After it was built, I drilled two small holes in the board, lined up with the mounting holes in a Radio Shack 4-cell AA battery holder (270-391). I then bolted the circuit board to the back of the battery holder, using #2 hardware and spacers.

This mounting system lets me stand the board vertically or lay it flat, depending on the type of experiment I am performing, and I saved the cost and fuss of putting the circuit inside a chassis box. This tool can perform several functions related to IR. For example, I recently purchased some IR detector modules from Electronic Goldmine. The company provided a hookup schematic, but no details on the unit's operating frequency.

I simply wired one of the IR detectors to the IR

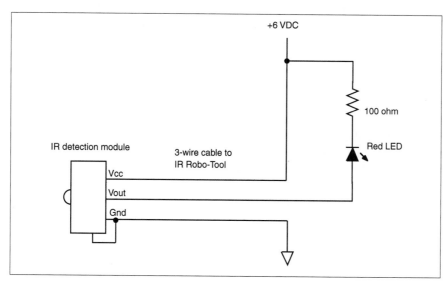

Figure 3. Schematic for IR Robo-Tool receiver.

Robo-Tool's input, connected an IR LED to the tool's output, then adjusted the 555's potentiometer until the red LED lit. This meant that the 555's output frequency matched that needed by the IR detector module. I then used my oscilloscope to measure the 555's output frequency. This showed that the Electronic Goldmine's IR detector (a Sharp GP1U01 unit) responds to 40 kHz, just as the Radio Shack module does. Incidentally, the Electronic Goldmine detector is far more sensitive than the Radio Shack device, and costs about one-third as much.

I don't know if Electronic Goldmine will have any of these units left by the time you read this, but contact them and at least ask for a catalog. They have good prices, plenty of surplus goodies, and excellent service.

Another use for the IR Robo-Tool involves testing different light sources. You can drive nearly any LED, regardless of color, with this tool. You just connect an LED of the desired color to the tool's output. Be sure to use the highest-output device you can find.

Detecting colors other than IR is almost as easy. Simply open up one of the IR detector modules and replace the sensor with an LED that matches the LED you hooked up to the Robo-Tool's output. For example, if you added a Radio Shack 276-087 to the tool's output, use another such LED to replace the IR detector's phototransistor.

Begin by carefully opening the metal case and removing the tiny PCB. Clamp the board in a bench vise or otherwise secure it firmly. Next, use a solder sucker and iron to remove the small dark phototransistor. Set this aside in your junk box, since it is a perfectly useable IR photodetector.

You'll have to get the LED's orientation correct when you install it on the detector's PCB. Examine the two solder pads on the PCB. The pad connected to the larger area of copper is the ground connection; insert the LED's cathode lead in this hole.

Insert the anode lead in the other hole, then solder and trim both connections. Carefully reinstall the PCB in its metal case, then hook up the detector module to the IR Robo-Tool's input. Now you can turn on the IR Robo-Tool, point the output LED at

the LED you added to the detector module, and the red indicator LED on the tool should light.

Note, however, that you will not get nearly the kind of range with visible LEDs as you do with the regular IR devices. I believe this is because the device normally hooked to the IR detector's input is a phototransistor or other amplifying device. The visible LEDs have no built-in amplification, so they cannot provide any gain by themselves.

Still, there are many robotics uses for short-range optical detectors. You could put a red LED emitter on one side of a gripper, and the LED detector on the opposite side of the gripper. When the beam is broken, the robot's computer knows that an object is now inside the gripper's area of action.

You could also use this type of sensor as a line detector in a line-following robot, or as a border detector in a Sumo robot. In both cases, the sensor will respond to a white-to-black transition. What's more, the sensor acts more like your eye does, letting you avoid the surprises that can come with guessing how IR sensors see the world.

But the biggest advantage to a visible-light object detector is that it won't interfere with an IR system running at the same frequency. You can run one 40 kHz IR emitter for long-range object detection, and a 40 kHz red emitter for gripper operation, and the two won't confuse each other.

Just wire the normal 40 kHz emitter circuit (such as those described in previous columns), using high-output red LEDs instead of the IR units specified. Then replace the IR phototransistor in one or two Radio Shack IR modules with the same LEDs. Now you simply place these visible-light emitters and detectors as needed on your robot frame.

The Radio Shack IR module invites even more experimentation. Readers with access to surface-mount technology (SMT) soldering equipment might try changing one or both of the capacitor chips on the detector's PCB.

You should be able to change the operating frequency of an IR detector module, using SMT caps pulled from surplus PCBs. IR detectors programmed to work at different frequencies (say, 40 kHz and 35 kHz) would let you have both object detector and IR communication systems running on the same 'bot, without interference.

I don't have the equipment to perform this modification, so I can't report on the proper cap values, or even if the idea works. If any of you try this, please let me know the outcome.

I'll close this section with one fairly obvious IR experiment. The IR detector side of the IR Robo-Tool responds to many of the surplus TV remote control units. For example, I picked up a couple of old controls at a local surplus store for about $1 each. I added a couple of AAA-size rechargeable alkaline cells, then aimed the control at the tool's IR detector and pressed a few buttons. The tool's red LED blinked each time, telling me that the surplus TV control emits a pulsed 40 kHz signal.

My control unit works from more than 15 feet away, and the combination has possibilities for arm-chair-based robotics. First, of course, you have to determine what signal you get for each of the different buttons. You would need to wire the output of the robot's IR detector to an input pin of the computer, then have software count the number and duration of pulses received when you pressed different keys.

After a little experimentation, you should be able to program your robot to respond to various keys on nearly any 40 kHz remote control unit. A robotic beer-fetcher is just around the corner!

The BOTBoard

Marvin Green, of the Portland Area Robotics Society (PARTS), has developed a 2 by 3 inch printed circuit board for building tiny 68hc11-based robots. His PCB, dubbed the BOTBoard, is etched and drilled for a 52-pin PLCC socket and contains traces for the usual crystal and passives needed to build single-chip 'hc11 computers.

Marvin set up his board so you can plug up to four hobby servos directly onto the board. Note that any servos you install on Marvin's board must have Airtronics connectors. The BOTBoard will NOT work with the Futaba J-connector that I use on the servo motors described in my recent articles.

He brought out all the 68hc11 PORTC signals to a 10-pin IDC connector pattern; most of the other 68hc11 I/O lines appear on a 26-pin IDC pattern.

Many of the board's other features stem from Marvin's previous work in building tiny robots. He added a 1 inch by 2 inch prototyping area, but designed the layout so that you can saw the prototyping area off if you need a really tiny computer board.

His circuit uses a MC34064 reset IC to protect the 68hc11's memory through power cycles. He even added pads for mounting a tiny SPST pushbutton switch to use for resetting the MCU.

I really like the documentation Marvin includes with his BOTBoard. You get the typical parts list, plus construction notes, list of suppliers (with phone numbers), and sample assembly language source for a couple of programs.

Marvin also supplies an application note describing how to build the needed RS-232 serial cable for the BOTBoard. Like the cable I described in a previous *Nuts & Volts* column, "Your First 68hc11 Microcontroller," Marvin's cable contains a MAX233 chip in the plastic hood of a DB-25 connector. This chip translates the PC's RS-232 signals into the TTL levels used by the BOTBoard circuitry.

You use PCBUG11 and the above serial cable to transfer your executable files from your PC to the BOTBoard's MCU. You can also use PCBUG11 to do limited testing of your code. PCBUG11 even includes a terminal emulator, so you can test those programs that need to talk to a serial terminal.

I will add one item I discovered while building my BOTBoard. Marvin used very tight spacing on the resistor layouts. You can fit the cylindrical 1/4-watt resistors into the holes so they sit flush on the PCB's surface, but the dumbbell-shaped dipped resistors don't quite fit. I had to form the leads on my resistors so they sit about 1/4 inch above the board's surface. You could also use 1/8-watt resistors, which will fit flush against the board, as well.

This board, like the single-chip CGN unit I've described before, "The Rapid Deployment Maze," can take any 68hc11 variant that fits in a 52-pin leadless chip-carrier package. This includes the 68hc811e2, which contains 2K EEPROM for program storage. Other variations of the 68hc11 contain up to 20K of EPROM (not EEPROM) in a windowed ceramic package compatible with the PLCC socket. These chips would give you plenty of code space, but you will still be limited to the 256 bytes of on-chip RAM. Contact your Motorola field applications office or distributor for details on the 68hc11 family of MCUs.

Marvin is offering his etched and drilled BOTBoard (no components) for $5.95 each (3 for $15.00) plus $3.00 shipping and handling. See Appendix A for details on ordering from Marvin.

For an additional $5, Marvin will include a floppy disc containing useful 68hc11 software and tools, including PCBUG11, Motorola's FREEWARE assembler, my tiny4th system, and several assembly language and tiny4th programs.

C for the 68hc11

I have just started using a newly released C cross-compiler for the 68hc11 chips. This program, called icc11, runs on the PC and compiles to 68hc11 assembly language compatible with the older Motorola AS11 FREEWARE assembler.

Richard Man, of ImageCraft, has made available for free what his company terms a "pre-release" version of a new C compiler. I am using pre-release version 0.30; the first real release of the product will be version 1.0.

Because of deadline pressures, I have only been able to spend a few hours working with icc11, so this will be a very preliminary review. Much of what follows is based on my limited experience with the product and on documentation supplied with icc11.

The icc11 system includes a C preprocessor, a C compiler, the as11 assembler, an assembly language source file containing the 68hc11 run-time and startup code, and a sample C program for a limited printf() function.

According to the documentation file, "The language preprocessor is close to ANSI C conformant

and the C language accepted by the language processor is ANSI C conformant."

The documentation file also notes that the icc11 system should run on DOS, OS/2, OS/2 VDM, Windows VDM, and "probably most DOS emulators on the popular Un*x and Macintosh machines."

I installed icc11 by copying the icc11.zip file into a special working directory named c:\icc11. I then used the PKUNZIP program to separate out all the component files.

Next, I used a text editor to examine the crt.s file. This file holds the assembly language source for the run-time executive. It contains several small subroutines that support 16-bit 68hc11 operations too large to be created in-line by the compiler's code generator. This file also contains the startup code for any icc11 C programs.

I added an ORG statement to the top of the file to declare the reset vector for my programs. I also changed the ORG statement for the main block of code to be $F800, consistent with my 68hc811e2 robot brains. Other than these two minor changes, I used the crt.s file as supplied.

I then used a text editor to create the supplied C test program, called bench01. See the accompanying listing (Figure 4). This program simply counts from 0 to $ffff, outputs a pulse on line 6 of port A, then starts over again.

I compiled the program with the command:

icc11 -l bench01.c

This created a listing file named iccout.lst and an executable file named iccout.s19.

I then hooked a small robot to my PC's serial cable, activated PCBUG11, and downloaded the file iccout.s19 into the 68hc11's code space. I switched a few jumpers, reset the MCU, and the program started running. By putting a logic probe on the pin connected to port A, bit 6, I could see a pulse each time the program finished its counting loop. The pulses appeared a little less than one second apart, a pretty respectable showing for compiled code.

```
/*
 * bench01.c
 *
 * First benchmark program to test
 * the icc11 C-compiler by ImageCraft.
 *
 * This program simply counts from 0
 * to 0xffff, then pulses bit 6 of
 * the 68hc11's port A line.
 *
 */

#define PORTA ((char *) 0x1000)

void main()
{
        int    i;
        do {
                *PORTA  =  0x00;
                for  (i=0;  i!=0xffff;  i++)  ;
                *PORTA  =  0x40;
        } while  (1);
}
```

Figure 4. Listing of bench01.c.

You can beat this performance easily in assembly language, but that isn't the point. Any compiler buys you (among other things) ease of code generation, usually at some cost in performance. icc11 turned out good-quality code very quickly, using a high-level language.

Another measure of a compiler's quality is the size of the created executable file. icc11's executable took up 224 bytes of code space when loaded into the 68hc11's memory. Most of this code space was devoted to run-time routines that my code never called. In fact, the active code for bench01.c is really just 35 bytes!

This crude first test shows that icc11 should offer a good platform for building small, powerful robotics programs. Since the 68hc811e2 contains 2K of EEPROM code space, you should be able to stuff quite a bit of C-code into your single-chip robots.

I couldn't help running a similar benchmark, using my tiny4th compiler. See the accompanying listing (Figure 5). When executed, this program cre-

ated pulses about 2.5 seconds apart. The executable file took up 765 bytes of code space.

Given the simple test file, these results are pretty much what I expected. I fully expect icc11 executables to always run faster than tiny4th code. As to code size, tiny4th programs will be smaller than icc11 executables for sufficiently complex programs. I just don't know where the breakeven point is in terms of source code size.

The bottom line is that the robotics experimenter now has two free methods of generating robotics code in high-level languages. You can get information on the icc11 compiler system by hitting the Imagecraft web site.

I should point out one restriction involving icc11. The license agreement in the supplied documentation stresses that icc11 is NOT public domain. Version 0.30 is a pre-release version of what might someday become a real product. Therefore, ImageCraft maintains that your license to use icc11 in its pre-release form expires if or when version 1.0 appears. No mention is made of any planned price for version 1.0.

```
(
( bench01.t4
(
( First benchmark program to test
( the tiny4th compiler.
(
( This program simply counts from 0
( to $ffff, then pulses bit 6 of
( the 68hc11's port A line.
(
( Time between pulses: About 2.25 seconds.
(

hex

1000 constant porta

: main
  0 1024 c!        ( shut off RTI )
  begin
    00 porta c!
    ffff 0 do
    loop
    40 porta c!
  repeat
;
```

Figure 5. Listing of bench01.t4.

In the Robo-News

The following report is taken from Internet releases published by Terry Fong of the NASA Ames Antarctica TROV project, and by Erik Nygren of the NASA Ames Research Center. Contact information for them follows.

In mid-October, NASA scientists began a planned two-month project in McMurdo Sound, near Ross Island in Antarctica. The project, dubbed TROV for Telepresence Remotely Operated Vehicle, will use a remotely operated, unmanned submersible to explore stretches of the ocean floor in McMurdo Sound.

The "remote" means really remote. Many times, the vehicle will be piloted by people (including high school students) in the Ames laboratory, located at Moffett Field in California.

A major project goal involves operating the vehicle in a "virtual reality" environment. Such a technique relies on a computer-generated model of the Antarctic terrain, and allows the pilot to steer the vehicle much like an aircraft in a video game.

The TROV receives power and guidance signals through a 1000-foot long tether containing electrical cables and fiber-optic lines. The vehicle sends video signals from two on-board cameras back through the fiber-optic cables.

This video is turned into stereo imagery and viewed with special LCD glasses. The stereo images provide a depth of field, so the operator can accurately use the TROV's on-board robotic arm and gripper for grabbing specimens or moving debris.

The remote-operation aspect of TROV will enable scientists who never leave California to search for and retrieve samples of all types from the Antarctic ocean floor.

NASA Ames will have an on-site team of project personnel to set up and maintain the TROV equipment, and to pilot the vehicle directly from the Antarctic site. The McMurdo site project leader is Dr. Carol Stoker. Also on-site are engineers Don Barch, Jay Steele, and Roxanne Streeter, and exobioligist Dale Andersen.

TROV team members at the Ames Laboratory in California include Butler Hine, Terry Fong, and Darryl Rasmussen. James Berry, a researcher for the Monterey Bay Aquarium Research Institute and a National Science Foundation (NSF) sponsored scientist, serves as chief scientist for the expedition. Another NSF scientist, James McClintock of the University of Alabama, will use the TROV to collect bottom-dwelling animals such as sponges for use in later studies.

The TROV station was declared operational and set up at its first survey site in Cape Armitage on Saturday, 23 October 1993. Dr. Stoker described the weather as "a balmy +10F...a shirtsleeve day which gives hope that summer has finally arrived."

Contact information:
Terry Fong terry@ptolemy.arc.nasa.gov
Erik Nygren nygren@athena.mit.edu

A Simple DC Gearhead Motor Controller

In this column from February 1994, I describe a couple of simple motor control ideas. One idea uses the PCB inside an R/C servo motor to drive a small 6 VDC gearhead motor; to your software, the gearhead motor looks just like an R/C servo. The other tip involves the TSC4427, a power MOSFET driver chip that has been out for several years.

I also describe a couple of tricks I use with Marvin Green's BOTBoard, originally mentioned in the January 1994 column. These tricks make the BOTBoard even easier to use, and let me run my favorite Futaba S148 R/C motors with the board, even though it was originally designed for another type of servo motor.

Finally, I give some details on the start of a large project undertaken by members of the Seattle Robotics Society, in conjunction with the Pacific Science Center (PSC) here in downtown Seattle. The project was ambitious, very large, and lots of fun; this column is your introduction to our efforts.

Not all hobby robots can use R/C servo motors for driving wheels. Sometimes you need to use faster motors, a different motor housing, or motors that run on a higher voltage. In these cases, you normally reach for a small DC gearhead motor, such as the Barber-Coleman motors I used in my earlier Sumo robot project. Surplus firms such as Herbach & Rademan or American Science & Surplus offer good assortments of such motors, and I rate their catalogs as must-have.

For small robots, you can usually find 12 to 24 VDC gearhead motors that draw around 100 mA and provide good torque at about 100 to 300 revolutions per minute (RPM). These motors make excellent robot drives. For example, Herbach & Rademan lists a Barber-Coleman motor (H&R P/N TM89MTR3021) that hits 300 RPM with 30 oz-in of torque, for just $15.50 each. See catalog #934 for details.

But once you get your motors, you still need to supply power and control their speed and direction. I'll discuss two methods of controlling small DC gearhead motors.

Servo Boards

The hobby R/C servo motors offer perhaps the simplest method of controlling a DC gearhead motor that runs on 5 to 6 VDC. Inside, an R/C servo is nothing more than a DC gearhead motor, a servo control board, and a potentiometer. I have already discussed modifying a hobby servo motor by replacing the potentiometer with two fixed resistors. Refer to Appendix B for details on modifying a Futaba S148 servo.

This modification leaves just a motor and a control board. You can carry this process a step further by replacing the hobby servo motor and its geartrain with a small DC gearhead motor of your own. Clamp the servo motor in a bench vise, then use a solder iron and a solder sucker to unsolder the motor from the small printed circuit board (PCB). Finally, connect leads from your selected DC gearhead motor to the motor terminals on the servo PCB. Polarity of the motor leads is not important.

This completes the electrical modifications. You now have a gearhead motor wired to a control PCB. The pair will likely not fit in the original servo housing, so you will need to come up with some type of mounting.

One idea is to hot-glue or epoxy the PCB directly to some part of the motor's case. This will yield a compact arrangement that keeps electrical noise from the motor to a minimum. Be sure you insulate the PCB so that no component leads touch the motor's housing.

When you have finished, you will have a DC gearhead motor that you can control just as if it were a modified hobby servo. In fact, you could replace the servos in Huey (described in "Allow Me to Introduce Huey") with your own heavier or faster motors. Simply plug the three-wire connector from your new motor's PCB into the appropriate header on Huey's circuit board.

Note that there are a few restrictions using this technique. The electronics from a Futaba S148 servo run on 4.8 to 6 VDC; you cannot use this PCB to control higher-voltage motors. The S148 servo PCB typically supplies about 100 mA of current to a running motor. I have not tested this technique under stall conditions, nor have I destructively tested a servo PCB to determine the maximum current it can supply. You can run motors that draw more than 100 mA if you like, but you are pushing the envelope.

Give this motor control technique a try on your next 'bot. For low-voltage motors, it offers the ease of servo control and the flexibility of using your own DC gearhead motors.

H-bridges

The usual technique for controlling the direction of a DC gearhead motor involves some form of an H-bridge. My early columns used an H-bridge made of a small relay to control direction and a power MOSFET to apply motor power.

You can also build solid-state H-bridges, using discrete transistors or dedicated ICs such as the L293D by SGS-Thompson. This latter chip provides two full H-bridges in a single 16-pin package; one chip can control two DC motors. Note that there is no single, perfect motor driver chip. There is also nothing stopping you from controlling motors with chips originally designed for other purposes.

I have mentioned Keith Payea, of the Seattle Robotics Society, in these pages before. Keith has introduced the club to several interesting ICs. One such chip is the Teledyne Semiconductor TSC4427 dual MOSFET driver.

The TSC4427 contains two electrical switches, built of medium-power MOSFETs, in an eight-pin DIP. The chip was originally intended to drive high-power MOSFETs using low current TTL or CMOS signals. Refer to the accompanying package outline and notes (Figure 1).

Rather than drive MOSFETs with this device, Keith used a TSC4427 to control each of the motors on his Albert robot. He simply wired the chips as H-bridges. Internally, the TSC4427 looks like two SPDT switches, each wired between Vs (the supply voltage) and ground. Each switch has an input line, labeled INA and INB in the diagrams. OUTA and OUTB serve as corresponding output lines.

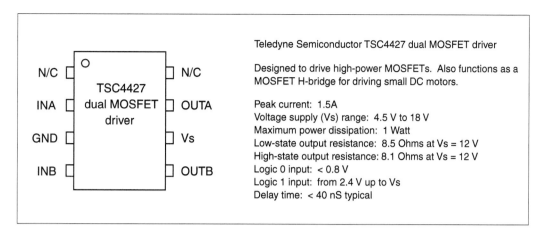

Teledyne Semiconductor TSC4427 dual MOSFET driver

Designed to drive high-power MOSFETs. Also functions as a MOSFET H-bridge for driving small DC motors.

Peak current: 1.5A
Voltage supply (Vs) range: 4.5 V to 18 V
Maximum power dissipation: 1 Watt
Low-state output resistance: 8.5 Ohms at Vs = 12 V
High-state output resistance: 8.1 Ohms at Vs = 12 V
Logic 0 input: < 0.8 V
Logic 1 input: from 2.4 V up to Vs
Delay time: < 40 nS typical

Figure 1. Notes on using the TSC4427 MOSFET driver.

You can connect any voltage from 5 to 22 VDC to the device's Vs pin. If you then apply a logic 1 to INA, the TSC4427 delivers the supply voltage to OUTA. Applying a logic 0 to INA causes OUTA to drop to ground. INB performs these same functions for OUTB.

Note that you don't have to work with just logic levels on the input. The TSC4427 considers any voltage from 2.4 VDC up to Vs to be a logic 1. This lets you control the chip with voltages from discrete circuitry, not just microcontrollers.

The chip uses CMOS technology and its outputs are MOSFETs. Thus, it delivers the entire Vs voltage to its outputs when activated. Contrast this with the 1.4 to 3 volts of saturation loss typical with bipolar driver ICs.

The TSC4427 does exhibit a fairly substantial resistance at its outputs. The Teledyne literature gives a low-state output resistance of about 15 ohms with a Vs of 5 VDC. My measured value weighed in at 14.5 ohms (See Figure 2). High-state resistance is listed as about one ohm lower.

This means that the output current will pass through an internal 15 ohm resistance (at Vs of 5 VDC). Any generated heat must be dissipated by the TSC4427 or the device will melt. The eight-pin plastic package can handle one watt of power. This limits your steady-state current through the chip to about 250 mA total, assuming Vs equal 5 VDC. However, the chip is rated to 1.5 A peak current.

Note that a dead short on an output (load resistance of 0 ohms) causes a current of 330 mA to flow through the device, given Vs equal 5 VDC. This is beyond the device's thermal capabilities, and if left in place long enough will destroy the chip.

As Vs increases, the output resistances drop substantially. At Vs equal 14 VDC, both output resistances fall to 8 ohms typically. This in turn lets you draw a steady 350 mA through the package.

The accompanying schematic and truth table show how to use a TSC4427 as an H-bridge (Figure 3). It doesn't get much simpler! Just hook up ground and Vs, wire the motor to the two output pins, and apply the proper digital inputs to INA and INB.

You can use the TSC4427 for driving a wide variety of devices. It was designed to switch high-capacitance loads such as power MOSFETs, at frequencies in excess of 1 MHz. It can withstand reverse currents of more than 0.5 A without latching-up, and will swing the output voltage to within 250 mV of either rail. This makes it ideal for driving solenoids, long cables or piezo devices, and for activating peripherals such as video cameras, headlights, or RF links.

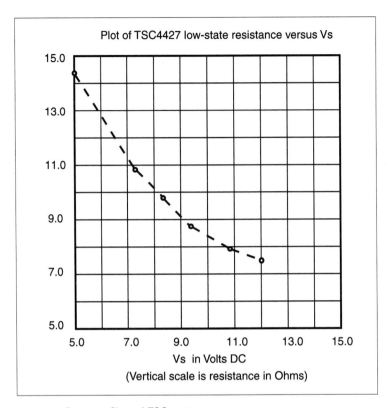

Figure 2. Chart of TSC4427 resistance measurements.

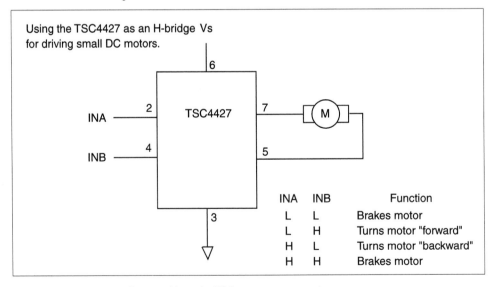

Figure 3. Using the TSC4427 as a motor driver.

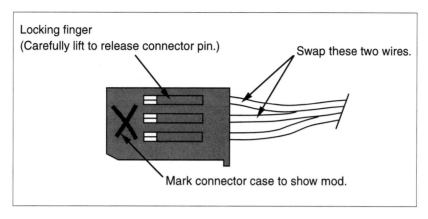

Figure 4. Modifying a Futaba J connector.

Best of all, you can buy this chip through DigiKey for just $1.37 each in singles. DigiKey also offers the TSC4426 and TSC4428 devices, closely related to the '27. The '26 features inverted outputs, while the '28 gives you one inverted and one non-inverted output. Remember DigiKey's minimum order when you call.

The BOTBoard

Last month I talked about Marvin Green's 68hc11 BOTBoard, a small PCB that lets you develop single-chip projects quickly and cheaply. I have already done one project with the BOTBoard ("Introducing the BOTBoard"), and want to pass along some thoughts.

Remember that Marvin is only offering an etched and drilled PCB; you have to supply all the other components. Still, at $5.95 each (plus shipping), I consider the BOTBoard a heckuva deal.

Marvin designed his board to accept up to four modified servo motors. You simply plug the servo's connector onto the selected three-pin header. Unfortunately for me, Marvin designed his BOTBoards to use Airtronics servo connectors. These are wired differently than the Futaba J connectors I use in my designs. The difference is significant; plug in the wrong motor and you will melt it.

I only use Futaba S148 motors with J connectors, so I spent some time modifying my motors to work with Marvin's board. The modification turns out to be quite simple. Refer to the accompanying diagram (Figure 4). You just need to reverse the red and black leads going into the J connector. Start by locating the tiny black plastic tab that locks the red wire into the connector shell. Carefully lift this finger up, just enough to ease the connector pin out of the shell. Do not lift too far, or you will snap the finger.

Use the same technique to release the black wire from the shell. Now reinsert both wires into the opposite positions in the connector. Push each pin completely into the connector shell until the matching finger locks the pin in place.

Finally, use a hot solder iron to score or mark a large X in both large sides of the connector body. This will serve as a reminder that the connector has been modified. Plugging this modified connector into a robot board designed for an unmodified Futaba connector will blow up the motor.

MODB Mod

Marvin designed his BOTBoard so the 68hc11 MCU always boots up in special bootstrap mode. Whenever the 68hc11 boots up in this mode, it automatically sends a special character out its serial transmit (Tx) line, then looks for an echo of that character on its serial receive (Rx) line. If the echo doesn't appear, the 68hc11 waits to receive a series of bytes on its Rx

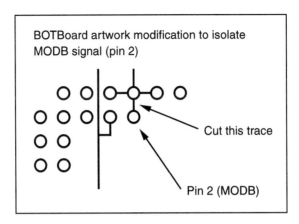

BOTBoard artwork modification to isolate
MODB signal (pin 2)

Cut this trace

Pin 2 (MODB)

Figure 5. Adding the MODB modification to a BOTBoard.

line. It then stores these bytes in RAM as a program, and jumps to the start of that program.

This works out well during program development, since your BOTBoard is automatically ready to talk to PCBUG11, Motorola's 68hc11 development system. If, however, the 68hc11 sees the echo of its special character, it assumes a program already exists in its EPROM or EEPROM, and automatically jumps to the starting address of that program.

Marvin intended that you run your working code by also booting up in special bootstrap mode. To do this, you need to first install a jumper across the serial port's RX and TX pins. Now, applying power brings the board up in special bootstrap mode. The shorting jumper across TX and RX insures that the 68hc11 will see an echo of its special character, forcing it to jump to the start of your program.

I have a few problems with this technique. First, it is quite easy to get the shorting block across the wrong pins on the serial port. Granted, the first time you do this and run down your batteries (or worse), you will likely get it right from then on.

But you still need this jumper in place any time you power up your robot. Not all robots are built so the PCB is easily accessible. If the jumper falls off during transport, your robot won't work until you find the darned jumper and put it back on the right pins.

Finally, most of my development robotics software uses the serial port for debug and control functions. I will load up a new program, boot the robot, then monitor the machine's behavior through a 9600 baud terminal program on my PC.

I am using a 68hc811e2 on my BOTBoard; this gives me 2K of code space for my programs. Since I write my code in tiny4th, I have plenty of room for text output, formatted numbers, and character I/O routines. Thus, I can build quite sophisticated debug software in just 2K of space.

However, the BOTBoard's use of the serial port forces me to first install a jumper, power up the BOTBoard, remove the jumper, install my serial cable, and then establish communication with my robot. I see this as too much fussing around.

The accompanying diagrams show a modification I have made to my BOTBoard that lets me use the serial port more effectively (Figures 5 and 6). The modification requires cutting a single trace on the BOTBoard, then wiring a 10K ohm resistor and a jumper into one corner of the prototyping area.

The trace cut isolates the 68hc11's MODB pin. After cutting the trace, locate a likely place to install the resistor and jumper block. I added my components near the ground buss on the prototyping area. Use 30 AWG wirewrap wire, cut and stripped to length, to make your connections. Use care when

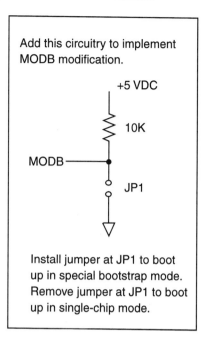

Add this circuitry to implement MODB modification.

+5 VDC

10K

MODB

JP1

Install jumper at JP1 to boot up in special bootstrap mode. Remove jumper at JP1 to boot up in single-chip mode.

Figure 6. Adding the MODB pullup resistor.

wiring to the MODB pin, as you can accidentally short to nearby pins. When installed, the MODB modification eliminates the need to add a jumper to the serial port when booting.

Now, you install a jumper on your new jumper block ONLY when you want to boot up and download a program into your BOTBoard. You remove the jumper anytime you want the BOTBoard to boot up and run your program.

This technique leaves the serial port completely free. Now, my robot code can begin accessing the serial port immediately upon startup; I don't have to worry about fiddling with the hardware.

What's more, the jumper only needs to be present during program development. I don't need to have the jumper in place during competition, leaving me one less potential problem to worry about.

Club Robotics

Seattle's Pacific Science Center (PSC) offers wonderful hands-on and interactive science displays for children and adults of all ages. I have spent many afternoons wandering through the Center and wishing I could somehow get involved in setting up an exhibit.

Recently, the PSC staff contacted the Seattle Robotics Society, asking if we would help them set up an exhibit of small, interactive robots that kids could somehow control. Obviously, the answer was a resounding "Yes!"

Our planned exhibit consists of a platform, about four feet by eight feet, and up to six mobile, autonomous robots. These robots will measure about eight inches in diameter, stand about one foot high, and roll around on three-inch wheels.

Each machine will contain modified hobby servos, IR object detectors, bumper switches, and will use a 68hc811e2 BOTBoard for a controller. The platform has been designed to help the visitors learn about robot behavior. Two large lights, one red and the other green, will shine inward from either end of the platform. The lights, mounted about one foot up on the inside wall of the platform, will serve as beacons for the roaming robots.

At either end of the platform, on the outside of the front skirt, will be mounted one large red pushbutton and one large green pushbutton. Pressing one of these buttons will activate the matching beacon at that end of the platform.

The robots will appear to gravitate toward the lighted beacon. Half of the robots will seek the red beacon, while the other half will only see the green beacon. As the robots move about the platform, they will use IR object detection and bumper switches to avoid other robots, the platform walls, and assorted obstacles scattered across the platform's surface.

Thus, the robots will display one of several different behaviors as they progress towards their goals. These behaviors might be called Seek, Avoid, and Flee, corresponding to seeking the goal, avoiding a nearby object, and fleeing from an object that touches the robot.

As each behavior activates, the robot will light a different LED on its display dome. This will give the visitors a visual clue as to why the robot has suddenly changed its behavior. Although the robots will appear to seek the red and green target lights, they will actually track on IR beacons mounted just below the lights. Beacons associated with the red light will emit an IR signal of one frequency, while the beacons for the green light will emit a different frequency signal.

Members of the SRS have taken on the design and construction of major elements of this exhibit.

Due to open 6 March 1994, the SRS exhibit will hopefully introduce the general public to amateur robotics and give the visitors a look at a fascinating new hobby.

If you find yourself in Seattle after early March, stop by the PSC and take a look at the robotics exhibit. And remember to join us on the third Saturday of any month; see the SRS web site for details. Visitors are always welcome...

A Gel-cell Battery Charger for Cheap

I wrote this column for the May 1994 issue of *Nuts & Volts* to fill a need: that of a cheap, easy to build battery charger. More and more SRS members were getting into gel-cell batteries, but chargers were outrageously expensive. Here is a dirt-simple charger you can build in an evening that will handle just about any 12 VDC gel-cell battery. And you can easily modify the circuit to charge batteries of other voltages.

I also covered a couple of "show-bot" ideas in this column. Any robot can use more lights, but the surface-mount (SMT) LEDs that were becoming more popular on the surplus market look way cool wired onto a robot. I also hacked one of the recently-announced Hallmark recording cards. I can't believe I actually spent $8 for a card back then! I must have been a real nerd...

Until now, my robots have been small, open-frame machines, powered by sets of C or AA batteries. The lack of covering lets me easily reach the batteries for recharging or replacement, and the batteries, while small, provide all the current my little machines need.

But building bigger 'bots usually means moving to larger batteries, such as the 12 volt gel-cells. And my new machine will have an outside skin, making access to the batteries difficult. In this case, I want to hook an external charger to the robot and recharge the batteries in place.

Gel-cells

You can group batteries of all types into one of two categories. Primary cells are those cells that give only one discharge cycle. This group includes cells such as mercury, lithium, silver oxide, and the early alkaline types. Secondary cells, however, can be fully recharged many times. This group includes cells such as lead-acid, NiCd, and the newer Renewal alkalines.

Gel-cells, or sealed lead-acid batteries, are secondary cells known for long storage life, plenty of power, and a high number of recharge cycles. They are commonly available in 6 volt and 12 volt units, though other arrangements are also possible. The sealed batteries are safe to use in any position, do not spill acid or other hazardous goo, and can be recharged in place without fear of dangerous fumes. Their compact size, rugged design, and long life make them good choices for self-contained power sources.

A battery's current capacity is usually expressed as a number of Amperes per hour (Ahr) into a load of fixed resistance. Typical ratings for gel-cell batteries run from two to six Ahr, though you can find brutes that provide 30 Ahr or more. The Ahr rating (also known as C) for a battery is typically the amount of current available during a 20-hour continuous drain. To calculate the battery's hourly available current for a 20-hour drain, simply divide the rating by 20.

For example, a battery rated at 2 Ahr will deliver 100 mA per hour for 20 hours. At the end of the 20 hours, a "12-volt" battery will be providing about 10.5 volts. Note, however, that this calculation is NOT linear with regard to current drain. Drawing two Amps from the above battery will fully discharge it well before one hour is up.

Charge and discharge currents are sometimes expressed as a multiple of C, the battery's Ahr rating. For example, a charge current of C/10 means a battery is being recharged at 1/10 of its current capacity. For a 4 Ahr battery, this works out to 400 mA.

In choosing a battery, you will usually know how much current you need to draw. What you don't often know is how long a given battery will last. Basically, there isn't a simple, accurate answer. One variable in calculating battery life deals with the definition of "fully discharged." A gel-cell battery is considered fully charged when each cell measures between 2.3 and 2.4 volts. As current is drawn from the battery, the voltage available at each cell begins to drop. At some point, the voltage at each cell drops so low that the battery is considered fully discharged.

Since a 12 volt gel-cell battery contains six cells, it is considered fully charged when it provides 6 * 2.3 volts, or 13.8 volts at its terminals. But "fully discharged" means different things to different people. The literature I have seen calls a battery fully discharged when its output drops to 10.5 volts. This gives a long life between charges. However, draining a battery this deeply can limit its life to about 50 recharge cycles.

Or, you can consider a battery as fully discharged at 12 volts, or 2.0 volts per cell. This will give you a shorter life between charges, but you will get many more recharge cycles. In reality, the "fully discharged" point may be set by your electronics. For example, you could have a regulator circuit that simply stops working properly when its input drops below 11 volts. And as I've already mentioned, you cannot simply divide your planned current drain into the battery's Ah rating and expect a meaningful answer. The relationship between current drain and battery capacity is best determined by consulting the maker's literature.

For a rough estimate, multiply your average current drain by 20, then look for a battery with that rating. Given a robot that draws 750 mA on average, you could start with a 15 Ahr battery. Remember, though, that this rating is based on discharging the battery down to 10.5 volts. Plan on recharging the battery before the 20 hours are up, or you will have to replace the battery after about 50 cycles.

Charge It

You should recharge gel-cell batteries with a constant voltage, current limited charger. Simply apply the proper charging voltage to the battery's terminals, making sure that the battery cannot draw more than a specific amount of current from the charger. After the battery has been sufficiently recharged, the recharger can provide a very small current of the proper charging voltage, to keep the battery fully charged.

For example, a 12 volt gel-cell battery is fully charged when it measures 13.8 volts, or 2.3 volts per cell. This is the voltage that the charger should provide. A deeply discharged battery can draw as much as 2C, or two times its Ahr rating, when first connected to a charger. This can be a large amount of current for higher-capacity batteries.

As the battery recharges, it will draw less and less current from the charger. Eventually, the battery will reach the fully charged voltage of 13.8 volts. At this point, the battery will only draw a tiny amount of current from the charger, usually from C/500 to C/1000. This reduced current is sometimes called a "trickle charge." Fully recharged gel-cell batteries can remain on a trickle charge for several years.

Generally, you can design a constant voltage charger to deal with the high initial current drain in one of two ways. One method calls for a charger that will supply up to 2C Amps of current for the intended battery. This could mean making a pretty hefty charger for some of the larger batteries.

Another idea is to build a charger that automatically limits the charging current to some lower amount, say one Amp or so. The second method has several advantages. It is cheaper, easier to build, smaller, and it can be used on batteries of nearly any rating.

As for disadvantages, it will take longer to recharge the higher-capacity batteries than the first type of charger. The accompanying schematic shows a LM317 used as a constant voltage, current limited gel-cell battery charger (Figure 1).

NOTE: This circuit CANNOT be used to recharge NiCd or Renewal rechargeable alkaline batteries!

The circuit is a straightforward LM317 regulator, adjusted to provide 13.8 to 14.4 volts DC at the output. These voltages correspond to a charge voltage of 2.3 to 2.4 volts per cell in a 12 volt battery. This charger needs a source of about 16 volts DC to properly charge a 12 volt battery. You can build a suitable power supply using a transformer, diode bridge, electrolytic capacitor, and (very important!) a fuse. Add in a chassis, power cord, and mounting hardware, and you have a good weekend project.

I chose instead to visit the local Radio Shack, where I picked up a "lump-in-a-line" power supply (273-1653) for $22. This unit is rated at one Amp, and comes complete with a panel-mounted circuit breaker. I call this a lump-in-a-line because it has a power cord coming out of one end and a second cord, fitted with a small connector, coming out of the other end. I like this arrangement, since I already have enough "wall-warts" plugged into my power strips.

There is one small problem with this supply, however. It doesn't supply enough voltage to use as purchased. It's within a volt of being enough, though. If only there was an easy way to milk an extra volt out of the supply...

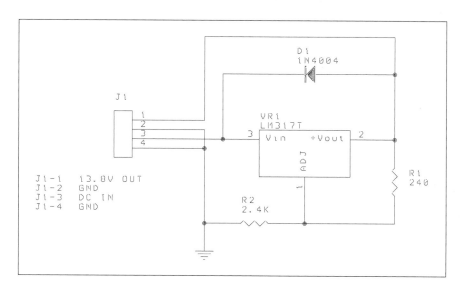

Figure 1. Schematic of LM317 battery charger.

I was discussing this project with Bob Nansel of the Seattle Robotics Society (SRS), and Bob explained that he had already built a battery charger, using this exact power supply. He then explained how to get the extra volt needed.

Open up the supply by removing the four screws in the bottom of the case. Inside is a transformer and a small circuit board containing, among other things, four 3-Amp power diodes. To get the needed extra one volt, carefully unsolder and remove these four diodes. Replace them with 3-Amp, 40-volt Schottky power diodes. The 1N5822, available from DigiKey and Active Electronics, costs about $1 each and works like a champ. Just make sure you install each new diode in the same orientation as its original counterpart. My power supply circuit board had the diode orientation silk-screened on the PCB, making the job easy.

Now build up the LM317 circuit as shown. Just about any type of construction method will work here. The circuit is so simple, I cannot imagine needing a PCB for it. Note that you do not need to use a heatsink on the LM317. If you work carefully, you can actually squeeze the entire LM317 circuit into the power supply chassis. If you try this, though, be very careful about accidental shorts in your wiring.

I wired my circuit up on half of a Radio Shack 276-148 dual mini-project board. I added a 4-position screw terminal block (276-1388) for connecting the input and output wires. The voltage adjustment resistor, R2, is fairly critical. You need to use a resistor as close to 2409 ohms as you can get. You have several approaches to take here.

You can use a 2.43K ohm, 1% precision resistor, available from places like DigiKey. You can also use a multi-turn potentiometer; simply measure between the wiper and one terminal, adjust the pot for 2.4K ohms, then put a drop of glue on the screw to lock it in place.

You can even use a 2.4K ohm, 5% resistor, provided you verify the actual resistance before installing it. I found several "2.4K" resistors that measured out at 2.6K ohms or more. This is enough difference to prevent the charger from working properly.

The schematic shows a 1N4004 power diode, connected across the LM317 to provide reverse voltage protection. You don't really need the 1N4004's 200 peak inverse volts (PIV) here; I just happened to have one handy. Anything from a 1N4001 on up should work fine. This diode protects the LM317 should you ever unplug the power supply while the charger is hooked to a battery. Under some circumstances, this could create a large enough reverse voltage to pop the regulator.

After you have built up your circuit board, you are ready to mount the board onto the power supply chassis. Start by covering the supply's metal faceplate (with all the lettering on it) with a piece of heavy insulating tape. Cover all of the faceplate completely, using at least one layer of tape.

Next, position the circuit board so it is centered over the taped faceplate, about 1/4 inch above the plate's surface, with the terminal block closest to the circuit breaker button. While holding the board in position, carefully glob some hot glue onto two of the board's corners, using enough glue to form a thick pad between the board and the chassis.

Hold the board in place while the glue cures; this should only take about a minute or so. When the glue has cooled enough, carefully glob hot glue onto the other two corners. Hold the board so it is parallel to the chassis surface as the glue cures.

Next, cut the output cord (not the power cord!) several inches from its strain relief in the side of the power supply chassis. The idea here is to leave just enough cord to reach from the strain relief to the proper terminals on the charger's connector block.

Strip and tin the ends on the shortened output cord, then slip them into the input terminals and tighten the screws. Watch the polarity on the wires; the lead with the white stripe on it goes to the positive input lead on the charger.

Now strip and tin the ends of the remaining length of output cord, then fasten them to the output terminals on the charger. Again, make sure you connect the wire with the white stripe on it to the positive output terminal.

Finally, dress the output wire so it lies against the strain relief on the output side of the power supply. Hold the output wire in place, then clamp it to the strain relief with a nylon wire tie.

Testing

You need to hook this charger up to a load that will draw at least 20 mA of current. This can be almost anything, such as a 12-volt relay or a light bulb. Connect the load to the charger's output cable, taking care to observe polarity precautions if your load is polarized. Next, plug the charger into the wall outlet.

You should measure 13.6 to 14.6 volts DC on the charger's output terminals. If you don't, you need to troubleshoot your work. One possible wiring error can be traced to the LM317's pinout. This device uses a pinout different from the common 7805 regulator. Refer to the accompanying diagram (Figure 2).

Another problem can lie with the resistor R2. Make sure it is as close to 2409 ohms as you can get, and certainly within +/- 100 ohms. Also check that D1 is properly wired; the cathode (banded end) must be connected to the LM317's input.

After you have cleared up any problems, and can see the proper voltage at the output terminals, your charger is ready to use. Incidentally, you can modify this circuit to charge batteries of other voltages, if desired. The value of R2 determines the charger's output voltage; use the following formula to determine the proper value:

$$R2 = R1 * ((V / 1.25) - 1)$$

where V is the desired output voltage. R1 is almost always 240 ohms, so this becomes:

$$R2 = 240 * ((V / 1.25) - 1)$$

For example, the calculation for a 13.8 volt output gives:

$$R2 = 240 * ((13.8 / 1.25) - 1)$$
$$= 240 * (11.04 - 1)$$
$$= 240 * 10.04$$
$$= 2409.6 \text{ ohms}$$

To build a 6 volt battery charger, you simply perform the same calculation, using a value of 6.9 volts for V (3 cells time 2.3 volts per cell).

The charger's input voltage needs to be at least 2.5 volts over the output voltage, but as close to that voltage as you can get. If you wanted to build a 6 volt battery charger, for example, you would want to look for a power supply that provides close to 6.9 plus 2.5, or 9.4 volts.

You might be wondering why a 12 volt power supply, modified to give 13 volts, can provide 13.8 plus 2.5, or 16.3 volts to run a 12 volt battery charger. The answer involves the loading that the power supply sees. Unloaded, this modified power supply provides 15 volts. If the charging battery draws a lot of current, the power supply output starts to sag closer to its advertised 13 (formerly 12) volts.

The LM317 needs a differential between input and output voltage (called the dropout voltage) of a certain minimum, depending on the current drawn. At lower currents, this dropout voltage can be as low as 1.5 volts at room temperature. As the current draw increases, however, this dropout voltage requirement climbs, reaching a top value of about 2.5 volts.

This leaves us with a conflict. As the current load goes up, the supply provides less voltage but

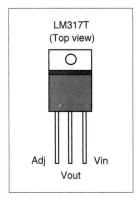

Figure 2. Pinout of LM317 (top view).

the LM317's dropout voltage requirement rises. Something has to give. What gives is the LM317's output voltage. The regulator always subtracts its dropout voltage from the input voltage, then works with what is left. If the result is higher than the desired output voltage, fine; the LM317 supplies the requested voltage. If the resulting voltage is not high enough, however, the LM317 simply provides whatever is left.

Back to our battery charger. Suppose we hook up a fully discharged 4 Ah battery to our charger. Initially, the battery will have a voltage of 10.5 volts and will try to draw as much as 8 Amps from the 13.8 volt charger. Immediately, the charger's output voltage will drop to 10.5 volts. The battery, which is already at this voltage, will draw almost no current, and the charger's output will start to rise.

The battery and the charger will quickly reach an equilibrium point. The charger will drop to a voltage and current that the battery finds suitable, and the charging will begin. As the battery charges, its internal voltage rises and the amount of current it draws at that voltage decreases. This in turn allows the charger's output voltage to rise.

This process continues until the battery reaches 13.8 volts. By this time, the battery is drawing so little current that the charger delivers the designed voltage at the trickle charge current. Note that the exact voltages and currents will vary, depending on components used and tolerances involved. But the above gives a picture of what takes place as the battery charges.

You can minimize this dropout voltage problem simply by supplying a higher input voltage to the LM317. I don't know how to easily get more voltage out of the Radio Shack supply, but you can explore a couple of other options.

I've already discussed one; build your own supply. I'm not going to get involved in specifying such a supply. If you want details on this, consult an Amateur Radio Relay League (ARRL) manual or Horowitz and Hill's excellent text, *The Art of Electronics*.

Another possibility involves scrounging the surplus markets for a more suitable, prebuilt power

supply. As a matter of fact, I located a 13.5 volt lump-in-a-line at a local Seattle surplus shop. The price was certainly right: just $8. Although it was a brand-new, working unit, it didn't have a circuit breaker like the Radio Shack unit. Still, I built a second regulator board, glued it onto the top of the new supply, and now I have two battery chargers.

Batteries

Now all you need are some gel-cell batteries. Several major suppliers provide batteries in a variety of physical shapes, weights, voltages, and current capacities. Panasonic, for example, offers a line of sealed lead-acid units called LCR. These are available from DigiKey; expect to pay about $30 for a 12 volt, 2 Ahr battery. Other companies that offer gel-cells include Gates, Yuasa, and Power-Sonic.

If you don't want to go mail-order for your gel-cells, look through the phone book for a local electric supply house. These cells are often used in burglar alarms, emergency lights, and security systems. You should be able to find a local source in town.

Talk to Me!

Before last Christmas, I saw a CNN spot on the Hallmark recordable Christmas cards. They cost about $15 at the time, and I ran right out to find one, but no luck. During a recent trip to a shopping mall, I dropped into a Hallmark store and asked about the recordable cards. Hallmark now sells them as birthday cards, for $8 each. Naturally, I picked one up.

The card's innards contain a 1.25 inch speaker, a separate piezo microphone, four tiny alkaline hearing-aid cells, and a 1 inch by 1 inch PCB with the recording chip on it. The card's plastic frame also contains some hardware and a couple of switches. One switch controls playback or record mode, the other switch turns the circuit on when you open the card.

The card will record a full ten seconds of speech when in record mode. You can then switch to playback mode, and the card will repeat your message as often as you press and release the little power switch.

The playback volume isn't very great, but that's no surprise, given the tiny speaker used. I imagine a boosting amplifier, perhaps based on an LM386, would provide plenty of volume. The sound quality was actually pretty good. Again, the speaker is no great shakes, so you can probably get much better performance with an outboard amp/speaker combination.

The small PCB contains two signal diodes and the speech recorder chip; the chip is hidden under a blob of black epoxy. The board also has a .022 µf ceramic disc capacitor and a 47 µf, 10 VDC electrolytic capacitor soldered to it.

Unfortunately, the board does not have any traces that appear to be address lines, so I don't know yet if you can record addressable units of speech, like you can with the Radio Shack IDS1000AP speech chip.

I replaced the piezo microphone with a small crystal mike from my junk box, and that improved the speech quality noticeably. I haven't had much time to spend with this circuit, but its 6 volt power supply and simple method of operation make it a candidate for low-power robots. Now you can customize your robot's introduction, collision annunciator, or people-detector.

Robo-lights

Robots somehow seem naked if they don't have blinky lights on them. But just slapping a few LEDs on a machine is boring. You need to try for something a little unusual or exotic in the lighting department.

I have been playing around with some red surface-mount technology (SMT) LEDs purchased from Electronic Goldmine. Since these are truly surplus, I don't have a part number (other than EG's, which is G2552), so you'll have to settle for a detailed description. Refer to the accompanying diagram (Figure 3).

The LED measures 0.125 by 0.094 inches, including leads! The leads are tiny metal ears that can be bent out slightly away from the LED's body. This will let you solder a short length of 30 AWG wirewrap wire to each lead. I connected one of these LEDs to a 3 VDC supply through a 100 ohm resistor.

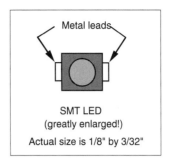

Figure 3. A typical SMT LED chip.

The resulting light is easily bright enough to see from across a typical living room. Note that from a distance, you can see the LED, but you cannot see the wires that provide the current. Thus, you see a pinpoint of pure red light that seems to hang in the air.

I could not find any marking for anode or cathode on these devices. Heck, I could barely see the LED itself. After attaching wires to the device, you will have to try applying power and checking the results, then recording which lead goes where.

Working with a device this small can present handling problems. I ended up putting a piece of double-sided tape on my work surface. Then I could stick the LED, top side down, onto the tape and carefully attach the leads. Use a small-tipped, clean soldering iron, of no more than about 25 watts. Work quickly, so you don't overheat the device.

These LEDs can be driven directly with a 68hc11 output line through a suitable dropping resistor. Without specs to design from, you'll just have to experiment. Try values from 150 to 240 ohms and check the results.

The insulation on a piece of wirewrap wire makes it seem fatter than necessary. If you really want to hide the harness on this project, use 28 or 30 AWG magnet wire. This wire uses enamel paint for insulation, making the wire about as thin as possible. Now just add a BOTBoard, a little code, and some imaginative packaging. You should end up with a real show-bot.

Electronic Goldmine also offers some green SMT LEDs, as well as SMT LEDs in other package styles. Call for a catalog.

Build a Switcher with the MAX642 IC

As this article from the January 1995 issue of *Nuts & Volt* clearly shows, not all ideas pan out. Still, the experience can be educational, and my run-in with the MAX642 switching power supply IC taught me a lot about using the one of the latest (at that time) chips. In the years since then, Maxim accelerated its research on switching power supply chips, and some of the newer offerings are simply awesome. Take some time to follow the tips outlined here, and load up on the latest information from chip makers such as Maxim and Motorola.

This article also describes adding another level to Max, my ongoing research robotic platform. I also discuss packaging a really nice black-and-white video camera, purchased from one of the larger mail-order surplus outlets. Finally, I close with tips on staying up with the industry and provide some information on the newly emerging blue LEDs.

Some project ideas work in general, but don't solve the original problem. Still, they demonstrate useful design ideas. Such is my recent adventure with the Maxim MAX642 power converter IC.

Third Floor

I have finished installing a third platform on Max, my experimental robot base. You can check some of the recent Nuts & Volts issues for construction details on building your own version of Max ("And Now, Here's...Max!"). I added this third platform to hold a tele-operation system consisting of a video camera, a TV transmitter, and a 9600 baud RF modem (Figure 1). I described the modem in last month's column and gave plans for mounting the modem's electronics in a Radio Shack construction box.

You can use the accompanying diagram of drilling details for the third platform to construct your own addition (Figure 2). As with Max' original two platforms, I made this new level out of 1/4-inch thick white Sintra plastic. You can also use acrylic or other plastic, or even 1/4-inch plywood. Measure and cut the material into an 11-inch square. Mark the

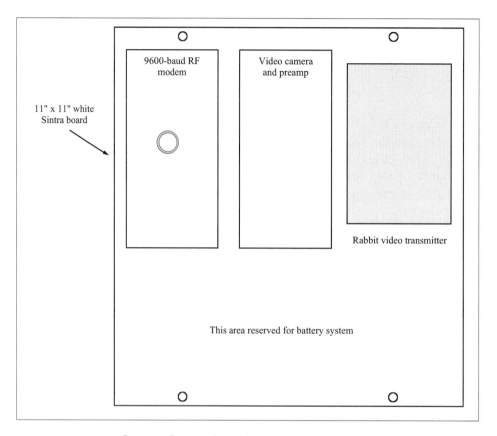

Figure 1. Diagram of Max's top-level layout (top view).

centerline, then locate and drill the required four holes, used for mounting this third platform to the existing second level.

You can also drill a second set of four holes, noted as optional on the diagram, if you intend to add a fourth platform to your Max base later. In fact, you can add several of these platforms to your Max base, by simply alternating the locations of your mounting hardware as you add the levels.

If you built a Max base from my previous articles, you will note that this third platform is slightly smaller than the original two. I intend to standardize on an 11-inch square for all platforms except the first, which holds the motors. Therefore, measure and drill your new platform first, then use it to locate and drill the new holes in your existing second platform.

Please refer to "And Now, Here's...Max!" for details on constructing the mounting hardware used to bolt together Max's frames. I used four 6-32 threaded rods, each five inches long, for adding this third platform. Note that you must cut the 5/16th-inch outside diameter (OD) brass tubing one inch shorter than the companion threaded rod, or four inches in this case. With the platform finished, I turned my attention to the video camera.

Adding Video

Some months ago, I had purchased a small black-and-white video camera from Herbach & Rademan. Their most recent catalog, issue 4 of 1994, lists this same camera as part number G3-032, priced at $169

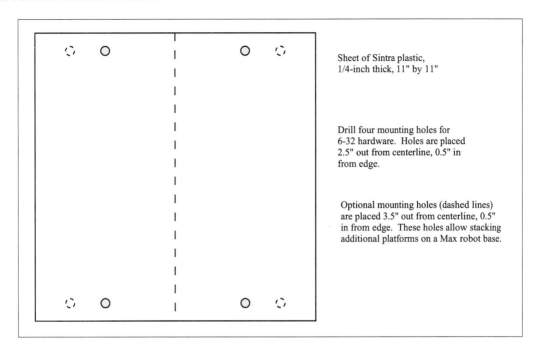

Figure 2. Drilling details for Max's third platform.

each. The specs for this unit match up well for tele-operations. The camera uses 125 mA at 12 VDC, and generates 1V composite NTSC video at 380 lines resolution. Best of all, it has a wide dynamic range and a rated sensitivity of 0.3 lux.

I chose this model because it has a square footprint, only 1.93 inches on a side. This makes it easy to mount in a small enclosure. I looked around for a suitable box, one that was readily available and easy to machine. I wanted something 2.5 inch on a side and about 1.5 inch deep, so I could put the whole camera, including lens, into the box. However, I couldn't find anything I liked.

So I instead went with a Radio Shack 270-238 aluminum project box. This case has a 3 inch by 2-1/8 inch end plate, suitable for mounting the camera. The box's length of 5-1/4 inch is far more room than the camera requires, but I intend to build a small audio preamp in the extra space, so my camera picks up audio as well as video.

On one end wall, I located and drilled 1/8-inch holes for mounting the camera's PCB. I also drilled a much larger hole, filed out to about 3/8-inch diameter, in the center of the square formed by the four holes, for the lens opening.

Next, I drilled holes for two RCA panel-mounted phono jacks on the opposite end wall, to provide connections for video out and audio out. I also drilled a small hole through which I intended to run a twisted pair of wires for the camera's power and ground.

I then used nylon 4-40 screws and spacers to mount the camera to the inside surface of the appropriate end plate. I stacked the spacers so that the end of the lens holder fits flush against the lens opening hole. This close fit reduces the vignette effect caused by the lens' wide viewing angle.

I bolted two RCA jacks into place in the opposite end panel. I also ran a short length of two-conductor 22 AWG wire, for +12 VDC and ground, through the previously drilled hole. I had already knotted this cable near the inside end, to serve as strain relief. I then filled the hole and coated the knot with hot glue, to prevent the cable from chafing when the camera was moved about.

Getting the camera's signal from the robot to a TV monitor requires a wireless link of some type, usually RF transmission. The Gemini Rabbit system is widely available, both in magazines and larger merchandise stores. You normally use a Rabbit system to send your VCR's output to another TV in your house, though it also works well as a link for a security camera or video baby monitor.

You can find the Gemini Rabbit video system for sale by many mail-order houses; at least one advertises in *Nuts & Volts*. The Rabbit system consists of a 900 MHz transmitter and receiver pair, with matching 18 VDC wall-wart power units and coax cabling.

The Rabbit transmitter accepts NTSC video and low-level audio signals, then transmits them as a single signal on a selected portion of the commercial 900 MHz band. The receiver turns this signal into RF on either channel 3 or 4. You can run this signal directly into a TV set, or into a VCR input if you want to record the event.

Both units are housed in rugged, black metal cases about one inch thick and three inches square, sitting up on rubber feet. They both have an on/off switch, power jack, and collapsible antenna mounted on the rear panel, and a red power-on LED located on the front panel. The transmitter has rear-panel RCA jacks for connecting audio and video signals, while the receiver's back panel holds a single coax connector for routing the received video to your TV set or VCR.

The transmitter's front panel also carries a three-position switch. The switch selects one of three transmission frequencies in the Rabbit's operating band, in case you use more than one Rabbit in your home. The receiver's front panel has a small knob for tuning, rather than a switch. You have to retune the receiver each time you change the transmitter's frequency selector switch. According to the manual,

you can use this same knob to adjust the receiver's output between channel 3 and 4.

You have to modify the Rabbit transmitter before you can use it on a robot such as Max. The conversion involves the transmitter's power supply. As purchased, the Rabbit transmitter contains a 7812 three-terminal regulator that drops the wall-wart's 18 VDC down to 12 VDC for normal operation. You have to remove this regulator so you can use the robot's battery supply to power the transmitter.

Begin by removing the transmitter's top case. The printed-circuit board (PCB) inside contains the 7812, mounted in the corner near the power switch. Carefully clip the three leads on the 7812 and discard it. Be sure to clip the two leads nearest the PCB's edge as close to the 7812's body as possible; you will need the extra lead length to make your own connections.

The middle lead of the three goes to ground; you will use this lead for the incoming ground connection. The lead closest to the PCB edge came from the 7812's output; you will connect your incoming 12 VDC here. The last lead originally supplied 18 VDC from the wall-wart. Connect a 0.1 µf disc capacitor between the ground and 12 VDC leads that were formerly part of the 7812. Also connect a 470 µf, 35 VDC electrolytic between these same leads. Be sure to observe polarity when wiring up the electrolytic.

Remove and discard the wires from the small three-pin power switch mounted on the back panel. Run a two-conductor wire through the hole in the center of the power jack, and solder one wire (preferably black) to the ground lead. Connect the other wire (preferably red) to the center terminal of the power switch. Connect the anode of a 1N5817 Schottky diode to the power switch's lead closest to the chassis' edge. Connect the diode's cathode to the 12 VDC lead that was formerly part of the 7812.

This completes the modifications to run the Rabbit transmitter on 12 VDC battery power. Reassemble the case. You need to be aware of some elements of the Rabbit transmitter's design when you work with it. First, the metal case shields the circuitry from interference that can degrade performance. The unit was assembled so as to provide the needed shielding, and you must take care not to disturb this shielding.

Two metal boxes on the circuit board are soldered to the case using large blobs of solder. Do not break either solder connection; they are part of the shielding. Also, a small tab has been soldered to the top of one such metal box, then bent up so the tab presses against the metal case when the case is fastened into place. Carefully bend this tab up slightly before you replace the metal case. This will force the connection back in place and keep the shielding intact.

Once reassembled, you can test your work by connecting 12 VDC to the power cable, observing proper polarity. If you get the connections right, switching the power switch on will light the front panel LED. If you accidentally get the power connections reversed, the Schottky diode will prevent damage to the Rabbit's circuitry. Simply swap the leads and try again.

Note that you don't have to use a Schottky diode; a 1n4001 will work as well. The Schottky diode has a lower voltage drop, though, letting you get good performance with slightly less battery voltage. In fact, you don't need a diode at all, if you feel confident that you will never accidentally hook up the power wrong. I prefer to use the diode, and not to tempt fate.

Power Source

The Rabbit transmitter and the video camera, both mounted on Max's platform, require a source of 12 VDC at about 250 mA. This current draw is based on my bench supply meter, not the most accurate of instruments. Still, it is a ballpark figure.

I ran my first test by wiring the camera and transmitter to the 12 VDC gel-cell batteries that power Max's motors. With Max stopped, the Rabbit sent a sharp, clear picture throughout my two-story house. As soon as I started Max running, however, the picture became useless. The motors dropped the batteries' voltage so low that both the transmitter and

camera stopped working properly. I'm sure the motors probably introduced some hash on the power lines as well, even though I added plenty of filtering when I installed the motors.

When I stopped Max again, a second or so would elapse before the picture snapped back. This performance was good enough to demonstrate that the video system worked, but not good enough to use in a contest or exhibition.

I thought about adding a second 12 VDC gel-cell battery to the top platform, dedicated to supplying current for the video system. But this would add considerable weight. I wanted a cheaper, lighter solution.

Maxim's MAX642

Maxim makes a truly wonderful array of specialty ICs. The company is probably best known in robotics for the MAX232, a chip capable of driving RS-232 signals directly from logic signals, without needing external +12 and -12 VDC supplies.

But Maxim also sells a large family of voltage converter chips for use in battery-operated equipment. ICs are available for generating a specific output voltage from almost any input voltage. These voltage converter chips are actually the core of a tiny switching power supply. You simply add an inductor and one or two other passive components, and your switching supply is done. The finished circuit usually runs at 85% efficiency or better. This comes in handy in a robot, since you want to waste as little of your valuable battery power as possible.

Each chip in Maxim's switching supply line addresses a different set of special needs, so you have to sort through them to find one that best matches your design. In my case, I wanted to convert 6 VDC to 12 VDC, with an output power of about four watts. Most of the Maxim chips are intended for the one- to two-watt range, too small to meet my needs.

But one chip, the MAX642, looked perfect. It contains an on-chip MOSFET driver circuit, letting it provide up to 10 watts of output power. The circuit in Maxim's application notes is so simple, and requires so few parts, that it easily fits on half of a Radio Shack 276-148 experimenter board. (See Figure 3.)

After I selected the MAX642 as the core of my new power supply, I called Maxim's office and asked for samples and literature. Two days later, a pair of MAX642s arrived in my mailbox, along with a set of application notes for designing power supplies.

The standard MAX642 circuit, capable of providing about 20 mA at 12 VDC, requires the IC, an inductor, and two capacitors. This basic circuit runs from an input voltage as low as 5 VDC.

To get more current from the MAX642, you must add an external n-channel power MOSFET such as the IRF530, and a Schottky diode. See the accompanying schematic for my +5 to +12 VDC up-converter (Figure 3). This schematic shows my first attempt at using the MAX642 in a power-boost configuration. I immediately ran into a problem finding the proper inductor.

Maxim's literature devotes several paragraphs and a chart to inductor selection, and you need to go over this material carefully when choosing the appropriate inductor for your design.

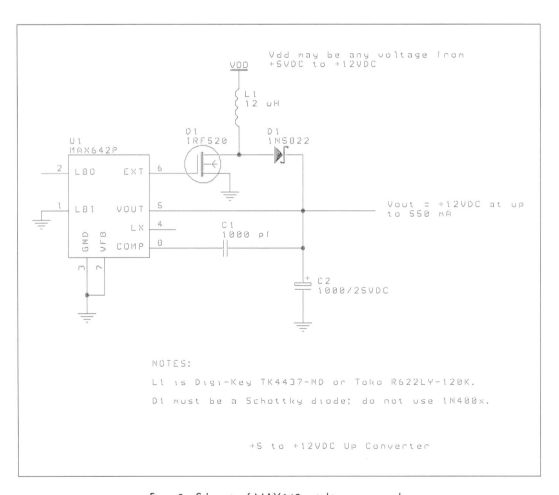

Figure 3. Schematic of MAX642 switching power supply.

In my case, I needed a 12 µH inductor with a coil resistance of 0.01 ohms. The first part is easy; Digi-Key offers a 12 µH inductor for $2 each. Unfortunately, the coil resistance for the Digi-Key part is slightly over three ohms. I ordered some anyway, and built one into my test power supply, just for the experience.

The higher coil resistance means the power supply cannot provide the 550 mA of current the design calls out. My test circuit was able to convert the 6 VDC from four Renewal C-cells into a solid 12 VDC, but only up to about 150 mA. When I tried to draw more current, the supply's output voltage dropped off sharply.

The MAX642 power converter circuit forms a switching power supply that runs at 45 KHz. This oscillator switches the high current running through the inductor, and can create some serious radio-frequency interference (RFI) if you don't use proper wiring layout.

The Maxim application notes discuss topics such as MOSFET and inductor selection, wiring guidelines, and bypassing requirements. They also describe some of the other features offered in this chip. For example, the MAX642 includes a Low Battery Output pin that your circuit can use to monitor the state of your input battery. If the battery's output drops too low for proper MAX642 operation, the open-drain LBO output switches to ground, signalling the condition.

Additionally, you can add a simple resistor divider network to the MAX642, in either basic or external MOSFET configuration, to generate output voltages significantly higher than 12 VDC. For example, the Maxim application notes provide a sample circuit for generating +50 VDC from a +12 VDC battery.

Running the MAX642 in my chosen configuration yields a power conversion efficiency of about 85%. To get 12 VDC at 550 mA out, I have to provide 6 VDC at about 1.3 amps. Since the Renewal C cells are rated at 4 AHrs, I won't get very much operating time before the batteries run down.

Bottom line: The MAX642 isn't quite the solution to my problem. Even if I added the proper in-ductor and created a robust 4 W supply, the current drain on the input battery wouldn't provide enough operating time per charge.

But you may well have an application needing higher voltages than your project has physical room to support. If so, consider the Maxim line of switching supply power converter chips. These chips are easy to use, very versatile, and Maxim provides some very good supporting literature.

Getting Samples

Some electronics companies strongly support engineers and designers, by making samples and literature easily available. Often, you only need to send a FAX or make a phone call describing the parts or documentation you need.

Companies such as Maxim and Motorola have been especially supportive of people looking for the right chips for their new designs. I have called Motorola's field application engineers (FAEs) more than once, and always been impressed with their knowledge and willingness to answer questions.

A few Motorola engineers are regular contributors to the comp.robotics newsgroup on the Internet. When questions arise about 68hc11 variations or circuitry, it's great to have someone from Motorola step in and provide a definitive answer. And it's especially nice to know that you can reach such expertise with email.

Maxim has taken the delivery of samples to a new level. The data books, free on request, include mail-in or FAX-in cards for free samples. Simply circle the part numbers of interest, fill out the rest of the form, and transmit it to Maxim.

You can get Maxim data books by calling your local Maxim distributor. You can also view any Maxim data sheet by hitting the company's web page. Note, however, that Maxim uses a different format for their data books than other companies. Maxim releases data books based on the year, so you need to know which year your target chip first appeared before you know which book to get.

At first, this seemed a little weird to me. But the Maxim 1994 book is nearly two inches thick, and having to rearrange all the pages to fit new products into a conventional style data book would probably prove too expensive.

Regardless, you can talk to the Maxim distributor to determine which data book you need for a particular chip. And you can always order individual application notes for most chips.

Keeping Informed

Maxim and many other major electronics developers advertise heavily in electronics trade journals such as *Electronics Design*. Articles and press releases in *ED* will keep you informed of the very latest in electronics. Many of the topics covered are so new that they currently exist only in the labs. The articles usually give clues as to how soon you can expect to see a commercial version of some hot new design.

For example, the 21 November 1994 issue of *ED* has a one-page article on new developments in blue LEDs. The current type of blue LED, made of silicon carbide (SiC), typically produces about 20 milli-Candelas (mcd) at a 20 degree viewing angle.

I've seen some of these LEDs in an indoor environment, and been underwhelmed. The light output is simply too low to be useful, especially running beside a 20-cent surplus LED putting out over 500 mcd.

According to *ED*, Nichia Chemical Industries of Japan has produced working lab versions of a gallium nitride (GaN) LED with blue light output in excess of 1000 mcd! The devices are still at least two years from commercial reality, according to a VP at Dialight Corp. Nichia doesn't even know yet how long the devices will last under normal use. Still, I found the news very exciting, and will look forward to more reports, both in the printed journals and on the Internet.

But the advertisements are by far the most entertaining part of trade journals such as *ED*. Companies with products across the spectrum of electronics offer free information, samples, technical literature, applications notes, and software, all to help you design their product into your new million-seller.

Often, you merely need to place a phone call and answer some questions, and your samples and literature are in the mail the next day. Even if you don't end up designing a best-selling whigmaleerie yourself, your discussions with others may be just the spark that leads to a big order for that supplier later on.

Like most trade journals, *ED* subscriptions are free to qualified readers. You demonstrate your qualifications by filling out and returning an application, available in any issue of the magazine. So scrounge a copy from a friend, fill out the subscription card, and start scoping out the opportunities for using the latest chips and parts. And you'll keep up with the newest technical advances at the same time.

Try This Junk Box Switcher Supply

I introduced readers to the Junk Box Switcher (JBS) in this column from the September 1995 issue of *Nuts & Volts*. Keith Payea, one of the Seattle Robotics Society's top electron wranglers, came up with the basic design after hearing numerous complaints about existing robot power supplies. With tweaks by Bill Bailey and other club members, the JBS finally took the form you see here. Of course, you can now find all kinds of way cool power supply ICs, notably from Maxim, but the JBS was a first for the club and it is still a clever circuit to study.

This column also describes 5th GEAR, the latest in the club's annual get-together under the trees. As always, it sparked plenty of ideas, most notably a design that the club has dubbed "Frank's thingy," in honor of Frank Haymes, the SRS member who first described it.

The Robotics Practitioner, mentioned in this and other columns, has gone the way of most other hobby robotics magazines. This must be a very difficult market niche to cover, despite the enormous enthusiasm shown by experimenters of all ages. *Nuts & Volts* has served up some excellent robotics articles in the past, my column not included, but *N&V* makes its money by catering to a large and diverse array of electronics fields. Robotics-only magazines have proven short-lived to date, and I wonder when the market will mature enough to sustain one or two focused magazines. Only time will tell...

Keith Payea, one of the electronics wizards in the Seattle Robotics Society (SRS), usually comes up with very cool robotics projects. This month, I'll describe one of his recent designs, a simple switching power supply.

And I do mean simple. You should be able to lash one of these switchers together in an evening, using parts from your junk box or a busted PC power supply. For lack of a better name, I'll refer to this project as the Junk Box Switcher, or JBS.

Inside the JBS

Keith designed this supply to meet requirements specific to robotics. It had to be cheap to build, and it had to provide an adjustable output voltage. He also wanted a supply that would provide its rated output, with little or no overhead. This means that you can set a JBS to supply +5 VDC, connect it to a 6 V gel-cell, and get +5V DC out, even as the battery's voltage sags due to usage.

A glance at the accompanying schematic will show how successful Keith was in keeping his design simple (Figure 1). Note that Keith's original circuit used a zener diode to select the output voltage level. The trimpot adjustment shown here was suggested by Bill Bailey, another of the club's electronics gurus.

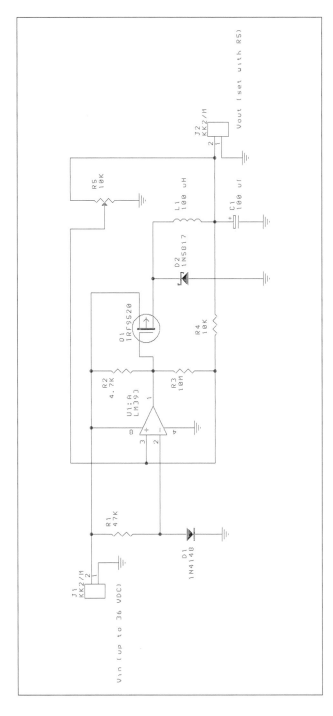

Figure 1. Schematic of the Junk Box Switcher.

```
DESCRIPTION            QTY    COMPONENT NAME(S)    DESCRIPTION

100 µf 25 WVDC         1      C1                   electrolytic cap
1N4148                 1      D1                   Silicon diode
1N5817                 1      D2                   Schottky diode
Molex KK male          2      J1, J2               2-pin connector
100 µH                 1      L1                   inductor
IRF9520                1      Q1                   P-channel MOSFET
47K                    1      R1                   1/4-watt resistor
4.7K                   1      R2                   1/4-watt resistor
10M                    1      R3                   1/4-watt resistor
10K                    1      R4                   1/4-watt resistor
10K                    1      R5                   10-turn trimpot
LM393                  1      U1                   Dual comparator
```

Figure 2. Parts list for the Junk Box Switcher.

The LM393 comparator, U1, samples the output voltage through the voltage divider formed by R5. This voltage is compared against the 0.6 VDC reference voltage at the junction of D1 and R1. If the sampled voltage is below 0.6 VDC, U1 turns on, bringing its output pin low. This turns on Q1, which switches the entire Vin voltage into L1. This voltage appears on the output pin, where it causes the sampled voltage to start rising. When the sampled voltage rises above 0.6 VDC plus a small hysteresis value, U1 turns off, raising its output pin to Vin through R2. This voltage also turns off Q1 and removes Vin from L1. Eventually, whatever load is hooked to the output will pull the sampled voltage back below 0.6 VDC, causing the entire process to repeat.

One element of this design should be immediately apparent; at any given moment, the switcher's frequency is unknown. U1 turns on and off at a rate determined by the load current, input voltage, setting of R5, value of L1, and the value of C1. Heck, the phase of the moon probably figures in, as well.

Since the switching frequency is unknown, there isn't any real concern over most component values. L1 can be nearly anything over 100 µH, and

C1 can be almost anything over 100 µf. You can use just about any P-channel MOSFET you want, as well.

However, you should choose a device with the lowest possible RDSon, as this parameter limits the amount of current you can draw from the switcher.

This limitation comes from the amount of power a MOSFET can dissipate. The typical TO-220 case can handle about 1 watt of power without needing a heatsink. This power, divided by RDSon, gives the square of the amount of current the device can handle.

For example, a JBS using a MOSFET with RDSon of 0.6 ohms can supply:

1.0 watts / 0.6 ohms = 1.67 Amps squared

or 1.3 Amps without needing a heatsink.

Be aware that this is a ballpark figure only. It gives you a rough guide on which to base your supply's current capacity, but it does not take into account some switcher characteristics. Note that you can boost the JBS' current capabilities by adding one or more MOSFETs in parallel to Q1, or by heatsinking Q1.

Besides the IRF9520 called out in the schematic, I have also used the IRF9Z30 device. This MOSFET sports an RDSon of only 0.14 ohms, giving you about four times the current capacity of the 9520. Price for the IRF9Z30 MOSFET is about $3 each.

The current rating of L1 will also limit the current supplied by the JBS. You need to use an inductor rated for about 50% more current than you intend to draw in normal use. For example, you should use a 3-amp inductor if you expect to draw two amps routinely from the supply. If you've ever leafed through a Digi-Key catalog, you know that such inductors aren't cheap. Where do you go, then, to find suitable inductors?

Many surplus stores and mail-order houses sell power inductors of various ratings. You can use just about anything over 100 μH, which simplifies the task of finding a useful inductor.

You can also tear apart a blown-out PC power supply. These are easy to come by, and offer scads of parts that you can use. You will find several inductors that should work in a JBS. Just pick one made of at least 18 AWG wire and give it a try. Besides high-current inductors, a junked PC supply will have a Schottky diode suitable for D2, as well as a high-value filter capacitor that you can use for C1.

Despite the intentional simplicity of this design, you do need to watch for a couple of possible problems. R3 determines how much ripple your output voltage shows. The specified value of 10 Mohms should yield a ripple of about 60 mV in a +5 VDC supply. Using lower values of R3 will produce greater levels of ripple.

If your project draws a lot of current, the switching frequency can easily exceed 1 kHz as Vin starts to drop closer to Vout. This increased switching frequency can cause C1 to overheat. If you expect to draw a lot of current from a JBS, use a high-quality cap for C1. Suitable parts include the Panasonic HF or HFU types, the Rubycon G2A series, or the Nichicon PX units.

You must use a Schottky diode for D2. This device acts as a catch diode for the inductor, and must handle high switching speeds. The breakdown voltage isn't critical, though it should be about 50% greater than the value you expect for Vin. The 1n5817

shown here has a PRV of 20 VDC, and should be good for use with a 12 VDC gel-cell battery.

You will probably want a multi-turn trimpot for R5. The adjustment range is very compressed at the upper end, and you will find the multi-turn's increased resolution helpful when setting the output voltage accurately.

The LM393 comparator that I used for U1 can handle Vin values of up to 36 VDC. This should be more than adequate for most robot needs. If you intend to use Vin greater than 36 VDC, you should select a different device for U1.

This circuit fits comfortably on half of a Radio Shack experimentor board (276-148). You can use just about any 2-pin connector for J1 and J2. If you want, you can just bring two twisted-pair wires onto the board for hookup.

Testing the JBS

After I built a working unit, I hooked it up to a 12-volt gel-cell and started playing around with it. Like most switchers, you have to add a load to the JBS's output before you can take accurate voltage measurements. In my case, I hooked the coil of a 5 VDC relay, with a coil resistance of 125 ohms, to the JBS's output terminals. I then set the output for a value of 5.0 VDC.

I left this setup running on my lab bench for a couple of days, taking periodic measurements of the battery voltage and the switcher's output voltage. As shown in the accompanying chart, the gel-cell's voltage dropped from 12.63 VDC down to 11.64 VDC, but the JBS's output only varied between 5.07 and 4.99 VDC (Figure 3).

I repeated the experiment, instead using a set of four Renewal alkaline C cells. This provided an initial battery voltage of 6.06 VDC. I retrimmed the switcher's adjustment, then repeated the measurements over a period of about 36 hours. This time, the battery's voltage dropped down to 4.86 VDC, yet the switcher's output remained as close to 5.00 VDC as possible. Refer to the accompanying chart (Figure 3).

In fact, the output stayed at 5 VDC until the battery voltage dropped below 5 VDC, at which point

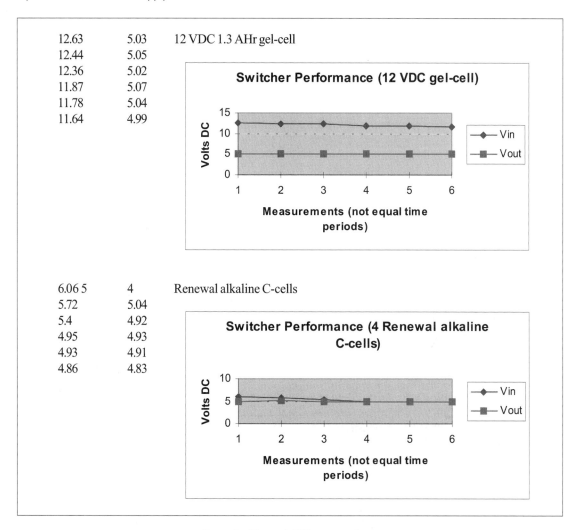

12.63	5.03	12 VDC 1.3 AHr gel-cell
12.44	5.05	
12.36	5.02	
11.87	5.07	
11.78	5.04	
11.64	4.99	

6.06 5	4	Renewal alkaline C-cells
5.72	5.04	
5.4	4.92	
4.95	4.93	
4.93	4.91	
4.86	4.83	

Figure 3. Charts of JBS output voltage.

the output voltage simply followed the input voltage. This exercise shows a strong feature of the JBS: it has essentially zero overhead voltage. This means you can get a regulated 5 VDC output from just about any voltage source of 5 VDC or greater.

Note that there is nothing magical about the output voltage; it doesn't have to be 5 VDC. You can set the output voltage to anything from 0.6 VDC up to the input voltage, and the JBS will try to maintain that output voltage.

You could, for instance, replace the trimpot with a large multi-turn pot and knob, creating a simple bench supply. Your JBS bench supply could run from nearly any source of DC voltage, such as a battery, a wall-wart, or even a solar panel.

If you use a battery as the DC source, you end up with an adjustable bench supply that runs from your car battery or an external gel-cell. The bench supply would be fully portable, letting you do robots on vacation.

A JBS would be so small that it would fit inside many battery-powered devices. You could then use a wall-wart to make an efficient line-powered supply for the device. I intend to do just this with my old Toshiba laptop, replacing the defunct NiCd packs and original AC supply.

At first glance, it might seem that you could do all of the above with a LM317 variable regulator and a few parts. You can, if you are willing to overlook the horrible inefficiency and high overhead requirements of such a supply.

Linear regulators such as the 7805 and the LM317 generate a specific lower voltage from a higher input voltage, wasting the difference as heat. For example, a 7805 hooked to a 12 VDC battery will provide 5 VDC output by converting the remaining 7 VDC, at whatever current your circuit draws, into heat. This means the best efficiency you can hope for with a 7805 in such a circuit is 5/12, or about 42%. The LM317 is no more efficient, it is just adjustable.

And these older regulators require a certain amount of "headroom" on the input voltage, typically about 2 VDC. Thus, you could not even use a 7805 regulator if your input voltage was less than 7 VDC, which eliminates the chance to use a 6 VDC gel-cell battery.

By contrast, the JBS needs no headroom, and will even follow the battery voltage down as it drops below the specified output voltage. This can be critical in those cases where your robot needs just a few more minutes of run-time, even though the batteries are nearly dead.

True, you can get very low dropout regulators today, but you still have to contend with the heat problem. As an extreme example, consider the power supply developed by one of the club's members. He was using a 7805 to generate 5 VDC, using 24 VDC as his supply voltage. This resulted in an efficiency below 21%; the remaining power was dissipated as heat. And a lot of heat at that!

When he tried substituting a JBS supply for his 7805, he was surprised with the difference. The MOSFET in the JBS ran at ambient temperature, with-

out a heatsink, yet the supply delivered all the power of his original unit.

Try wiring up a JBS for your next robo-project. I think you'll like the difference a switcher makes.

5th GEAR

I just got back from three days in the woods, vacationing with other members of the Seattle Robotics Society (SRS). We ate lots of food, consumed a brew or two, and talked at length about things robotic.

The SRS holds these annual events, called GEARs, each year around the end of July. A GEAR (Great Escape and Retreat) gets you away from work and chores, and gives you ample time to kick around ideas for that new 'bot or compare notes with friends.

I find the GEARs a perfect place to recharge my own batteries, to come up with new projects and to get revved up again. Long-time readers of this column will recall that 3rd GEAR spawned my 68hc11 tiny4th, while 4th GEAR was the springboard for Max's motor adapter.

This year proved to be just as rewarding. Keith Payea, Frank Haymes, and Bill Harrison and his daughter, Isa, joined Linda and me on Whidbey Island at our usual campground. Bill and Frank brought lots o' robots to work on and talk about.

During the discussions, Frank described a simple device that he had been waiting to see developed. He envisioned a 68hc11 board, such as Marvin Green's BOTBoard, permanently connected to a PC's serial port. The PC would issue commands to the 68hc11, which in turn would twiddle its output ports, or read its input ports and relay the results to the PC. Thus, the 68hc11 would become an I/O expansion port for the PC, hooked to its host via an RS-232 cable.

Frank couldn't believe that such a device didn't already exist, and asked me about it. Not only had I never heard of such a device, I was shocked that I hadn't already come up with such an obvious gadget myself. I mean, think about it. You could use the 68hc11 as the control end of a tethered robot. The PC would do the intelligent part, while the 'hc11 moved the motors or read the sensors.

And the PC wouldn't even have to be a PC. You could use a Mac, a TI 99, or anything else with a serial port on it. And you could even write the PC's program in just about anything you liked; Forth, Basic, C, or Pascal. You might even be able to cobble together a text file, and just send it down the serial line using the DOS COPY command.

Frank's concept, which we promptly dubbed the "thingy," could well remove the burden of learning almost anything about the 68hc11's innards. If the 68hc11's firmware were properly designed, a user would only need to know what tokens to send to the 'hc11 and what responses to expect.

We spent some time discussing what features to support in the 68hc11, and how best to make them available to the host's program. After a few hours of thrashing, we came up with a good starting set. I immediately grabbed my notebook and a pen, and started scribbling down 68hc11 assembly language code.

As of this writing, I have about 75% of the first version coded. By the time you read this, the project should be finished. I don't even know what I will call the program, but I will make source code available on my web site. Check this column next month for more details on the thingy code.

Author's note: I never finished the original thingy project, but the whole concept mutated into the RoboScrn project, described in "Tacklebot, a Backyard Explorer." I always liked the name of the thingy project, though.

Frank had filled his car with robot goodies for this event, and we all spent most of Saturday at his campsite, tinkering and talking. Frank set up a vintage Radio Shack laptop, on which he is developing the Basic code for his current robot.

Frank powered his laptop from a set of gel-cell batteries, as the NiCds had long since given up the ghost. Seeing his power setup sparked a discussion about imbedding a JBS inside my Toshiba laptop, and that led to talk about adding heavy gel-cells to a truck's electrical system. This would let you take your lab and robots on the road.

I was particularly intrigued by Frank's Meccano construction set. This is a metal-piece build-it sys-

tem, similar to an Erector set. Frank showed me a nice motor system you can buy at most toy stores. It consists of a 6 VDC motor in a special housing, with a small selection of brass pinion and drive gears, and some mounting hardware for bolting the motor to other Meccano elements. The motor system costs about $20, and deserves a look-see if you need a drop-in motor design for your next project.

You can also find details on the Meccano construction set by leafing through the Metal Construction Set FAQ in comp.robotics. Jeff Duntemann maintains this collection of tips and pointers, and it should help get you started with this intriguing robotics resource. Note that the FAQ covers several constructions sets besides Meccano, including Erector and Temsi.

Speaking of drive systems, Bill Harrison showed off a robot frame he had made, using key elements from a torn-apart RC car. He had dismantled the car and pulled out the front axle and steering system, as well as the rear axle and differential. He then mounted these elements to a frame of his own construction, using the car's design as a guide. He also added a small servo motor in the steering mechanics, so he could control the steering with a BOTBoard or other small computer.

The end product is a strong, light-weight robot frame that steers like a model car, yet has plenty of room for mounting other robot goodies. Considering how many RC cars turn up at garage sales, this is one way you might make a cheap but strong frame of your own.

Bill also showed off his little walking robot, which uses just one servo motor! The robot stands about five inches high, and can not only walk forward, but turn right or left as well. It uses a Basic STAMP for a brain, and even includes IR object detection.

I think Bill plans to publish details on his robot in a future issue of *The Robotics Practitioner* magazine, so I won't go into any more of the design here. Certainly all readers of this column should also subscribe to TRP; it belongs on your magazine shelf right next to your copies of *Nuts & Volts*.

The group closed out Saturday night with our customary potluck dinner, then sat around a crackling fire and talked well into the evening. The discussions covered many aspects of hobby robotics, and I'm sure will lead to new projects and club events.

As usual, this GEAR provided me with material and ideas for several future projects and articles.

There is something about this annual campout, with the Olympic mountains in view across Puget Sound, that helps recharge all of us. The fresh air, chill mornings, and campfire cooking seems to be a natural tonic. I'm already looking forward to next year.

Son of
BOTBoard

Marvin Green, one of my robot buddies from Portland, Oregon, has come up with several powerful tools for the amateur robot builder. This column from the February 1996 issue of *Nuts & Volts* describes his BOTBoard-2 (BB2), a larger and more powerful version of his 68hc11 BOTBoard project. Several of us in the Seattle Robotics Society built BB2s into our robots, and we had more than one Saturday session devoted to this board. The software and hardware hacks I mention here are the result of some long and very enjoyable hours of work on the BB2.

Ever on the lookout for ways to build better motor assemblies, I one day ordered a couple of Tamiya motor gearboxes from Edmund Scientific. Easy to build, well constructed, and reasonably priced, the Tamiya products deserve the attention of any of the more mechanically challenged of us.

Finally, I describe a couple of catalogs that you really need to check out. Most readers will have heard of Mondotronics by now, but if you haven't seen an American Science and Surplus catalog yet, you are in for a treat!

Marvin Green, who developed that staple robotics tool, the BOTBoard, just released his next generation robot brain dubbed, appropriately, the BOTBoard-2. I've been using one of his prototype boards for some time now, and thought I'd pass along my impressions.

Like its predecessor, the BB2 uses a 68hc11 and supporting circuitry. The BB2, however, supports 32K of expanded memory, including battery-backup if you decide to add static RAM. The BB2 also contains artwork for a MAX232 RS-232 level-shifter and a pinout for a 9-pin serial connector. This really makes it easy to hook the board to your PC or Mac; you don't have to build an external level-shifter.

Of course, you can use the normal BOTBoard 4-pin serial cable on the BB2, if you choose. Marvin included pads for mounting the usual 4-pin Molex KK-style connector used on the BOTBoard. Those of you who already have a BOTBoard serial cable with built-in adaptor can use the same cable on your BB2.

Other carryovers from the BOTBoard include four servo ports, on-board reset switch, and connectors for the 68hc11's port A and port E pins. Note, however, that the BB2 uses a pair of 10-pin IDC connectors for these ports, rather than a single, larger connector as on the BOTBoard.

Marvin also added several extras, essential to robot development. The BB2 contains a 1 inch by 5 inch prototyping area, large enough to hold some substantial circuitry. What's more, he arranged the prototyping area so pairs of pads are joined with traces, creating a versatile experimenter's layout. This really simplifies wiring extra circuitry into the prototyping area.

Since the BB2 uses the 68hc11 in expanded mode, many of the I/O pins become unavailable, as they now carry bus signals for the external memory. To help you recover the missing I/O lines, Marvin has added partial decoding for addresses from $4000 to $7fff. His board divides this address range into four equal parts, and generates separate *READ and *WRITE signals for each of the four regions. For example, you get a pad that provides a low signal whenever your program reads an address in the range $4xxx, and a separate signal for writes to $4xxx.

Marvin brought six of these signals and the eight-bit data bus out on a 16-pin IDC connector. This lets you build a custom expansion board that simply hooks onto the BB2, either directly or using a ribbon cable.

Other goodies you can add directly on board include a piezo speaker for sound effects and an LED under software control. Marvin's layout also provides a 10-pin IDC connector for the SPI bus and a separate connector for wiring in a battery-backup supply.

He also includes pads for a four-position DIP switch. Two of these switches control the MODA and MODB pins, so you can change 'hc11 configurations without having to move jumpers around. The other two switches are uncommitted; you can run wires from their pads to whatever I/O lines or signals you need.

Finally, the advanced hacker will find spare pads provided on all vital 68hc11 signals, in case you need to do a mondo hack. For example, all of the multiplexed bus traces contain spare vias for making connections with wirewrap wire.

The BB2 comes with a very nice technical booklet, which includes the full schematic, a parts list (with vendor names and typical prices), and good information on how to wire up the many BB2 options. It provides details on adding extra 8-bit input and output latches, sample code for moving servo motors, and a schematic for building a BOTBoard-style RS-232 serial cable.

Bottom line: This is a well-designed board for the advanced robotics hobbyist, especially if you intend to use RC servo motors to drive your 'bots. I personally think the blank board is worth much more than the $18 (or $45 for three boards) that Marvin is charging. Be sure to add $3 to your order to cover the cost of shipping and handling. Oregon residents already know they don't have to add sales tax, 'cause there isn't any in Oregon...

If you want to take a closer look at the BB2 materials, check out Marvin's web page.

Hacking the BB2

Naturally, I can't leave well enough alone. As soon as I got my BB2 up and running, I started hacking on it. I had wired it up to hold 32K of static RAM, complete with a backup lithium battery, but I wanted to do more.

First up was some software. I used a 68hc11e1 for my MCU; this device has 512 bytes of EEPROM and 512 bytes of on-chip RAM. The extra RAM is always nice; it makes a good place to stash the stack. But the EEPROM has real possibilities. It runs from $b600 to $b7ff, and provides enough room for some sophisticated assembly language code. In my case, I added a S19 loader.

Those of you who read "Remote Reloads with 811bug" will remember the S19 loader I installed in BYRD, my Back Yard Research Drone robot. The loader I wrote for the BB2 is similar, and provides a smooth way to install and test new programs. In fact, the loader is so powerful that I no longer even use PCBUG11. I just start up a comm program, such as Crosstalk, and download my software.

The only problem I had to solve involved starting the loader, called E9BUG, from reset. The 'hc11e1 is normally used in expanded mode, and it is difficult to pass control to $b600 to start my S19 loader. So I start the 'hc11e1 in special bootstrap mode, by setting the DIP switches appropriately. The 'hc11 comes out of reset by sending a break signal out the serial port's (SCI) transmit line. If it sees a break on its receive line, the 'hc11 responds by jumping to $b600.

The only trick is getting the break character to the SCI receive line. While most comm packages let you send a break signal, which will probably do the job, I chose a simple hardware mod instead. I just hooked a SPST pushbutton switch across the SCI Rx and Tx lines at the 68hc11. Note that I wired this switch to the pins as they come from the MCU, NOT after the signals have been converted to RS-232 levels.

Now the reset sequence is just slightly more complicated than normal. I hold down the SCI button, press and release the reset button, then release the SCI button. Bingo; my S19 loader starts up and I'm ready to send my software. This works because the closed pushbutton switch routes the break signal that the 'hc11 sends back to the 'hc11 receive line. Thus, the 'hc11 gets a break signal immediately, and transfers control to $b600, starting up E9BUG.

Of course, E9BUG has to figure out when you've released the pushbutton; otherwise, it won't be able to recognize legal commands from garbage. So it sends out a stream of nulls ($00) until it fails to see a null on its input port. This means you have released the pushbutton, breaking the connection between the 68hc11's transmit and receive pins.

Naturally, your comm software is also seeing this stream of nulls, but most comm programs will ignore them. If your program has problems with this, just change the E9BUG source file to send a different test character.

Once my E9BUG monitor has control, I can download any Motorola S19 object file by simply doing an ASCII send operation. Most comm programs support this, and can even suspend transmission of one line until the receiving computer (in this case, the BB2) acknowledges receipt of the previous line. If your comm program has this support, just set it so it waits for a carriage return (CR) before sending the next line of a file.

E9BUG remembers the execution address of the last S9 record it receives, so starting your downloaded program requires nothing more than sending a lowercase-x after you finish downloading. E9BUG

will immediately jump to the execution address from your last downloaded file.

Note that this works with S19 files created with the Motorola FREEWARE asmhc11 assembler, and with files compiled by SBasic and tiny4th, since they use the asmhc11 assembler in the final step of compilation. It will NOT work, however, with the older as11 assembler, since that assembler does not preserve the execution address in the S9 record. If you are still using the as11 assembler, I urge you to dump it and move to the more flexible asmhc11 program. My SBasic and tiny4th distribution files contain the asmhc11 assembler; you can get these files from my web site.

E9BUG does have a command for supplying your own execution address. You simply enter a lowercase-j followed immediately by a four-digit hexadecimal address. For example, to set an execution address of $8000, you would enter:

j8000

Note that E9BUG does not accept any intervening spaces, nor does it allow backspaces to correct mistakes. If you make a mistake, simply press RETURN, then enter the command again.

Note that the j-command only sets the execution address; it does not transfer control to the address you provide. After you have entered the correct j-command, you use the x-command, described above, to start execution at that address.

E9BUG's simple command set opens up many possibilities. It lets you download S19 files quickly, without having to resort to PCBUG11. It lets you provide an execution address of your own, or use the address contained in an S9 record. Finally, it allows you to jump to any of several routines that you might already have stored in the BB2's on-board battery-backed RAM.

But the BB2 only has one 28-pin memory socket, usually filled with a 62256 32K static RAM chip. While this is plenty of space for programs, it can prove inconvenient during development. Development code tends to run away during tests, which usually trashes the RAM holding the code under test. This naturally forces another reload. You can remove the BB2's write-enable jumper after each download, preventing writes to memory after the runaway, but this is an extra step that I'd personally like to avoid.

Besides, what you really need during development is a solid monitor program, tucked away in EPROM where it can't be corrupted, and where it can't interfere with the program you're developing. Fortunately, it's easy to find a good 68hc11 monitor program.

Motorola's BUFFALO monitor has been around for years, and the source code is readily available at a number of FTP sites. BUFFALO offers a variety of powerful functions, including built-in assembler and disassembler, breakpoints, and single-step execution.

But BUFFALO was written to work in a far more complicated environment than a BB2, so it contains several features you won't need. I spent a couple of weekends editing the source code for BUFFALO 3.4, and stripped out all the extraneous code. I also cleaned up the menu text quite a bit.

The resulting BUFFALO monitor takes up about 7K of code space, and can fit anywhere in the BB2's address space. It does not take over the vector area, as the original BUFFALO does, and I placed the RAM variables that my new BUFFALO needs in the upper section of the 'hc11e1's 512-byte RAM space, to free up as much lower RAM as possible for development programs.

So now I have this cool monitor program, but nowhere to put it, since the BB2 only has the one 28-pin socket and it already contains a RAM chip. I

needed to add a second 28-pin socket, with address decoding for a 27c64 EPROM chip.

This hack actually proved quite easy to do. To add a second 28-pin socket, you will need the following components:

1   14-pin solder-tail IC socket
1   28-pin solder-tail IC socket
1   74hc20 IC
     some 30 AWG wirewrap wire
     some 1/16" heatshrink

NOTE: You MUST use a 28-pin socket with leaf-type pins; you CANNOT use a machine-pin socket for this mod!

Begin by locating a place in the prototyping area for the 14-pin socket. I mounted my socket immediately below the silk-screened legends for C15 and C16, near the MAX232 pad layout. I positioned the chip with pin 1 toward the C15 legend and pin 14 toward the C16 legend.

Use wirewrap wire to connect the 74hc20 as shown in the accompanying schematic (Figure 1). The pad traces on the prototyping area will make this job easier. Leave the connection from pin 8 of the 74hc20 off for now; we'll get to it later.

Note that four of the 74hc20's connections run to I/O select pads provided by Marvin for device decoding. As shown, this mod uses 6000R, 6000W, 7000R, and 7000W. This combination means that the new 28-pin device will reside in the address space from $6000 to $7fff, and that the device is read/write capable.

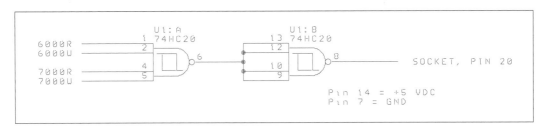

Figure 1. Schematic of BOTBoard-2 memory modification.

You have some flexibility here. For example, if you know you will always put read-only devices such as EPROMs into this new socket, you can omit the connections to 6000W and 7000W; just wire those pins on the 74hc20 to the corresponding xxxxR lines. This leaves 6000W and 7000W free for later use with output-only devices such as 8-bit latches. You can also move the new 28-pin socket to $4000 through $5fff, by using the 4000R, 4000W, 5000R, and 5000W pads instead.

After you have wired the 74hc20 to suit your needs, it is time to install the new 28-pin socket. Begin by selecting a known good 62256 32K static RAM chip. Be sure you have tested this chip in your BB2, using a program like PCBUG11, and know it works properly. What you are about to do to it will make it very difficult to replace, should the chip prove defective later on.

Push the chip's pins into a piece of black anti-static foam, and heat up a fine-tipped soldering iron. Fit the 28-pin leaf-pin socket over the pins of the RAM chip, so the socket's pin 1 matches up with the RAM chip's pin 1. Make sure that each socket pin is centered on and touches the matching pin of the RAM chip.

When properly positioned, you should have what looks like a very tall socket, with a RAM chip nestled underneath. Make sure the underside of the socket sits flush against the top of the RAM chip.

Carefully bend pins 28 and 20 of the socket outward, away from the corresponding RAM pins. You want to bend them out just enough to break electrical connection, but not enough to damage the pins. Carefully solder all socket pins EXCEPT 28 and 20 to the corresponding RAM pins. Work quickly on each pin, and make sure you get a good solder flow between the two pins.

Cut a 3-inch piece of 30 AWG wirewrap wire, strip 1/8 inch insulation from each end, and carefully solder one end to pin 28 of the socket. Make sure you don't accidently solder this pin to the underlying pin 28 on the RAM chip! Run a small piece of heatshrink over the entire pin and shrink it into place. Perform the same step above for pin 20 of the socket.

You are now ready to put the RAM chip into the 28-pin socket on the BB2. Visually inspect the RAM chip's pins, and clean off any deposits you might have picked up during soldering. Align the RAM chip properly in its socket, then carefully push the chip into its socket. Power up the BB2 and verify that the RAM chip still works properly.

Connect the wirewrap wire from the new socket's pin 28 to a convenient pad on the BB2 that holds +5 VDC. DO NOT connect this pin to the battery backup circuitry! If you don't have an expansion connector wired into EXP1, you can use pin 1 of this pad layout as a source of +5 VDC.

Connect the wirewrap wire from the new socket's pin 20 to pin 8 on the 74hc20's socket. If necessary, you can route this wire through a convenient pad in the prototype layout, in order to reach pin 8.

Insert a 74hc20 IC into the 14-pin socket, and insert a 28-pin 8K memory chip into the new 28-pin socket. Power up the BB2 and verify memory at the addresses you have selected for the new socket. Note that you can use a 6264 RAM chip or a scrap 27c64 EPROM for this test, since all you really care about is whether the chip responds to the memory addresses. Even if you wired your socket up for read-only operation, you can still verify that some type of data exists in the address range.

You have just added a separate 8K memory socket to your BB2. You can burn a copy of my modified BUFFALO monitor into a 27c64 and install the chip into the new socket. Once installed, you just use E9BUG's j- and x-commands to set up the execution address and start running BUFFALO. Note that if you wired your 74hc20 for different addresses than I show here, you have to edit the BUFFALO source file so it assembles at your socket's address. Just change the equate for ROMBS, near the top of BUF-FALO, to the proper address and reassemble.

One nice feature of the BB2 and BUFFALO combination is the ability to single-step through your code. To do this, you must provide a connection between the 68hc11's XIRQ and OC5 pins. On the BB2, you can do this by first installing a 2-pin header in the XIRQ pad on the SPI layout and the pad immediately below it. Then run a wire between this empty pad and the pad labeled 3 on the PORTA layout. To enable single-step, simply put a jumper on these two pins; removing the jumper disables the single-step. Note that with the jumper in place, you will not be able to run a servo motor connected to servo header S4.

Tamiya Kits

Tamiya offers a small selection of very well-made kits for building sturdy, high-torque gear boxes of various reductions. One of these kits could serve as an excellent starting point for a robotic sculpture or demonstration.

For example, I just finished assembling a Tamiya Worm Gear Box. The gear box, including a small 3 VDC motor, could serve as the core to a single motor drive system. By adding a second gear box, I could make a dual-drive system for a small robopet.

I bought my gear box kit from Edmund Scientific for $16.95 plus shipping, and have to say I was impressed with the kit's quality. It reminded me of the model airplanes I built as a kid; snap the pieces apart from their "tree" carrier, then fit them together.

But these plastic pieces were very well designed, and fit together perfectly. What's more, elements such as the axle and gear holder were made of high-quality metal, with a smooth finish and good machining. The gear box kit comes with enough plastic gears to generate one of two gear reductions, either 336:1 or 216:1; I chose to build the 336:1 box.

You also get a small bag of metal screws, nuts, and other assorted hardware, including a small Allen wrench for tightening the metal grub screw when mounting the gear holder onto the axle.

The kit contains a spare axle and mounting hardware, so you can add your own wheels and have most of a single-motor car all done. It also has various control horns for hooking up your own control rods, pulleys, or cams.

It even includes a tiny tube of gear grease, with more than enough lubricant to complete the kit. The instructions warn you to use only the supplied grease, to prevent chemical damage to the plastic gears.

The instructions consist of a single sheet of paper, printed with clear pictures of each step and very terse instructions in both Japanese and English. My gear box kit went together in about an hour, and I only had one problem. That involved pressing a small metal shaft into the helical gear. You need to get a little rough with this step. I ended up using a hammer to tap the shaft GENTLY into the gear. Be sure to tap the shaft all the way into the gear. When I tried to fit the helical gear into the housing, the exposed end of the shaft bound in its fitting, preventing the gears from turning smoothly. I spent a few minutes spinning the gear by hand, and carefully crimping the tip of the shaft to remove some of the edges. Eventually, I got the gear shaft to spin freely in the housing.

I then assembled the housing, using the supplied metal screws, and snapped the 3 VDC motor in place. When I applied power from a pair of C cells, the output shaft started turning at about 30 RPM. Tamiya designed this gear box to make it as easy as possible to build into larger frames, such as robots and toy cars. This attention to detail shows up in several ways. The output shaft is threaded at either end, with grooves for E-rings and holes for locking shafts. The included control horns mount easily to the shaft, giving you plenty of options for using the power this unit generates.

Overall dimensions are quite compact. The output shaft is 3.25 inches long and sits perpendicular inside the gear box housing, which measures 3 inches long, including motor, by 1 3/8 inch across at the mounting ears. The housing is less than 1.25 inches high, and the whole unit weighs less than two ounces.

I find the Tamiya Worm Gear Box to be a rugged, well-designed drive system that is reasonably priced. True, you could buy a plastic toy car and hack the gear system out of it, but for little more than the cost of the toy, you can use instead a professional-looking, well designed gear box.

Edmund Scientific also sells two other Tamiya gear boxes. The High Power Gear Box kit yields gear ratios of 41.7:1 or 64.8:1, while the High Speed Gear

Box sports 11.6:1 or 18:1 ratios. If you need a gear box for that next robot, Halloween project, or kinetic sculpture, check out the Tamiya gear boxes. I certainly like what I've seen.

Lagniappe

If you haven't seen the latest Mondotronics catalog, send some email and let them know you want one right away. Mondotronics is fast becoming the one-stop source of complete amateur robotics items such as books, kits, videos, and software. You can buy many of Scott Edwards' BASIC Stamp boards, plus Marvin Green's BOTBoard and finished machines such as the PC Rover.

Roger Gilbertson started Mondotronics years ago by offering the Shape Memory Alloy Muscle Wires, and still sells some very good project kits using SMA wires. He also carries a wide assortment of plastic robot kits, including the Tamiya Gear Box kits. And the catalog is colorful, densely packed, and quite weird in places; very entertaining.

I recently saw an American Science & Surplus catalog for the first time in a couple of years. Offering without a doubt the oddest assortment of gadgets, geegaws, and whigmaleeries imaginable, this company belongs in every robot hacker's list of resources.

The catalog sports everything from ammo boxes to parking meters, glow-in-the-dark human skulls to rubber iguanas, gearhead motors to pencil sharpeners shaped like a human nose. If the Mondotronics catalog is weird, the AS&S catalog is totally bizarre. Get one, you'll love it.

More (and More) LEDs

Infrared (IR) plays a big part in amateur robotics. It serves as a cheap, powerful, easy-to-build foundation for many object detection systems, and some people in the Seattle Robotics Society, notably Gary Teachout, have even created object ranging systems with it. This column, from the November 1995 issue of *Nuts & Volts*, shows a simple two-IC circuit that lets you add up to eight IR emitters to your robot, without tying up a bunch of I/O lines.

This column also contains recommendations for a PCB artwork package and a board house that will etch and drill your new creations. The DOS version of Protel's Autotrax package is still around, and some of us in the SRS still use it, though other low-cost alternatives exist today. Alberta Printed Circuits continues to offer their low-cost, quick-turn service, which is a real boon for the hobbyist on a budget. The GC Prevue package described in this column has been updated to a Windows version that gets rave reviews from others in the club.

Peter Dunster of Australia, perhaps the most distant member of the SRS, has been selling a well-designed system of robotic electronics, based on the 68hc11f1 variant of Motorola's powerful microcontroller family. Besides shipping his circuit boards worldwide, Peter kindly donated the artwork for his boards to the community.

I still get asked to recommend the best programming language to use when writing robotics software. As this column makes clear, there is no single best language. Your choice will depend on many factors, and I devote some space to discussing possible candidates and the reasons for choosing one in particular.

I usually use modified Futaba S148 RC servo motors for my little robots, such as Huey and Arnold. The servos are easy to use, but mounting a wheel onto one of the servo motor's control horns has always seemed more painful than necessary.

I like using the Dave Brown Lite Flite wheels, available at most hobby stores. These wheels have a wide, soft foam tire on them, with a two-piece hard plastic hub. But the hub has five evenly spaced radial ridges on it, which make aligning the hub and the servo's control horn difficult. Plus, I used to insist on using two #2 sheet-metal screws to hold the hub to the control horn. The uneven hub surface

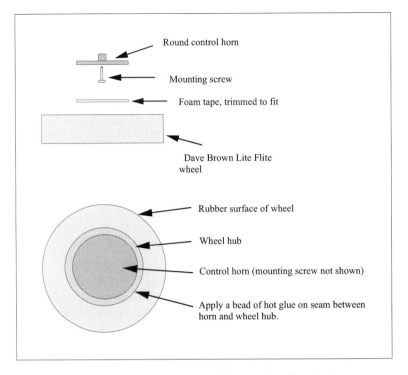

Figure 1. Mounting a servo control horn to a Lite Flite wheel.

guaranteed that the wheel would mount crookedly, giving the finished robot a bow-legged gait.

My latest technique results in a clean, solid coupling between the wheel and control horn, without using screws or machining. It only takes about five minutes per wheel, and gives a quality-looking finish. The technique relies on two of my favorite tools, a hot-glue gun and a roll of double-sided foam tape. Refer to the accompanying diagram of the mounting technique (Figure 1).

Begin by selecting a control horn of the proper size. The Futaba servo package, available from Tower Hobbies, usually includes three different control horns. Use a round control horn that is at least 1/4 inch smaller than the diameter of the wheel's hub. For example, I used a control horn 1-3/8 inches in diameter with a 3-1/4 inch wheel. Since the hub for this wheel measures 1-5/8 inches, I had exactly 1/4 inch of hub overlap to use.

Begin by cutting two pieces of foam tape, each slightly longer than the diameter of the control horn. Stick the tape down onto the wheel hub, covering as much of the hub as you can. Use an Exacto knife to trim the foam tape away, exposing 1/4 inch of hub. Use the knife to punch a hole in the tape at the center of the hub, where the axle was designed to go.

Next, put the control horn mounting screw into the control horn, with the threaded end of the screw sticking out through the mounting rim of the horn. Peel the protective paper from the foam tape, then carefully position the control horn onto the foam tape. Take care to align the control horn properly; you should see an even band of hub around the horn's rim. Make sure the tip of the mounting screw still sticks up above the mounting rim.

Press the control horn firmly in place, then run a bead of hot glue completely around the circumference of the control horn. Use plenty of glue, and make sure you get a thick, smooth bead that touches both the wheel hub and the control horn in all places.

That's it. The whole process takes about five minutes, with no tricky alignment. Makes for a good-looking product, and my robots are starting to drive straighter lines. I've already built up a bin of wheels, ready to mount onto whatever robot needs them. A variation of this technique lets you use the four- and six-point control horns, also shipped with the Futaba servos. Just cut a single, smaller piece of foam tape, and put it over the center hole of the hub.

The tape should be only about 3/4 inch square. This will leave plenty of exposed hub when the control horn is pressed onto the tape. Now, run a bead of hot glue around each arm of the horn, making sure the glue flows into the gaps between the arm and the wheel hub.

You can use this tape-and-glue technique for wheels other than the Dave Brown units. In fact, you can probably tape-and-glue a control horn to just about anything. Just remember to leave a hole in the middle of the mounting surface, and be sure you insert a mounting screw into the control horn before taping the horn in place.

More IR

Regular readers of this column should be familiar with using a modulated beam of IR light as an object detection system. I won't go into all of the gory details, but instead refer you to "Allow Me to Introduce Huey," and to the bible of amateur robotics, *Mobile Robots*.

I recently started a robot that requires multiple IR object detectors. I quickly realized I would need at least three IR LEDs, each emitting a 40 kHz signal on demand. I could have built up a second 74hc04 circuit from the *Mobile Robots* book, but frankly, I wanted a more general solution to the problem of multiple IR sources.

My solution uses two ICs, a CMOS 555 and a 74hc138, to provide up to eight 40 kHz IR sources. Refer to the accompanying schematic for details (Figure 2). U2, a CMOS 555, generates a constant frequency square wave at pin 3. Trimpot R2 sets the

output frequency, and as shown spans a range of just a few kilohertz to well above one megahertz.

The output square wave runs at exactly 50% duty cycle; that is, pin 3 stays high for exactly the same amount of time as it stays low for each cycle of the signal. This 50% duty cycle is crucial to reliable operation of any IR object detection system using the Radio Shack IR detector modules. If the duty cycle gets too far above or below 50%, the system's performance is severely degraded. Note that R2 only controls the frequency of the 555's output; it has no impact on the duty cycle.

The circuitry shown for U2 will not work with an NMOS device, and it will not work with a Motorola 1455 CMOS timer. I have used a Texas Instruments chip (TLC555C) successfully, as well as a National LMC555 device.

The output from U2 is routed to the enable line, pin 6, of U1, which is a 74hc138 demultiplexor. I have wired the 74hc138 to function as a selectable data router. If pins 4 and 5 of U1 are both pulled low, the device is enabled. If these lines are pulled high, the device is disabled, and all outputs remain high.

When enabled, the '138 uses its three address lines, pins 1, 2, and 3, to determine which of eight output lines will carry the inverse of the signal on pin 6. For example, if pin 1 is high and pins 2 and 3 are low, then output Y1 (pin 14) will show the inverted state of pin 6.

Since I've routed the 40 kHz signal to pin 6, any enabled output line will show a 40 kHz signal and the disabled output lines will remain high. The enabled output line will provide a current path from +5 VDC, through limiting resistor R1 and the connected LED. The enabled LED will turn on when its associated output line goes low, and will turn off when that line goes high. The current through any LED is limited by R1. You can select any value you like, but you should limit current to no more than 20 mA, to stay within specs on the 74hc138.

Your computer controls this circuit using the signals on the six-pin Molex KK-style connector J9. Select lines SEL0, SEL1, and SEL2 determine which LED lights, while a low on *ENABLE is required to

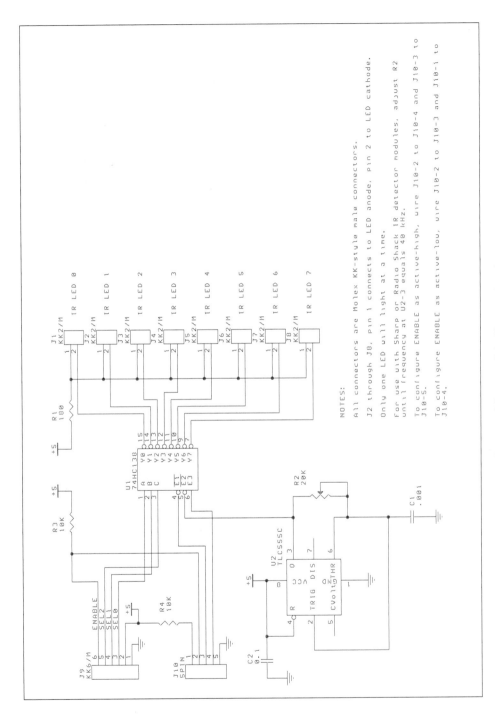

Figure 2. Schematic of the IR source board.

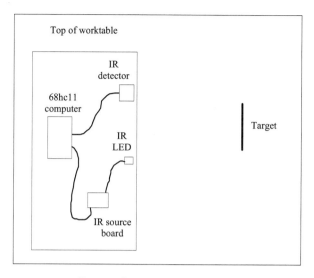

Figure 3. Set-up of IR test platform.

enable the circuit. In my prototype, I hard-wired *EN-ABLE to ground right on the IR source board, and used just a five-pin cable back to the host computer. This means that one output on the board is always supplying a 40 kHz signal.

I also installed a red LED as IR LED 7, wiring the LED directly onto the circuit board. My software writes 111 to the three select lines by default, which means that this red LED is normally lit at half-brightness. This default condition lets me know that I have power on the IR source board, and that the computer select lines are controlled. If the computer needs to turn on a different LED, perhaps to test for an object's presence, it briefly selects the proper LED, performs the test, then selects LED 7 again.

Still More IR

After building the prototype circuit for the above IR source board, I started rooting through my junk box, looking for suitable IR LEDs. I had always used some surplus detector/emitter units, purchased from a local store, but my stash of those was empty.

I did find a large wired assembly of LEDs and IR detectors, purchased some time ago from Electronic Goldmine. Just for grins, I pulled one of the LEDs off

the assembly and plugged it into LED port 0 of my circuit board. I wrote a little SBasic code for a 68hc11 board, connected up the IR source board, and set up a test platform (Figure 3).

I started by taping one of the LEDs to the edge of my work table, with the LED's lens pointed out away from the table. I also taped an IR detector module to the table's edge, placed about five inches from the LED, also placed with the sensor facing out away from the table.

The program causes the LED to broadcast a continuous stream of 40 kHz IR. The code then samples the IR detector's output every half-second, and lights a bi-color LED, based on the detector's signal. Normally, the bi-color LED remains red, signalling that no 40 kHz IR has reached the detector. If, however, I place a suitable object in front of the emitter/detector arrangement, the reflected IR pulses will register in the detector's output and the code will turn the bi-color LED green.

The IR LED from Electronic Goldmine, a TIL-31B, gave a dependable signal out to a distance of 19 inches, in contrast to the nine inches of the unit I used to use. For this test, I used a white sheet of paper, positioned perpendicular to the LED's beam, as a target. This test platform has proven valuable in studying the performance of IR object detection. I was easily able to determine the angle and relative intensity of the beam projected by the TIL-31B, and to test the IR reflectance of different surfaces.

Electronic Goldmine was still offering the IR detector/emitter assembly in their Summer 1995 catalog. For $1, you get a wire harness containing four IR emitters, four IR detectors, a red LED, and connectors. The catalog included a price for 1,000 pieces, so you can probably still get these: EG item number is G911.

I also ordered the bi-color LED, described above, from EG. The T1-style device contains three leads: a common cathode, a red anode and a green anode. I hooked each anode through a 220 ohm resistor to the output of a 74hc device, and connected the cathode to ground. Now, bringing one of the chip's outputs high causes that half of the LED to light. Lighting both halves at once gives an orange LED. Add in blinking of one or both colors, and you can create a wide variety of signals out of just one LED and two control lines.

Prototype PCBs

While developing my small printed circuit boards, I've come to rely on Alberta Printed Circuits for prototyping. AP Circuits offers one-day turnaround (!) on two-layer boards, with some restrictions, for a very reasonable price. And they aren't kidding about one-day turn. If they have your data in hand by 8:00 AM Alberta time, your order is done and on its way to you the next morning.

You get a limited set of features for this price, of course: no silk screen, no solder mask, and a limited set of drill sizes. But this price does include plated-through holes, and you can always do information text in copper.

AP Circuits wants your board files in Gerber format, and I have settled on the Protel Autotrax board layout package for my work. The Seattle Robotics Society made a group purchase of the Protel package some time ago, and it has become a sort of standard for us.

I have not yet been able to substantiate a rumor that Protel has placed the final version of this package on their BBS, free for downloading. I do know that Protel stopped selling the package shortly after the SRS made their buy.

AP Circuits moves the data through the process so quickly that you had better be darn sure you have everything right before you pull the trigger; you won't have much chance to stop them after they start. To help verify your design before you ship it, AP Circuits provides a very cool tool, called GC-Prevue, downloadable from the APC web site.

GC-Prevue lets you load in all the drill files and Gerber plot files for the various layers of your board, then view the PCB that would result from those files. It shows all drill holes in true diameter and placement, and all traces are shown true width.

You can zoom in and out on your design, move from place to place, and even use the mouse to measure actual distances between points on your circuit board. It took me a little time to get the hang of GC-Prevue's user interface; I've seen more intuitive designs. But after reading the skimpy and poorly written docs and running a couple of sessions, I finally got up to speed. It helped having access to Dan Mauch, the club's CNC expert, to answer some of my questions.

GC-Prevue can even save all of your PCB files into a single, larger file called a workgroup file. Later, you can run GC-Prevue and restore this single file, and your entire design is restored, just as you left it. Even better, AP Circuits accepts this single file as your design set. This reduces those awful errors in which you send all but one crucial file to the fab shop, then get back a whole run of boards you can't use.

This last feature is so handy! I can look over my design as much as I want, then send the workgroup file out and know, without a doubt, that the board won't come back screwed up because I left out a file. More likely, it will come back screwed up because I messed up the design, but at least I'll know all the files were there.

AP Circuits and GC-Prevue are a dynamite combination. Hit APC's web site for their new-user documentation. I find their prices good and the turnaround unbeatable. Remember that many of the prices quoted are in Canadian dollars, so you US readers will get a price break on the exchange rate.

Robots in Oz

Peter Dunster of Australia, developer of a versatile 68hc11f1 computer board, brought his family on a short visit to Seattle recently. Pete, Dan Mauch, and I spent a fun day trolling the local surplus shops and swapping stories about robots we have built.

Pete marvelled at the number and variety of surplus shops available to us here in the Seattle area. Granted, few places in the States have anything like Boeing Surplus (need an inch-thick sheet of titanium?), but I don't consider Seattle's surplus outlets all that great when compared to, say, San Jose.

But apparently, hobbyists Down Under have almost nowhere to go for finding electronics scrap or junked printers and whatnot. So they have to make do with limited mail-order access to expensive components from a narrow range of offerings.

Even so, Pete has designed a marvelous set of PCBs for robotics, thanks to support from his employer, the University of Wollongong. His F1 board sports a 68hc11f1 MCU and two sockets for a total of 64K of memory. The 'f1 MCU is similar to the standard 'a1 chip, but it has extra I/O lines and some cool programmable chip select lines.

Pete also offers an add-on PCB called the motor-driver board. You can plug this board directly onto the expansion connector of an F1 board, then hook motors to the driver board. The motor-driver board controls about any type of motor you can imagine:)servos, steppers, and DC gearheads.

And, you can download the Protel .PCB files for a number of other add-on boards, such as a 64-line digital I/O board. Pete has made the designs of these add-ons available on his FTP site, free for the taking. I encourage anyone looking for a high-powered, well-designed robot controller to check out Pete's work.

By the way, Dan Mauch serves as the US distributor for Pete's F1 board and motor-driver board,

along with other goodies Pete offers. Contact Dan for prices of the various boards by sending him email.

What Language?

I get so many letters and email asking me what language to use for programming robots that I thought I'd address the question publicly. The answer is: It depends.

It depends on the size of your robot's memory. If you are using an unexpanded 68hc11a1, you only have 512 bytes of memory. Therefore, the size of the executable program becomes paramount, and you may be limited to using languages you don't understand or don't use easily. If, however, you have 32K or 64K of memory available, you have more freedom on your language choice, and can go with one you enjoy using, even if it creates larger code than other languages might.

It also depends on the execution speed your final program will require. If your code needs to position multiple stepper motors, sample A/D ports, and scan an array of switches, you may have to pass up some languages as being too slow for the job.

And it depends on your wallet. If you have bucks to spare, you can pick up a commercial development package, with cool debug support and extensive libraries of developed code. Most hobbyists, however, must stay with cheap or free software, which can limit choices.

Finally, it depends on your own skill level. Guys who write software for a living usually are fluent in one or more languages, and have more choices available. If you have never written software before, you will have to deal with the learning curve problem, and some languages are easier to learn than others.

Given all of those depends, I'll provide a list of recommendations. I'm sure that some readers will disagree with some of my ideas, and that's OK, but this is my column and these are my recommendations. In all cases, I assume you are using a 68hc11 MCU in your robot brain.

For robots with 512 bytes or less of code space, I suggest 68hc11 assembly language or SBasic. Assembly language is way more complicated and difficult to learn than SBasic, but experienced programmers will create much smaller code with assembly than with SBasic.

For assembly language, you can pick up a copy of Motorola's free 68hc11 assembler, asmhc11. I use this assembler exclusively, and even distribute copies of it with my SBasic and tiny4th systems. You can download your own copy from Motorola's FREEWARE web site. Note that the documents with asmhc11 are rudimentary at best; they assume you know assembly language already. If you don't, check the comp.robotics.misc FAQ for pointers to books on 68hc11 assembly language.

If you're just starting out, begin with SBasic; you will get your programs running sooner with SBasic. Also, SBasic code runs about half as fast as well-crafted assembly language, which is usually plenty fast for most robot software.

For robots with 512 to 2K bytes of code space, you can choose between the above two languages, as well as a couple of C compilers or tiny4th. icc11 is ImageCraft's 68hc11 C compiler. Those in the SRS who have used icc11 like it a lot, and it comes highly recommended. Another C compiler that gets a lot of positive press in comp.robotics.misc is Dave Dunfield's Micro-C. This package costs $100, and Dave has created an extensive suite of support software and tools. I have not used either icc11 or Micro-C, but am passing along both because I have heard lots of good things about them, and nothing bad.

I wrote tiny4th a couple of years ago, and distribute it free from my web site. tiny4th translates programs written in a subset of the Forth language into tokenized assembly language, which is then assembled by the Motorola asmhc11 assembler. If you know or want to learn Forth, tiny4th makes a great way to get up to speed. Note that tiny4th's programs run about six times slower than assembly language, due to the tokenizing.

For robots with 2K to 64K of code space, you can use nearly anything. All of the above options will work, plus you can try MIT's Interactive C, or IC. As its name implies, IC is interactive. You can run it on your robot's computer, and enter C statements from your keyboard. The statements are executed immediately, giving you instant feedback on your system. For a more complete discussion of IC, check your copy of *Mobile Robots*.

Those are my recommendations, along with one more. You cannot write robot code if you don't try. Start with something simple, such as SBasic, and write a program. Keep it simple, maybe just flashing an LED. If that works, move on to something more complex.

If that doesn't work, or when you start writing code that doesn't work, don't be discouraged. Software may be the most difficult aspect of robotics, but it can also be the most rewarding. Keep trying, and use email and the Internet news groups as a source of help and guidance.

Software is a necessary evil when it comes to amateur robotics, and I say this as someone who earns his living writing software. I've seen more than one promising hobbyist stopped in his tracks by the need to write software; don't let it stop you.

And Finally...

There is more to life than building robots. When it's time to set aside my soldering iron or keyboard, I usually reach for a good book, and that book doesn't even have to be about robots.

Beryl Markham's autobiography, *West with the Night*, is, as Ernest Hemingway wrote in a review, "really a bloody wonderful book." Beryl grew up in British East Africa in the early 1900s, and went on to become a celebrated aviatrix, horse trainer, and adventurer.

More than that, she writes like an angel, of a life beyond anything most of us could dream about, let alone experience. The stories of her childhood, spent hunting in the African plains with only a spear or raising a new-born Abyssinian colt, are marvelous reading. And her beautiful prose haunts you long after you finish the book. Highly recommended.

Design of a Simple Line-following Array

My June 1996 column for *Nuts & Volts* covered an assortment of topics. I started by describing a line-following pod I had developed for use in an SRS event. Building a line-following pod using point-to-point wiring quickly becomes tedious, and after about the third one, I was ready for a PCB. The small board I designed holds five photodetectors and five red LEDs, arranged in a suitable pattern. This section also covers a few pointers on line-following pod design.

Next, I devote a bit of space to my history as an embedded firmware designer. Each of us is a product of people we've known before, and I owe much of my current skills to those who helped start the microcontroller and microcomputer hobby over two decades ago. This column includes a "thank you" to those experimenters who donated so many hours of time and labor, that all interested in the hobby could benefit.

I conclude this column with a step-by-step explanation of how to go from an SBasic (SB) source file to a working program in your robot. I cover details such as the command line options for the SB compiler, running Motorola's pcbug11 utility, and setting up your 68hc11 microcontroller board so it can load your program and run it after reset. This type of "getting started" information is vital for anyone beginning a 68hc11 project, and having it available in one column will, I hope, make the process easier.

This time, I'm going to devote my column to robotics odds and ends; those little tidbits of information that don't make up a full column, but deserve discussion.

Line-following Sensor

I've designed a simple, five-sensor line-following array that can be used on nearly any robot. The board measures 1.5 by 2 inches, and provides layout and traces for five LEDs and five matching photodetectors. The circuit is quite simple, consisting of five current-limiting resistors for the LEDs and five pullup resistors for the photodetectors. See the accompanying schematic for details (Figure 1).

I finally sat down and designed this board, rather than hand-wire yet another sensor pod. The physical arrangement works very well with a 1/4-inch black line, which is customary in Seattle Robotics Society contests. The layout features three emitter-detector pairs arranged in a straight line across the direction of travel, with another pair at either end of the three, offset by 45 degrees. This pattern can provide right and left leading or trailing sensors, depending on how you mount the board.

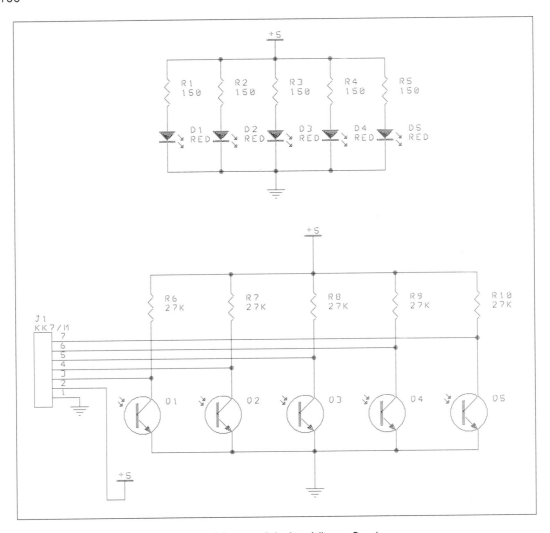

Figure 1. Schematic of the Line-following Board.

The circuit provides analog signals out for all five photodetectors. You can run these signals into five A/D channels, such as those available on a 68hc11, or you can design your own circuitry to convert the analog outputs into a digital form, for hooking to a parallel port.

The circuit isn't critical with regards to sensors or emitters. I've had excellent results using the photodetectors pulled from the Electronic Goldmine wiring harness (G911). EG has offered this harness for months, and still quotes a price for 1000 pieces, so I figure they've got a bunch on hand. Each harness contains four IR emitters, four IR detectors, a red LED, and three or four 0.1" female AMP connectors. Quite a bargain for only a buck per harness. I just pull the photodetectors off their wire leads, clean them up a bit, and wire them into the board.

For emitters, I like to use ultrabright red LEDs in water-clear lenses. I've had problems making IR emitters work in line-following contests, and finally given up on using them. IR tends to go through paper and bounce off the floor, or reflect strongly off of a dark line that doesn't look like it should reflect. With red LEDs, you can be pretty sure that you are seeing what the photodetector sees, making it easier to align your sensor pod.

SBasic and KIM-1

As of late April, I've released version 1.1 of my 68hc11 SBasic compiler; you can find the latest distribution files on my web site. This version cleans up some problems with code generated for array references and the ADDR() function. I also added the SWAP command, for exchanging elements on the SBasic data stack, and I cleaned up some minor errors in the manual.

I continue to get email from readers thanking me for SBasic, and asking about sending donations or shareware fees. As I've explained to at least one reader, I placed both SBasic and tiny4th in the public domain, because I've long felt that software should be free, or at least minimally priced.

I started using computers back when they were known as microcomputers. My first machine, which I still have, was a KIM-1. This 6502 board was built with NMOS and 74LS parts, so it took at least an amp of current at 5 VDC. It sported a whopping 1K bytes of RAM and used a 16-key keypad and a six-digit LED readout for I/O. For mass storage, you could connect up a cassette tape recorder and save your programs to tape.

At first, I had to code my programs in machine language (read: hex digits) and enter them by hand, one byte at a time, via the keypad. This wasn't too much work if the program ran properly, but it was a pain in the wazoo if the program crashed. A crash meant I had to enter the entire program again, with changes, and test it once more.

I honestly think that the pain resulting from poor work habits made me a better software designer. The price for firing off a program without double-checking for errors was simply too great. If I did my job well, I was rewarded by being able to play chess or

Hunt the Wumpus for the rest of the evening; get it wrong, and I lost a lot of valuable time cleaning up the mess. By dint of continual negative reinforcement, I changed my style to minimize errors.

Writing out programs for the KIM-1 also forced me to improve my documentation habits. Since I didn't have an assembler when I started out, I would print, by hand, all of my source code in school notebooks. For each instruction, I would write the address it would occupy in memory, the machine code needed, and the assembler syntax for that instruction. I included a comment for each and every line, since I found I had problems remembering exactly what some instructions did after a few days.

To this day, I still comment almost every line of code, even if I'm using a higher-level language such as C or Forth. This habit has paid dividends many times over, in ease of maintenance and in passing my code on to others.

This was a tedious way to code, but I became very familiar with the 6502 opcodes and machine characteristics. Hand-coding machine language programs probably seems like a huge waste of time, given today's powerful compilers and debuggers, but I don't know of a better way to learn the fine details of a microcontroller.

After I added some memory to my system, bringing it up to 5K bytes total (!), I started looking around for some more powerful software. That's when I discovered Tom Pittman's Tiny Basic. TB let you write programs in BASIC, with 26 variables (A through Z) and subroutines and other really cool features. I'm sure this seems odd to someone who has just installed Visual Basic 4.0, but TB opened some very big doors for those of us in the early days of the hobby. We could write REAL programs now, using a REAL language, and I for one enjoyed it thoroughly.

And best of all, Tom practically gave away his software; he only wanted $5 for the software and a manual. I got a copy of TB from a friend, so I simply sent Tom a check for $5, with a letter explaining that he didn't need to send me a copy of the program, but thanks for writing TB. I later found out that this was a pretty common practice among KIM programmers, and I've always wondered how many copies Tom sold versus how many he received money for.

Many times since, I've thought back on how much I owe people like Tom Pittman. Using Tom's $5 creation, I was able to leverage my time and creativity, to write larger and better programs, and to explore areas of microcontrollers previously beyond my reach. I don't know if Tom was by nature a generous person, or if he simply accepted the fact that hobbyists in those days didn't have much money. Regardless, Tiny Basic gave me power way beyond its cost. I turned my experiences with the KIM-1 and Tiny Basic into the foundation for my current career as an embedded control software designer.

Legend has it that a noted science fiction writer once thanked Robert Heinlein for his help and inspiration, and asked Heinlein how he could ever pay him back. Heinlein reportedly said, "You can't, so you pay it forward; you help someone else."

That, in a nutshell, explains why I don't charge anything for SBasic or for tiny4th. Think of these programs as a repayment to Tom Pittman, Jim Butterfield, Peter Jennings, and others who helped the early KIM-1 users get started.

A Wheel Adapter Design

I'm working on my first stepper motor robot design right now, and I'll have full details out sometime soon.

For now, however, I want to describe a little mechanical gem that Dan Mauch and I developed in the course of starting on my stepper robot.

The R/C servos I normally use have a very nice design when it comes to mounting wheels onto them. You just add some double-sided foam tape to a wheel, slap it onto the servo's control horn, then screw the control horn onto the servo shaft. No muss, no fuss, and you can move the wheel and its control horn from servo to servo.

But steppers don't come with such a cool wheel mounting system already in place. Since my mechanical skills aren't too good, I turned to Dan for some help. Between us, we came up with the adapter shown in the accompanying diagram (Figure 2).

Dan started with a piece of 5/8-inch round aluminum stock and cut it to 3/4-inch length. He drilled a hole halfway down the length of the adapter using a #29 bit, then tapped for an 8-32 thread. He swapped ends of the adapter and drilled a hole halfway down the length of the adapter using a 5mm bit. He then drilled and tapped for an 8-32 set screw on the side of the adapter, about 1/4 inch from the unthreaded end as shown.

Then Dan got a little fancy, and milled the threaded end of the adapter to take a press-fit 3/8-inch inside-diameter (ID) fender washer. He matched a washer to the milled end, clamped the pair in his bench vise, and leaned on them. When he took the

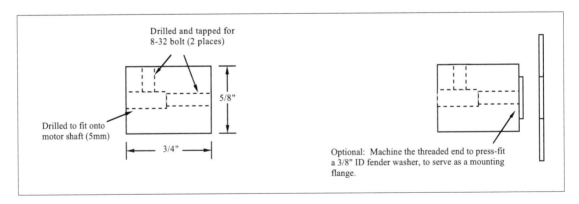

Figure 2. Wheel adaptor design.

piece out of his vise, he had an adapter complete with a large mounting flange on the threaded end. Dan assures me the whole operation takes very little time, and the end result is one of the most versatile wheel mounting systems I've ever seen.

I couldn't wait to get a pair of these adapters home and start playing with them. I grabbed an 8-32 bolt 1 inch long and screwed it through the axle bore on a Dave Brown Lite Flite wheel. I use these wheels on all of my machines, and an 8-32 bolt will cut into the wheel's plastic hub for a locking fit. I then screwed the protruding end of the bolt into the threaded end of the adapter, and I had a perfectly matched unit, ready to bolt onto the stepper's shaft.

The diagram shows a 5mm bore for the motor shaft, but you would use whatever drill bit size was proper for the motor at hand. Thus, you could make a version of this adapter for use with gearhead motors. Similarly, you could use this adapter to mount pulleys onto high-speed motors as part of a speed-reduction belt drive. You can also make these adapters tapped for other bolts, such as 6-32, 4-40, or metric. These other sizes would let you use smaller wheels and attachments if necessary.

If anyone at your club has the proper machinery to make these adapters, see about having some made for the whole club. If necessary, contact someone at work about having a few made. They will make the task of mounting wheels to motors so much easier.

By the way, if you haven't used Lite Flite wheels on your robot yet, give them a try. The wheels are about 3/4 inch wide, with a dense, black foam-rubber tire and a spiffy gray plastic hub. They are usually available at larger hobby stores in a variety of sizes, and you can buy them mail-order through Tower Hobbies.

RF Modems

The comp.robotics.misc newsgroup and the various listservers catering to robotics and the 68hc11 usually carry requests for some type of RF or IR modem. The need for such devices, especially in robotics, is so great that I try to keep an eye out for opportunities,

despite the time-lag involved in magazine publication.

Halted Specialties, a prominent advertiser in *Nuts & Volts*, offered a set of Air-Share RF modems recently. I picked up a pair of modems, with battery adapters and a wall-wart power supply, for $49.95. Unfortunately, HSC only had about 100 sets available, and they ran out shortly after I told the rest of the SRS about the deal. So why am I mentioning it now? Two reasons, actually.

First, you have to keep your eyes open. Various RF modems do surface from time to time, and the convenience and flexibility they give to any robot is worth the effort needed to track them down. Second, if you keep after companies such as HSC, Timeline, and others to offer such devices, they in turn will spend more time and effort finding them. After all, the surplus houses need to find popular products to sell, or they won't make money. Let them know what you want, and you'll find they will happily work with you.

The Air-Share modems use 900 MHz narrow-band FM, on one of three frequency pairs, to provide serial communication at up to 115 Kbaud. The units do not provide any packetizing or error detection; think of them as an RF wire. Though they seem to be fairly accurate in benign environments, you will definitely want to use some type of file transfer protocol when exchanging data using these units.

In fact, my modems arrived in a sealed bag with the Traveling Software logo on it, so they were evidently surplus from an RF version of the popular LapLink software. Just for kicks, I bought an older version of LapLink from a software surplus house and used it to run these modems between two PCs; works great. But back to the modems ...

The system includes two battery adapters that let you run a modem from a 9 VDC battery. Each modem has a short DB-9 serial cable on it, and one of the modem control signals acts as a power-on line. When the host computer or robot pulls the line high, the modem switches on and begins transmitting. With the modem off, current drain is in the microamp region; when on, the device draws about 70 mA.

As of this writing, HSC thought they might be able to get in a few more units. Contact them to see the current status, and to let them know what kind of product you want to see for sale.

68hc11 and Robots

I notice that too often, a beginner thinks he has everything needed to start working on a robot. He's built a frame, scrounged some motors and electronics, and a BOTBoard or other 68hc11 computer. He even has a small program written in SBasic, ready to go. But the final step, that of moving the code into the BOTBoard so it can run, eludes him. How, in detail, does one move an S19 object file from a PC into a BOTBoard, and make it run?

I'll assume here that you are using a BOTBoard with a 68hc811e2. This chip contains 2K of EEPROM, usually addressed from $f800 to $ffff. This addressing scheme works perfectly for the BOTBoard and SBasic; you just put your program's code at $f800, the stack at $00ff, and the variables at $0000.

To serve as an example, I've written a small SBasic program that prints "Hello, world!" out the 'hc11's serial port (SCI). I'll take you through the steps needed to move this program into the 68hc811e2 and see it run (Figure 3).

Begin by compiling the program with the various segments set up as described above. To do this, use the command line:

sbasic hello.bas /v0000 /s00ff /cf800 >hello.asc

Use your text editor to verify that the output file, HELLO.ASC, does not contain any SBasic errors. Next, invoke the asmhc11 assembler to generate the final S19 file:

asmhc11 hello

You should now have a finished object file named HELLO.S19, ready for downloading into your BOTBoard's 68hc811e2. The first step in this process involves a program named pcbug11, available free from Motorola's FTP site. If you don't already have a copy of this program, get one. pcbug11 is an indispensable tool in 68hc11 development.

Hook up the serial cable between your BOTBoard and PC, strap your BOTBoard for special bootstrap mode (MODA and MODB both tied low) and reset your board. Now invoke pcbug11, providing it with the proper arguments for a 68hc811e2 on whatever COM port you are using. For example, if your PC is hooked to the BOTBoard using COM2, you would enter:

pcbug11 -a port=2

pcbug11 should present a blue screen with a command prompt at the bottom. If you get an error message in a black box at the center of the screen, you have a problem with the serial hookup. Verify that you are using the correct COM port, and that you have the proper connections on pin 2 and pin 3 of your serial cable.

If you get an error when you first invoke pcbug11, you can retry the connection without exiting pcbug11. First, press ENTER until you see the pcbug11 command prompt. Then reset your BOTBoard again and enter the pcbug11 command **restart**. pcbug11 will try again to set up your BOTBoard for communication.

When you have pcbug11 talking to your BOTBoard, you need to enter a few commands to prepare your 68hc811e2 for downloading. Enter the following sequence:

control base hex
ms 1035 10
eeprom f800 ffff

The **control** command simply tells pcbug11 that you will use hexadecimal numbers in all commands. The **ms 1035** command turns off the block protection bits for all of the on-chip EEPROM, allowing you to change it. The **eeprom** command tells pcbug11 that any writes to the memory from $f800 to $ffff will be writes to EEPROM. pcbug11 must know this address range, since it handles writes to EEPROM differently from writes to RAM.

With these commands behind you, you are finally ready to transfer your HELLO.S19 file into

```
`
`    hello.bas          Send  a  message  every  once  in  a  while
`

include   "regs11.lib"

declare   n                        ` temp storage
declare   wait                         ` timer variable
const   DELAY = 255                    ` timeout (1 sec)


interrupt   $fff0                  ` RTI vector
if wait <> 0                           ` if timer not yet 0,
    wait = wait - 1                ` decrement it
endif
pokeb   tflg2, %01000000           ` rearm RTI
end                                    ` end of RTI ISR


main:
pokeb   baud, $30                  ` 9600 baud
pokeb   sccr2, $0c                 ` turn on rcvr and xmtr
pokeb   tflg2, %01000000           ` arm RTI
pokeb   tmsk2, %01000000           ` permit RTI interrupts
interrupts on                          ` permit system interrupts

wait = 0                               ` start with 0
n = 1                              ` counter starts at 1
do
    if wait = 0                    ` if timed out...
        wait = DELAY                   ` reload delay value
        print n;                       ` show counter
        n = n + 1                      ` bump counter
        print "Hello, world!"    ` show message
    endif
loop                                   ` do forever
```

Figure 3. Listing of HELLO.BAS program.

the 68hc811e2's EEPROM. You do this with the command:

loads hello

This command loads an S19 file, assumed to reside in the same directory as pcbug11. If the file resides in a different directory, simply specify the full pathname, as in:

loads c:\mydir\hello

pcbug11 will begin moving the file into EEPROM, giving you a progress report as it does so. When pcbug11 finishes the download, you need to enter the command:

verf hello

This forces pcbug11 to read the file a second time, verifying the contents of the 68hc811e2's memory. If the verification works, your program is loaded and ready to run. If the verification fails, you might not have entered the **ms** or **eeprom** commands above properly.

Now you can enter the following pcbug11 command:

term

to activate pcbug11's 9600 baud terminal window. With the terminal window active, strap your 68hc811e2 for single-chip mode (MODA low and MODB high), then press reset. You should see the "Hello, world!" message on the terminal screen. Press the ESC key to leave terminal mode and return to pcbug11's command mode.

Using pcbug11 to move programs into an expanded 68hc11, one with external battery-backed RAM, is only slightly more complex. Be sure to use the proper option for your 68hc11 variant when invoking pcbug11. For example, use **-e** for chips such as the 68hc11e9. Consult the pcbug11 manual for help, or just enter pcbug11 without options at the prompt, and walk through the questions the program asks. After you enter the **control** statement above, you need to replace the **ms** command with:

ms 103c f5

This command switches on the 68hc11's external bus, giving pcbug11 access to external RAM. Since you will not be writing your program into EEPROM, you don't need to enter the **eeprom** command. After you finish downloading your program into RAM, remember to strap the 68hc11 for expanded mode (MODA high and MODB high) before resetting it.

All of this may look complicated, but you will quickly get the hang of the process. pcbug11 supports a macro system that lets you hide most of this, but it's good to know how the chips and pcbug11 work together. Besides, you might end up trying to use pcbug11 on a buddy's system, and your macros may not be available.

Stepper Motor Basics

In this column from the July 1998 issue of *Nuts & Volts*, I spent some time with stepper motors. Most hobby robots use a DC motor, such as a hobby servo motor or a surplus gearhead motor. The switch to stepper motors isn't difficult, and this article will get you started.

I begin with a quick run-down on how stepper motors differ internally from the more common DC motors. Then I describe the two most common types of stepper motor driver circuits, the L/R and the chopper drivers. Each has its strong points, though the chopper driver gives such superior performance that serious robot builders should avoid the L/R circuit entirely.

Next, I go into detail on a chopper driver circuit that Bill Bailey, one of the SRS' electronics gurus, put together. This circuit now forms the core of some very fine robots, and Bill has done the club a real service by making his PCB and design available for all.

Finally, I develop a short piece of SBasic code to show how you might control such a chopper driver board in software. The code is straightforward, and my explanation should get you started on your own routines.

I've never built a stepper motor robotics platform before, but there's never any time like the present. So I asked others in the Seattle Robotics Society (SRS) for some pointers, spent a few evenings in the workshop, and built up the two-tiered machine you see here (Figure 1). I enjoyed the construction, and picked up a few tips that I'd like to pass along.

I'll leave a discussion of stepper motor theory and design to others, and focus instead on building the actual platform. You can find excellent information on steppers through many channels, including a Web search using Alta Vista, or a browse through back issues of the popular electronics magazines, including *Nuts & Volts*. Also check the application notes from major chip makers such as National Semiconductor and SGS-Thompson. These app notes usually include some excellent material on the physics of steppers, as well as info on designing driver electronics.

I approached my platform design from the premise that stepper motors are very much like the usual DC gearhead motors I've used so often before. You bolt them to the chassis, wire them up to a driver board, write some software to give them the electrical signals they want, and your robot goes. As it turns out, this is pretty close to correct, but the

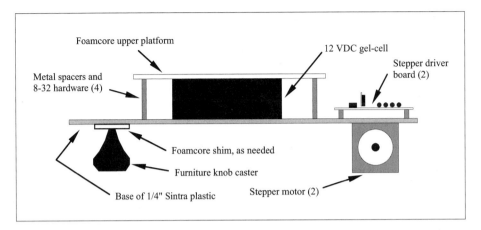

Figure 1. Layout of a stepper-motor robot.

differences between steppers and DC gearheads proved interesting.

Drive That Motor

A stepper motor, such as the Vexta PXB44H that I used, consists of a magnetized rotor element that spins within a cylindrical field of alternating electromagnetic coils. When one coil is magnetized, the rotor moves slightly to align itself to the field. If that coil is de-energized and the second coil energized, the rotor again moves slightly, or steps, to realign itself. If the motor's coils are energized rapidly and in sequence, the rotor will rotate around its axis in a series of steps, with each step corresponding to the distance between coils.

This very simplistic explanation shows the major difficulty in using stepper motors. You don't just apply current and watch the motor spin; you need some fairly complex driving circuitry. Luckily, owing to the widespread use of steppers, companies such as SGS-Thompson have developed ICs that ease the design of stepper driver boards.

Stepper driver boards come in two major designs, the L/R type and the chopper type. Both designs try to deal with the same problem when supplying voltage to a stepper's coils; namely, forcing the maximum rated current into the coils as quickly

as possible. The faster you can get the maximum current into the coils, the sooner you will have full torque available. But the inductance (L) and resistance (R) of the motor's coils combine to place an upper limit on how quickly current can build up at a given voltage.

I'll use the Vexta motor above as an example. The motor runs at a rated 6 VDC. The two coils are center-tapped (the motor has six leads), and the resistance across each full coil is 15 ohms. This means each full coil, when driven by 6 VDC, will pull 0.4 amps.

The motor will show maximum torque when the coils carry the rated current. However, sticking 6 VDC on the coils just locks the motor in place. Remember, to move a stepper you have to energize its coils in a sequence. This implies a sequence of square waves, with 0 VDC (and hence, zero torque) when a coil is off, and 6 VDC (and maximum torque) when a coil is on.

This brings us back to the L and the R. Current does not build up instantly in a coil simply because you switch on the juice. The time, in seconds, that it takes the current to reach its maximum is determined by the L/R constant, where L is the coil's inductance in henrys and R is its resistance in ohms. For a full explanation of this relationship, refer to any copy of the *ARRL Handbook*, a reference book long valued by the amateur radio community.

For now, I'll stay with a general discussion of the problem. You want to bring the current in the motor's coils to the rated value as quickly as possible, to get the maximum torque, but the L/R time constant gets in the way. You need a driver board that somehow gets past this L/R time constant, or at least helps you mitigate its effects.

The L/R driver tries to offset this time constant by increasing the R value. If each coil measures out at 15 ohms and you were to somehow double that, the time constant would suddenly become L/2R, or half its original value. This would let you reach the rated current in the motor's coils twice as fast. Rewinding the motor's coils with smaller wire wouldn't be practical, so most L/R drivers simply stick a hurking power resistor in series with each coil.

Assume you add a resistor equal to the coil's resistance in series with each coil. For my example, this would take a pair of 15 ohm resistors. The added resistance means that in a holding situation, with a coil fully energized, the coil will only see half the applied voltage. So right off, we know that we have to double the driving voltage if we want the coil to ever see its rated voltage. Thus, we would have to supply 12 VDC to our L/2R driver board to run a 6 VDC stepper motor.

Next, we need to look at the power rating of the resistor. In worst case, with the coil fully energized, the resistor will carry the coil's rated current of 0.4 amps. This means the resistor must dissipate I*I*R watts, or (0.4 * 0.4 * 15) watts, or 2.4 watts. Since each stepper motor has two coils, you need two resistors, which will dissipate a total of 4.8 watts.

Because the stepper coils have the same resistance of 15 ohms, each coil will also use up 2.4 watts. This means that a L/2R drive shows 50% efficiency, with half of the applied power going into the coils and the other half into the air as heat. As you boost the resistance of the series resistor, the effiency drops accordingly. An L/4R driver board in our example would use 45 ohm resistors, require 24 VDC to reach rated current, and dissipate a whopping 7.2 watts per resistor. This yields an efficiency of just 25%.

This characteristically low efficiency makes the L/R design a true bow-wow for robotics. Most robots have enough problems hauling batteries around to let them run at all, without having to carry extra juice just to run the onboard room heater.

Besides the obvious problems with efficiency, this example points out another weakness to the L/R design. The current eventually applied to the coils depends directly on the voltage you use. Put another way, your L/R design must specify both the series resistor value AND the applied voltage. This becomes obvious when you think of the resistor and the coil forming a divider. You have to balance the resistance and the applied voltage so the motor's coil ultimatly sees the rated voltage and current when fully energized.

So why do so many L/R designs still exist? Because they're cheap. About all you really need is a set of power transistors or MOSFETs, the series resistors, and some type of electronics to create the correct sequence of steps. If your design can use a wall outlet for power, all the better. Not having to worry about the battery drain just leaves you with the problem of venting all the built-up heat.

Chopper drives take a more intelligent approach to the problem of getting the rated current into the coils quickly. Here, the idea is to dump a high voltage, well beyond the rated voltage, into the coil and monitor the coil's current. Whenever the coil's current reaches some set limit, usually below the rated value, the chopper drive momentarily shuts off the voltage but continues to monitor the current. When the current drops below a lower bound, the driver turns the voltage back on and the process starts over. This repeated cycle of sample and switch lets the chopper drive maintain full current in the coils longer than would be possible with a simple L/R drive. This in turn lets you run a stepper faster, with more high-end torque.

The increase in speed and torque over an L/R drive can be considerable. National Semiconductor's application note AN-828 cites a typical comparison of torque between a standard L/R drive and a chopper drive based on the company's LM18200 chopper IC. Using a motor rated at 18 oz-in of torque, the torque using an L/R drive dropped to 10 oz-in at a step rate of about 400 steps per second. Compare this to the chopper drive, where the torque didn't drop that low until the motor hit 2000 steps per second.

A chopper drive usually consists of logic circuitry to create the proper stepping pulses, a current monitor for sampling the current flowing through the motor's coils, and a set of high-current switches. These usually consist of an H-bridge made of power bipolar or MOSFET transistors, though many companies now offer a single package containing one or two high-current H-bridges.

One of the older chopper drive systems on the market is the SGS-Thompson L297/L298 combination. The L297 stepper controller chip comes in a 20-pin DIP format. It accepts a set of digital signals from some host device and generates the proper switching signals for a high-current driver. The L297 works well with the L298 dual high-current H-bridge driver, which comes in a 15-pin Multiwatt power package. This driver can handle up to 46 VDC and a total of 3 amps of current (intermittent).

Real Electronics

A single L297/L298 combination can drive one stepper motor. The two chips plus supporting circuitry fit confortably on a PCB 2.75 by 2 inches. I know this because Bill Bailey, one of the SRS stalwarts, designed just such a circuit board. Rather than design my own, I used his in my latest robot. Refer to the accompanying schematic of Bill's chopper driver board (Figure 2).

You can think of this chopper design as a black box. You plug a stepper motor into one end, add voltage, and hook up a computer to the other end. The computer provides a set of signals to the board's inputs, and the board in turn switches current into the motor's coil as required. Refer to the accompanying drawing of the L297/L298 chopper driver (Figure 3).

The host computer, a 68hc11 in my case, must provide a small set of control signals. The CLOCK signal acts as a step command; each low-going pulse causes the L297 to advance the motor one step in the current direction. The CW/*CCW signal determines the direction of rotation; a high on this line causes the motor to step in one direction, while a low makes the motor step in the opposite direction. The ENABLE signal controls the overall operation of the H-bridge. If ENABLE is high, the bridge drives current into the selected coils. If ENABLE is low, however, the H-bridge removes all current from all coils. This latter condition can serve as a power-saving feature, since you can shut off the current when the robot is at rest.

The HALF/*FULL line determines whether the stepper moves in half-step or full-step increments. Using half-step mode lets you double the resolution of your stepper system, though your motor will run half as fast for the same pulse rate.

The schematic shows a 1K trimpot, R4, which you use to set the maximum current that the chopper will deliver to a coil. You simply hook a voltmeter to pin 15 (VREF) of the L297, then read the voltage as you adjust R4. The voltage you read, in millivolts, corresponds to the amount of current the chopper will deliver to each coil, in milliamps. Thus, setting VREF to 200 mV causes the driver to maintain 200 mA of current in any energized coil.

Note that the L298 power H-bridge only uses the motor's voltage; it does not need a source of 5 VDC. You can drive the L298 with anything up to 46 VDC, though you may have to provide a heatsink at higher currents.

You do need, however, to provide 5 VDC to the L297, since it is basically a logic device. Bill has added traces for a simple 78L05 voltage regulator supply that derives 5 VDC from the motor voltage, though you can also provide 5 VDC through the host computer connector, J2. If you do use the 78L05 regulator, make sure you keep the motor voltage at 12 VDC or below. Higher voltages will cause the 78L05 to dissipate too much heat, and it will thermally shut down.

Bipolar H-bridges, such as the L298, can suffer from thermal runaway and burn up if you pull excessive current. I strongly suggest you install a fuse in

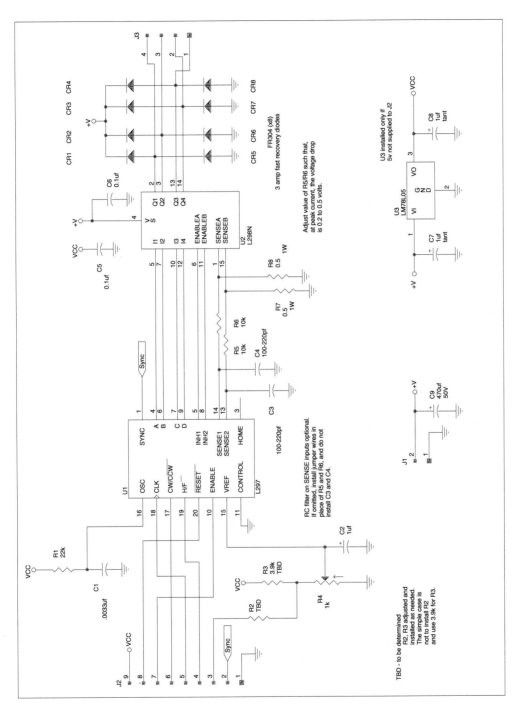

Figure 2. Schematic of chopper driver circuit.

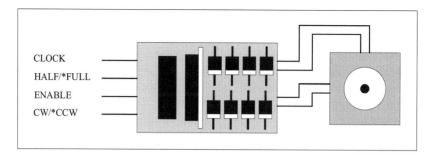

CLOCK

HALF/*FULL

ENABLE

CW/*CCW

Figure 3. Hooking up the chopper driver board.

the motor voltage line coming into the board to protect the electronics. The fuse rating isn't too critical; just use one about twice your intended current draw. In my case, I had set both motor driver boards for about 300 mA, so I was drawing 600 mA per board or a total of 1.2 amps worst case from my 12 VDC battery. I chose a 1.75 A fuse, inserted it into a Radio Shack automotive in-line fuse assembly, and wired the harness into the power system of my robot.

You need to pay attention to some of the components when building your own L297/L298 stepper driver board. Most important, be sure to use Schottky diodes for the snubbers; these are CR1 through CR8 in Bill's schematic. I used 1N5822 devices, as suggested by Bill, but other 3-amp, 50 V fast-recovery diodes will work as well.

Also, the two sense resistors, R7 and R8, should be non-inductive devices. This means you cannot use wirewound resistors. Bill's circuit calls for 0.5 ohms each, but he notes that 0.47 ohms will work as well.

After you have built your driver board, you can hook up a small 555 oscillator to generate a stream of square pulses at about 200 Hz. Temporarily tack jumpers to the input lines of your driver board to get this configuration:

J2-1 Ground
J2-2 N/C
J2-3 N/C
J2-4 Ground (HALF/*FULL)
J2-5 Pin 3 of the 555 (CLOCK)
J2-6 Ground (CW/*CCW)

J2-7 Vcc (ENABLE)
J2-8 Vcc (RESET)
J2-9 See text (VCC)

If you installed the optional 5 VDC power supply shown in Bill's schematic, do not connect an external 5 VDC source to pin J2-9; that pin will already have 5 VDC on it. Instead, hook J2-7 and J2-8 directly to J2-9, to tie those signals high.

Turn on the driver electronics, then adjust R4 for a reading of 200 mA at pin 15 of the L297. Remove power, install a stepper motor on the connector J3, then turn on power to the motor board and to your 555 circuit. You should see the motor spinning smoothly. If you used a trimpot in your 555 circuit, you should be able to alter the pulse rate and see a corresponding change in the motor's step rate.

Now, Some Software

Time to add some software, so you can put some intelligence behind your motors. I used one of Marvin Green's BOTBoards, complete with a 68hc811e2 MCU, to act as my robot's brain. I wrote all of the stepper driver code in SBasic, and the complete stepper library only takes up about 512 bytes of EEPROM.

The heart of the stepper code is the two interrupt service routines (ISRs) used to generate step pulses on output pins OC4 and OC5. I will describe how OC4 works; OC5 behaves identically (Figure 4).

This simple routine begins by clearing the RMASK bit in the TFLG1 register. Clearing this bit

```
interrupt    $ffe2                        ' right motor ISR
rtocisr:
pokeb    TFLG1, RMASK                     ' clear the interrupt flag
if rscnt <> 0                             ' if something to do...
      poke    RTOC, (peek(RTOC) + rsdly)        ' set timer for next pulse
      if rscnt <> $ffff        ' if not running forever...
            rscnt = rscnt - 1 ' count this step
      endif
      pokeb    PORTB, (peekb(PORTB) or RCLK1)        ' bring clk high
      pokeb    PORTB, (peekb(PORTB) and RCLK0)       ' now bring it low
endif
end
```

Figure 4. Listing of ISR for stepper motor control.

rearms the TOC4 timer, acknowledging the interrupt that got us here in the first place. If you don't clear this bit, the interrupt will re-trigger as soon as you try to leave the routine, putting you in an infinite loop.

The code next checks the variable RSCNT, which holds the right-motor step count. This variable always holds the number of steps remaining before the current motor motion completes. If RSCNT does not already contain zero, then the code assumes another step is required. It prepares the step by first updating the timeout value in RTOC, which holds the timeout count for the next pulse. The code does this based on the value already in RTOC, which is just an equate for the 68hc11 TOC4 register. That register already holds the timer count that got us into this routine originally, so adding a delay to that value and storing the sum back into RTOC makes sure we get another interrupt at the right time. The delay we add to RTOC has already been stored in variable RSDLY, which stands for right-motor step delay.

The code then tests to see if RSCNT holds the value $ffff. This is a special value that means "move forever." If RSCNT holds $ffff, the ISR does not decrement RSCNT. Thus, RSCNT stays non-zero and unchanging, so the right motor will spin continu-

ously. If, however, RSCNT does not hold $ffff, the ISR decrements it. This insures that eventually RSCNT hits zero and the motor pulses stop.

Finally, the time comes to provide the motor pulse. The code does this by toggling a selected bit in output port B. The constant RCLK1 defines an OR-mask for setting this bit to logic 1 and RCLK0 defines an AND-mask for clearing this bit to logic 0. The two **POKEB** statements each read the current value of port B, modify the proper bit, then write the new byte back to port B.

When control exits this routine via the **END** statement at the end of the ISR, the 68hc11 will execute an RTI instruction, which returns from the interrupt back to the mainline code.

This ISR forms the bulk of the stepper motor code. All the other code in the stepper library exists only to support this routine in some way. For example, the routine InitSteppers sets up the TOC registers for the right and left ISRs, and initializes the variables used by these routines. Similarly, StartSteppers activates the motor interrupts so the first pulse can happen.

I've tried to provide good comments to all of this code, to make it easy to follow. This code should serve as a good starting point for more sophisticated routines, and is worth study. I'll make a copy of this code available on my web site.

Final Thoughts

I haven't gone into much detail on my stepper robot platform, since I wanted instead to focus on the electronics and software. I used a pair of surplus Vexta steppers, rated at 0.8 amps at 6 VDC, as my drive motors. You can find suitable steppers in just about any mail-order surplus catalog.

I powered the robot using a single 12 VDC, 1.3 AHr gel-cell battery for the motors and a set of four AA Renewal cells to run the 68hc11 MCU board. I used a separate MCU supply to keep as much stepper motor electrical noise out of the MCU's circuitry

as possible. I also hooked up a charging jack on the gel-cell, so I can recharge the gel-cell in place, without having to pull wires and mess with connectors.

I took the easiest possible way out for mounting these motors onto my Sintra plastic base. I simply stuck them in place using double-sided foam tape. The tape has plenty of grip; in fact, I'm reluctant to change the arrangement, since it's such a pain pulling the motors off the base.

As for mounting the wheels onto the motor, I used the wheel mounts described in last month's article "Design of a Simple Line-following Array." Dan Mauch, one of the club's machinists, helped me with the design, and it really simplifies wheel mounts. Take a moment to leaf back through the last article for details.

A First Look
at the 68hc12

I love doing projects like the one in this column from the September 1996 issue of *Nuts & Volts*. Sitting around the robot lab with your buddies, hacking a chip that almost no one has ever seen before; the hobby doesn't get much better than this. Kevin Ross, Marvin Green, and I were some of the first people outside of Motorola to see the newly developed 68hc812a4, Motorola's first offering in the 68hc12 family of microcontrollers.

By pooling our efforts, and with help from Motorola's Jim Sibigtroth, we were able to put together the whole package in just a few days. We had the assembler, we had the prototype PCB, we had the interface electronics, and we had the PC-based development environment. The project was an intense hoot from the beginning, and our work (play?) opened the door to a high-powered MCU for use in hobby robotics.

I have since ported both tiny4th and SBasic to the 68hc12 family, and these chips continue to be strong robot brains. Though there will always be a place for the older 68hc11 in hobby robotics, I'm sure that future development will shift to the faster, more powerful 68hc12 devices. More and more companies are offering development boards and tools for these chips, and I suggest that you check them out when beginning your next design.

In mid-June 1996, Motorola announced its latest family of microcontrollers (MCUs), built around the newly-designed CPU12 core. The company targeted the first chip in this group, dubbed the 68hc812a4, for sampling by early July. Billed as a migration path from the 68hc11 to the 68hc16, the 68hc12 is packed with features ideal for robotics. Since most of the Seattle Robotics Society (SRS) members use 68hc11's in their robots, there was immediate interest in the new chip.

This month, I'll describe the journey that Kevin Ross, Marvin Green, and I took in developing a low-cost hacker tool set for the 68hc12. I'll also introduce you to several valuable sources of info on this chip, and give you a brief outline of the power this new device can provide in your robot designs.

A Look at the 68hc12

I don't want to spend a lot of space on describing the 68hc12; for full details, contact your local Motorola distributor and ask for a copy of the *MC68HC812A4 Technical Summary* (MC68HC812A4TS/D). While you're at it, get a copy of the *CPU12 Reference Manual* (CPU12RM/AD), which serves as a programmer's reference manual for using the instruction set.

If you can't wait, or don't have access to a Motorola distributor, you can download .pdf files of both documents from links reachable from Motorola's FREEWARE web site. Just browse through the 68hc12 links and you'll get there. Please note that these are very large files; the file for the *Technical Summary* holds 180 pages of text and drawings, and is an exact duplicate of the printed document distributed by Motorola.

If you have an Acrobat reader (also available from the Motorola web site), you can actually view this document on your screen before printing it. This makes a very handy reference source, though I had some problems with my version. It acted as if it was using an imported font, rather than a font already on my machine, and the characters within the font got mixed up. This resulted in plausible but wrong numbers for items such as pin numbers or timing info. The printed document, however, came out perfect.

I applaud Motorola for making these documents available in this format, especially as they followed the formal announcement of the chip by only a day or so. The documents contain enough information to start designing a 68hc12 system, even though at the time the real chips weren't available and the printed docs weren't shipping.

The first difference you notice between the 68hc812a4 and its older cousin, the 68hc11, is speed. Not only does this chip run on a crystal of up to 16 MHz, its internal processor clock runs at half the crystal frequency, rather than using a quadrature system like the 68hc11. So right off, you're looking at a 4x speed boost by going to the newer device.

The 'a4 carries much more on-chip memory than the 68hc11 variants, as well. This first chip contains 1K of static RAM and 4K of EEPROM, plenty of room for fairly strong projects. It also contains an expanded set of 68hc11 peripherals, so you veteran 'hc11 users will feel right at home moving up.

For example, the 'a4 carries two asynchronous serial ports (SCI) and a synchronous serial port (SPI). Both have been enhanced over the same peripherals found in the older 'hc11. The 68hc12 SPI contains additional logic supporting a bidirectional serial in-

terface. This means that a 68hc12 can send and receive data using only a clock line and a single data line, rather than the two data lines previously required. Note, however, that it appears all SPI transfers are still forced to use 8-bit exchanges.

Motorola also enhanced the two SCI channels over that found on the 68hc11. These SCI peripherals can perform parity calculations in hardware, and use separate modulus counters for baud rate generation. The latter gives you much more flexibility in defining baud rates. Oddly enough, given a crystal of 16 MHz, the fastest baud rate this chip can support is 38.4 Kbaud.

The 68hc11's timer/counter system has always been one of its strongest points. Long-time readers of this column know well how to use it for generating pulse-width modulation (PWM) signals with zero software overhead. But the basic 68hc11 only supports this feature on its four output-compare (OC) channels, while the remaining three channels can only handle input compare (IC) functions.

The 68hc812a4 has eight timer/counter channels, any of which can be configured for IC or OC operation. You can still slave any of these OC channels to timer 7, which acts as a master timer channel. This would let you run seven hobby servos with no software overhead.

In the timer/counter area, Motorola has really beefed up the versatility of the pulse accumulator. It now uses a 16-bit counter and you can insert a /256 or a /65536 prescaler if desired. On the downside, the 'a4 still only carries one pulse accumulator.

The 'a4 has an improved real-time interrupt (RTI) system as well. Many of my projects use the 68hc11's RTI system to control down-counting timers, for timing motor motions or actuator delays. But using the typical 8 MHz crystal on a 68hc11, the fastest RTI signal I can get is 4.1 msecs. This is fine for most uses, but sometimes I need to divide a second into finer slices. The 'a4 provides seven different RTI rates, and using a 16 MHz crystal can get the RTI time slice down to 1.024 msecs.

The 'a4 also has an 8-channel A/D system much like that of the 68hc11. In the 'a4, these A/D channels provide 8-bit resolution, but the I/O registers for providing the A/D data all occupy two bytes. I asked a Motorola representative if this meant that future versions of the CPU12 family would offer 10-bit or higher resolution on the A/D ports, but he replied "No comment." I noted, however, the faintest of smiles on his face, so I'm looking for more resolution down the road.

Motorola did address one complaint I've long had with the 68hc11 A/D system. The 'a4 lets you read all eight A/D channels with one command, and to pull the results from eight separate registers. This eliminates the convoluted folding of eight A/D channels into four result registers, as used in the 68hc11.

The CPU12 family also offers features not seen on the generic 68hc11. For example, the 'a4 MCU has up to 24 I/O lines with key wakeup interrupt ability. Key wakeup means that the MCU can generate an interrupt if the state changes on one of these special input lines. You can use this wakeup interrupt feature to make a keypad that doesn't need to scan the matrix continuously; it just wakes up whenever the user presses a key.

The 'a4 also supports a non-multiplexed address and data bus, for memory expansion. The CPU12 architecture permits addressing of up to 5 Mbytes of external memory, using a windowing system. This is not as clean a layout as that used by the 68hc16, since some of the larger address space gets folded into windows in the chip's 64K linear address space. Still, it eliminates all of the hardware gyrations used by most 68hc11 large-memory designs.

Along with the increased address space, the 'a4 has seven programmable chip select lines. These I/O lines provide chip select signals for whatever address space you declare, simplifying the addition of external memory or I/O devices.

Motorola has abandoned the 68hc11's bootstrap loader in favor of a background debug module (BDM) similar to that used on the 68hc16 and 68332 chips. The BDM provides the main entry way to the 'a4, providing support for downloading programs, dumping memory and registers, and doing single-step, trace and breakpoint operations. The CPU12 BDM system is very different from that used in the other two chips, and it presented a substantial obstacle in getting our first project running; more on this later.

Hidden inside the CPU12 are even more surprises. Motorola has made substantial improvements to the programmer's model, adding some very powerful indexing modes to not only the X and Y registers, but to the S (stack) and PC (program counter) registers as well. These enhancements should help quiet the grumblings from those programmers who moved to the 68hc11 from the 6809 and felt the pain of the 68hc11's feeble indexing modes. Of course, software types are rarely satsified, and many will ask why Motorola didn't throw in a second stack register or another accumulator. Still, the CPU12 represents a big upgrade to the 68hc11 family, and I'm looking forward to working with the new model.

Also, Motorola added support for fuzzy logic and digital signal processing (DSP) on this chip. Several special CPU12 opcodes permit you to set up a fuzzy logic engine in only 50 or so instructions. The docs go into considerable detail on how to set up and use these instructions, and Motorola has provided application notes that give much more information. The DSP instructions, while lacking the strength of a full-power DSP controller, will get you started on DSP programming and will solve some problems previously beyond the reach of the 68hc11.

In all, this first member of the CPU12 family is a large win for those who have outgrown the 68hc11. Readers of this column should prepare for a shift toward 68hc12 projects, though I intend to keep supporting my 68hc11 tools such as SBasic.

The First 68hc12 Project

Motorola's announcement fired up a lot of people in the SRS and on the Internet. Marvin, Kevin, and I immediately grabbed the available docs and started checking out what was obviously the future in small robots. Each of us focused on a different aspect of this new chip.

Marvin Green, developer of the 68hc11 BOTBoard and BOTBoard-2, used his documentation to start the design of a single-chip PCB, sort of a BOTBoard for the 68hc12. Kevin Ross, with his expertise in PC-based software and his considerable Windows experience, took up the challenge of a PC-based front end for the BDM. And I zeroed in on PC-based tools, such as an assembler, for writing new 68hc12 programs once Marvin and Kevin finished their tasks.

We quickly realized that we just didn't have enough information to complete our individual tasks. Using the Internet, I was able to locate a Motorola engineer, Jim Sibigtroth, who proved to be a valuable 68hc12 resource. Jim is a senior member of the technical staff at Motorola's Advanced Microcontroller division in Austin, Texas, and he recently published a lengthy article on Fuzzy Logic for the 29 July 1996 issue of *Electronic Engineering Times*. When Jim read my plea for assistance on the 68hc12 development project we had begun, he offered his help.

Motorola, via Jim Sibigtroth, donated the source code for the in-house absolute assembler used to develop test code for the first 68hc12 chips. The code was originally written for use on a Unix system, but I was quickly able to port it over to the DOS environment. I cleaned up a few bugs, removed an Easter egg or two, and prettied up the output it generates. Thanks to Motorola's support, I now have a working absolute 68hc12 assembler.

You can use this tool, known as as12, to create Motorola .s19 object files from 68hc12 assembler source files. Because this is an absolute assembler, you don't get some of the fancy features of the expensive relocating assemblers, such as the ability to make linkable modules. But as12 does offer many strong features lacking on the older Motorola shareware assemblers.

For example, as12 supports the #define, #include, and #ifdef constructs familiar to C programmers. as12 also provides an extensive macro capability, useful for defining a set of commonly-used functions.

I didn't lack for sample assembler language source code to use in testing as12. The CPU12 chips are source-compatible with 68hc11 assembly. Note the term "source-compatible," not "object-compatible." You can assemble a 68hc11 source file and get code that will run on a CPU12 device, but you cannot take a 68hc11 executable file and expect it to run on a 68hc12 chip.

This happens because Motorola kept the assembler mnemonics the same for both chips, but revamped much of the underlying opcodes. For example, the 68hc11 instructions dealing with the Y-register all begin with a $18 pre-byte. Thus, a LDY #$a5 assembles into the bytes $18 $CE $00 $a5, where the instruction LDX #$a5 becomes $CE $00 $a5. This means your 68hc11 programs pay a one-byte penalty for each use of the Y-register.

On the 68hc12, however, Motorola rearranged the opcodes so the penalty disappears. Instructions using the X- and Y-registers assemble to the same number of bytes. One obvious fallout of this rearrangment means that 68hc11 object files won't run on the 68hc12.

You can check out the new as12 assembler by accessing my web site. Note that this is a .zip file for running on a PC. Those of you with Macs, take heart. Porting of this assembler to the Macintosh environment is underway, and a Mac version of as12 should be ready by the time you read this.

My next task was to get my hands on some actual 68hc12 chips. After a few telephone calls to Motorola, I reached their samples desk. I signed up for two samples of the 'a4, which arrived by mail a few days later. Interestingly, these chips carried the part number PC68HC812A4. The PC prefix marks these as prototype devices, with designs even more unstable than the XC, or experimental, chips. This means these chips are using silicon dies with a design that may well be unreliable or poorly documented. As it turns out, the PC chips I received had some serious design flaws in them, which hampered our efforts. But we only discovered that later.

Now, however, I was the proud owner of two 68hc12 chips, each housed in a 112-pin thin quad flatpack (TQFP) package. This package design uses leads only 0.5 mm wide on 0.65 mm centers. Marvin's

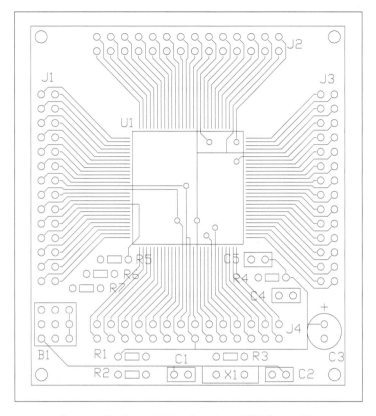

Figure 1. Top layer of Marvin's 68hc12 PCB (2:1 scale).

efforts to design his board were slowed because we just couldn't get definitive information on lead arrangement and spacing. Eventually, however, he received enough information to complete the layout for holding the chip (Figure 1).

Marvin also added traces for the necessary resistors, capacitors, and other passives to use the 'a4 in single-chip mode. He included the same 3-pin power connector and 4-pin serial connector on this board as used on his BOTBoard products. He even added four layouts for 28-pin dual-row headers, used to bring each pin of the 'a4 out to the board edge for probing or connection.

This last feature really makes his board design flexible. On one of his prototype boards, I installed four 28-pin dual-row male IDC headers. I then built up a Radio Shack experimenter board with matching

female connectors. Now I can just plug my 68hc12 PCB into a matching experimenter board, moving the MCU from board to board as needed. Or, you could solder the female connectors into the 68hc12 board. This would let you plug wires into the connector pins and out to a prototyping board, so you could quickly build up test circuits around a 68hc12.

Marvin ran his ideas past Kevin and me many times, and between the three of us we eventually hit a design that Marvin liked. He then packaged up his design and shipped it to Alberta Printed Circuits (APC) for etching. APC, as usual, did a first-rate job and had the boards back to Marvin in a couple of days.

While Marvin was getting his PCB design finalized, Kevin was up to his armpits in the BDM interface. His task would have been far simpler if Motorola had stuck with the BDM design used in the 68hc16 and 68332 devices. Those chips use a 3-wire interface that contains a clock line, a data-in line, and a data-out line. The clock line, controlled by the host, means that the PC talking to the MCU dictates the rate at which the data transfer occurs. Thus, you have no problems with synchronizing the timing between the host and target systems.

For some reason, Motorola dumped this design on the 68hc812a4, opting instead for a 1-wire interface. Here, the target and host exchange data over a single wire. The timing for the exchange is dictated by the target MCU, in that the host must use a data transfer rate that is based on the MCU's crystal. Unfortunately, the host does not have visibility as to the transfer rate, since there is no clocking signal in the BDM interface.

After much discussion among us, Kevin chose to build a BDM interface around a PIC 16c84. He needed to use a 20 MHz PIC device, to keep up with a 'a4 using a 16 MHz crystal. We had built our first boards using 8 MHz crystals, however, so Kevin could get by with a 10 MHz PIC, available from Digi-Key for less than $10. Refer to the accompanying schematic of Kevin's PIC-based BDM interface (Figure 2). Note that the schematic calls out a PIC 16c61, which is basically a one-time programmable version of the 16c84.

The PIC 16c84 is highly cool, since it comes with 1K of EEPROM to hold your code. It doesn't have a lot of on-chip peripherals such as serial ports, but the bloody thing runs so fast that you can bit-bang serial data with little trouble. Plus, Microchip's web site is crammed with technical info, application notes, and source code for plenty of finished projects and subsystems. You can also find the latest versions of Microchip tools, such as the MPASM PIC assembler.

Kevin designed his BDM interface to run from the PC's serial port. The first cut at his PIC electronics used 5 VDC and a 4-wire cable to the target 68hc12 board. This cable supplies connections for ground, +5 VDC, BDM data, and a reset line. The latter wire is needed because Kevin's PIC uses it to reset the target board and control what mode the target enters after reset.

As you can see from the accompanying schematic, there just isn't very much to this BDM interface board. Kevin added some LEDs for status display, a MAX232 to provide the needed RS-232 level shifting, and a DIP switch to select one of several operating modes on powerup. For example, one switch tells the PIC that the target system runs with either an 8 or 16 MHz crystal.

Most of the hard work in Kevin's design took place in the PC and PIC software. Kevin created a very nice Windows-based program that acts somewhat like a modem program, exchanging data and commands serially with the PIC BDM interface. As the PIC receives each command, it translates it into the proper BDM sequence and sends it to the target. If the PIC needs to handle a response from the target, it collects the data and translates it for passing back to the PC host.

Putting It All Together

Our mutual efforts converged one Friday afternoon at my home in Bothell, Washington. Marvin had driven up from Portland with several copies of his first PCB, and Kevin had brought much of his software design tools over from nearby Duvall. With our gear spread out on a table in my living room, we began the task of integrating all our efforts.

While Kevin worked on completing the PIC software, Marvin started building up one of his 68hc12 boards and I began breadboarding a PIC BDM interface. It took a couple of hours, but I finally had the electronics ready to fire up. Marvin, in the meantime, had all of his 68hc12 board soldered except for the one device none of us wanted to touch, the MCU itself. The fine pitch on its leads just looked like too much for us to handle, given our relatively crude tool set.

We solved this problem with typical software ingenuity; we went out for Chinese food. By the time we returned, we had a plan of action. And since these were my chips, I was elected to start the soldering operation.

I began by spreading a very thin film of Kester paste flux on the traces destined to hold the 68hc12

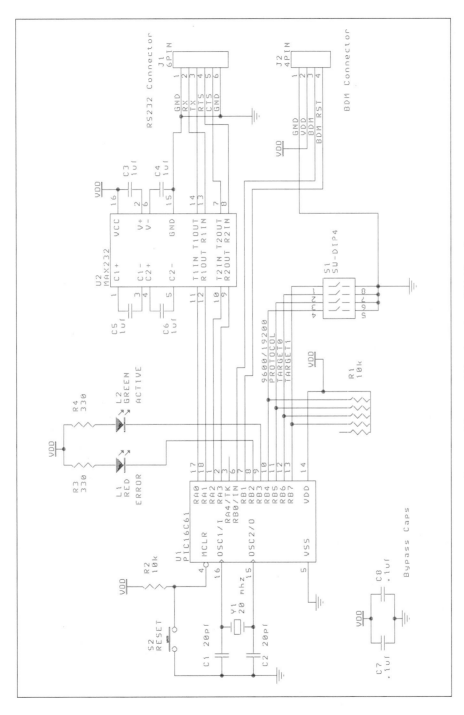

Figure 2. Schematic of Kevin's 68hc12 BDM board.

chip. I intentionally used the thinnest possible film, since I only wanted enough paste to stick the chip into place. I then positioned the chip onto the board, taking care to align the dot on the chip's case with pin 1 on Marvin's board.

If you build such a board of your own, pay close attention to this alignment procedure. The text on the chip's surface, giving the Motorola logo and part number, gets printed in a random orientation, and you cannot use the printing for chip alignment. Always use the alignment dot when placing the chip on the board.

Next, I worked very carefully to shift the chip on the trace layout, aligning the chip's leads with the matching traces. This took a little time, as the flux only keeps the chip from sliding but lets you push it around fairly easily. Once I had the board resting on my work surface and I was satisfied with the alignment, I was ready to solder the leads.

I cleaned the tip of my grounded soldering iron, then added a thin film of solder to the tip. Be sure to use a grounded-tip iron to minimize risk of static damage while soldering the chip in place. Then, using only the soldering iron and no other solder, I gently touched the trace at one of the chip's corners. The solder plating on the trace melted and fused with the film of solder on the iron's tip, then wicked underneath the pin. When it cooled, the lead was soldered. I then moved to the diagonally opposite pin and repeated the operation. This left the chip tacked in place at two corners, essentially immobilizing it.

Take care during these steps not to touch the chip or its leads. If you do, you risk misaligning the chip and soldering it in the wrong place at the same time. Once soldered by two or more adjacent pins, you will have a tough time removing the chip and repositioning it.

With the chip locked in place, I started at a third corner and began soldering the remaining pins. I still soldered only to the traces, letting the melted solder wick underneath the pins to complete the connection. We took turns on this phase and when done, I went over all connections with a magnifying glass to check for opens or shorts. Satisfied that we had

actually soldered this baby in place accurately, we were ready for some power.

We connected Kevin's BDM electronics to the PC and to the 68hc12, then applied power in the form of four Renewal C cells. Using a logic probe, we were able to verify that the MCU's clock was operating properly. This sparked a lot of grinning and congratulations, since it meant that Marvin had designed the correct trace layout for the many power and ground pins, as well as the passives needed to get the chip running.

When Kevin entered a command on his PC screen and actually got a response from the BDM interface, there was more stomping around and high-fiving. We were having a great time celebrating these small steps, because it had taken so much hard work just to get this far.

However, we eventually sobered up when we realized that the PIC BDM interface was not returning the proper responses to our commands. Evidently, the MCU was either not running or the PIC interface had some kind of timing problems. We spent an hour or more trying different fixes, but we finally called it quits around 2:00 AM that Saturday morning.

During the next week, Kevin worked with Keith Payea, another SRS stalwart, to verify the timing on his PIC interface. He cleaned up a few small problems, but still could not make the MCU respond properly. In desperation, I once again turned to the Internet, posting a request for help to the 68hc11 listserver. And once again, Jim Sibigtroth came through. Thanks to his intimate knowledge of this chip's inner workings, he was able to solve our problem. It turns out that "first silicon," that is the experimental devices we were using, had a flaw in the BDM logic. This flaw inserted an additional /2 in the BDM timing chain, making the chip's interface run at half the advertised speed.

Kevin added code to his PIC firmware to compensate for this timing flaw and the MCU "woke up" and began talking to the PIC via the BDM interface! Immediately, Kevin was able to write data to RAM, to download simple programs, and to single-step through those programs. We had done it! Using nothing but cheap hacker tools and the club's expertise, we had put together a 68hc12 development system.

What's Next?

This kind of project just grows without bounds. Marvin is already making plans for selling his 68hc12 PCB; contact him directly for pricing and availability. Kevin has also talked about selling PIC chips already programmed with the firmware for his BDM interface. As I've already mentioned, you can get a copy of the current version of the as12 assembler from my web site.

In the meantime, I've started work on converting tiny4th to run on the 68hc12 MCU; the code should be finished by the time you see this. Check on my web site for a copy of the new compiler. Porting SBasic to the 68hc12 will take considerably more time, but I will work on it whenever I can.

And then there are all the robots to build using this new chip. The small physical size and huge I/O capabilities mean that you can pack a lot of robot smarts into a very small package. Previous designs that called for one or two added chips can now be done using just this one device. And Marvin's 68hc12 BOTBoard is only slightly larger than its 68hc11 cousin, so you won't pay a large price in real estate for using the newer device.

If you think that 4K of EEPROM might limit you in your designs, then stick around until the fourth quarter of 1996, when Motorola is scheduled to ship first silicon on the '-b32 version of the 68hc12. This device will contain 32K of flash EPROM. As I understand, it is targeted to come out in an 80-pin quad-flat pack, which will make it even smaller than most 68hc11s. The smaller lead count means that Motorola will sacrifice some of the I/O; check the Motorola web sites for details on this new chip.

Check Out This New 68hc12

My first look at Motorola's new 68hc12 offering, the '912b32, really got me pumped. Here was a single chip containing a fast processor, 32K bytes of program space, 1K bytes of data space, and lots of I/O. I couldn't wait to start hacking on it, and this article from the March 1997 issue of *Nuts & Volts* describes how I modified SBasic so I could use it to generate 68hc12 code.

Once I had upgraded SB, I used it to write two small programs for the 'b32. The first program, speed12.bas, is a simple empty loop that blinks an LED periodically. I wrote the same program for the 68hc11. Comparing the performance of these two systems was a real eye-opener; the 68hc12 runs rings around the 'hc11.

I also developed a program for using the sophisticated pulse-width modulation (PWM) subsystem built into the '912b32 device. This four-channel I/O subsystem handles all aspects of a PWM generation with no CPU overhead; you essentially get free motor control signals. This means you can do your motor control software without concern about CPU overhead impacting any other part of your design.

I predicted then that the '912b32 would make a strong impact on hobby robotics. By now, a number of companies have offered 'b32 development boards, and the toolset for working with this chip gets

stronger every day. If you're looking for a small but powerful robot brain, check out this Motorola chip.

Motorola's latest variant of the 68hc12 hit my workbench just before Christmas, and I've had several weeks to play with it. The 68hc912b32, I can safely predict, will cause a huge sensation in the amateur robotics community.

This 80-pin TQFP chip uses a 16 MHz crystal to generate an internal system clock of 8 MHz, giving you an immediate 4x speed boost over the older 68hc11. Add the improved addressing modes and spiffed-up instructions, and your 68hc11 code will run up to 10 times faster on a 'b32. Even better, the 68hc12 opcodes are source-compatible to the 68hc11, so you won't have to edit the opcodes in your assembler source files.

Three members of the Seattle Robotics Society were privileged to test-drive both the 'b32 chip and a new evaluation board (EVB), courtesy of Motorola. Kevin Ross focused his efforts on connecting his newly designed Background Debug Mode (BDM) board to the 'b32. Marvin Green used his 'b32 chip as the core for his newest BOTBoard-style prototyping tool, a small single-board computer (SBC) for the 'b32. And I spent my quality time porting SBasic to the 68hc12 so I could write actual 68hc12 software and test my programs on a working EVB.

Inside the 'b32

Like its older cousin, the 'a4, discussed in this column some months ago, the 'b32 sports many of the newly designed I/O subsystems common to the CPU12 chipset (see "A First Look at the 68hc12")-. Besides the CPU12 core, with its enhanced addressing modes, fuzzy logic instructions, and 20-bit arithmetic unit, the chip handles 8- or 16-bit expansion busses, has eight channels of fast 8-bit A/D, eight 16-bit timer/counters, a 16-bit pulse accumulator, hardware-generated pulse-width modulation (PWM), an SCI for asynchronous communication, an SPI for synchronous I/O, and CPU12-style BDM.

But the real killer feature of this chip, as far as amateur robotics is concerned, lies with the on-board memory resources. The "9" in the chip's part number (68hc912b32) indicates that the device carries flash EPROM. In fact, it holds 32K bytes of flash EPROM, plus 1K of static RAM and 768 bytes of the more common EEPROM. This means the chip can handle very large programs without having to use an expanded bus, with the attendant loss of valuable I/O lines (Figure 1).

Even cooler, the 32K flash EPROM, which sits at the top half of memory, comes preloaded with a 1K byte bootloader. This program lets you download S19 records via the SCI serial port at 9600 baud, using a conventional communications program such as ProComm. Thus, getting your program into the 'b32 takes little more than resetting the board, then doing an ASCII file download from ProComm. As simple as pcbug11 is for the 68hc11, the 'b32 bootloader is simpler still. And the 32K flash is protected with a lockout, so your development code can't trash memory if it "runs away" during test.

The bootloader itself sits in the very top 1K of memory and is protected with a second level of lockout, so you cannot accidentally overwrite it with a file transfer or code runaway. If you really do want to overwrite the original bootloader with your own code, you can do so via the BDM. Likewise, you can use the BDM port to restore the bootloader if you so choose.

But those are possibly the only times you would really need to use the BDM. In general, you can do everything you need to do with the 'b32 using nothing more than a serial connection through the SCI. This is a huge boon to experimenters, opening up

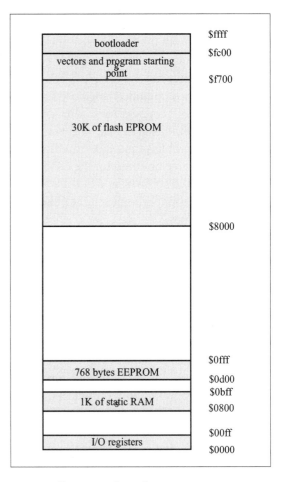

Figure 1. 68hc912b32 memory map.

access to this way cool chip without the expense and fuss of an additional BDM board.

Motorola also sent me a prototype of their new 'b32 EVB, part number M68EVB912B32. This board measures 3.5 by 5 inches and contains a good-sized prototyping area, a MAX562 RS-232 level-shifter, and the 'b32 chip proper. The rest of the board space holds the needed passives and pullups to drive a 'b32 in single-chip mode, plus an assortment of jumpers. It also has a fancy lever-type connector for +5 VDC, a 2-pin header for the +12 VDC needed to program the flash EPROM, and a 9-pin RS-232 connector for hooking to your host computer.

Sources inside Motorola tell me that the board should be ready for release by the time you read this, with a projected price tag of less than $100. I consider this an excellent price for the quality design I've got sitting on my workbench, and wouldn't hesitate to buy one. Contact your local Motorola distributor for pricing and availability.

SBasic and the 'b32

I wanted to jump in immediately and begin writing code for this chip, but the only software tool I had was my as12 assembler. Granted, this freeware assembler, available from my web site, is good enough for simple programs. But I didn't want to get bogged down in assembler programming, I wanted to play. And I play best in SBasic.

So I spent a precious week or so editing SBasic so it would generate output code for either the 68hc11 or the 68hc12. I didn't go so far as to change the code generator so it would use the nifty new opcodes. I settled instead for rewriting the generator so it would output 68hc11 assembler source compatible with my as12 assembler. This takes advantage of the source-level compatibility between the two devices.

You can get a copy of my latest SBasic compiler, version 2.0, from my web site. You control which chip the compiler uses as a target by means of a command line option. To compile for a 68hc12, invoke SBasic with a command such as:

sbasic foo /m6812

This compiles the file foo.bas and generates code for the 68hc12. If you leave off the /**m** option, you get code for the 68hc11 by default.

With that effort behind me, I was ready for my first test. Since the first test I do with any new chip is a standard, blink-the-LED speed test, I hooked a 470-ohm resistor to the anode of an LED, wired the LED's cathode to ground, and ran a wire from the other resistor lead to I/O pin PB0 on the 'b32 EVB. This gives me a simple LED indicator that lights when I bring PB0 high and goes dark when I bring PB0 low.

Next, I had to dig through the preliminary docs on the 'b32, supplied by Motorola with my EVB. The 68hc12 differs greatly from the 68hc11 with regard to programming the I/O registers, and I spent quite a while poring through the many pages.

The 68hc11 only has 64 I/O registers, many either reserved or not used. I know this seems like a lot of I/O registers to twiddle if you want the chip to perform, but for most programs you will use a very small subset of this full collection. Contrast this with the 256 I/O registers used by the 68hc12, and you get a crude measure of the increased sophistication in this chip.

For example, port B on the 68hc11 is an output-only port. To affect the outside world using port B, you simply write a value to register PORTB, and the I/O lines change accordingly. On the 'b32, however, port B is bidirectional, so you must first modify register DDRB, the port B data-direction register, to assign each pin as either input or output. If you intend to use some of these lines for input, you next want to alter the pullup control register (PUCR) to enable or disable the port B internal pullups. Likewise, you need to select either reduced or full drive level for port B output pins, using the RDRIV register. Finally, you can write a value to register PORTB to change the outside world.

Having gathered all this information together, I was ready to write my first 68hc12 SBasic program. See the accompanying listing of speed12.bas (Figure 2).

```
include    "regs12.lib"

declare    n

org    $8000

main:
pokeb  copctl, $08            ' disable COP resets and clock
poke   sc0bdh, 52            ' 9600 baud
pokeb  sc0cr1, 0             ' 8 bits, no parity, 1 stop
pokeb  sc0cr2, $0c           ' enable xmtr and rcvr
print
print  "speed12"

pokeb  ddrb, 1               ' make PB0 an output
pokeb  pucr, 2               ' enable port B pullups
do
       pokeb portb, peekb(portb) xor 1
       for n=0 to* $fffe
       next
loop

end
```

Figure 2. Listing of speed12.bas.

This program, though small, shows many aspects of 'b32 software. I compiled it with the command line:

sbasic speed12.bas /s0bff /cf700 /v0800 /m6812 /i >speed12.asm

which sets the stack at $0bff, the variables at $0800, and the code at $f700. The code setting, in particular, is important. The bootloader in my 'b32 EVB takes control immediately after reset and tests the state of two I/O lines. These lines, tied to jumpers on the EVB, determine the bootloader's function. When the bootloader detects that the EVB is strapped to run a user's program, it automatically jumps to $f700 after reset. Thus, you have to set the start of your code at $f700 if you want it to run.

This command line also shows the new /i option. Normally, SBasic automatically adds the reset vector (and any other interrupt vectors you select) to the 68hc11's vector area at $ffc0 to $ffff. Since that area of memory is locked out in the 'b32 and cannot be modified with a simple program download, I added this option to supress generation of interrupt vectors. This option comes in handy when I need to install vectors in another address area, as I will when working with the 'b32.

SBasic compiles all programs to begin execution at the address given by any /c option. If you choose, you can add **ORG** statements inside your SBasic program to move later code to a different address, as I've done in speed12. This combination of the **/cf700** option and the **ORG $8000** statement lets me put the start of my program where the bootloader wants it, yet store the bulk of the program down at the beginning of the 'b32's flash EPROM, where I want it.

Like all SBasic programs, my code actually starts at the required label MAIN. Here, I begin by disabling the COP watchdog. I could have chosen to service the COP repeatedly, preventing it from timing out and resetting my program, but I wanted this to be a simple test so I just disabled the COP.

Next, I modify the three main registers associated with the 'b32's SCI port. My initializations set the EVB for 9600 baud, 8N1, and enable the SCI. Then I print a greeting to the serial port so I know the program at least got that far. I follow this by setting up PB0 as an output pin and enabling full output drive.

Now I'm ready to begin toggling PB0 so my LED changes state. I used an infinite **DO-LOOP** structure around a **FOR-NEXT** loop that counts from 0 to $fffe, unsigned. Thus, the LED will change state each time the loop counts past 65534. It's important that my **FOR-NEXT** loop end at $fffe and not $ffff. SBasic **FOR-NEXT** loops terminate when the index variable **exceeds** the limit value, not when it reaches it. Using a limit value of $ffff would actually create an infinite loop, since the index can never exceed $ffff.

After compiling the program, I hooked my EVB to the PC's serial port, fired up ProComm, strapped the EVB for bootloader mode, and hit the reset switch. When I got the bootloader's prompt, I used the E command to erase the flash EPROM, then used the P command to prepare the EVB for program download. I next used ProComm's ASCII file transfer to send the file speed12.s19 to the EVB. After I got confirmation from the bootloader that the program was written properly, I restrapped the EVB to run my program, then hit the reset switch again; the LED began blinking.

I immediately noticed that the LED was blinking very quickly, changing state roughly seven times per second. I compiled essentially the same program for a 68hc11 and moved it into a BOTBoard holding a 68hc811e2 and 8 MHz crystal. The LED on this board changed state slightly faster than once per second. Not a bad speed boost!

Time for Some PWM

But robots need more than time-wasting loops, and I wanted to take a look at the 'b32's brand-new PWM subsystem. The next robot on my horizon will need some PWM support for motor control, and I was curious to see how the new chip handled this task. My first experimental code appears in the accompanying listing of pwm12.bas (Figure 3, at end of article).

This example is longer and more complex than speed12 above, but it is much more instructive. pwm12 sets up four independent PWM outputs, all running at the same frequency but with different duty cycles. Before I walk you through the setup, please note that the final waveforms are generated entirely in the CPU12's hardware subsystem. This complex waveform generation does not impact the CPU at all.

My program begins by declaring some needed variables, then using the **ORG** statement to move the code to $8000, as done above in speed12. Here, however, I added the second argument, **CODE**, to the end of the **ORG** statement. This option is new for version 2.0, and fills a subtle yet critical need when writing code for this early version of the 'b32.

The 'b32's bootloader overlays the entire interrupt vector table, including the reset vector. Therefore, if your program wants to use interrupts of some kind, as pwm12 does, you have to use the jump table provided in the bootloader. This jump table reroutes all interrupts except reset through a table of addresses starting at $f7a0. Each entry in this table is three bytes long, allowing your program to insert a JMP-extended instruction to the location of your interrupt service routine (ISR).

My program uses the real-time interrupt (RTI) to decrement a variable once per interrupt, making that variable a down-counting timer. But to pass control to my ISR, I have to set up a JMP to RTIISR where the interrupt is serviced. The block of code following the RTIISR label services that interrupt, and the label RTIISR gives SBasic a way to refer to the address of the ISR.

But back to the **CODE** option in the first **ORG** statement. This option tells SBasic to mark its internal code section as now starting at $8000, rather than the $f700 as originally called out in the command line option. Any subsequent references to the **CODE** section in my program will refer to addresses in the

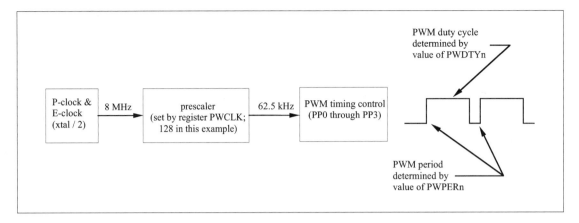

Figure 4. Timing diagram of PWM signal generation.

$8000 range. This becomes critically important later, as you'll see. Without this **CODE** option, SBasic would still use addresses in the $f700 range as the code section.

The RTI ISR code is fairly simple. It rearms the RTI, then tests the variable WAIT to see if it has reached zero; if so, the routine returns immediately. If WAIT has not yet reached zero, the RTI ISR code decrements WAIT and again tests for zero. If WAIT has now reached zero, the code calls the HEART-BEAT subroutine to provide a short pulse at PA0. Regardless of the value in WAIT, the RTI ISR code now returns control back to the main code.

The main code, starting at label MAIN, turns off the COP, sets up the RTI subsystem for an interrupt once every 1.024 msecs, then sets up line 0 of port A and the SCI. Next, the code calls HEART-BEAT to provide an initial output pulse and to make sure PA0 returns to a known state. It then issues a sign-on message, presets variable WAIT, and enables interrupts so the RTI can actually occur.

The next chunk of code initializes the PWM subsystem to provide pulses on all four PWM port output pins, labeled PP0 through PP3 (Figure 4). These four pins are controlled by two identical PWM subsystems; the first handles PP0 and PP1, while the second runs PP2 and PP3. You can set up PP0 and PP1 as two separate 8-bit PWM counters, or

concatenate them and treat them as a single 16-bit PWM counter. You have this same option with PP2 and PP3. My code writes a value of $3f to register PWCLK so I can use all four PWM outputs as separate 8-bit counters. This value also assigns a clock prescaler value of 128 to the two PWM subsystems. Since the PWM clock is derived from the CPU12 P-clock, running at half the crystal frequency, this yields a PWM basic clock rate of 8 MHz divided by 128, or 62.5 kHz.

Writing a value of $00 to register PWPOL selects the proper clock source, described above, to the two PWM subsystems. It also defines the PWM outputs as active high (resting low). My code then writes $02 to register PWCTL; this enables the pullups on all PWM output port pins.

Next, I have to set the actual PWM period and duty cycle. Since I'm using all PWM channels in 8-bit mode, the period of each PWM waveform is controlled by a separate 8-bit register. I'll describe the arrangement for PP0; the other three channels operate identically.

The clock for PP0 is fed by the 62.5 kHz clock stream selected by my previous write to PWCLK. In left-aligned mode, used here, the period of PP0's PWM cycle is always one more than the 8-bit value written to register PWPER0. My code writes a $ff to PWPER0, which means the PWM period for PP0 is

256 cycles of the 62.5 kHz clock stream, or 4.1 msecs; this works out to a 244 Hz cycle rate.

I then assign the duty-cycle of PP0 by writing another 8-bit value to PWDTY0. Since I had already assigned this channel as active high, this value determines how long the PWM output stays high during each PWM cycle. As with the PWM period, this value is measured in cycles of the 62.5 kHz clock stream. In my example, I used a count value of $40, which is 25% of a full-scale value of $ff, and yields a 25% duty-cycle. You can see that I also assigned duty-cycles of 50%, 75%, and nearly 100% to the other three PWM channels.

Then I enable the entire PWM subsystem by writing the proper bit pattern to register PWEN. This final write activates all four PWM channels so the desired waveforms appear on the assigned output pins.

After activating the PWM subsystems, my code enters an endless loop that simply tests the WAIT variable, updating it as necessary. The value written to WAIT determines how much time elapses between each heartbeat pulse; the value I chose (977) provides a one-second delay, given the RTI interrupt rate of 1.024 msecs.

The only task remaining must be handled by the SBasic compiler, and that involves setting up the jump vector so control will reach the RTI ISR whenever an RTI interrupt occurs. This task is accomplished with the block of code following the **ORG $f7e8** statement. After changing the ORG to $f7e8, where the bootloader expects to find my jump vector for an RTI interrupt, I use an ASM block to assemble a JMP instruction to label RTIISR.

Following the assignment of the JMP opcode, I must make sure I use an **ORG CODE** statement to switch SBasic back to the main code section. After SBasic finishes compiling my program, it will append any needed library files at the current location. If I don't change back to the code section, SBasic will end up adding the library code to the interrupt vector area, which is incorrect.

Finally, we get to the real reason for using that first ORG CODE statement. If I had not added the CODE option to that statement, SBasic would now revert to its original code section, which started at $f700 as called out in the command line. SBasic would thus add the 68hc12 library code somewhere in the $f700 area, potentially overwriting the jump vector table at $f7a0 and above. This in turn would cause a nasty crash when I tried to run the program. The bottom line is that you need to be aware of SBasic's behavior and of the 'b32's requirements whenever you design code for this new chip.

Having said all this, I'll point out that Motorola is redesigning the bootloader, and the final version of their firmware will likely behave somewhat differently. In particular, there is talk of going to a two-byte jump table, rather than a three-byte table. This will significantly change the way you set up an interrupt vector for the 'b32 in SBasic, making it more like the traditional 68hc11 technique of a single INTERRUPT statement with an address argument. Consult the Motorola docs for the bootloader that comes embedded in your chip for full details.

That's a Wrap

This completes my first look at what I'm sure will prove a very popular chip. The 68hc912b32 offers 68hc11 code compatibility in a small chip with plenty of program space. The I/O goodies, including the PWM subsystem described above, make it a natural for small robots. The embedded bootloader, which lets you download executable files with commonly available tools, will simplify your test and development cycles. And with SBasic's capability to generate 68hc12 assembler source files, you can quickly get your new programs up and running.

I'll keep you posted on further developments regarding the Motorola 'b32. Contact your local Motorola office, or a distributor such as Future/Active, for price and availability of both the 'b32 chip and its EVB. And be sure to check the Motorola web site now and then for new product information.

```
include    "regs12.lib"

const    ONE_SEC = 977

declare    n
declare    wait

'
'    Declare the location of the main code. Change the code
'    section also.
'

org    $8000 code

heartbeat:
pokeb porta, 1
pokeb porta, 0
return



'
'    Define the RTI interrupt service routine. Note that
'    the INTERRUPT statement does not have an address argument.
'    The actual vector will be set up in code later on.
'
interrupt
rtiisr:
pokeb rtiflg, $80                          ' rearm the RTI
if wait <> 0
      wait = wait - 1
      if wait = 0
            gosub    heartbeat
      endif
endif
end



'
'    The main program
'
main:
```

Figure 3. Listing of pwm12.bas.

```
pokeb copctl, $08              ' disable COP resets and clock
pokeb rtictl, $81              ' 1.024 ms, RTI enabled
pokeb rtiflg, $80              ' rearm RTI
pokeb ddra, 1                     ' make PA0 an output
pokeb pucr, 1                     ' enable port A pullups
poke  sc0bdh, 52               ' 9600 baud
pokeb sc0cr1, 0                ' 8 bits, no parity, 1 stop
pokeb sc0cr2, $0c              ' enable xmtr and rcvr
gosub   heartbeat              ' set up heartbeat line

print
print "pwm12"

wait = ONE_SEC
interrupts on

pokeb pwclk, %00111111         ' A/128, B/128
pokeb pwpol, %00000000         ' start all pwm channels low
pokeb pwctl, %00000010         ' active pullups, full drive, left-
                               ' aligned
pokeb pwper0, $ff              ' period count for pwm0
pokeb pwper1, $ff              ' period count for pwm1
pokeb pwper2, $ff              ' period count for pwm2
pokeb pwper3, $ff              ' period count for pwm3
pokeb pwdty0, $40              ' duty cycle for pwm0
pokeb pwdty1, $80              ' duty cycle for pwm1
pokeb pwdty2, $c0              ' duty cycle for pwm2
pokeb pwdty3, $fe              ' duty cycle for pwm3
pokeb pwen, %00001111          ' enable all pwm channels

do
     if wait = 0
          wait = ONE_SEC
     endif
loop


'
'    Declare the interrupt vector jump table here, rather than
'    at the end of the code. This keeps the library routines from
'    being compiled in the vector area.
```

Figure 3 continued.

```
'

org $f7e8
asm
      jmp   _rtiisr                    use the SBasic label rtiisr
endasm


'

'   Finally, return to the code section so the library routines
will
'   end up in the correct location (following the main program, not
'   in the vector area).
'
org code

end
```

Figure 3 continued.

Part IV
Mechanics

Many of my columns touched on mechanical subsystems of one kind or another, but this section of four articles featured mechanical elements. The first article describes a group project designed by those of us at a club campout; I provide enough information to get you started on a similar robot of your own.

Next up is Max, a research platform that I designed so I could throw different experiments onto a rolling robot. This kept me from having to build another robot everytime I wanted to try out another idea. Max is easy to build, and I give you all of the details for the various structures. You will find Max discussed in several of my articles, but this was the introduction.

My Dancer robot used an open-frame design of brass rod and copperclad board. Robots don't need to look like flat pieces of plastic that roll around on the floor, and you can apply the techniques I use on Dancer to design your own kinetic sculpture.

The last article in this section describes several techniques for adding optical encoders to R/C hobby servo motors. These motors are the best way to get started building robots, since they are cheap, modular, powerful, and easy to install. Adding encoders gives you terrific control over your robot's speed and distance, opening up new areas of robotic application.

A Basic Robot Design

My column from September 1994 covered the fourth annual Great Escape and Retreat, or GEAR. This robotic campout is always a favorite of the Seattle Robotics Society membership, and has spawned some very clever projects and designs. There's something about sitting around the campfire with a bunch of fellow robot-builders that just inspires everyone. This year, the GEARheads did some major groundwork on designing a simple, practical, easy-to-build mechanical base for a 15-20 lb robot.

The parts for this design are either available through mail-order or can be machined with a minimum amount of skill and fuss. In this column, I cover motor selection and designs for the wheel adapters and motor mounts.

Marvin Green, long an SRS stalwart, showed up this year with a solar-powered robot that kept us all fascinated for hours. It ran around on the asphalt road in the campsite, using only the sunlight for electricity. Marvin also brought a prototype for his latest robot frame, a two-tier circular design complete with a clear plastic dome and a bumper skirt.

Finally, we all spent some time getting a very humbling lesson in design efficiency. Watching ants go about their daily tasks makes you realize how far behind Nature

you are in your robot designs. Try to imagine how you would create an ultra-light, extra-strong intelligent machine that can do its bit to sustain an entire colony of similar machines, without external direction, and do this in an environment filled with dangers and obstacles of all kinds.

I've said this before, but it bears repeating. The strength of a group such as the Seattle Robotics Society (SRS) lies in its diversity. The SRS brings 50 to 60 members to each meeting, and includes expertise in all fields relevant to robotics. Thus, a small group of SRS members, in a suitable setting, can generate more well-designed, hammered-out ideas than any single member could in several times the time. As a case in point, I offer our Great Escape and Retreat, or GEAR.

Each year, a few of us gather at the South Whidbey State Park for three or four days of car-camping, eating, beach-strolling, and robot designing. The setting is quiet, the scenery lush, the weather ideal, and and we spend our time discussing things robotic. A GEAR is named for its sequence since the first year, in 1991. This year was 4th GEAR.

The date of a GEAR mutates. Last year we met in early August, this time we met the weekend after the 4th of July. But the GEARs always start on a Thursday, for those who can make it, and generally breaks up early Sunday morning. We have even considered a Winter GEAR, since winters in Seattle can be dreary. How better to spend a rainy weekend than building 'bots with your buddies? So far, however, we haven't found the ambition or the meeting place. Perhaps this year...

The attendance is usually sparse; a good turnout is six members, with families. But the quality of the designs produced by these GEARheads creates enough energy among SRS members to ratchet everyone up a couple of notches. Last year, 3rd GEAR produced the first designs for what became tiny4th.

Bob Nansel, Marvin Green, Keith Payea, and I strolled the beach and hiked the trails, talking about how we could cram the output of a high-level compiler into 512 bytes of memory. By the time we headed home, I could see that 512 bytes was too small, but 2K was certainly doable. So Marvin, Bob, and I hammered out the next stage on the deck of the ferry boat taking us all back to the "real world." Within a month or so, I had a working version of the tiny4th compiler up and running. The rest, as they say, is history. See "My Tiny Forth Compiler" for details.

4th GEAR

This year, we focused on the mechanics of a robot base. Many in the SRS (notably Marvin Green) build very small robots, using R/C servo motors and 3-inch or 4-inch hobby wheels. But there is a strong interest in building larger 'bots, as well. For example, I can't get Huey, my RoboPet, to tote a beer. I need a bigger machine.

Now, everyone who starts out in robotics usually starts out BIG. Grab a car battery and some windshield-wiper motors, get some angle iron for mounting brackets, and graft on a couple of lawn mower tires. The machine quickly becomes too heavy to lift, too big to move, and often ends up in the garage for a few years.

It isn't that this approach is wrong. It's just that many of us cannot make the vital machine parts necessary to keep the design small. So we make do with what we have. The design gets a little bigger here and a little heavier there, and before you know it, the machine is too darn big.

What the GEARheads agreed on was designing or selecting modular components for motors, wheels, and mechanical linkages. We wanted to create the mechanical equivalents of Marvin Green's BOTBoards; a common design that each builder could start from, without needless reinventing.

The Motors

The motor seems to be the starting point of all the problems. Finding a widely available source of a suitable high-torque DC gearhead motor has proven difficult. Some of the surplus markets, such as Herbach & Rademan, carry a limited line of Barber-Coleman gearheads. But I've found the shaft speed too fast, or the torque too low, or the motor not beefy enough.

Ideally, we wanted to settle on a single motor suitable for building a 15 to 20 lb robot. We needed a large enough shaft that we could bolt some hefty tires onto it. The motor had to be low cost, easy to mount, draw low current in run mode, and be available from a large mail-order supplier.

Keith Payea recommended the motors he built into his robot Ernst. Ernst wasn't available for examination, as he isn't an ATR (all-terrain robot), so Keith described the Dayton 2L011 motor he used. It fit all of our requirements, especially the wide availability. You can order these motors from Grainger's, the nation-wide industrial supply house. The motors, brand new, cost about $21 plus shipping. They are rugged units, supplying 10 in/lbs of torque on a 5/16th-inch diameter, 7/8th-inch long flatted shaft. The motor housing includes four 10-32 studs already installed for easy mounting. The motor and gearbox are replaceable units, should the need arise.

Consult your yellow pages for the Grainger's nearest you. You really need one of their catalogs. It measures 2.5 inches thick, is printed on very thin paper in small type, and is crammed with every possible industrial gadget or commodity you could imagine. The prices range from pretty good to kinda high: a great resource.

The Wheels

After the motor comes the wheels. Most large hardware stores carry a good selection of replacement lawn mower wheels. These run from 5 inches to 7 inches in diameter, usually 1.5 inches wide, with a 0.5 inch bore. The low-end wheels, with plastic frame and molded rubber or plastic tires, cost about $5 each.

After much discussion, we all agreed that a six-inch diameter wheel probably offered a good starting point. When connected to the 2L011 motor, which turns at 50 RPM when driven by 12 VDC, this wheel would move a robot at about 9 inches per second, a very respectable speed.

More important, however, was agreeing on a common axle bore. This became apparent as soon as we started talking about adapters to mate the wheel to the motor shaft. For various reasons, the mechanically ept among us proposed a half-inch bore for the wheels.

The Adapters

All that remained was a mechanical method for mounting the wheel onto the motor shaft. This design occupied most of our time. The servo motors used on the smaller robots have a very nice feature: the R/C motor's control horn. We usually bolt this control horn to a wheel, then bolt wheel and control horn to the motor's shaft, using the screw supplied with the servo motor. This design lets us quickly change wheels on our robots, should the need arise. A couple of seconds with a small Phillips screwdriver, and the wheel comes off. Push the replacement in place, tighten the screw, and your robot is ready to go again.

Ideally, we wanted a mounting method for the larger robots that would let us replace wheels easily if necessary. After several rounds of "Well, what

about this...," we settled on a threaded 1/2-inch diameter adapter, drilled down the center for a 5/16th-inch shaft. We would then drill a second hole through the adapter's shoulder and tap it for a setscrew. The setscrew would hold the adapter to the motor shaft. A properly sized nut would bolt the wheel onto the threaded adapter.

We originally envisioned this adapter as being turned down from round steel stock. Mike Birdwell, one of this year's GEARheads, created a prototype adapter for us to show off at the next SRS meeting. After further discussion at the meeting, we decided to simply drill the two needed holes into a standard 1/2-inch steel bolt, available from the local hardware store. Refer to the accompanying details of the wheel adapter (Figure 1). Dan Mauch, the club's CNC guru, turned out a few of the newly-designed adapters on his metal lathe. He gave me a couple to try out on some motors I had purchased from Grainger's.

The adapters are, in a word, awesome. I sandwiched the plastic wheel between two 1/2-inch internal-tooth star lockwashers, then bolted the whole unit together onto one of the adapters, using a stock 1/2-inch steel nut. Tightening the assembly together is easy, since both the nut and the adapter have hex heads on them. Once tightened, the wheel and adapter form a single unit.

This means I can change a wheel or get to the motor's front surface by simply loosening the setscrew. If I build the frame properly, I could even change wheel sizes, should that prove necessary.

This adapter typifies the classic robotics construction problem. I have until now avoided building large machines, since I had no easy, low-tech way to fasten a large wheel onto a large motor. I couldn't buy what I needed from a catalog, since I didn't really know what I needed. And even if I did know what I needed, and could find it in a catalog, it probably wouldn't prove to be the right solution for readers of this column. After all, I try to design my machines so others can find the needed parts easily and duplicate my efforts.

Now, I grant that most of you don't have a metal lathe standing idle in your garage; actually, neither

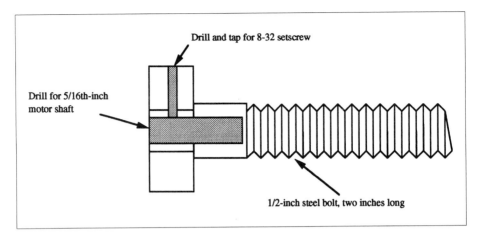

Figure 1. Wheel adapter details.

do I. But I do know people who have such, and are willing to do a little metal work in exchange for software help. This trading on strengths is a hallmark of the SRS and clubs like it. So talk among your robotics friends and your acquaintances, to find someone with a metal lathe willing to do you a good turn See what you can trade for about 15 minutes of machining time.

You can always contact the local high school, vocational school, or trade school for help in machining these adapters. And remember that local businesses usually have someone willing to do a little side work. Seek them out, tell them about your hobby, and ask for some help. You might need to come up with a few bucks for their time, but the quality of the finished product will be worth it.

Additionally, there is talk within the SRS of perhaps creating a box of parts needed to assemble motors and wheels for our 20-lb. robot. The talk is just that right now; everyone is too busy to take on a large task such as coordinating the construction and sales of a robot kit.

But we know there is likely a need out there for a uniform motor and wheel assembly. Perhaps the SRS, or some other robotics club, would be willing to make and sell these adapters for those who simply can't find access to a lathe.

The Mounts

I need to finish up with a few comments on the motor mounts. The 2L011 motor mounts perfectly onto a 3 inch by 3 inch surface. Refer to the accompanying drill guide (Figure 2). I bought a length of 3-inch angle aluminum stock at a local metal surplus shop. Dan Mauch cut me two 3-inch lengths on his CNC metal band saw, then drilled mounting holes in one of the faces, using his CNC drill press.

He also drilled two holes in these same faces, large enough for the motor shaft. He placed these holes so that the motor could be mounted facing "right" or "left." This lets me build a right-side and a left-side assembly for my machine, starting with the same mounting brackets One bracket, two uses: keeps your inventory low.

More GEAR

We didn't spend all our time designing robots at 4th GEAR. We also found time for short hikes through the Pacific Northwest woods, and for watching the sun set over the San Juan Islands at the mouth of the Puget Sound.

Eating, and preparing to eat, consumes a lot of GEAR time. Each year includes a friendly rivalry

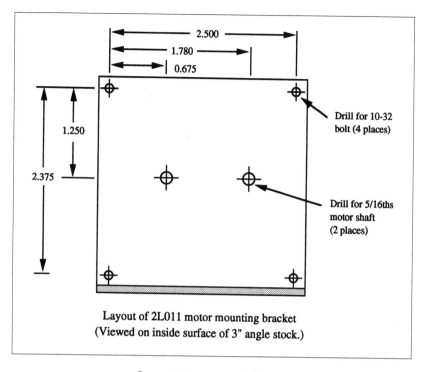

Layout of 2L011 motor mounting bracket
(Viewed on inside surface of 3" angle stock.)

Figure 2. Motor mount drill guide.

among the caveman chefs in the club. Saturday night's potluck dinner featured enchilada cassarole, grilled steaks, pasta salad, and other goodies. We usually eat well at the GEARs, and I often weigh more afterwards than before.

The camp's ranger, Richard, has seen us arrive each year for the last four years, but had not seen any actual robots. Despite our assurances that we do, in fact, build robots, I sensed his skepticism. Oh, he was pleasant enough, but I knew he really wanted to see proof of our robot-building claims.

This year, Marvin Green brought a solar-powered platform he had built. It didn't have any computer on board; it was little more than a motor, solar panel, and wheels. But it made quite an impression on all of us, including Richard. Marvin set the 'bot down in the sunshine on the camp road, and it motored off. No batteries, no steering mechanism, no brains; but it moved, quietly and efficiently.

Marvin built his solar platform using one of the Panasonic Sunceram II solar cells available from Digi-Key. These are very high output solar panels, small and lightweight, and reasonably priced. For example, Digi-Key offers a solar cell measuring only 1.5 by 3.2 inches that produces 3.2 volts at over 40 mA in direct sunlight; the cost per catalog #944 is only $3.77 each in singles.

Marvin used a beautiful, high-efficiency gearhead motor for his platform. He told us the motor only draws two mA when powered by his solar cell. Naturally, we all wanted similar motors, but getting them may prove difficult. Marvin bought the motor from a small surplus shop in Portland, so this is not a widely-available item.

He added an electrolytic capacitor across his solar panel, to serve as a temporary battery. When the platform rolled into the shade, the capacitor supplied power to move the machine forward an extra foot or so. This was sometimes enough juice to move the machine back into the sunlight.

While I'm on the subject of solar panels, I'll mention Ken Birdwell's contribution to the GEAR show-and-tell. Ken had purchased a panel almost one square foot in area from Electronic Goldmine. This unit produces six volts at 400 mA under direct sunlight.

The panel is encased in a tough TEFZEL plastic, claimed by EG to be suitable for long-term outdoor use. You can bend this solar panel to at least a six-inch diameter, so it is quite flexible. This would make a great panel for wrapping around the body or dome of an outdoor robot. Note, however, that this panel is fairly pricey; EG lists it at $65. Still, it provides plenty of current for recharging on-board gel-cell batteries.

Marvin also showed off his prototype for a generic robot platform. He had built a beautiful little round base, just large enough to hold two R/C servo motors for driving the wheels. The base included a clear plastic ring, about one inch tall and 0.25 inch thick, that encircled the frame and acted as a bumper skirt. The ring's inner surface rested against three microswitches and was supported in position by rubber bands.

Pressing inward against any outer surface of the ring resulted in the closing of at least one microswitch. The skirt design was clever, simple, and foolproof. Best of all, the skirt was cheap. Marvin simply went to a local plastics shop and had them cut some one-inch lengths of 4-inch plastic tube. He then drilled four holes in the skirt, properly placed, to thread two rubber bands through. Finally, he fastened the ends of the rubber bands to posts on the underside of the robot frame.

Marvin's complete robot frame fits comfortably in the palm of my hand, yet contains two servo motors, the bumper system, a BOTBoard, sensors, and a four-cell AA battery holder. It is a wonderful piece of design work, and could serve as the core for classes, science fair projects, or research programs.

Marvin already intends to sell just the bare frame (no electronics, sensors, batteries, or switches) as a

kit. Contact him through his web site if you have questions on cost, availability, or features.

And I encourage any of you who can make it to Portland to attend one of Marvin's Portland Area Robotics Society (PARTS) meetings. Marvin builds righteous robots, and studying his projects always gives me ideas for my own machines. PARTS meets the first Saturday of each month; contact Marvin for time and place.

Nature's 'bots

Sitting in the shade of a Douglas fir and waxing robotic, several of us noticed a red ant lugging a dead bee back to the nest. We started discussing this ant's task in robotics terms.

The ant moved by backing up, clamping the bee with its mandibles and dragging the load behind it. The bee was easily three or four times the size and weight of the ant, and the terrain was very rugged, in ant-scale terms. But the ant simply plugged along, backing up and over "boulders" easily three times its height, and dragging the bee over the boulders with it.

The ant's path took it directly to the nest, though there was no trail visible to our senses. The path was so direct, in fact, that the ant made no attempt to ease its journey, choosing instead to barge over any obstacle, regardless of height.

At one point, a second ant wandered by. Ant 2 promptly grabbed the bee and began helping ant 1 drag the bee back to the nest. The two ants worked well together, never seeming to disagree as to path or technique. Eventually, ant 2 released the bee, wandered away, then seemed to forget about its partner and went off to take care of other business. Ant 1, meanwhile, continued to lug the bee home.

It took the ant many minutes to cover the four feet or so to the entrance of its nest. Throughout this journey, we kept up a running stream of questions and comments about what we were watching. How did the two ants coordinate their behavior? How did the first ant "decide" to grab the bee and start homeward? How did the ant change its grip on the bee to ensure it could carry the bee? How did the ant even recognize the bee as something worth dragging such a huge distance?

Then we started thinking of the ant's performance in terms of robotics. How many neurons was the ant using? How much software and hardware would it take to build an autonomous "food" gatherer? How was the ant's behavior coded internally, and how would we code such behavior in our own robots? How would we build a machine capable of dragging an object weighing three times its own weight?

We didn't come up with a robot ant, but we did get a greater respect for Nature's designs. For all we

have been able to achieve in making smaller, faster, cheaper robots, we are still a long way from an autonomous machine that can do even this small part of an ant's daily chores.

Back at the Lab

I used the brackets and adapters Dan Mauch had made to create a prototype motor platform for a 20 lb robot. Refer to the accompanying layout diagram (Figure 3). The base uses 1/4-inch thick white Sintra plastic, cut to a square 11 1/2 inches on a side. Sintra is a

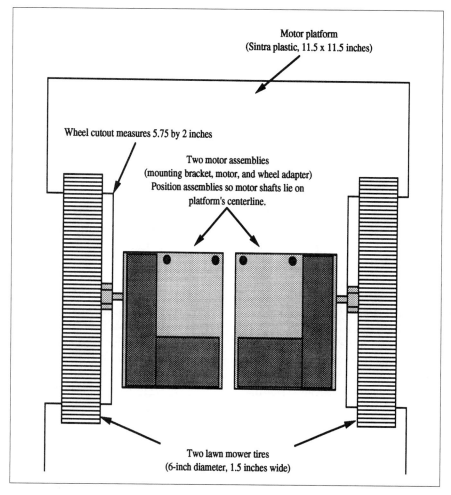

Figure 3. Diagram of the 20 lb robot design.

light, foam-based plastic that cuts easily, provides good strength, and accepts cyano-type glues such as Super Glue or Locktite 406. Prowl through the scrap bin at your local plastics shop for appropriate pieces.

Note that you can use other base materials as well. Double-sided copper-clad, used to make printed circuit boards (PCBs), makes a good base material. You'll want to use some fairly thick stock, ideally made of a fiberglass substrate, for mechanical strength. Other suitable base materials include acrylic or thin plywood.

I cut out rectangles on opposite sides of the square, to allow clearance for the two wheels. Make sure your cutouts are deep enough to allow for the 1/2-inch nut that fits on the end of the adapter to lock the wheel in place.

I then drilled appropriately placed holes in the Sintra base for mounting the two motor assembly brackets. Take care to align the motor's output shaft on the base's centerline when placing your brackets. Be sure to use large washers when bolting the brackets to a Sintra base. The plastic will mush in (a technical term) when you tighten the nuts down; the washers spread the force across a greater surface area and minimize damage to the Sintra base.

Though not shown in the diagram, I added two furniture knobs at the leading and trailing edge of the platform, to act as slides or casters during motion. I then mounted two 12-volt, 1.9 AHr batteries on the platform, holding them in place with double-sided tape and a harness made of 22 AWG wire.

I tested the platform by wiring the motors in parallel, wiring the batteries in parallel, then momentarily applying juice to the motors. The platform runs like the proverbial bat out of Hell. Seeing the phrase "9 inches per second" does not do justice to the speed this platform reaches. The first software I add to this beast will be pulse-width modulation (PWM) speed control.

This prototype platform is only one method for using these motor assemblies. Dan is already talking about cutting some channel stock of suitable width to make U-shaped boxes. Drilling the proper holes and mounting the gearhead motors inside the channel would create a single assembly containing both motors. Building a robot base would then involve nothing more than bolting the motor box to a flat platform.

Another option involves mounting a single motor assembly, made of the L-shaped brackets described above, on the floor end of a table leg. If you used four motor assemblies, one on each leg, you would get a motorized table capable of carrying 20 lbs or more.

I believe this assembly gives the SRS a mechanical module equal in power and versatility to Marvin Green's BOTBoard. We are slowly reaching the point where we can build robots by adding together existing, tested modules. And several SRS members are already starting on the next group of modules: power management and sensors. Watch here for details.

Miscellaneous

Kevin Ross is the new president of the Seattle Robotics Society, based on recently completed elections of club officers. Kevin developed the software for the Pacific Science Center robots, described in (see "Designing an Interactive Robot Display"). I wish Kevin the best of luck as the club's new president. I was elected to the dual office of vice-president and treasurer. Dan Mauch became the club's new secretary, an officer we have been sorely missing. Dan immediately proved his worth at this post by taking real meeting minutes. I'm sure all of us in the club look forward to Dan's contributions during the next year.

And Now, Here's... Max!

This column, published in October of 1994, introduces Max, a two-tier robotic platform that can carry about 20 lbs of experiments. I based Max's design on the ideas from 4th GEAR, discussed in the previous column, "A Basic Robot Design." The heavy-duty motors, strong motor mounts, and large wheels make Max a real mover, and the flat upper surface is ideal for carrying custom electronics, video cameras, or beer.

This column also details the core of a pulse-width modulation (PWM) motor drive program. A PWM program is essential for slowing down a robot of Max's weight and speed; fortunately, such a program is really pretty simple to write. Here, I blend tiny4th and 68hc11 assembly language to show how quickly you can get such a program running. And the technique used here will work with other types of motor drivers and with other microcontrollers.

Finally, I include a review of a wonderful PC-based 2D drawing package called TopDraw. This column covers version 1.0a, but I kept getting upgrades until the company finally stopped supporting the product, at version 3.0. TopDraw was a first-rate product, well worth the $40 shareware fee, and I used it in pretty much every article I did from this column forward. I realize that praise and good reviews don't pay the rent, but I hope the designers of TopDraw see this one day and understand in what high regard I hold their product.

In "A Basic Robot Design" I discussed my high-power robot base, built from a pair of stock Grainger's DC gearhead motors and SRS-designed mounting brackets and wheel adapters. In this article, I'll continue that discussion, and introduce you to Max.

First off, I made a couple of minor modifications to the prototype motor base. I used a 1/2-inch Nylok nut on the end of each wheel adapter. I went to Nylok nuts because the regular nuts kept spinning off the threaded adapter shaft. This, even with two star lockwashers and lots of elbow grease to hold the nuts in place.

The second mod concerns filtering out motor-induced voltage spikes on the wiring. I added a three-capacitor filter to each motor; see the accompanying filter schematic for details (Figure 1). Without the filter, the controlling computer would flip out after a few minutes running time. With the filter, the computer has not had any problems after several days of testing.

225

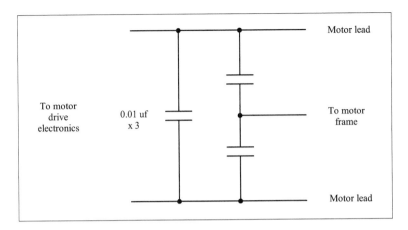

Figure 1. Schematic of motor noise filter.

The Base

I built this motor base to serve as the foundation of an experimental robot platform. I wanted a platform that could carry about 20 lbs total weight, so I could really load on the equipment. The heavy tires, large motors, and two 1.9 AHr batteries make for a substantial base. But the motors, batteries, and wheels take up all the base's horizontal area. I realized immediately that I needed to add a second story.

The second platform measures 10 inches by 11 inches, and is cut from 1/4-inch white Sintra plastic, just like the motor base. I located the center of this rectangle and drilled a 1/2-inch hole to provide a routing point for the motor cabling. I had to mount the second platform five inches above the base, to allow clearance for the wheels. I usually use 4-40 spacers and hardware on my smaller 'bots, but this called for something heftier.

I decided to use 10-32 hardware, since that was what I used in building the motor base. But I couldn't find any five-inch threaded 10-32 spacers, so I had to improvise. Refer to the accompanying detail on spacer construction (Figure 2). Each spacer is made from a length of 5/16th inch OD brass tube, cut to five inches long. You can probably get satisfactory results using a hacksaw and patience, but I really recommend using a pipe cutting hand tool.

This consists of a clamp with a twist grip; the clamp has a cutting blade in the center of its jaws. You simply clamp the tube in the tool's jaws, then alternately rotate the tool around the tube and tighten the clamp. Each rotation cuts a little deeper into the tube, until finally the cut is complete and the tube separates at the scoring.

These are available at most large hardware stores for about $15, and do a wonderful job, giving you a straight, clean cut. Spend a few extra bucks and get the kind that automatically flares the cut for you. You will love the smooth finish and quality look this tool gives your project.

You will need the hacksaw (and patience) though, for cutting the 10-32 threaded rod to length. Each spacer needs a six-inch length of rod. Again, this rod is a common item at large hardware stores; a two-foot length will make exactly four bolts for the base's spacers. If you work carefully, you shouldn't have any trouble getting a clean enough cut to thread a 10-32 nut over. Alternatively, you can use bolt cutters, provided the jaws don't bung up the threads at the cut ends.

I located and drilled holes in the motor base, then located and drilled matching holes in the new platform. I installed the four threaded rods in the motor base, adding nuts, washers, and lockwashers to the underside on each rod. Next, I ran a nut all the way

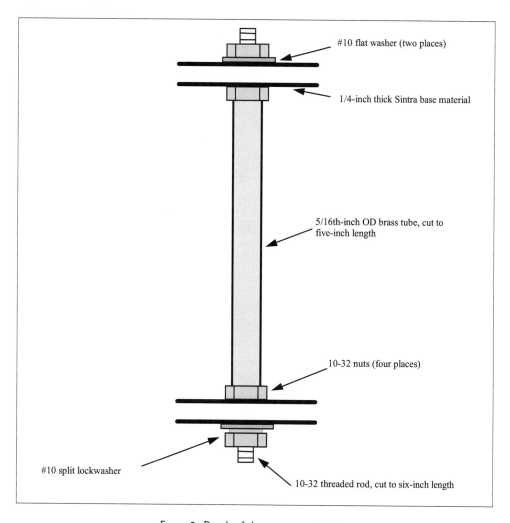

Figure 2. Details of the spacer construction.

down each rod, tightening it in place on the top side of the motor base. This left each threaded rod fastened firmly in place. I slipped a brass spacer over each rod, followed by a nut to clamp each spacer in position.

Finally, I fitted the new platform in place, then added a flat washer and nut to each threaded rod protruding through the drilled holes in the plastic. Tightening these nuts down gave me a rugged, two-tier robot base that I can easily lift by grabbing the top platform, yet is sturdy enough to carry large loads.

With the second platform in place, I laid out the electronics I wanted to install. Refer to the accompanying diagram of the second platform layout (Figure 3). This platform carries a four-cell C-size battery holder, providing 6 VDC for the computer and electronics. This helps isolate the computer's voltage from the motors' voltage, reducing the electrical noise on the computer's power lines.

I also added a relay-based dual DC motor controller board, handwired on a Radio Shack 276-168A

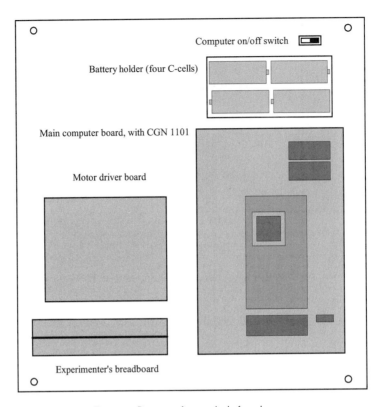

Figure 3. Diagram of second platform layout.

experimenter's board. This is a standard circuit, and I usually have a couple of finished boards kicking around. Refer to the accompanying schematic for wiring details (Figure 4).

For a computer, I used a handwired 68hc11 system, based on a CGN 1101 module. I built the circuit by wirewrapping the components onto a Radio Shack 276-147 experimenter's board. Since this board measures more than 4 inches by 6 inches, I had plenty of room to add 8K of static RAM, 8K of EEPROM, and an 82c55 parallel I/O chip.

There is no complete schematic for this computer. I built it using the tables and circuits featured in previous articles, notably the July 1994 issue, "68hc11 Memory Expansion." Using the CGN 1101 module really speeds up construction and minimizes errors. You can find a review of this CGN board in the July 1994 column.

Note that there is nothing stopping you from mounting a 68hc11 BOTBoard on this platform and running the robot with it. I foresee needing more memory than the 2K of EEPROM available in a 68hc811e2, so I chose to go with an expanded-mode system of my own design.

Finally, I added a small solderless breadboard so I can quickly build and test circuits. Mine is from CS Specialties and measures 4 inches long; you should be able to find similar breadboards from several mail-order firms. You can also use the Radio Shack 276-174 if you change the mounting arrangement to allow for the larger size.

I call this new robot Max. Future articles will describe experiments based on Max, so you might want to check last month's article, "A Basic Robot Design," for more details on the motor base construction.

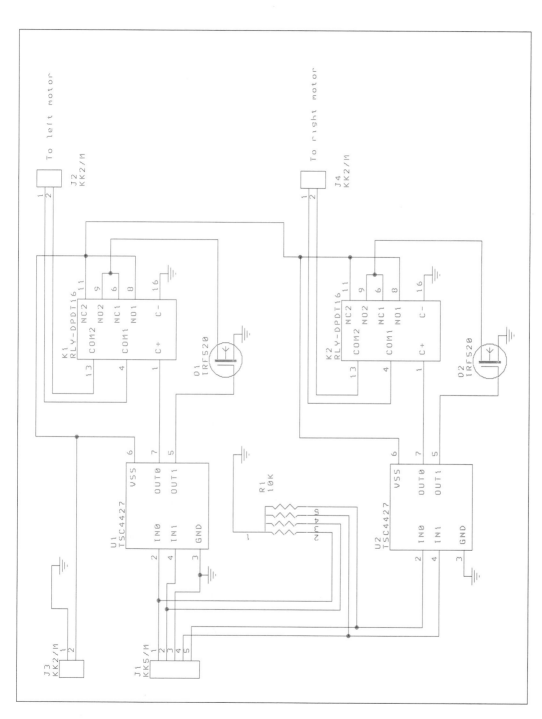

Figure 4. Schematic of relay motor driver board.

Slow Down!

As I mentioned last month, the motors and wheels used on Max give a top speed of nine inches per second. That is on the upper edge of manageable, and the robot's weight of seven pounds makes for a dangerous combination. This puts pets, babies, furniture, and unwary guests in harm's way.

The first block of code I wrote for Max provides pulse-width modulation (PWM) speed control of the motors. Refer to the accompanying listing (Figure 5, at end of article). I don't normally include listings in my articles, as they take up a lot of space, but this code is vital to Max's proper operation. Also, the concepts behind this code can be applied in many other areas of robot software. In all, I think this is space well spent.

Note that this is not a complete program. You cannot type all this in and expect it to compile, let alone run. This code is simply to illustrate some concepts, and was taken from a larger, working program. The PWM code is a mix of 68hc11 assembly language and tiny4th. You can rewrite the tiny4th code in assembly, if you prefer; the tiny4th source can serve as a guide.

Basically, this code uses a periodic interrupt on the 68hc11's TOC2 to time the PWM signals to the right and left motors. On each interrupt, control transfers to the routine TOC2_ISR, defined with the tiny4th **CREATE** word and written in assembly language.

When control reaches TOC2_ISR, the code immediately calculates the "time" of the next PWM signal, based on the constant PWMDELAY defined in the tiny4th source. This constant contains the number of 0.5 μsec tics between each PWM update. In the listing, this value is set to $1388, or about 2.5 msecs. The code then rearms the TOC2 interrupt.

Next, the code shifts the contents of variable RSPEED one bit to the left. The bit shifted out of this appears in the carry (C) bit of the 68hc11's status register. It is also rotated back into the new least-significant bit (LSB). If the shifted bit was a zero, the code prepares a mask to turn off the right motor. If,

however, the shifted bit was a one, the prepared mask will turn on the motor.

The same operation is performed again, this time using the contents of variable LSPEED. The two motor control masks are ORed together, then written to the motor port. This final step actually turns each motor's MOSFET on or off, based on the corresponding shifted bit. Finally, the code executes an RTI to return from the TOC2 interrupt.

This means that your software can control the motors' duty cycle simply by writing 16-bit values into variables RSPEED and LSPEED. The motor control circuitry will turn a motor on when a one appears in the PWM pattern, and turn the motor off when a zero appears. Refer to the accompanying diagram of typical PWM patterns (Figure 6). Each one or zero in the pattern data corresponds to a time-span of 2.5 msecs.

Using this type of PWM pattern, your software could run a motor at full speed by writing the value $ffff to the proper speed variable. Similarly, the code would write $0000 to a speed variable to turn that motor off. The value $5555 (or $aaaa) contains a pattern of alternating 1s and 0s, corresponding to a 50% duty cycle.

The listing for TOC2_ISR shows how to access tiny4th **VARIABLE**s from within a block of **ASM** code. The leading underscore ('_') character in front of a **VARIABLE** name causes the compiler to pass the internal name of that **VARIABLE** to the assembler, so the assembler can properly resolve it. Without the leading underscore, the compiler would pass the name unchanged to the assembler, which would not be able to resolve it and would then report an error.

Also note how I used tiny4th's **ASSIGN** word to set the address of the TOC2_ISR routine as the TOC2 interrupt vector. When the 68hc11 processes a TOC2 interrupt, it jumps to the address stored in $ffe6 and $ffe7. The **ASSIGN** statement at the end of the listing generates the correct assembly language instructions for performing this setup.

The block of code at SPEED_TABLE shows one way of assigning values to RSPEED and LSPEED. Here, a fixed set of possible speeds are stored in a table as 16-bit entries. Software can index into this

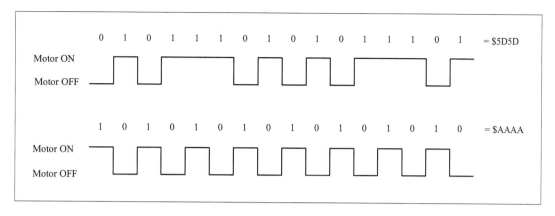

Figure 6. Typical PWM patterns.

table to fetch a selected value and write that value into RSPEED or LSPEED, as needed.

The comments in SPEED_TABLE give the percent of duty cycle represented by each entry. Note that these values are duty cycle, not actual speed. The actual speed will generally differ from the duty cycle, due to mechanical elements such as wheel inertia.

Changing the value of PWMDELAY will change the pulse width used. Smaller values will result in shorter pulses and smoother motor movements, but will cause the code to spend proportionately more time in the PWM service routine. Larger values, however, will let the MCU spend more time running your main program, but will result in rougher motor movements. Try some modifications and gauge your results.

You will also need to add a small block of code that starts up the PWM system, following power-up. This code appears in the accompanying listing for initializing the PWM system (Figure 7). Basically, this code simply preloads the TOC2 register with the current value of TCNT plus the number of counts in constant PWMDELAY, setting the time at which the next TOC2 interrupt occurs. The code then arms the TOC2 interrupt and exits.

When the sytem counter counts up to the value written to TOC2, the TOC2 interrupt triggers and control passes to TOC2_ISR, where the next pulse is

created. TOC2_ISR then calculates the time of the next pulse, rearms the TOC2 interrupt, and returns.

These two chunks of code form a complete PWM system for any robot that uses a motor controller similar to the one described in my earlier articles. Note that this code does not handle motor direction control, only speed. Feel free to build on the code given here, and to add your own code for moving different PWM values into RSPEED and LSPEED. Your code might use the state of various sensors to change speed and direction, or signals received from an external beacon might trigger changes. Regardless, these few lines of code should form a solid core for your future experiments.

TopDraw

In the past two years, I have searched for a cheap, high-quality PC-based drawing program that I can use for making diagrams and simple schematics. I didn't think I was asking for much. But every shareware package I tried had serious flaws in the user interface, didn't provide what I felt were minimum features, or flat didn't work.

So I was forced to use my trusty old Macintosh Plus, running MacDraw II, to do my artwork, while I did all my text on my 386. After two years, I had about given up on finding a cheap PC-based package that did nothing more than duplicate MacDraw II.

```
: init-motors
   40   tflg1   c!          \ clear any toc2 interrupt
   tcnt  @                  \ get time now
   pwmdelay  +              \ add delay
   toc2  !                  \ save as time for first delay
   40   tmsk1   c!          \ arm toc2 interrupts
;
```

Figure 7. Code to initialize the PWM system.

Then I ran across TopDraw at a local software store here in Seattle. For about $6, I got to test-drive a Windows drawing package that is everything I have been looking for, and more. TopDraw's user interface is clean and well-designed, the tool set is limited but very powerful, and the program has every type of fill, line, and texture method I could ask for.

Pans and redraws are fast, even on my 40 MHz 386. The zooming technique is different from MacDraw II's, but effective and easy to use. Best of all, this package offers something that has proven oddly difficult to find in other programs: the ability to create and use libraries of predefined drawing elements.

TopDraw calls these shape sets, and you can have an unlimited number of them. So I can make a shape set of schematic symbols, a shape set of robot elements, and shape sets for anything else I use more than once. Each element in a shape set may be placed, stretched, rotated, and grouped, just as any drawing element can. What's more, TopDraw graphically displays the elements in a shape set before you select, so you can immediately see what you are about to choose. The graphical thumbprints also include a text description, provided by you when you created the element originally.

Another TopDraw feature I really like is the node tool. This lets you edit individual segments of any line, breaking the line or gluing two lines together as needed. You can pull and stretch lines, deforming them as you want. TopDraw also has a very powerful undo feature. Not only does the undo list appear bottomless, the Undo menu selection always tells you just what type of edit you are about to undo.

The text tool lets you use all your Windows fonts, in all available sizes and styles. It even supports justification, double- and triple-space text, and centering. When you click on a drawing element to select it, a group of handles appears around the shape. Most of these handles provide the normal stretch and squeeze functions. But the transposition handle is quite powerful and smooth.

Simply put your mouse cursor on the transposition tool, hold down the left mouse button, and start moving the mouse. Your shape will smoothly rotate around its centerpoint. When you have the shape at the desired angle, let up on the mouse button and the shape is redrawn at the new angle. If you need even more control, a menu selection brings up a full screen of Windows control elements for moving, rotating, and otherwise manipulating the shape.

I am using TopDraw version 1.0a, so I expected some bugs. Surprisingly, I have only discovered one, and I have done a lot of work with this program. The one bug surfaced when I tried to add some text inside one square of a grouped set of squares. The text appeared properly on the screen, but I could never get the text to print out on my laser printer. I worked around the problem, and Top Software, the creators of TopDraw, have probably found and solved this already. So I don't consider this bug anything serious.

I can't say enough good about TopDraw. It is far beyond anything I hoped to see in shareware. I consider Top Software's registration fee of $40 to be reasonable. If you need a Windows drawing package, I urge you to pick up a copy of TopDraw. Highly recommended.

```
hex

variable    rspeed                      \ holds right motor PWM pat-
tern
variable    lspeed                      \ holds left motor PWM pattern
variable    toc2_temp                   \ temp variable for toc2 isr

1388    constant    pwmdelay                \ pwm updates every 2.5 msecs

porta   constant    mtr-port

   20    constant    rmtr.pwm    ( bit mask for driving right motor )
   08    constant    lmtr.pwm    ( bit mask for driving left motor )

create    speed_table
8888    ,                  \ 25%
9249    ,                  \ 33%
ad6b    ,                  \ 62%
eeee    ,                  \ 75%
ffff    ,                   \ 100% of full speed



create    toc2_isr
asm
*
*   This code uses toc2 to create PWM pulses for controlling the
motor
*   speed.
*
        ldx           #iobase           point x at io regs
        ldd           #_pwmdelay        get pwm delay value (tics)
        addd          toc2,x            add to latest toc2 value
        std           toc2,x            save as new toc2 value
        ldaa          #$40              get mask for toc2
        staa          tflg1,x           rearm toc2 interrupt
        ldd           _rspeed           get right motor PWM pattern
        asld                            move next bit to carry
        bcc           toc21             branch if top bit was cleared
        orab          #1                roll top bit around to low bit
toc21
        std           _rspeed           and save rotated speed
```

Figure 5. PWM code listing.

```
        clra                            assume need to turn motor on
        asrb                            restore carry bit
        bcs        toc22                branch to turn motor on
        ldaa       #_rmtr.pwm           no, get mask for right motor
toc22
        psha                            save motor mask on stack
        ldd        _lspeed              get left motor PWM pattern
        asld                            move next bit to carry
        bcc        toc23                branch if top bit was cleared
        orab       #1                   roll top bit around to low bit
toc23
        std        _lspeed              and save rotated speed
        pula                            assume need to turn motor on
        bcs        toc24                branch if so
        oraa       #_lmtr.pwm           no, or in mask for left motor
off
toc24
        staa       _toc2_temp           save in temp location
        ldaa       _mtr-port            get current port value
        anda       #$d7                 drop only motor bits
        oraa       _toc2_temp           or in motor bits
        staa       _mtr-port            update motor port
        rti                             and outta here
end-asm

ffe6   assign    toc2_isr
```

Figure 5 continued.

Build an Open-frame Robot Body

My February 1995 column for *Nuts & Volts* described a new robot, named Dancer. The frame was built entirely of brass rod and blank copperclad circuit board stock. These two items, commonly available from mail-order outlets and hobby stores, make excellent construction materials for robot builders. You can cut them easily, and fastening them together takes nothing more than a heavy-duty soldering gun.

I also touched on some concepts behind line-following contests. These events are easy to stage and make real crowd-pleasers. You can build a simple line-following robot in just a few evenings, and watching your creation work its way around the map will give you a real sense of accomplishment.

I close with a look at a truly versatile robotics gizmo, the Leatherman's Super Tool. My wife, Linda, gave me one for Christmas the previous year, and I've worn it ever since. The prices for these tools have dropped dramatically in the last couple of years, and a true robo-nerd should carry one at all times.

I was in the mood to build a new robot frame. Max, discussed in several previous columns, makes a great platform for large experiments, but Max is, well, large (see "And Now, Here's...Max"). I needed a smaller, lighter frame to serve as a base for some sensor experiments I had in mind. Much of this need was driven, as always, by developments within the Seattle Robotics Society (SRS).

The SRS members have shown renewed interest in line-following robots, and we held a club-wide competition recently. My venerable CBE-1, built nearly three years ago, did manage to win the contest. But the gleam in the eyes of the other contestants made it clear that CBE-1's days as champion were numbered.

CBE-1

I'll briefly describe CBE-1, so you get an idea of how I put it together. This is mostly historical information, as I don't intend to build a CBE-2. Incidentally, CBE stands for Crude But Effective, a comment made about my machine's design shortly before it went on to win the first of many line-following contests.

I built CBE-1 around a skeleton made of three pieces of bookshelf frame. You've probably seen this frame material in the larger hardware stores. It usually comes in three- or four-foot long sections, and has pre-drilled holes spaced at about 3/4-inch intervals along its length. The material is made of U-shaped aluminum or soft steel stock about 1/2 inch wide.

I started by cutting one piece, about 12 inches long, to serve as the central brace for the robot. I then bolted two other pieces, each about ten inches long, to this brace. I mounted the two shorter pieces to the main brace at their centers, crosswise to the main brace, so the unit formed a plus-sign with two horizontal lines instead of just one.

I then bolted two Maxon gearhead motors, one at either end of the two-piece central brace. The motors fit between the two braces, held in place with circular clamps. Above the motors, I added two four-cell D-size battery holders. I also added two furniture knob casters, one front and one rear, from the single main brace. I used a plastic flower pot as the cover, held in place with spacers and hardware to the main brace. Additional hardware held the electronics to the top of the flower pot.

The electronics consists of a motor controller board and a 68hc11 computer board. The motor controller uses two DPDT relays and two power MOSFETs to provide direction and pulse-width modulation (PWM) speed control of the motors. The computer board contains 32K of EPROM and 32K of battery-backed static RAM. Finally, I mounted a line detector pod, consisting of three cadmium-sulfide (CdS) photocells and three LEDs, to the front underside of the robot, just behind the front caster.

But CBE-1 was one of a kind. I can't get any more of the motors, I don't want to build another complex 68hc11 computer, and I can't find the Forth compiler or source I used to create CBE-1's code.

Dancer

So I decided to start over completely. The new machine, called Dancer, uses several construction methods I've tried out in recent designs. Dancer features an open frame made of stout brass rod and double-sided copper-clad printed-circuit board (PCB) stock. These materials are cheap, easy to find, and can be machined with simple tools such as drills, files, and wire cutters.

My earlier robots all had a large horizontal platform, which supported the electronics and batteries. Dancer's frame has only one small vertical element, set at the back and holding the batteries and modified R/C servo motors. Refer to the accompanying top and side views of Dancer's wire-frame design (Figure 1).

Begin by bending two lengths of 3/32-inch brass rod to the indicated shape. You can find this rod at most large hobby stores; price is about 89 cents for a three-foot length. Trim the rod to length using heavy steel-jawed nippers or stout wire cutters. The brass is soft, but it will probably damage fine diagonal cutters normally used for cutting wire.

Next, cut two pieces of 1/16-inch brass rod, each 3-1/4 inches long. I'll describe how to connect one of these pieces to one of the heavier frame sections; follow the same steps for the other piece.

Using heavy pliers, wrap one end of the thinner rod around the upper horizontal element, approximately as shown in Figure 1. Crimp this wrap down securely, then coat the junction with solder flux paste, heat with a solder gun (100 watts or more) and flow rosin-core solder into the junction. You can find the paste and the solder gun at most hardware stores. Most soldering irons suitable for electronics work don't generate enough heat for this type of soldering, and do a lousy job.

After the junction has cooled, wrap the other end of the thinner rod around the lower horizontal element. The two horizontal elements should be spaced about 2-1/2 inches apart by the vertical element. Crimp the wrap, coat it with paste, then solder the junction. Trim the excess thin brass rod from both ends. Repeat the above steps to add a vertical element to the other frame section.

The frame sections are designed to hold a circuit board built on a Radio Shack experimenter board (276-158A). If you choose to build your frame for the same size board, you need to space the two frame sections properly. Refer to the accompanying details of additional hardware (Figure 2).

The brass screws for holding the circuit board must be soldered onto the frame elements such that

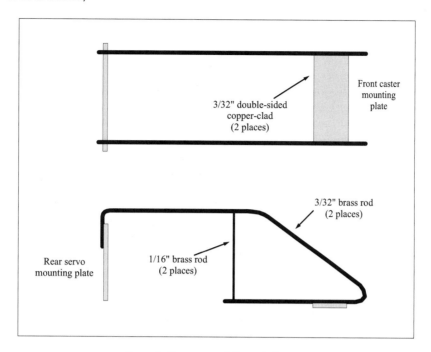

Figure 1. Two views of Dancer's frame.

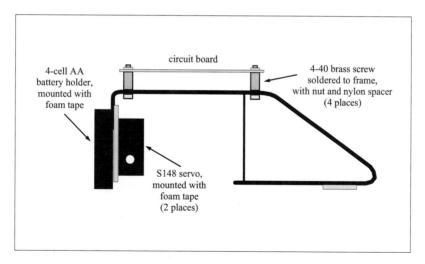

Figure 2. Details of additional hardware.

the screws form a rectangle 2.5 inches wide by 3-11/32 inches long. The simplest way to align these screws properly is to use a blank experimenter board as a guide. Place the two frame pieces side by side, locate a point near the rear edge of the frame, and mark that location on both frame pieces. Using the blank board as a guide, make a second mark near the front bend of the each frame for the second set of mounting hardware. The first and second marks on each frame element should now be 3-11/32 inches apart.

Now you need to attach 1-inch long brass 4-40 screws, available at most hardware stores, to each of the frame elements at the marked locations. Start by stripping the insulation from a piece of solid copper wire about two inches long. Wrap the wire around both the screw and the frame element, so the screw is held snugly against the inside edge of the frame at one of the marks. Carefully check that the screw is vertically aligned, relative to the rest of the frame element.

Coat the junction between the screw and the frame element with flux paste, then solder the screw to the frame. Repeat this operation to solder a second brass screw at the other marked location on this frame element. Then solder two screws to the other frame element. Make sure you solder all screws to the inside edge of the frame elements.

Now you should be able to mount the experimenter board to the four screws, using nylon spacers and 4-40 nuts. Double-check the screw locations to ensure that they are vertical and that the ends of the frame match up. Leave this circuit board mounted to the frame elements for the remainder of the frame construction; this ensures that your frame will stay aligned.

If a screw isn't positioned properly, reheat the junction and realign the screw. If necessary, you can also rebend parts of the frame elements, though the brass will fatigue easily and may break after repeated bending.

Make the rear servo plate from a piece of double-sided PCB stock, cut to a rectangle 3-1/4 inches by 2-1/2 inches. Position the plate as shown in the details (Figure 2), so that the longer side lies horizontally between the frame elements. Mark the mounting locations, then solder one frame element to the

plate. Solder the second frame element in place, then reheat the first junction if necessary.

Cut a second piece of double-sided PCB stock to a rectangle 2-1/2 inches by 1 inch. This will serve as a mounting plate for the front caster. Drill into this plate whatever holes you need for your selected caster. Since I usually use a plastic furniture knob as a caster, I just drilled a single hole in the center of the plate.

I also drilled two holes, each 1/8 inch in diameter, in the rear edge of this plate. I positioned the holes properly so I could later attach one half of a Radio Shack experimenter board (276-148). This board's small size makes it an ideal development platform for sensor pods and other experiments. After you have drilled the needed holes, position the caster plate as shown and solder it to the underside of the frame. Finally, attach your caster.

This completes the construction of Dancer's frame. Now all you have to do is add the motors, batteries, and electronics. Dancer is built to use modified Futaba S148 R/C servos. You can buy these servos at most hobby stores. You can also order them from Tower Hobbies, usually at prices well below most stores. Tower sells these servos for about $15 each. I have provided instructions for modifying these servo motors for robotics use in Appendix B.

After modifying the motors, mount them to the inside face of the servo plate, using double-sided foam tape. You can find this tape at most hobby stores and large hardware stores. It holds the motors firmly in place, yet you can (with effort!) remove the motors to replace or realign them if necessary. Be sure to clean the plate's surface with alcohol, then dry it thoroughly, before putting the tape in place.

The outside face of the servo plate holds a four-cell AA-size battery holder (Radio Shack 270-391). Mount this battery holder in place, also using double-sided foam tape. Be sure to clean the surface of the servo plate before sticking the tape in place.

I made wheels for Dancer by fastening a five-point Futaba control horn to each of two 3-3/4" Dave Brown Lite Flite wheels. You can get the wheels at any large hobby store, or through the mail from Tower Hobbies. The control horns are supplied with the

Futaba servo motors, and the tiny sheet metal screws are available at any hobby store.

Dancer's frame design makes a light, stylish robot. Since the large wheels help hide the batteries and motors, the robot appears to consist of a circuit board, wheels, and some brass rod. And the open frame design is amazingly strong and rugged. Best of all, you can modify the design at any time, with nothing more than some brass hardware, solder, and a soldering gun.

I built Dancer's 68hc11 computer board by wiring a CGN 1001 68hc11 module into a Radio Shack experimenter's board. I have reviewed a couple of CGN boards in previous columns (see "Quick and Easy 68hc11 Expansion"). If you haven't tried a CGN module before, I strongly recommend you give the company a call and ask about their products. I have been impressed with the quality and design of all their products that I have tried so far.

I also added a handful of male Molex KK connectors to the experimenter's board, to support various robotic subsystems. A pair of three-pin connectors let me plug in the Futaba J-connectors from the servo motors. I used a two-pin connector for the wiring from the AA-battery holder. I used a four-pin connector for the serial data hookup to my PC, needed when I download any programs into the 68hc11. I also added three four-pin connectors to analog ports AN0, AN1, and AN2 on the 68hc11. These let me quickly add custom analog sources to Dancer for experimentation.

Finally, I added a special five-pin connector, wired to +5 VDC, ground, and analog ports AN4, AN5, and AN6. This connector supports a simple three-element robotic eye I am currently developing.

I also added the obligatory power switch, reset switch, power-on LED and dropping resistor, and a large electrolytic filter capacitor. And I still have enough room on the board to add a considerable amount of circuitry.

Line-following

The SRS members have renewed their interest in line-following for several reasons. First, it is a cheap event

to hold. You just need a roll of white paper, such as butcher paper, about three or four feet wide. You also need a black marking pen that can draw a 1/4-inch wide line.

Just roll out a suitable length of paper onto a hard, flat surface such as a tile floor. Most of our events use a strip of paper 10 to 12 feet long, though you can certainly run longer if you like. We generally apply some strips of wide masking tape on the edges of the paper, to keep it flat on the floor during the event. All that remains is drawing the line to follow. We use a set of guidelines drawn up for a previous event, that helps insure all robots will face a high-quality layout.

The line will be at least 1/4-inch wide along its entire length. The line will contain no intersections or branches. The line may contain 90-degree turns, either as a rounded arc or as a sharp angle. Turns made of an arc will use a radius no smaller than six inches.

Wherever any segment of the line passes near to another segment, the two segments must be separated by at least six inches of white space. The starting point of the line must consist of a straight line segment no less than 12 inches long. The starting line will be marked by a faint pencil line, drawn perpendicular to the contest line, six inches from the end of the contest line. The contestant's robot must be placed so that no part of its body lies past the starting line before the contest begins.

The contest run is started when the time-keeper signals. Note that some robots might have a built-in delay after power-up before they actually begin moving. In such cases, the time-keeper will start the clock when the robot begins moving.

The contest run ends any of several ways. A robot that leaves the maze paper will be flagged as did-not-finish (DNF) and will be given a time for that run equal to the maximum allowable time, usually 10 minutes. A robot that leaves the line, then rejoins the line at a point significantly farther along the line than where it left, will also be declared DNF.

A robot that is touched during the run by the builder or anyone else, in an effort to correct or improve the robot's performance, will be declared DNF. A robot that travels the length of the maze line will be timed at the instant the front point of its body reaches the end point. The time-keeper will announce the running time, ideally to a tenth of a second.

Note that sometimes a robot will get turned around and head back to the starting point. This is not cause for disqualification or penalty, other than the extra running time that such a move creates. Often, such a robot will travel to the starting point, turn around, and restart the contest run. Its time continues to run, and is not restarted simply because the robot crossed the starting line a second time.

This type of line event requires very little sophistication so far as robot design is concerned. In fact, the only difficult part of the line event described above is the right-angle turn. A robot can successfully make its way from one end of this line layout to the other, using nothing more than two light/dark sensors. Mind you, it won't necessarily be fast or efficient, but it will get through.

A robot using this type of sensor pod often employs a bang-bang type of line-following algorithm. This means that the robot goes forward until one of the sensors, spaced about three line-widths apart at the front end of the robot, runs over the line. The robot then immediately turns in that direction, until the same sensor sees, then doesn't see the line. This means the line is again between the two sensors, and the robot switches to moving forward.

I call this a bang-bang algorithm because the robot is continually hitting the line with first one sensor, then the other, as if it were banging off a wall. This algorithm, while useable, suffers from a serious flaw. The robot's computer never knows for sure when the robot is actually on-course; it only knows for sure when the robot has strayed off-course.

This weakness can lead to tacking, where the robot moves down a straight section of line by alternately hitting the line with first the right, then the left sensor. This tacking motion can cost valuable time in a contest run. In fact, each line layout should have at least one very long straight stretch in it, to put such designs to a test.

CBE-1 used left and right sensors, as well as an on-course line sensor. This latter sensor detected when the line was directly below the robot. In such cases, CBE-1 went full-speed ahead, knowing that it wouldn't have to slow down and turn immediately. Often, CBE-1 would be out-maneuvered by other robots in the turns and arcs, only to gain back all the lost time in the straight sections.

The next step up in line-following complexity uses a line layout that contains branches and intersections. Such a layout would usually have the finish line at the end of a branching line. This means a robot would have to use some type of a "wall-following" algorithm to explore the maze. Robots that could explore the maze, then determine the solution and run it properly in a second heat would have a big advantage.

The most complex maze includes branches that loop back onto themselves or onto the main line. Solving this kind of maze requires both mapping and measuring. A robot must not only determine where each intersection occurs, it must also know the distance between nodes so that it can weed out loops and determine the shortest distance to the finish line.

Line-following contests make super club events. You can build a line-following robot easily and, at least for the first type of line layout, the software is very simple. These events are fun to watch, and even more fun when you have a machine running in them.

Super-Tool

I am writing this on the day after Christmas. My wife, Linda, never knows what robotics-related gifts to give me, even though I try to provide specific details on my gift list. I guess the technology is just too daunting. This year, however, she came up with the perfect gift. She gave me a Leatherman Super Tool.

The Super Tool is sort of the ultimate Swiss Army knife. I have the larger version, which measures 4-1/4 inches long and a little over one inch wide. It consists of two halves, hinged at one end, that open up into a pair of needle-nosed pliers. The pliers contain a wire-cutting edge and a larger, open section for gripping large nuts and bolts.

The handles of the pliers contain fold-out tools of all types. One handle sports a file, a razor-sharp knife, large and medium flat-blade screwdrivers, and an awl/punch. The other handle contains a serrated knife, a can and bottle opener, a Phillips screwdriver, a wood saw, and a small flat-blade screwdriver.

Other parts of the Super Tool serve as a nine-inch ruler and an electrial connection crimper. The whole thing folds up into a small tool that fits into a leather sheath attached to my belt.

I'll be honest with you. When I first opened this gift, I was somewhat disappointed. After all, I am a software-type, and I usually view tools as simply another means of removing blood from my body. But our CD player had gone belly-up a few weeks ago, and I had bought a new unit for Christmas, which meant the old one was now a source of robo-parts.

Just for grins, I set the old player on the dining room table, opened up the Super-Tool, and started in.

What a blast! I had the defunct player torn down in a matter of minutes. And I had all the screwdrivers, pliers, and knives I needed, right there in the Super Tool. This experience made me think back on all the times I had been at a competition or an SRS meeting, when some robot-related gizmo broke down. My tools were at home, and no one had the screwdriver or pliers I needed. Boy, I sure could have used something like the Super Tool then.

Bottom line: Take a trip to your nearest cutlery store and check out a Super Tool. Linda says these come in at least two different sizes, and they are highly prized by electricians and engineers who do field work. I haven't had the guts to ask how much it costs.

Every part of this tool is made of very high-grade, corrosion-resistant steel. The Leatherman company guarantees the Super Tool for 25 years, and I have no doubt this device will easily last that long.

Adding an Encoder to a R/C Servo

My March 1996 column for *Nuts & Volts* covered a topic of major concern to robot builders: how to add shaft encoders to hobby robot motors. To keep my design as universal as possible, I chose to work with the Futaba S148 hobby servo motor. I patterned my design on work done by SRS member Mark Castelluccio, whose servo-based 'bot scored an impressive victory over stepper-driven robots in the club's Dead-Reckoning contest.

Adding a shaft encoder to such a motor requires some careful work, so you might consider picking up a couple of spare motors for this project. But the final product, a servo motor that can reliably move a robot a specific distance, is a big win and opens up all kinds of possibilities for navigation.

This column also goes into some detail on Mark's original design, and on a similar design developed by Marvin Green. I also show an SBasic program that uses the encoder information to control the servo's motion. You can use this SB program as the core for your own experiments. The notes I've supplied on how the encoder software works should prove helpful.

I've had a lot of fun building robopets and other small machines, but they've all had a common lack; none of them included wheel encoders. Now, that hasn't actually been much of a problem. Line followers don't need wheel encoders to run simple mazes; they react to the line in front of them. Robots for the Rapid Deployment Maze don't need encoders, either; the operator controls them with a joystick.

Even Huey, my original robopet, does just fine without encoders. The subsumption architecture design of Huey's software means that it reacts to its environment, so path planning and navigation aren't essential. Note, however, that encoders could be added, and then Huey's code could be modified to show other behaviors.

Wheel encoders open up a whole new world of experimentation for the amateur robot builder. With encoders attached to your machine's drive system, your software can measure distance of travel, detect wheel slippage, and use dead-reckoning to make reasonable guesses as to present location.

Despite all of the advantages offered by wheel or shaft encoders, I haven't used them on a machine yet, for a simple reason. I hadn't seen an encoder that I felt was suitable for designing into a robot for this column.

Sure, I've seen some real deals on the surplus market: encoders with gobs of resolution for pennies on the dollar. But if I built a machine using such a device, would the thousands of *Nuts & Volts* readers be able to duplicate my design? Probably not. And generally, mounting an encoder on a shaft or motor requires some form of machining or special adapters. These devices would likely be unavailable to readers as well.

So for a long time I had to do without encoders, waiting for a design that matched some pretty tough criteria. It had to be cheap, reliable, easy to construct, and it had to work with as wide a variety of motor systems as possible. Failing the last, it had to fit onto a motor or wheel system that was as widely available as possible.

As usual, the Seattle Robotics Society (SRS) provided the answer I needed. Mark Castelluccio showed off a robot some months ago, driven by hobby servo motors. His machine competed in the Dead-Reckoning contest, an event that requires accurate drive systems. Since "hobby servo motors" and "accurate drive systems" don't usually go together, I was surprised when his robot took first place against some stepper-based designs.

Mark said that he had installed a home-built encoder system inside the servo motor housing itself. I really like this concept; the encoder goes where the motor goes, and you don't have to worry about special shaft adapters or drive systems.

Based solely on Mark's statement that he had installed the encoders, but never having seen his implementation, I developed the following technique. Marvin Green, designer of the BOTBoards, tackled the same problem and came up with a different solution, which I'll describe briefly. Finally, I'll give you a quick rundown on the technique Mark actually used, so you can evaluate all three and pick the one that seems the best to you.

My design fits inside the case of a Futaba S148 hobby servo, one of the more popular servos used in hobby robotics. The servo with encoder looks and works just like a servo without the encoder, except that it has an extra three-wire cable coming out of the side of the case.

The only special component you need to make this mod is a Siemens SFH-900-1 IR emitter/detector unit. Granted, most of you may not have such a device in your junk box, but they are actually quite easy to get, and very cheap to boot. Just pick up your Electronic Goldmine catalog and look up their part number G3355. EG sells these devices, brand-new, for $0.50 each. They also list a price for 1000 pieces, so I figure they have a bunch on hand. You will also need some simple tools, such as an electric drill and a 1/8-inch bit, some ribbon cable, some 1/16-inch heatshrink, and a hot glue gun.

Begin by removing the four Phillips screws on the bottom of the servo motor case, then removing the top of the case (the part that has the Futaba name embossed on it). You need to locate and drill a hole in one side of this case top. When you are done, whichever side you use will have a blob of hot glue stuck on it, and it will have a short length of three-conductor cable coming out of it. If you intend to mount this motor on its side, take the effects of this mod into account when you choose which side of the case to drill. The mod will work with either side, but subsequent motor mounting could get tricky.

The mod requires you to mount the IR emitter/detector (E/D) inside the servo's case, so the component sits about 1/4 inch above one of the gears and looks down on the gear's surface. This means that the placement of the hole you are about to drill is critical; you don't have a lot of room to work with inside the reassembled motor case. The hole must be placed exactly 1/4 inch up from the bottom edge on the top of the case (the open section that fits over the servo PCB), and exactly 3/4 inch from the wall next to the Futaba legend. See the accompanying hole placement guide for details (Figure 1).

Mark the hole's location, then clamp the case top in a bench vise or other holder. Carefully drill out the hole, then clean and deburr the inside of the case. Remove any loose plastic shavings so they don't get into the gear mechanism later.

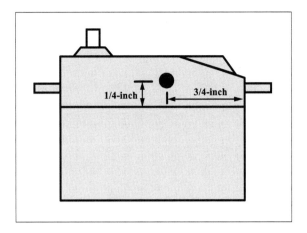

Figure 1. Hole placement details for modifying a servo motor.

Now temporarily insert the E/D through the hole in the case from the inside, with the leads sticking out. Make sure you have the E/D's active elements, coated with a deep blue epoxy, pointed downward, toward the open part of the case top.

With the bottom edge of the E/D flush against the inner wall of the case top, carefully bend the device's leads so they fit flat against the outer wall. You should bend the leads toward the open part of the case top, so they extend beyond the edge of the

case. Tape the leads in place, using a small piece of masking tape. This should leave the E/D positioned inside the case, with the leads taped flat against the case (Figure 2).

Carefully put the pieces of the servo back together. You don't need to install the four screws; you are simply checking for fit and alignment. Make sure that the E/D doesn't interfere with rotation of the servo's gears in either direction. You can test this by clamping a control horn or wheel onto the

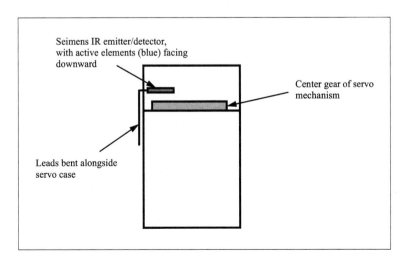

Figure 2. Internal view of servo case, with emitter/detector installed.

motor, then gently turning the motor shaft back and forth. If necessary, reposition the E/D, using the masking tape to hold the device in place.

When you are satisfied with the alignment, separate the case top from the rest of the servo assembly. Heat up the hot glue gun, then apply a large bead of hot glue to the hole on the outer wall, filling the hole and covering no more than 1/8th-inch of the device's leads. Remember that you still have to solder wires to the leads; make sure you leave yourself with enough length on the leads. After the hot glue sets, put the servo back together without using the screws, and verify that the motor still turns freely. If necessary, reheat the glue and reposition the E/D.

Separate the top case again and prepare a length of three-conductor ribbon cable. You can use a piece as long as the servo motor cable, if you like. Strip 1/16th-inch of insulation from each wire at one end of the cable, then tin the three wires. Separate the three wires for about 1", then slide a 1/4th-inch length of heatshrink over the center wire and move the heatshrink down the wire, safely away from the bare end of the wire.

Carefully solder each wire to a pin on the E/D. Make sure you connect the center lead of the ribbon cable to the center lead of the E/D. After the leads are soldered in place, slide the heatshrink up over the connection to the center lead, then shrink it in place using the tip of your soldering iron.

With the wiring complete, apply a blob of hot glue to the E/D's leads to complete fixing it in position. Make sure you apply glue to both the case material and the leads, and be sure to cover as much of the leads as you can. Note that you will still have bare leads protruding beyond the lower edge of the case top.

Now set aside the case top and pick up the gear assembly. Note the large center gear, sitting flat against the gear platform, positioned midway between the outer two gears.

Find a 45-degree arc section of the upper surface of this gear that is as free of gear grease as possible. If necessary, use a cotton swab dipped in rubbing alcohol to clean off a 45-degree section. Take care to remove only the smallest amount of grease necessary, and do not disturb the grease on any other gears in the assembly.

Using a black Sharpie marker, carefully fill in the cleaned 45-degree section of the gear surface. Apply one coat of ink, let it dry, then apply a second coat. When dry, this ink will serve as the encoder signal for the E/D. Carefully rebuild the servo motor and tighten the four screws to complete the assembly. This completes installation of the servo motor encoder system.

The encoder you have just added provides 45 pulses per each wheel rotation. While this isn't up to the hundreds or thousands of pulses per rev available from commercial units, it only set you back a buck or two. What's more, you can move this motor from one robot to another without having to deal with mounting problems beyond where to put the motor. Next up is testing the encoder and adding some simple electronics.

Electronics

The E/D detects the darkened section on the servo gear by the amount of IR light reflected off the gear's surface. When the IR emitted by the LED half of the E/D falls on the white, untreated surface of the gear, a large amount of IR reflects back to the detector. This in turn creates a low resistance across the phototransistor and a correspondingly low output voltage.

Conversely, the darkened section of the gear reflects much less IR. This causes the phototransistor to show a large resistance. By adding a resistor pulled up to +5 VDC, you can use the resulting voltage across the phototransistor to determine when the dark section of the gear falls underneath the E/D.

Finding the proper pullup resistor takes a little tweaking. The resistor's value will vary from servo to servo, depending on many conditions. It is affected by the alignment and orientation of the E/D, by the density of the ink you put on the gear, and even by the type of pen you use when you mark the gear.

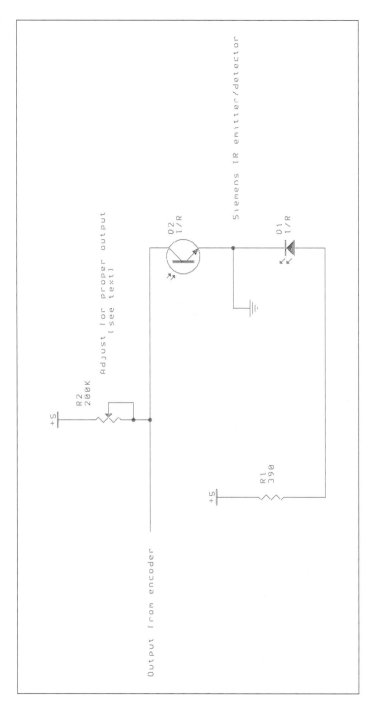

Figure 3. Schematic of encoder test circuit.

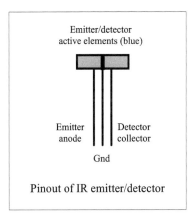

Figure 4. Pinout of emitter/detector device.

Fortunately, you can build a simple test circuit on a Radio Shack plug-in experimentor's board, and quickly find the best resistor value to use. Refer to the accompanying test circuit schematic for details (Figure 3). Also refer to the accompanying pinout diagram of the E/D device. (Figure 4). Always begin your tests with a current-limiting resistor of 390 ohms. You may need to change this value later, but it will work in most cases. This value is far larger than what I usually use for driving LEDs, but the reduced current is important. If you use too much current to drive the IR emitter, you end up with IR bouncing all over the inside of the servo's case, giving you false readings.

Set the trimpot to its midrange, then apply power and monitor the voltage at the junction of the detector's collector and the trimpot. Very slowly rotate the motor's drive shaft until you get as high a reading as possible; this shouldn't require more than 90 degrees of turn. Adjust the trimpot until you measure a voltage of at least 2.7 volts. If you get a reading above 3.0 volts, adjust the trimpot to bring the voltage down below 3.0 volts.

Now slowly turn the motor shaft until the voltage drops dramatically. This means the darkened section of the gear has moved underneath the E/D. The voltage should be below two volts, ideally below 1.5 volts. If it is not, adjust the trimpot until the

voltage falls below 1.5 volts. Repeat this rotate and tweak operation until you read a voltage above 2.7 volts for the light portion of the gear, and below 1.5 volts for the dark portion of the gear. When you are satisfied, disconnect the trimpot and measure the resistance between the two terminals you had connected to the circuit. This value represents the optimum pullup resistor value for that motor.

You will have to perform this same setup with any servo to which you add an encoder. It's a good idea to mark the pullup value for each motor somewhere on the motor's case, or on the cable coming from the motor's case. You will need this information later, when you hook up multiple motors in a robot.

You can now add circuitry, similar to the test circuit, to your robot computer boards. Add one set of the circuitry for each encoder you intend to use. Note that you don't have to find the exact value for the pullup resistor, as determined by your setup. Just use the nearest common value. For example, if your test shows a 107K resistor is optimum, you can use a 100K or 110K without any problems.

If you choose, you can also leave the trimpot in place when you add the encoder circuitry to your robot. You would need to readjust the trimpot if you change motors, but that would only take a few minutes, and would only need to be done once.

In some cases, you may not be able to get the above cited voltage range from your encoder. If so, you have a couple of options available. The easier option involves the darkened area of the gear. Go over the arc you marked out again, making it as black and evenly-coated as you can. In extreme cases, you might try cutting out an arc-shaped piece of black construction paper and actually gluing it to the gear's surface.

Less easy but also worth trying is to vary the LED current-limiting resistor. If you cannot get the encoder's output voltage to fall below 1.5 volts, try increasing the current-limiting resistor to 470 ohms. If the output voltage won't go above 2.7 volts, try decreasing the resistor to 270 ohms. If possible, you need to use the same current-limiting resistor for all of your encoder circuitry. This will make it easier to move your encoder-equipped motors between robots.

I hook the output of the encoder circuitry to one of the three input-capture pins on the 68hc11. These are port A pins 0 through 2, though some 'hc11 variations let you configure other port A pins for this function, as well.

Software

Now that you have the electronics and mechanics taken care of, you need a little software. The accompanying listing provides an SBasic program for moving a motor a prescribed number of pulses (Figure 5, at end of article). With it, you can determine how well your encoder system works. Plus, the program can serve as the starting point for developing your own encoder-based programs.

The program uses encoder interrupts on input IC1 (port A, pin 0) to increment a counter; each interrupt bumps the value in variable MTR_CNTR by one. The program also expects the servo motor to be hooked to OC2 (port A, pin 3). This software sets up the interrupts needed to handle the servo motor and the encoder, then sits in a loop, waiting for input from the user via the serial port (SCI). You can enter several one-letter commands to control the servo motor. Additionally, the software reports vital information back, concerning servo pulse width used to control direction and speed of rotation.

For example, you can enter a single number from one to nine. The software will turn the motor that number of revolutions, based on 45 counts per wheel rotation. After the prescribed number of rotations, the software will stop the motor in place.

Other commands available include a space for stopping the motor immediately, plus and minus to increase or decrease the motor's speed, 'f' and 'b' to set the direction either forward or backward, and 'z' to zero the encoder pulse counter MTR_CNTR.

While the motor is turning, the software also displays the current value of the servo timer, TOC2. This directly controls both the speed and direction of rotation. The values shown in the listing work for my motor, though you may need to tweak them for yours.

As you review this program, notice the simple code used to track encoder pulses, and how I use the information stored in MTR_CNTR to control the program's flow. You would need to use similar code in your robot to do navigation or path-planning. I have also included the listing for an SBasic file called servo.bas (Figure 6, at end of article). This library file contains code for setting up interrupts needed to control servo motors. This code shows how easily you can develop powerful modules in SBasic, then reuse your code in other projects.

Other Ideas

After I developed my encoder system, I got an up-close look at what Mark did to install his encoders. Basically, he had used a Dremel motor tool and a tiny saw blade to remove the plastic walls inside the servo motor case, where the servo potentiometer fitted originally. In the hollowed-out space Mark fitted a dime-sized piece of perfboard containing one IR LED, two separate IR detectors, and resistors similar to those used in my circuit.

Using two detectors gives 90 pulses per revolution, and Mark was able to align the detectors so that he actually got quadrature signals from his encoders. This means his encoders can not only determine rotational speed, but direction as well. Also, because all of his circuitry fitted inside the case, Mark was able to steal +5 VDC from the servo electronics. Thus, he only needs three wires for his encoder system: ground and two encoder outputs. I encourage readers with the proper tools and ambition to consider Mark's mod when deciding how to add an encoder. Though his technique isn't for everyone, the added capabilities may be what your project needs.

Marvin Green developed a method midway between those used by Mark and me. Marvin also carved out the plastic walls inside the motor housing, but he installed just a single Siemens IR E/D. The Siemens E/D units really lend themselves to this type of mod, and I'm sure many of you will be able to fit them into encoders of your own design.

You can use one of these techniques or create your own, but putting the encoder mechanics directly inside the servo case is a big win. You eliminate perhaps the biggest hurdle in optical encoders on robots, the uneven results caused by fluctuations in ambient light. Whether it's direct sunlight sneaking into the detector or noise from overhead fluorescents, optical encoders can prove hard to design. The dark, light-tight interior of a servo case is a perfect mounting place.

Well, that's it for this month. I'll close with a comment regarding software. I've avoided including software listings in the past, since it tends to fill up the article and I wasn't sure how much value it gave. But I've received a lot of support for my SBasic listings, and will try to make them a regular feature in the column. Apparently, many readers enjoy actually seeing how to solve some tough programming problems, and use my examples as jumping-off points for their own designs.

```
'
'  encoder.bas Program for testing encoder-equipped servo motors
'
'  This program allows the user to enter one-character commands
from
'  a serial port and control the behavior of an encoder-equipped
'  Futaba S-148 servo motor. Available commands include speed and
'  direction control as well as number of rotations.
'
'  This program assumes the servo motor is hooked to TOC2 and the
'  encoder output line goes to IC1. You can change the
'  InstallServo call below if you want to use a different TOC.
'
'  You must compile this program with SBasic11 version 1.0a or
higher.
'
include "regs11.lib" ' define the 68hc11 regs
include "servos.bas" ' use the servo library
const DELAY = 60 ' update rate (60 * 4.2 msecs)
const PULSES = 45 ' pulses per revolution
const FWD = 1
const BKWD = 2
declare n
declare wait
declare mtr_cntr
declare target
declare running
declare fspeed
declare bspeed
declare direction
```

Figure 5. Listing of encoder.bas, a program for moving hobby servo motors.

```
interrupt $fff0 ' RTI isr
if wait <> 0 ' if timer not yet at 0...
wait = wait - 1 ' drop it
endif
pokeb tflg2, %01000000 ' rearm RTI
end

interrupt $ffee ' ic1 isr
mtr_cntr = mtr_cntr + 1 ' count this pulse
pokeb tflg1, $04 ' rearm ic1
end

main:
pokeb baud, $30 ' 9600 baud
pokeb sccr2, $0c ' turn on SCI
mtr_cntr = 0 ' init the pulse counter
pokeb tmsk1, $04 ' allow interrupts on ic1
pokeb tflg1, $04 ' clear flag for ic1
pokeb tctl2, $10 ' interrupt on rising edge only
wait = 0 ' clear the wait timer
pokeb tmsk2, peekb(tmsk2) or %01000000
pokeb tflg2, %01000000 ' start RTI system
gosub InstallServo, toc2 ' install a servo on toc2
interrupts on ' allow interrupts
target = 0
running = 0
fspeed = $f800
bspeed = $f200
direction = FWD

print
print "encoder.bas"
print
do
if wait = 0 ' if time to update...
wait = DELAY ' rearm wait timer
printu "\rmtr_cntr = "; mtr_cntr; " ";
printu " target = "; target; " ";
if direction = FWD
printx "fwd speed = "; fspeed; " ";
```

Figure 5 continued.

```
else
printx "bkwd speed = "; bspeed; " ";
endif
endif
'
' Check for a key from the user; if key is pressed, do the com-
mand.
'
' You can add extra commands by inserting your own CASE statements
' in the large SELECT structure below.
'
n = inkey() and $ff
select n
case 'z' ' zero the pulse counter
mtr_cntr = 0
endcase
case 'f' ' go forward
poke toc2, fspeed
direction = FWD
endcase
case 'b' ' go backward
poke toc2, bspeed
direction = BKWD
endcase
case ' ' ' stop the motor
poke toc2, 0
endcase
case '+' ' increase speed
if direction = FWD
fspeed = fspeed - $40
else
bspeed = bspeed + $40
endif
endcase
case '-' ' decrease speed
if direction = FWD
fspeed = fspeed + $40
else
bspeed = bspeed - $40
endif
endcase
```

Figure 5 continued.

```
case '1'  ' set pulse count = n * 45
case '2'
case '3'
case '4'
case '5'
case '6'
case '7'
case '8'
case '9'
target = (n - '0') * PULSES
running = 1
endcase
endselect
'
' This code checks the motor. If it is running, the code
' then tests to see if the pulse counter has hit the target.
' If so, this code stops the motor and resets the target value.
'
if running <> 0
if mtr_cntr >= target
poke toc2, 0
running = 0
target = 0
endif
endif
loop
end
```

Figure 5 continued.

```
`
`  InstallServo set up a timer output channel to control a servo
`
`  Invocation: gosub InstallServo, toc
`  or: n = usr(InstallServo, toc)
`
`  When called as a function (USR format), InstallServo returns a 0
`  if the setup worked, or -1 if the toc argument was invalid.
`
`  Upon exit, the specified TOC is set to 0, effectively stopping
the
`  motor.
`
InstallServo:
select pick(0)
case TOC2 ` if this is TOC2...
push $c0 ` push the proper TCTL1 value
push $40 ` push the proper OC1D/OC1M value
endcase
case TOC3
push $30
push $20
endcase
case TOC4
push $0c
push $10
endcase
case TOC5
push $03
push $08
endcase
drop 1 ` drop invalid TOC argument!
return -1 ` report a failure
endselect
pokeb oc1d, peekb(oc1d) and (pick(0) xor $ff) ` clear bit on TOC1
pokeb oc1m, peekb(oc1m) or pull() ` connect to TOC1
pokeb tctl1, peekb(tctl1) or pull() ` set bit on TOC
poke pull(), 0 ` clear TOC for now
poke toc1, 0 ` TOC1 hits at 0
return 0 ` show all went well
```

Figure 6. Listing of servo.bas, a program for initializing servo motor control.

Part V
Robotics Projects

Here are several robots that you can build for your own use. I include a machine for running through a table-top wooden maze that the SRS calls the Rapid Deployment Maze (RDM). You can also find a two-servo platform that runs a marble through a maze by tilting the platform back and forth.

I built BYRD out of a roll-around beer cooler, and had great fun using its remote-control feature to drive it around in my back yard. The beer-cooler concept gives you plenty of construction space, and you can get some nifty designer colors on your big robot frame if you go this route.

The Time-Speed-Distance (TSD) rally format has been a favorite among sports car drivers for years. I've tailored that contest for robots, and did an article that covered such concepts as rally layout, instructions, traps, and time checks. I even included a sample layout, complete with quizzes for testing your own rally skills. Building a robot for this event would prove a serious challenge, but running such a contest would be great fun.

For those interested in precision mechanical motion, check out my article on the stepper-based robot design. It includes a discussion of the SRS' Dead-Reckoning event, which puts a premium on mechanics and software design. Speaking of software, this article includes SBasic code to help you get your 'bot up and running.

I took a swing at building a tiny robot, and ended up with Hercules. By today's standards, Hercules isn't really very small, but my motor skills aren't very good, either. Still, the techniques used to build this little machine show how easy it is to construct a small robot, and you can probably get your own Hercules going in just a weekend or two.

The Mars Pathfinder took the hobby robotics community by storm, and Tacklebot was my response to the landing on Mars. I built Tacklebot for exploring my back yard, and using a tacklebox for the frame solves a number of design problems from the get-go. You can have your own version running in just a few evenings, and spend the next several weeks tinkering with the design.

The Mini-Sumo robotics contest offers robot builders a serious challenge; to design a small robot that can successfully find a competing robot and push it out of the Sumo ring. I did an article that introduces you to the world of Mini-Sumo, and even includes SBasic code with the core set of features you will need.

I did several columns on my attempt to build a robot to compete in the annual Trinity College Fire-fighting contest. I never did get the robot done, due to various distractions, but this inaugural article describes the event in detail, and should get you started on a machine of your own. Hopefully, you'll make more progress than I did.

The Rapid Deployment Maze

As a crowd-pleasing demo event, you'll be hard pressed to match the Rapid Deployment Maze (RDM). Two competitors try to steer their robots through a maze, each attempting to be the first to push a ping-pong ball through a common goal. RDM robots are simple to build; it shouldn't take more than a few evenings. The power supply for the event will take even less time to construct. I can speak from experience about running a robot in the event; it is a real hoot!

This column, published in September of 1993, also marked the first time I used one of the CGN 68hc11 modules. Each board features a complete 68hc11 system on a single wirewrap footprint, ready to hook into your larger design. At $19 each, they saved a lot of time and made an excellent addition to the robot builder's hardware toolset.

I have spent the last year trying to simplify my robot construction. I wanted to reach the point that I could design and build a working robot in one week. The first three articles in this book are the result of my efforts. A single-chip 68hc11 computer board, mated with a pair of converted hobby servo motors, forms the core of a small robot.

You can then build a customized frame from printed circuit board (PCB) stock and brass rod, add four NiCd batteries and a little software, and produce a working robot for science fair projects, research, or entertainment.

I encourage you to refer often to these three. In future projects, I will mention these articles in describing a robot's design. You will need to check these past articles for details on construction.

CGN

I use a small 68hc11 single-chip computer for most of my robots. The circuit, supplied in previous articles, is quite simple and fits on a small Radio Shack experimenter's board. But after building three or four of them, even wiring this simple circuit became unacceptably tedious.

CGN offers several small PCBs that dramatically simplify construction of 68hc11 projects. These boards contain a 52-pin PLCC socket and the needed components to build either a single-chip or expanded-mode 68hc11 system.

Connections are brought out on wirewrap pins. You simply install the CGN board onto a larger piece of perfboard, using wirewrap wire to hook up connectors, sensors, and other components. The CGN 1001 board carries the 52-pin socket, an 8.0 MHz crystal, a 34064 reset controller, and assorted passive components. You only need to add a 68hc11 chip, and +5 VDC and ground connections to have a working computer. The 1001 measures 1.75 by 1.75 inches, and is about 1.25 inches thick, including wirewrap pins.

The 1001 brings all 52 68hc11 pins out on wirewrap posts. Refer to the accompanying schematic for the few connections needed to make a 68hc11 robot controller (Figure 1). The 52 pins on this board correspond functionally to the 52 pins on a 68hc11a1. This means that pin 28 of the CGN board supplies the OC2 signal, just as pin 28 of a 68hc11a1 does.

Therefore, you can think of this board as a "super" 68hc11 chip. You wire it into your circuit just as you would use a real chip. You just don't need to wire in the boring support circuitry.

CGN even supplies a small wirewrap ID guide. This is a paper overlay, pre-printed with pin numbers and names, that slides down over the wirewrap pins on the underside of your project board. It lets you see at a glance the location and function of each wirewrap pin as you wire your board.

I paid $19 plus shipping for my CGN 1001, and figure I saved at least that much in construction time. Note that you cannot build as small a computer board with the CGN module as you can by custom wiring your own board. For one thing, the CGN module's wirewrap pins will make your finished board much thicker than if you built a point-to-point version.

But your $19 buys you reduced construction time, more reliable construction, and freedom from chasing down parts. CGN also sells the 1101 module (about $35; call for exact pricing). This board contains surface-mount components for running the 68hc11 in expanded-address mode. You can then connect up to 64K bytes of RAM/EPROM to your computer board, for those really big robotics projects.

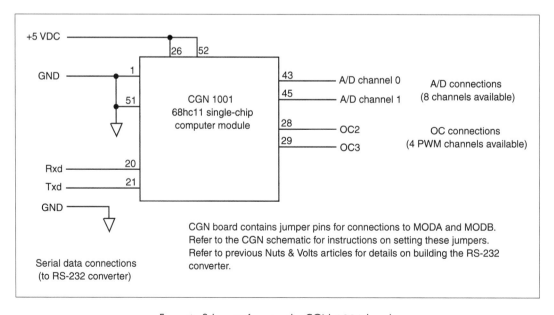

Figure 1. Schematic for using the CGN 1001 board.

As with the 1001, all connections are brought out as wirewrap pins. On the 1101, however, the pins supply a fully demultiplexed set of address and data lines. I am already planning a future article using this board.

Note that using the 68hc11 in expanded-address mode costs you a lot of the chip's I/O capabilities. However, the Rug Warrior (described in *Mobile Robots*, by Jones and Flynn) uses the 68hc11 chip in this mode and it still contains a rich array of sensors and actuators.

CGN also sells a variation of both of these boards that contains on-board RS-232 conversion. I still prefer to have the RS-232 conversion done by off-board circuitry, ideally built into a DB-25 connector shell. But CGN offers you an alternative, for just a few bucks more than the basic 1001 or 1101 modules.

CGN also offers the Motorola 68hc811e2 chip. This version contains 2K bytes of EEPROM, rather than the 512 bytes in the standard 68hc11a1. Consequently, there is a greater demand for the chip and it has proven quite difficult to get. I purchased two chips from CGN for about $20 each. Note that this is a reduced price, available only when you order 1001 or 1101 boards.

I was impressed with the quality of the CGN boards, and by the prompt delivery. Note, however, that CGN does not accept credit card purchases. I used COD and the shipment arrived about three days after I placed my order.

Incidentally, the order was short two modules (I had ordered four units). I phoned the company, explained the problem, and the missing two units were promptly shipped out. A good company offering a quality product at a fair price; recommended.

The RDM

Members of the Seattle Robotics Society developed an event called the Rapid Deployment Maze (RDM) contest, to fill a need for demonstrations. The RDM platform fits on a large table, folds up into a two-person carrier, and can be easily set up in schools, libraries, or malls whenever a quick robotics demo is required.

The contest involves two small robots, connected via ribbon cable tethers to a power supply and controlled by human contestants. Each contestant tries to maneuver his robot from the starting point to a ping-pong ball, then wrestle the ping-pong ball through the maze to a common goal in the maze's centerline.

The RDM platform consists of a 4 foot by 4 foot wooden maze, hinged along the centerline and fitted with clasps so it can be folded up and latched into a single suitcase-type form. Refer to the accompanying drawing for details on the maze layout (Figure 2).

Each corridor of the maze measures about 7.5 inches across. This limits the size of the robot that can run the RDM contest; a robot too large won't be able to turn around inside a corridor. Along with the maze platform, the RDM uses a plastic toolbox, 16 inches long by 7 inches wide, fitted with two 12 VDC power supplies, an AC switch, and four DB-9 connectors. Two 18-foot long, 9-wire ribbon cables, with a male and female DB-9 connector at opposite ends, serve as tethers between the power supplies and the two competing robots.

The controls used to steer the robots are patterned after the Atari joysticks. I used five industrial-grade pushbuttons for each controller, for the FWD, BKWD, RIGHT, LEFT, and CMD buttons. The first four buttons are arranged on the controller's panel in a diamond shape (with FWD at the top), while the CMD button sits just above the FWD button.

I mounted the switches on blue plastic Radio Shack project boxes that are just the right size to hold in my hand. The boxes are made of ABS plastic, which can be easily worked with a file, drill, and Exacto knife. I brought the switch connections out to a 6-foot long cable made of 26 AWG stranded wire. The exact wire size isn't critical, but the cable should be very flexible so that it doesn't hamper movement during the contests. Note that I originally tried to use Atari joysticks as controllers, but the switches inside the Atari devices are just not reliable enough for this purpose.

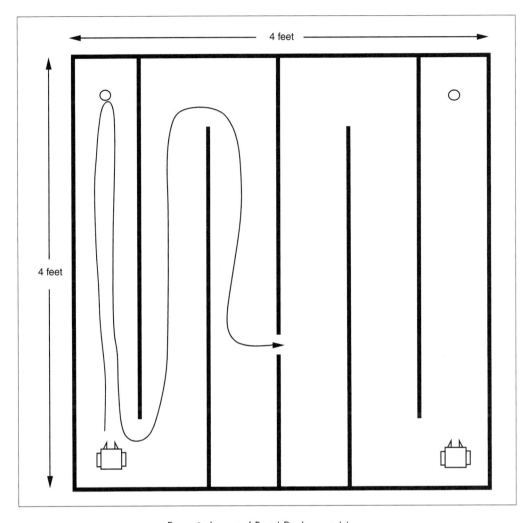

Figure 2. Layout of Rapid Deployment Maze.

Wire these five switches as shown in the accompanying schematic (Figure 3). The pin numbers supplied in the diagram refer to the DB-9 connectors. Note that the cable from each controller should end in a DB-9 female connector.

Next, select a plastic tool box for use as the RDM power box. It should be large enough to accomodate two surplus 12 VDC power supplies, each capable of providing about 2 amps of current. Note that you may choose to wire these up your-

self, if you like. I just went down to the local surplus shop and bought a couple of 3-amp units for about $17 each.

Mount the power supplies inside the tool box. Also mount an appropriate on/off switch, fuse block, and 120 VAC cord connector on one end panel. I used a PC-style connector block for the latter. This lets me unplug the unit's power cord from the outside of the box, then put the cord inside the power box for transport or storage.

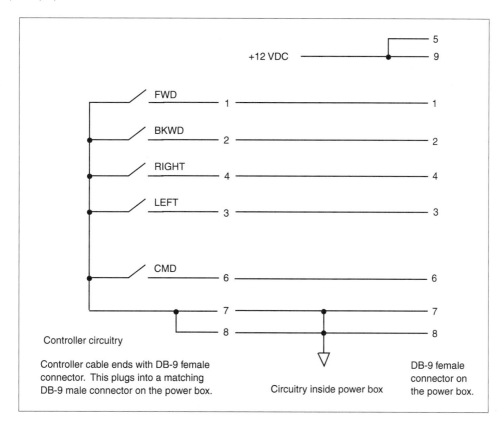

Figure 3. Schematic for RDM control cable.

Wire the power supplies, power switch, AC connector, and fuse block. Use stranded 18 AWG or heavier wire for these connections, and double-check your wiring. This section will have 120 VAC connected to it during operation, so use proper care to avoid exposing dangerous voltages.

Cut appropriate holes in the opposite end panel of the tool box for two DB-9 male connectors and two DB-9 female connectors. Refer to the accompanying suggested layout drawing (Figure 4).

Wire these DB-9 connectors per the accompanying schematic (Figure 3). Each 12 VDC power supply connects to only one of the DB-9 female connectors on the power box's panel. This isolates the two competing robots. Finally, cut two 18-foot lengths of 9-conductor ribbon cable. Clamp a DB-9 male IDC

connector on one end of each cable. Clamp a DB-9 female IDC connector on the other end of each cable. Wire these connectors for straight-through operation; that is, wire pin 1 of one connector to pin 1 of the other, pin 2 to pin 2, etc.

Plug the power box's AC cord into a wall outlet and turn on the power switch. You should measure 12 VDC at the appropriate pins of each DB-9 female connector on the side panel of the power box.

Next, connect both tethers to the DB-9 female connectors on the power box. You should measure 12 VDC at the appropriate pins of the DB-9 female connectors at the ends of both tethers. This completes construction of the RDM power box and controllers. Now all you need is a robot to run in the maze.

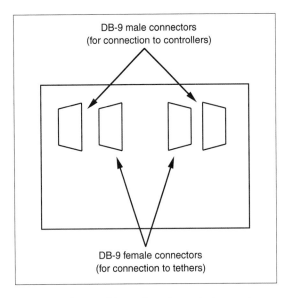

DB-9 male connectors
(for connection to controllers)

DB-9 female connectors
(for connection to tethers)

Figure 4. Suggested layout for side panel of power box.

RDM Rules

To compete in the Seattle Robotics Society RDM contest, a robot must fit within a cube measuring five inches on each side. Each robot will be measured prior to starting a contest run. Since the robot will draw power from the tether, it doesn't need to carry batteries; this helps reduce the size of the machine.

Each robot must have a DB-9 male connector on it, for connection to the RDM tether. A robot may draw up to an amp of current through the tether.

A robot receives its control signals as switch closures on the appropriate tether lines. Note that these signals are simply closures to ground; the robot's circuitry must supply pullups to 5 VDC to turn these closures into digital signals.

This obviously means that the robot's circuitry must also supply some means of converting the tether's 12 VDC to 5 VDC for powering its onboard logic. Note that if you use a 7805 regulator for this purpose, you will have to provide a hefty heatsink.

RDM contests consist of best two-out-of-three runs between two competitors. The event's judge

signals the start of the event by switching on the power to the tethers and announcing "Go!" Each competitor must drive his machine up to the ping-pong ball, then move the ping-pong ball back through the maze to the opening in the center wall of the maze. The first competitor to break the plane on his side of the center wall with his ping-pong ball is the winner of that run.

The rules state that a robot must fit within the five-inch cube before the event starts. Once power is available, however, a robot may reconfigure itself as needed. Note that any such changes must be done only by the robot's mechanism and only as triggered by signals received through the tether.

No robot may carry any extra power source of any kind; all power must be derived from the 12 VDC supplied through the tether. Most robot designs use the CMD signal to activate some type of grabber mechanism. You could also use this signal to throw some type of obstruction onto your opponent's playing area. Note, however, that any deliberate attempt to damage your opponent's machine is forbidden; violators will be disqualified from the event. Similarly, the judge may disqualify any competitor whose robot is deemed a danger to spectators or other contestants.

My RDM Robot

I ran an RDM robot in the recent Robothon Northwest, taking second place. My machine was based on a handwired single-chip 68hc11 computer, as described in previous articles. (See "Inspiration and Implementation" and "Your First 68hc11 Microcontroller.") I hooked two converted hobby servo motors to OC2 and OC3; this gave me motive power. I built a frame from blank PCB stock, using 4-40 hardware to support the 68hc11 circuit board off the frame. I used the blank PCB stock as a heatsink for a 7805 regulator, which in turn supplied power for the 68hc11 circuitry. The 7805 never got hot, despite running for an hour or more at a time (people really love to play with these machines!).

I mounted the hobby servos to the underside of the PCB frame at the back end, using double-sided tape and nylon wire ties to clamp them in place. I attached 2-inch wheels to the servos' control horns, as outlined in "The Basics of Hobby Robotics." I used a red plastic drawer knob, purchased from a local hardware store, as a front skid. I mounted the knob to the underside of the PCB frame, using appropriate spacers and hardware.

I mounted a third hobby servo (this one was not converted) to the top of the PCB frame, again using double-sided tape and nylon cable ties. I positioned this servo so that the surface of its control horn was aligned with the side of the PCB frame. Refer to the accompanying diagram (Figure 5).

Using two 2-56 machine bolts, washers, and nuts, I bolted a short length of 1/16-inch brass rod to the control horn. I had previously shaped the brass rod into a two-prong fork for capturing a ping-pong ball. I then connected this unconverted servo motor to OC4 on the 68hc11 computer board. I also wired the five control signals from the RDM tether to selected pins on port C of the 68hc11 board, using a 10-pin header and matching 10-pin IDC ribbon cable.

All that remained was the software. Basically, the software simply watches the appropriate inputs on port C, looking for a low on one of the lines. If it sees a low on only the FWD, BKWD, RIGHT, or

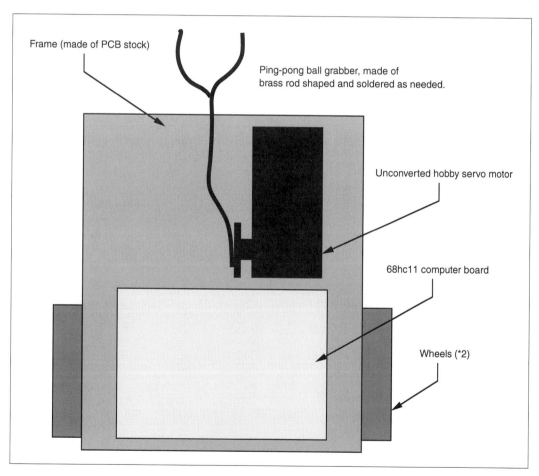

Frame (made of PCB stock)

Ping-pong ball grabber, made of brass rod shaped and soldered as needed.

Unconverted hobby servo motor

68hc11 computer board

Wheels (*2)

Figure 5. Layout of my RDM robot.

LEFT lines, it steers the robot accordingly. If the software sees a low on the CMD line, however, it behaves differently. In this case, a simultaneous low on the FWD line moves the grabber to its lowest position, snaring the ping-pong ball. Similarly, a simultaneous low on the BKWD line moves the grabber to its highest position, so it is out of the way of the ping-pong ball.

One interesting feature of the design concerns the use of an unconverted servo motor. If a servo has been modified, writing a value into its TOC register translates into a direction and speed of rotation. If a servo has not been modified, however, that same value translates into a position. Thus, the equates I used for speed and direction for the wheels' servos serve as position equates for the grabber servo.

The software is small and simple, and loads quickly into the 68hc11's EEPROM, using pcbug11 as described in previous articles (See "An Intro to 68hc11 Firmware.") Remember to set the RXD and TXD lines properly before you download and before you try to run the code.

I really enjoyed building this little machine and watching it perform during Robothon. Judging by the fun others had playing with it, the RDM contest has a bright future.

> Boy, was this project a blast! I built BYRD, my backyard research drone, from a LARGE beer cooler on wheels, and with help from a couple of guys from the Seattle Robotics Society, got my tele-operated machine going in record time. My December 1995 column for *Nuts & Volts* provides plenty of details on the subsystems I built for BYRD.
>
> Perhaps the most interesting element of BYRD's design was the firmware that I stashed in the top 2K of EEPROM in the 68hc811e2 MCU I used. The firmware, coupled with the RF modems I had picked up earlier on the surplus market, let me do a full program download and reboot, even if BYRD was outside or in the next room. This feature was way cool, and sped up my code development something wonderful.

Build BYRD, a Back Yard Research Drone

I now have my all-terrain robot (ATR) up and running. I've named the machine BYRD, for Back Yard Research Drone. BYRD will serve as the core to several upcoming articles, and you can build your own BYRD for about $200, depending on your junk box and access to tools.

Much of the following descriptions will refer to robot subassemblies discussed in previous *Nuts & Volts* columns. I built BYRD around a Rubbermaid wheeled beer-cooler, purchased from a local hardware store. Regularly $80, I waited for a big sale and picked one up for considerably less.

The cooler is quite light, but very rugged and nearly waterproof. The molded plastic construction leaves lots of places to fasten robo-thingies, and you can easily hot-glue, bolt, or foam-tape your add-ons wherever you need. Besides being perfectly designed for robot retro-fit, this cooler really looks, well, cool. The body is bright green, the top is grey, and it has a large, purple plastic handle for pulling; the handle folds down onto the top and locks in place when not needed.

The cooler stands about 22 inches tall, and sports four 7-inch hollow plastic wheels. The front wheels are actually independent casters, which

Figure 1. A portrait of BYRD.

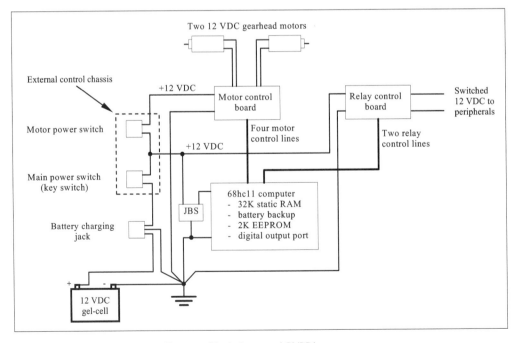

Figure 2. Block diagram of BYRD's system.

means that one of the hardest aspects of robot frame design is already done for you. The original back wheels are free-spinning, mounted onto a single axle that clips into a groove along the back of the cooler. Dan Mauch, one of the Seattle Robotics Society's top machinists, pulled the back axle off, then mounted a motor support beam along the cooler's underside. This beam holds a pair of Dayton 2L011 12 VDC gearhead motors. I fastened two 7-inch diameter lawn-mower wheels onto 1/2-inch motor shaft adapters, then fastened the shaft adapters onto the two motor shafts. The finished frame tilts forward at a rakish angle, giving BYRD a devil-may-care profile.

Once I had the motors and wheels mounted, I was ready to start on the interior. The cooler has a huge cavity, large enough to hold ice and a couple of six-packs. Mine now holds a pile of electronics. Refer to the accompanying system block diagram (Figure 2).

Electrical System

I started with a pair of 6 VDC, 10 AHr gel-cell batteries, wired in series to give 12 VDC. The wiring harness mates to the batteries using standard female spade lug connections. I used 18 AWG stranded wire for all high-current wiring. The ground connection runs to a single large lug, which serves as a common ground return for all electronics. A common ground point plays an important part in keeping down electrical noise within the robot.

This ground point appears in the system diagram as a single large ground symbol. Note how all subassemblies have their grounds returned to this one point.

The battery's positive lead runs directly to a SPDT charging jack, mounted inside the robot's cavity. The charging jack passes power to the rest of the system until I plug in a battery charger. Then, the jack disconnects all the other electrical systems, and routes the charging power to the battery. I used a Radio Shack 274-1565 DC power jack, covering the wire connections with heat-shrink. I didn't bother mounting the charging jack onto the robot's frame; instead, I just let it flop around inside the cavity.

There are no exposed voltages, and I don't mind lifting the hood to plug in the charger.

The battery voltage from the charging jack next runs to a key switch, mounted in a metal chassis on the outside top of the robot. This key switch, a Radio Shack 49-515, sports a weatherproof spring-loaded cover. The key switch lets me securely turn off all power to the robot, so I can let it sit unattended without worrying that someone might start it up on their own. I also mounted a red LED and a 1.2 K dropping resistor in the control chassis, wired to the switched side of the key switch. This gives me a visual clue when main power is turned on.

I ran an 18 AWG wire from the switched side of the key switch back to the main electronics. All electronics except the motor control board get 12 VDC whenever the key switch is turned on. I also mounted a large SPST toggle switch in the control chassis, wired into the switched side of the key switch. The other lug of this switch provides power to the motor control board, via an 18 AWG wire. This toggle switch acts as a "panic" switch, allowing anyone to quickly shut off motor power should BYRD suddenly lose control. I also installed a red LED and dropping resistor in the motor voltage line, so I can tell when the motor voltage is on.

I use two battery chargers with BYRD. One is a standard wall-wart unit, based on a LM317 circuit from an earlier column, "A Gel-cell Battery Charger for Cheap." This device can recharge BYRD overnight, and I usually let BYRD trickle-charge when I'm not using it.

The second charger uses a large solar panel that I had lying around from somewhere. It provides 18 VDC in full sunlight, and I can use this charger outdoors during the summer. I built the solar charger by hot-gluing the solar panel onto a large piece of industrial-grade foam-core, then connecting it to a small circuit board containing my usual LM317 battery charger circuitry. The whole panel fits neatly inside BYRD's cavity, ready to use whenever the sun comes out. Which, living in Seattle, isn't often, but it's the thought that counts. Actually, I'd be better off building a charger around a waterwheel or a treadmill for slugs...

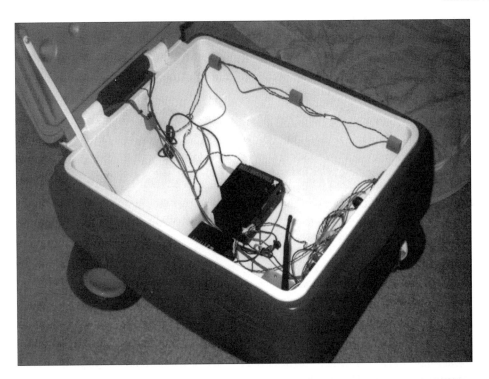

Figure 3. A look inside BYRD.

Main Computer

BYRD uses a 68hc11 board with 32K of battery-backed static RAM as its main computer. I knew from the beginning that the limited code space on a BOTBoard was not going to be enough room for my planned projects. I did, however, install a 68hc811e2 MCU in the board, so I could stash some special code in the 2K on-chip EEPROM; more about this later. The '811e2 chips are proving tough to find, but they are worth the effort. Going price is about $20 when you can get them, but many places are quoting multi-month shipping delays.

This board uses the '811e2 in expanded mode, so most of the MCU's I/O lines are lost to address and data busses. Since I needed some extra output lines, I wired a 74hc574 into the board's prototyping area, to act as an 8-bit output latch. Signals from this latch serve as motor control lines and relay control lines.

I used a Junk Box Switcher to convert the battery's 12 VDC down to the 5 VDC needed by the computer board (see "Try This Junk-Box Switcher Supply." The JBS runs completely cool, costs almost nothing to build, and fits in a corner of the frame's cavity, safely out of the way. I'll pass along a couple of tips to those of you who have built JBS boards. In some cases, you may find the MOSFET runs very hot. This means the device is running in its linear mode; that is, it isn't switching properly.

To correct this, replace the 10 Mohm resistor with a 1 Mohm device. You might see increased ripple, say about 75 mV in a 5 VDC output, but the MOSFET should begin operating in the switching mode. If even this doesn't do it, you might have too much inductance on the input side of the JBS. At one point, I had installed a 500 μH choke on the JBS' input. That increased inductance was enough to send the MOSFET into linear mode.

Motor Control Board

The motor control electronics consists of a hand-wired Radio Shack experimenter's board (276-168A) containing two TSC4427 MOSFET drivers, a pair of

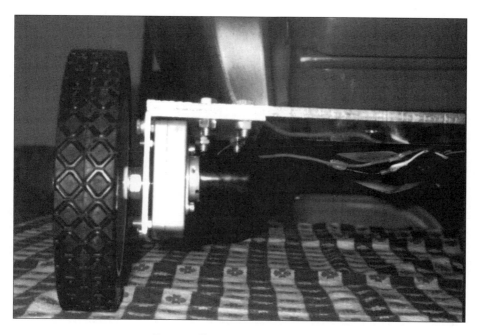

Figure 4. Close-up of the motor mount.

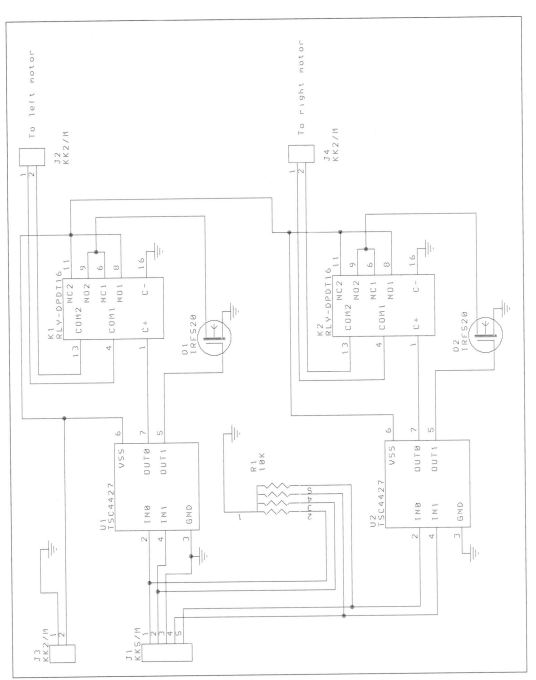

Figure 5. Schematic of BYRD's motor control board.

12 VDC DPDT relays, two N-channel power MOSFETs, and assorted passives (Figure 5). The relays are good for about 2 amps each, more than enough for the selected motors. You can get the TSC4427, or its equivalent, through Digi-Key or other large mail-order houses. This 8-pin IC uses digital level inputs to control higher voltage devices such as power MOSFETs. Each TSC4427 sports two switches, controlled by signals IN0 and IN1; a high on an input connects Vss to the corresponding output.

I have used the TSC4427 before in several circuits. It is cheap, easy to get, and nearly bullet-proof. The chip is rated to 600 mA, far more than a relay coil will draw, and the TSC4427's internal circuitry acts as a snubbing diode, surpressing the coil's back EMF.

You can find suitable relays for this board through any large surplus parts house. I used the miniature devices that plug into a 16-pin DIP socket; these go for a couple of bucks each just about anywhere. Make sure you get 12 VDC relays, since that's the voltage the TSC4427 supplies to the relay coils.

Just about any high-current N-channel MOSFET will work in this design; I used some IRF520s from my junk box, but similar devices will work as well. B.G. Micro carries a good line of N-channel MOSFETs, including the BUK-455-50, with an Rdson of 0.038 ohms; price is under $1 each.

Be sure to include the four pulldown resistors, R1, shown in the schematic. These resistors ensure that the TSC4427s' input lines all see a valid logic 0 whenever the cable to the computer board is disconnected. This stops the robot from "running away" if you pull the signal cable while you still have motor voltage on. I show a value of 10 Kohms, but just about anything from 4.7K to 20K will work as well.

Relay Control Board

The relay control board is a simplified version of the motor control board (Figure 6). It lets the 68hc11 turn 12 VDC peripheral devices on and off, using digital output lines. This comes in handy when you want to control a video camera, a tape recorder, floodlights, a siren, or other actuators. A single relay control board provides two switched 12 VDC output lines. The lines are switched on or off by two relays, similar to those used in the motor control board. These relays are in turn controlled by the output lines from a single TSC4427 MOSFET driver.

The two output connections on the relay control board are open-circuit if the corresponding TSC4427 input line is low; an output goes to 12 VDC if the input line goes high. Note that I have only wired up half of each relay in this circuit. You could easily wire up the second half of a relay, if you needed to switch two different devices at exactly the same time.

Serial Communications

I designed BYRD as a tele-operated robot; it exchanges signals with the operator via the 68hc11 serial port. While you could run a long wire between your host computer and BYRD, I chose instead to connect the two using a pair of 9600 baud RF spread-spectrum modems. I discussed these modems in an earlier column, "Wiring Up an RF Modem Link." The original modems, made by Proxim, are no longer available, but you can probably find similar RF modems by searching through the mail-order houses.

BYRD's RF link makes a huge difference in ease of use: no more fussing with serial cables whenever I need to reload BYRD's software. I don't need to pull the electronics out of the chassis; I don't even need BYRD in the same room or the same building. I just fire up my comm program and press a key; if BYRD is powered up and within 500 feet, it responds.

Part of this power comes from the firmware I've loaded into the 2K EEPROM onboard the 68hc811e2 on BYRD's computer board. The firmware, which I call 811bug, includes a full S19 loader, compatible with the Motorola S19 object records generated by SBasic, tiny4th, and the Motorola assemblers.

Following powerup, the 68hc811e2 jumps to 811bug and waits for a character from the serial port. If it sees a 'S' as the first character of a new text line, 811bug assumes that the entire line contains an S-record of some kind. 811bug's S19 loader handles the three most common S-records in an object file. It

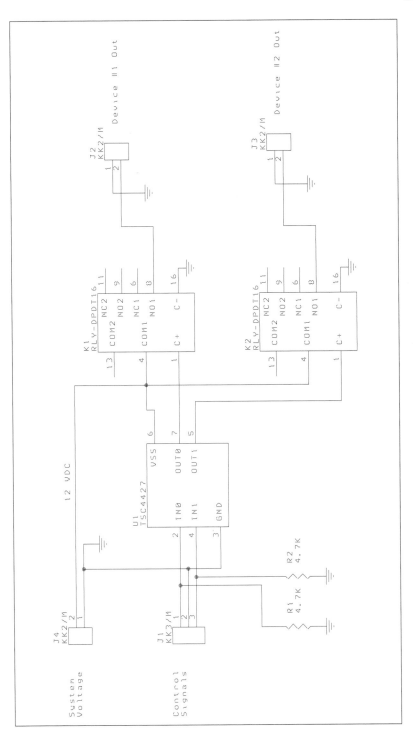

Figure 6. Schematic of BYRD's relay control board.

ignores the S0 header records. It processes S1 records by storing the supplied bytes in the appropriate memory addresses. Finally, it saves the execution address in an S9 record for later use.

This means that downloading a new program to BYRD is as simple as sending an ASCII file from my comm program. I don't even need to enter any kind of "load" command; I just start the file on its way, and the S19 loader handles the rest.

After I have loaded a complete S19 object file into BYRD, I send the character 'x' to BYRD to start execution. 811bug uses this character to signal a jump to the execution address contained in the most recent S9 object record.

Since 811bug resides from addresses $f800 to $ffff, it takes over the 68hc11 vector area. Therefore, I included a jump mechanism that routes almost all 68hc11 vectors through a dedicated area of on-chip RAM. This means that you have to modify your software if you want to use interrupts in programs downloaded with 811bug.

Specifically, most vectors automatically jump through dedicated addresses in the range $b0 to $ec. Each RAM vector becomes a three-byte sequence of the form:

jmp start

where START is the address of the 811bug firmware. To use the RAM vectors in your program, simply overwrite the 16-bit value of START at the appropriate spot in RAM with the address of your interrupt service routine (ISR). Do NOT overwrite the JMP opcode, or your program will die when the first interrupt occurs.

For example, the real-time interrupt (RTI) vector normally resides at $fff0. The 811bug firmware stores a $00d7 at this address, since the RAM vector for the RTI has moved to $00d7. 811bug then stores a JMP START sequence at address $00d7, before it loads any S19 records. This means that any unexpected RTIs will simply restart the 811bug firmware.

To use an RTI ISR in your program, your code must overwrite the value in $00d8 (NOT $00d7!) to contain the address of your ISR code. If you are writing in SBasic, you can perform this function with a single line of code. Assume your RTI ISR code begins at label RTIISR. Simply include the line:

poke $00d8, addr(RTIISR)

in the main code of your program. After setting up the RTI subsystem and turning on interrupts, your RTI ISR will begin fielding and servicing RTIs properly.

811bug does not treat all interrupts this way. In particular, the Illegal Opcode and COP Failure interrupts are hardwired to jump back to the start of 811bug. These two vectors serve as insurance; should your code go completely bananas, there is still a chance that 811bug can recover control.

I will post the source for the latest version of 811bug on my web site. I wrote 811bug to assemble with the Motorola asmhc11 FREEWARE assembler; you can get a copy of this assembler by downloading either tiny4th or SBasic from the same site.

Feel free to modify the source file and reassemble as needed. I have included a few lines of assembly that are specific to my version of BYRD; you will likely need to doctor 811bug for your platform.

811bug includes a few features and options beyond those discussed above. I have added plenty of comments and documentation within the source file, so you shouldn't have too much trouble figuring out how 811bug works and where to add your own enhancements.

I will save a discussion of BYRD's operating software for next month, (see "Remote Reloads with 811bug"). I will say that to date, the entire control program is written solely in SBasic; it contains not a single line of assembly language. This, despite using multiple interrupts and PWM speed control of the drive motors.

My thanks to Dan Mauch, Marvin Green, and Keith Payea for their help in building this prototype BYRD robot; had a lot of fun, guys.

Rally 'Round the 'Bot, Boys!

The robot rally project described in this column from the April 1996 issue of *Nuts & Volts* would make a super Science Fair or club contest. I patterned this event on the Time-Speed-Distance (TSD) rallies so popular nationwide. These rallies are great fun to run in real cars, and the concept scales down to robots very well. I begin this column with a quick overview of how a TSD rally works, and I even include a fairly large TSD layout for robots that you can use to practice on.

Next, I show you how various rally instructions would work, using the practice layout. I cover the notion of time traps and the odometer check. Then I describe how to convert some of the basic TSD elements to use in a robot event. This includes passing the rally instructions to each robot, the types of sensors that would be needed to support these basic elements, and a few tips on writing a robot rally.

I conclude this column by describing other, more sophisticated rally elements that you might include in your rally layout. Adding any of these elements means that competing robots must have a stronger set of sensors than those needed for the basic rallies, but these new elements will be fun to add and they give the rallymaster more options in writing the rally instructions.

In this article, I'm going to extend the idea of a line-following robot to its extreme, and introduce you to a contest idea that will test even the best robot designs. You should be able to adapt some of my earlier designs for this contest, or come up with suitable robots of your own.

TSD Rallies

When Linda and I lived in Phoenix, we competed in car rallies through the city streets, usually on the first Friday night of each month. Unlike the Monaco-style speed rallies, these events depend on following instructions and solving problems. To compete, your two-man team has to drive a fixed course in a set amount of time, so these rallies are known as Time-Speed-Distance, or TSD.

Porting this contest idea to robotics is a natural, and I hope it will catch on as a group and spectator event. I am still finishing up a suite of tools to develop such robots, and I should have an article about a working rally robot in an upcoming issue. For now, however, I'll outline what a TSD robot should do and how to set up a contest.

A TSD rally for humans starts with a set of instructions. The instructions for simple rallies consist of turns, stops, and speed changes at designated landmarks. Unless directed otherwise, a team

R	Right turn
L	Left turn
SS	STOP sign
SI	Signalized-intersection; an intersection controlled by a traffic light.
Opp	Opportunity to turn; an intersection with more than one exit from your direction of travel.
Last Opp	An intersection with exits to the right and left, but no exit straight ahead.
CAST	Change Average Speed To; assigns an <u>average</u> speed for all subsequent rally instructions
Pause	Stop in place for the assigned number of seconds
On course	Indicates a rally landmark, but otherwise requires no action

Figure 1. Glossary of rally terms.

starts the rally with instruction #1, and ends at the last instruction.

Each instruction consists of an action to perform and a landmark at which to perform that action. For example, the instruction "Right at 3rd SS" means turn right at the third STOP sign. Refer to the accompanying glossary of rally terms (Figure 1).

The person who developed the rally instructions, known as the rallymaster, has designed the course to cover a precise path. Thus, he can measure the course in advance, and knows the distance the course covers to within 1/100th of a mile. This accounts for the "Distance" part of TSD. The rally instructions always assign an average speed, so the rallymaster knows exactly how fast you should be traveling at any point in the course. This accounts for the "Speed" part of TSD.

The instructions also contain numerous pauses to allow for traffic conditions, and the rallymaster knows the sum of all these pauses. In short, the rallymaster knows exactly how much time you should take to arrive at any given point in the rally, to within a second. This accounts for the "Time" part of TSD.

You begin a rally at the starting point, where your team receives several pages of terse rally instructions. Your car is assigned a precise time to leave the starting point, a time noted by the crew manning the start point. You and your partner must now drive the rally course, based on the instructions provided.

The rallymaster knows precisely the course, the instructions, and the time required to execute those instructions. Thus, he can calculate exactly when your car should arrive at any point in the rally. Your progress through the rally course is measured at specially manned checkpoints, placed at strategic points along the route. The rallymaster, in laying out the course, assigns these checkpoints to serve as reference marks, to gauge your team's performance.

As your car pulls into a checkpoint, the crew notes your car number and the precise time of arrival. They subtract your actual time of arrival from your expected time, and assign your team one penalty point for each second in difference, up to some maximum (usually five minutes). Note that arriving one minute early gets you just as many penalty points as arriving one minute late.

Such a contest moves very easily to the robotics arena. Just substitute a black line 1/4-inch wide for the city streets, a white surface for the adjoining land, and a very simple text file for the printed rally instructions. With nothing more than black lines on a white surface, and a minimal set of rally instruc-

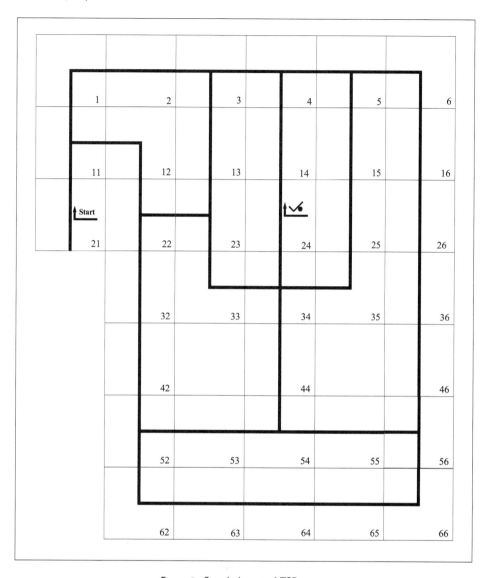

Figure 2. Sample layout of TSD maze.

tions, you can create some very difficult robot rallies. Refer to the accompanying sample rally layout (Figure 2).

Note that I have provided a number for each square in the layout. These are for reference, and would not appear in a real robot rally. Similarly, the light square outlines are not actually part of the lay-

out. I will discuss the starting line in square 21 and the checkpoint in square 24 shortly.

The Ground Rules

Before starting out on a simple rally, your robot needs to know a couple of ground rules. First, your robot

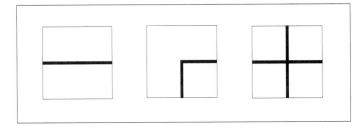

Figure 3. Examples of "straight as possible" routes.

must always travel in a forward direction, taking a course known as "straight as possible," or SAP. Usually, your intended course is obvious. For example, if it enters square 2 on the left edge, SAP means it must continue forward and exit on the right edge.

Similarly, SAP forces your robot to continue straight ahead as it enters square 3 from the left edge. Thus, with no rally instruction to the contrary, your robot will exit square 3 on the right edge.

SAP also controls travel as your 'bot enters square 6 from the left edge. Even though this square contains a bend in the road, the bend is not a right turn. This is because the SAP rule directs your robot in the straightest possible course. Since there is no choice but to make the bend and continue downward to square 16, your robot cannot consider the bend to be a turn (Figure 3).

But notice what happens when your robot enters square 23 from the left edge. This intersection is known as a last opp, or last opportunity to turn. Here, SAP is of no help; there is no "straight as possible" way to leave this intersection. The rallymaster's instructions must provide guidance in

the form of a right or left turn instruction. See the accompanying discussion in Figure 4.

These examples provide the first ground rule: Unless directed otherwise, always proceed on a course as straight as possible. Implied in this rule is the convention that bends in the route forcing your robot to turn either right or left are not actual turns, and cannot be counted as such when executing an instruction.

Based on this first ground rule, start at square 21, moving upward, and execute the following rally instruction:

1.   R   last opp

which directs you to turn right at the last opportunity to do so. What square are you in when you execute this instruction? How many squares have you traversed by the time you execute this instruction? (Do not count square 21, and do not count the final square.)

By following the SAP rule, you should have traveled completely around the outer perimeter of the rally

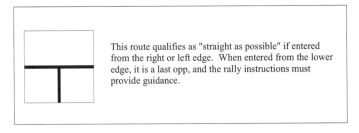

This route qualifies as "straight as possible" if entered from the right or left edge. When entered from the lower edge, it is a last opp, and the rally instructions must provide guidance.

Figure 4. "Straight as possible" or last opp?

map, entering square 11 from the right edge, where you execute the instruction. This leaves you headed toward the upper edge of square 11. In making this journey, you should have traversed 22 squares.

The next ground rule concerns counting landmarks while executing an instruction. An instruction can require you to perform an action based on having passed some number of landmarks, such as right turns. All such instructions assume that the count starts at 0 when the instruction becomes valid. For example, the instruction:

L   **3rd opp**

means to turn left at the third opportunity to do so. Your robot must begin counting left turns when this instruction becomes valid. Any left-turns seen or made prior to this instruction do not count.

To see how this works, again start at square 21, moving upward, and execute the following rally instructions:

1.	**On course**	**R opp**
2.	**R**	**4th opp**
3.	**R**	**last opp**
4.	**R**	**last opp**

What square are you in when you execute instruction 4? How many squares have you traversed by the time you execute instruction 4?

You should have executed instruction 1 at square 11. You should have then continued to square 56, where you executed instruction 2. Instruction 3 gets executed at square 52, and instruction 4 at square 11. This requires you to traverse a total of 20 squares, not counting 21 and 11.

But suppose you had not started your R opp count at 0 when you began instruction 2? If you mistakenly counted the R opp at square 11 as part of the four needed by the instruction, you would have turned early, at square 5. This would have caused you to execute instruction 3 at square 3. You would have then continued around the outer perimeter, executing instruction 4 at square 11.

A robot that made this mistake would have traversed a total of 31 squares, or 11 more than the correct route required. Since it would have been moving at an assigned average speed, it would have taken considerably longer to arrive at square 11 than if it had followed the correct route. Thus, it would pull into a checkpoint late, and receive a penalty for the late arrival.

Note that the mistake did not affect the sequence of instructions. Your robot would have executed instruction 4 at square 11, which is correct. It would simply have taken longer to get there, since its mistake threw it momentarily off course.

This set of instructions is called a trap. All TSD rallies contain traps of some type, to challenge the teams to stay alert and to test their interpretation of the instructions. Traps in well-written rallies will not cause a wrong decision to leave a team hopelessly off course or lost.

Distance and Average Speed

This brings us to the last major topics: distance and average speed. From the instant your rally team leaves the starting line to the time it enters the last checkpoint, it must always travel at an assigned average speed. This speed should always be well within the legal speed limit. The rally master sets the average speed with a CAST (change average speed to) instruction.

Note the constant use of "average" when discussing speed. Your rally team is not compelled to always drive the assigned speed. Instead, the average speed lets your team track how close to "on-time" it is at any point in the rally. To do this, your car must be equipped with an accurate odometer.

The car's odometer measures distance traveled. Rally odometers are very accurate, and a target of much pampering among true TSD devotees. The rally team uses the odometer to record the distance traveled in executing each instruction. This distance, divided by the assigned average speed, yields the amount of time that should have passed in driving to any point in the course.

Since driving conditions vary, a well-written rally always includes plenty of pause instructions. These let a team make up lost time due to slow traffic, confusion, missed turns, or other natural hazards. These pauses, added to the calculated travel time, produce the rallymaster's target time for any part of the rally.

To get a better understanding of how distance and average speed work on a robot rally, consider the following rally. The robot must leave the starting point at square 21, maintaining an average speed of two inches per second. Each square is exactly nine inches on a side. Assume the following instructions:

1.	R	2nd opp
2.	L	2nd opp
3.	L	2nd opp
4.	CAST	3
5.	L	
6.	PAUSE	15
7.	R	last opp

Instruction 7 will carry your robot into the checkpoint from the lower edge. What should be its total travel time upon crossing the checkpoint?

Note that instruction 4 does not include a landmark. Thus, it is executed immediately when it becomes active. This means your 'bot would have executed instructions 1 through 3 at the initial average speed of two inches per second, then changed immediately to an average speed of 3 IPS, and maintained that speed until it reached the checkpoint. Similarly, instruction 6 is also executed immediately when it becomes active.

The proper course works out as follows:

Execute instruction 1 at square 3; total travel is 36 inches at 2 IPS, or 18 seconds.

Execute instruction 2 at square 5; total travel is 72 inches at 2 IPS, or 36 seconds.

Execute instruction 3 at square 3; total travel is 18 inches at 2 IPS, or 9 seconds.

Execute instruction 4 at square 3; total travel is 0 inches at 3 IPS, or 0 seconds.

Execute instruction 5 at square 34; total travel is 36 inches at 3 IPS, or 12 seconds.

Execute instruction 6 at square 34; total time is 15 seconds.

Entering the checkpoint; total travel is 9 inches at 3 IPS, or 3 seconds.

Total time to reach the checkpoint: 93 seconds

This completes an overview of the TSD rally as it applies to robots. True TSD rallies, as run by humans, are far more complex, and a true test of a team's problem-solving abilities and its friendship. If your area supports a TSD rally club, I urge you to contact the club and check them out. The rallies are a blast to run, and you can make some good friends along the way.

Robot TSDs

To translate a TSD for use in robotics, you have to make a few changes. I suggest you set up the course using nine-inch flat white floor tiles, available from any large hardware or tile store. The floor tiles let you use a fixed size and shape for each component of the rally layout, and helps keep the line and background uniform.

I recommend using a matte press-on line available from most art supply stores, as the "road." Use a line that is flat black, with a matte finish, 1/4 inch wide. You can press this line in place, then spray on a matte finish plastic spray to seal it in place and protect against moisture.

You could simply draw a black line on each tile, using a Sharpie or other indelible marker, but this usually gives uneven results. It's hard to make the line a consistent width and blackness; I suggest staying with the press-on artwork.

Notice that you only need a few distinct patterns to duplicate the rally layout I've shown. If you make several tiles with the patterns shown in squares 1, 2, 3, and 34, you've covered them all. You can draw the starting line and the checkpoint line in light pencil at the time you lay out the rally. The robots will always be stationary as they are positioned at the starting line, and they won't stop as they go through the checkpoint, so you don't need special tiles for either.

Be sure to carry plenty of extra blank tiles along. You will need these to set up a perimeter around the layout, to provide space on the inside of turns, and to allow for robots overshooting a turn.

If some of the robots running the course will be hefty, you might want to put some double-sided adhesive tape (not foam tape) to the underside of each tile. This will hold them in place as the robots start up and turn.

Manning the starting line is no big deal. Just get someone with a stopwatch to say "Go!" when he presses the watch's button. The checkpoint crew should also start a stopwatch at the same time, then stop the watch as the robot runs past the checkpoint line. The starter can serve as a backup timer to the checkpoint crew, calling out his time to a notetaker as the robot crosses the checkpoint line.

Note that you can have more than one checkpoint in a rally layout. You can even have more than one robot running the course at the same time, if your rallymaster did a _really_ good job of designing the course. This gets very tricky, however, and you're probably better off just running one robot at a time.

As rallymaster, you must take care to measure the prescribed course accurately, and to verify all running times and distances. Be sure to include plenty of pause instructions in your rally, and to throw in a CAST instruction now and then to test a robot's wheel encoders. If possible, have others double-check your layout and instructions, "walking through" the instructions several times to look for typos or dead-end traps.

Now for the rally instructions that get passed to each robot. These take the form of a simple text file. Before each run, the starter connects the robot to a laptop and downloads the file into the robot's memory

for storage. When a robot starts its run, its software parses the rally file and executes each instruction as directed. Note that you could also publish the rally instructions in advance, provided the rallymaster carefully matched the instructions to the layout. Obviously, the rally layout is kept secret until the event.

The format of the rally file is very simple, so the robot's software doesn't have to be too sophisticated. Refer to the accompanying rally file format (Figure 5, at end of article). A rally file may contain blank lines, which should be ignored wherever they occur. Any line containing an asterisk in column one is a comment, and should also be ignored.

The rally instruction always begins in column one. It generally consists of a single letter and two ASCII digits, though some commands may use more letters or digits. The units for an argument depend on the instruction, and should be obvious. The following examples should prove helpful:

R01	**Right at the next opp**
L00	**Left at the last opp**
C35	**CAST to 3.5 IPS (note decimal point)**
P10	**Pause 10 seconds**

Note that this is only the beginning of robot rally possibilities. Advanced robot rallies can have multiple instructions active at the same time. Thus, your robot could be counting right turns as part of one instruction, while looking for a last opp to satisfy another, simultaneous instruction.

Other additions include special markings on the side of the road that designate STOP signs, railroad tracks, or even street names. The latter could be implemented with barcodes, allowing instructions such as "R onto Lark Ln." (That has a nicer ring to it than "R onto 1792350" or "R onto Palmolive Original Scent Dishwashing Liquid," assuming you get your barcodes off of discarded grocery items.)

You can even implement traffic lights fairly easily. Just suspend a very bright, tightly focused IR LED about 18 inches above an intersection. When the "light" is red, the LED emits a 40 kHz signal, which would be picked up by an upward-looking Radio Shack IR sensor on the top of the robot. You could have a second, red LED wired up so the spectators and judges could tell when the traffic light was red. You would have to modify the rally instructions to allow a pause for the SI, so the robot could recover whatever time was lost waiting for the stoplight. And you would have to include a severe penalty for "running" the light; disqualification or maximum time seems appropriate.

Oddly enough, a railroad crossing is quite easy. Just mount a length of light wood dowel or brass rod onto the control horn of a hobby servo, then fix the servo in place alongside the "road." Tape a piece of white paper or fabric to the wood dowel so it hangs down when the dowel is horizontal, forming a large flag. Robots equipped with a single forward-looking IR object detector would sense the lowered flag and be forced to stop while the "train" goes by.

I can see that one could easily get carried away with this concept. How about billboards? A side-looking barcode reader, properly designed, could spot signs at the side of the road. At the very least, you could have a "sign" that consisted of a blinking green or red LED, mounted so it is visible only from a certain location on the layout. This would make possible an instruction such as "R onto GREEN St." A piece of colored plastic film in front of a photoresistor should do the trick as a sensor.

You could make a STOP sign out of a short piece of metallic tape, available at art supply or hobby stores. Just put a piece of tape on the right-hand side of the road, a fixed distance out to the side. A robot could use a reflective IR or visible light sensor to look for the tape. A sudden increase in signal indicates your machine has "seen" the sign and must come to a complete stop for some fixed amount of time. Note that this sensor would not mistake a section of the road for a sign, since the road is darker than the background, and will return a reduced signal.

One refinement that certainly belongs in an advanced robot event is the odometer check. This is always the first several instructions of the rally, designed to let the robot calibrate its internal odometer. After executing the odometer check instructions, the robot reads an instruction that provides the actual distance that should have been covered from starting.

The software can compare this expected distance to the recorded distance, then calculate a correction factor. This factor can then be used to correct distances covered, so the robot can accurately maintain its average speed. Like real rallies, the odometer check must not contain any traps. It should be a straightforward route of several feet (for robot rallies) so the robot can record a fairly long run. Note that unlike real rallies, the robot rallies are laid out on precise squares. This lets smarter software automatically resync the robot on each detected turn or bend in the road.

The simple rally file format I've described here cannot handle some of these additions, and that's OK. As more sophisticated rallies evolve, you can modify the format appropriately. If I have to update the format to handle my own rally improvments, I'll keep you posted.

I haven't yet talked about the type of robot you'll need to run such a contest. You can begin with one of my earlier line-following robots. Note that because distance plays such an important part in this event, you will need some type of wheel encoder system.

On the software side, your robot program must be able to record and process the simple rally file. It must also handle the bookkeeping chores of recognizing and counting turns and intersections, timing pauses, and tracking which instructions have been executed and what to do next. Certainly a BOTBoard-2, running some SBasic code, can handle this job.

This Robo-Rally event strikes me as a contest that would make an excellent centerpiece to a high school Science Fair or Technology Challenge, or as a public demonstration event for any robotics club. It is very simple in concept, yet will provide plenty of challenge to most hobbyists. In fact, you might want to team up with someone else when you tackle this project. Then, like your counterparts in a human TSD rally, you could enjoy the excitement of competing against other teams.

Rally File Format

*	Asterisk in column one marks a comment line; all text in the line is ignored.
Rnn	Turn right at the nn'th opportunity. nn = 00 means last opp.
Lnn	Turn left at the nn'th opportunity. nn = 00 means last opp.
Pnn	Pause for nn seconds immediately.
Cnn	Change Average Speed To n.n inches per second.
Onnn	Odometer check; distance covered since starting is nn.n inches.
NRnn	On course at nn'th right turn. nn cannot be 00.
NLnn	On course at nn'th left turn. nn cannot be 00.

All commands start in column one. Commands are case-sensitive, and must be entered as shown. Each number field contains exactly two digits (three for the Odometer command). Use a leading 0, if necessary, to fill out a number field. Blank lines may be included, and are ignored. Each line ends with a carriage-return (CR) character; it may also include a line-feed (LF) character as well, in the form CRLF.

Figure 5: Format of rally instruction file.

The Dead-Reckoning Event

My November 1996 column for *Nuts & Volts* describes a stepper-motor based robot that I built for competition in the Seattle Robotics Society's Dead-Reckoning (DR) contest. This event places a premium on base construction, software design, and drive-train implementation. The DR contest makes a great club event, one that the crowd can understand and appreciate.

I begin this article by describing my search for a lightweight power supply, and why I ultimately settled on the new National Semiconductor SimpleSwitcher® product. This little board saved me many ounces on my final design, and I've since used it in many other projects.

Next, I go through the steps (sorry!) in setting up the motors. I used Bill Bailey's chopper drive PCB to power my two Vexta stepper motors. Along the way, I learned a thing or two about driving two motors off of the same clock line, and how to handle some of the power issues involving steppers.

I close with a discussion of the 68hc11 SBasic program I wrote to control my DR stepper motor 'bot. The software isn't too difficult to follow, and I've made the code available on my web site so you can customize it, if you like. I built the code so it uses a motion table made with DATA statements, to define each sequence the robot must perform during the event. This

technique carries over to other stepper-based designs, and should prove instructive to anyone designing stepper motor software.

Building a Dead-Reckoning Robot

The Seattle Robotics Society (SRS) runs several contests throughout the year, and one of the more unusual is the Dead-Reckoning (DR) event. The DR contest offers a formidable challenge using very few rules, and can be run with anything from a simple robot to a sophisticated machine. In this column, I'll take you through the contest rules and my design for a viable DR robot.

In running the Dead-Reckoning event, your robot must begin from a marked starting point on a flat surface. When the judge says "Go!," you start your machine on a prescribed but unmarked course. The course consists of a two meter run forward, a 45-degree turn clockwise, then another two meter run forward. The course's endpoint is marked for human recognition, though not for robot recognition. Your

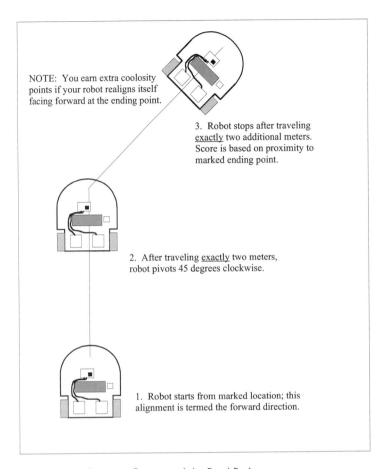

NOTE: You earn extra coolosity points if your robot realigns itself facing forward at the ending point.

3. Robot stops after traveling <u>exactly</u> two additional meters. Score is based on proximity to marked ending point.

2. After traveling <u>exactly</u> two meters, robot pivots 45 degrees clockwise.

1. Robot starts from marked location; this alignment is termed the forward direction.

Figure 1. Overview of the Dead-Reckoning event.

robot's performance is judged by how close it stops to the endpoint (Figure 1).

Your machine must carry a pointer or marker of some kind, so the judge can verify your robot's starting and ending position. Though not required, you will probably want to include some type of pointing or aligning mechanism, so you can properly align the robot at the starting point. Some of the SRS machines have put in excellent performances, and any robot entered in one of these events had better hit the endpoint within 1/4" or it won't have a chance for first place.

Gearhead motors, equipped with suitable encoders, can perform very well in this event, but find-

ing a suitable motor isn't easy. One of the most ingenious designs was fielded by Mark Castelluccio, who managed to imbed quadrature encoders inside the housing of a Futaba R/C servo motor. His machine hit the endpoint with amazing accuracy; Mark's software even inserted a 360-degree pivot in the middle of the run, just to show off a bit. I've described a few designs for embedding encoders in these servo motors in previous columns; refer to the article "Adding an Encoder to an R/C Servo" for details.

But stepper motors remain the most obvious drive system for the DR event. I based my DR robot design on Bill Bailey's stepper motor driver boards, showcased in the column, "Stepper Motor Basics." Rather than duplicate that schematic here, I'll refer you to that column instead.

The LM2595 Switcher

My robot design uses a 12-volt gel-cell battery to supply power for the stepper motors. To drive the on-board electronics, I originally used a set of four C cell Renewal alkaline batteries. The design worked, but I was never happy with the weight. It just seemed like too much bulk to supply the few milliamps needed by the 68hc11 BOTBoard and other circuitry. And the extra battery pack was, well, inelegant. I was about ready to drop in a Junk-Box Switcher when I ran across an advertisement in *Electronic Engineering Times* that provided the perfect answer.

EE Times, as you can guess, caters to the electronics engineer, and carries a huge assortment of articles on high-end engineering topics. You can always find the latest skinny on new chips, new network protocols, new technology, new legislation affecting the communications and electronics industry, and other cutting edge topics. The newspaper, published weekly, is free to qualified engineers and managers. They even have a web-based version you can check out.

But a major appeal to me is that the big chip vendors carry a lot of advertisements in *EE Times*, and sometimes they announce some real deals. For example, the August 26, 1996 edition ran an offer by National Semiconductor that was too good to refuse.

National has long carried a great line of linear components, and every well-stocked designer's shelf should have a bunch of National Semi data and application books. Lately, National has been pushing devices such as the LM2595 switching power supply controller. This device comes in a 5-lead TO-220 case, and forms the core of what National calls the SimpleSwitcher® power supply. Using just this device, two capacitors, an inductor, and a Schottky diode, you can build a switching power supply for converting anything from 7 to 40 VDC input down to 5 VDC output at up to one amp of current.

To entice engineers into trying the device, National offered to send readers a working power supply built with this chip, along with a pack of data sheets AND a floppy disc of design software, free. All you had to do was call National Semiconductor and ask for a LM2595 demo package. So I did.

My demo board arrived within a week, and I was impressed. The board measures only 1-3/8 inches by 1-1/2 inches and has a small male header for the few electrical connections needed. The switcher runs at 150 kHz and sports thermal shutdown and current limit protection. Using 12 VDC on the input, the LM2595 reaches 85% efficiency, which means you'll get the most out of your batteries.

The board doesn't have mounting holes, so I just placed it where I wanted it on my robot base, then fastened it down with a bead of hot glue on each of the PCB's corners. I soldered wires directly to the proper ground, voltage in, and voltage out pins, and even added a red LED and 820 ohm dropping resistor between ground and the output pin, as a power-on indicator.

One plus for this power supply was immediately apparent. I saved nine ounces of overall weight by using the switcher to replace the original set of four C-cell Renewal alkalines. I spend a lot of time on my designs trying to reduce weight, and dropping over half a pound with this PCB was a big win.

The National package includes a set of very complete data sheets for the LM2595 and other devices in the SimpleSwitcher family. Each data sheet contains a full range of graphs and design guidelines, to help you design your own switcher. You get tips on selecting the inductor and a list of recommended vendors. You even get a PCB layout, so you can build additional boards. This is a first-rate evaluation package and I really like the LM2595. Recommended.

A Bit About the Base

I built my DR base using my favorite material, Sintra plastic. This is a foam-based plastic, available in sheets at most large plastic houses. By scrounging the scrap boxes at local stores such as TAP Plas-

Figure 2. A shot of the DR robot, showing the two stepper control boards (left), the 12 VDC gel-cell battery, and the 68hc11 BOTBoard that drives the machine. I'm using 2-3/4-inch Dave Brown Lite Flite wheels. The robot measures 12 inches long by 8 inches wide.

tics, I've built up a reserve of Sintra in assorted sizes and colors. I like using Sintra because even the 1/4 inch-thick sheets cut very easily, the material takes simple machining well, and you can even tap it for screws. To glue Sintra, use a free-flowing Acrylic cement, available from TAP Plastics or other large suppliers.

I cut my base to be 12 inches long and 8 inches wide, with a rounded front end and cutouts at the back to accomodate inset wheels. I've pretty much standardized on the Dave Brown Lite Flite wheels, a foam-type unit with two-piece Delrin snap-apart hub. The axle bore is exactly the right size to thread a #8 screw through, which in turn matches with the direct-drive wheel adapters that the SRS' Dan Mauch made for me. The wheel and adapter make a perfect unit for the business end of my 'bot.

I added my usual drawer knob as a front caster. I started with a large, black plastic knob. I then cut a piece of foamcore to a square about 1-1/2 inches on a side and drilled a 1/4-inch hole in the center of it. I ran the knob's screw through this hole, then into the knob, cinching the knob tightly against the foamcore. Finally, I used two pieces of foam tape to stick the knob and foamcore to the proper location on the underside of the robot's base.

Fine-tuning the Steppers

For motive power, I hooked up a couple of Vexta PXB44H-02AA steppers to my base. The Vexta motors are rated at 6 VDC, 800 mA, and provide plenty of torque for moving the base. I strapped the stepper driver boards for half-stepping, giving me 400 steps per revolution. I then wrote a small SBasic program for providing a stream of pulses to the motors.

But when I turned on the power so I could talk to the BOTBoard's 68hc11 in bootstrap mode, the motors locked in place, then the 1.75 A fuse I had installed in the power circuit popped. When the 68hc11 comes up in bootstrap mode, your software isn't running and you have no control over the I/O port lines. The stepper driver boards were getting full current and blowing the fuse before I could download my program and gain control of the lines.

This led to two changes. First, I installed a second SPST switch for the motor voltage, in series with the switch already used to control the power for the 68hc11. Thus, I could turn on the MCU but leave the motors unpowered. Next, I reset the current control on both boards. I had originally set it for 500 mA per winding, which was over the fuse limit; it was also over the free-air dissipation rating of the L298 driver chips, so they would have gone into thermal shutdown, anyway. I trimmed the drivers for 200 mA per winding, which proved to be plenty of current for my needs.

But now a new problem surfaced, one that had me stumped for a couple of hours. The motors moved, but the motion was ragged and had little torque. At first I suspected a blown driver transistor in one of the L298 chips, but each driver board, run separately, worked fine, so the L298s were good. Finally, I got in touch with Bill Bailey and Dan Mauch, the club's main stepper experts. They got me headed in the right direction.

The L298 has an internal oscillator for timing operations with the stepper motor. You can run several L298s off of the same oscillator if you wire them together properly. You must first select a single L298 as the master, and disable the oscillators on all other L298s. This only involves tying pin 16 of each device to ground. Then, you simply wire the SYNC pins on all of the L298s together. This is pin 1 of the chip, or pin J2-2 on Bill's PCB.

My problem occurred because I had wired both of the SYNC pins together, but not disabled one of the oscillators. Since the oscillators on both boards were running, they would occasionally get out of sync, causing the ragged performance and reduced torque I saw. I simply tied one of the pin 16s to ground and both motors began running beautifully.

In tuning the steppers, I needed a quick and dirty way to vary the frequency of pulses my software issued to the stepper driver board. I could have

Figure 3. Here is a closeup of the robot, showing the two toggle switches for controlling power to the motors and the BOTBoard. The National Semiconductor LM2595 switching power supply sits between the switches and the gel-cell battery. To the right of the switches, I've added a power jack for charging the gel-cell. The D cell in the foreground gives an idea of the robot's size.

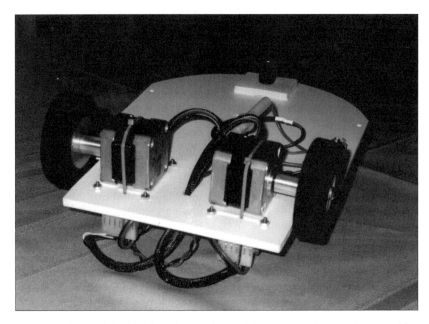

Figure 4. The underside of the DR robot, showing the two Vexta steppers that provide power. I fastened the motors in place using foam tape, then added a nylon wire tie to each for extra security. Note the direct drive mounting system; I used shaft adapters machined for me by the SRS' Dan Mauch.

used a serial port connection to the 68hc11's SCI and some SBasic code to read incoming characters, but I chose instead to add a frob knob. The term comes from MIT's engineering labs, and refers to a general-purpose analog control or dial.

Given the 68hc11's eight A/D converters, adding a frob knob takes nothing more than a 10K-ohm trimpot between +5 VDC and ground, with the wiper hooked to one of the A/D input lines. At selected points in your software, you read the current value of the A/D channel, then convert the reading into the appropriate range and pass it to your test software.

In my case, I wired the frob knob to PE0. My conversion software translated the voltage on the A/D port into a pulse duration count, which then determined the step rate for the motors. The frob knob let me alter the step rate on-the-fly, smoothly and efficiently; a simple idea that can come in handy in many designs.

One last change to my system involved mounting the heavier components. I usually favor double-sided foam tape and hot glue; they are cheap and light, and they don't require great mechanical dexterity, a real plus for most software types. But this robot was going to get some rough treatment, and the bulky motors and battery required sturdier fasteners than just foam tape. So after I got everything taped down exactly where I wanted it, I marked and drilled holes into the Sintra base, then ran a single nylon wire tie around each major component. The adhesive on the foam tape keeps the large objects from shifting or rotating in place, and the wire tie keeps them from lifting off the tape, even if I pick up the robot by the battery or the motors.

The DR Software

In theory, the dead-reckoning software seems quite simple. A couple of quick calculations give you the number of steps to make to cover a prescribed distance. For example, my robot uses 2-3/4-inch diameter wheels, which means a wheel measures 2.75 * 3.14159 inches in circumference, or 8.64 inches. Given 39.36 inches per meter, this works out to 4.55 full revolutions. Since my robot uses half-stepping, it needs to issue 400 steps per revolution, or 1820 steps per meter. For two meters, this becomes 3640 steps.

But you cannot just fire off 3640 steps at any old rate and get consistent stepper performance. You will find that stepper systems have a range of pulse rates that work better than others, and some pulse rates flat won't work at all. The major influences over stepper performance are the mechanical resonance and inertia of the system.

You can get a graphic demonstration of these influences by using the frob knob to vary the starting pulse rate for your motors. Beginning with a rate of about 200 pulses per second, repeatedly start the robot from a full stop, increasing the step rate slightly each time. You will not only find an upper limit beyond which the robot simply won't move, you likely will also find a few settings in between where the motors chatter or skip a few pulses before starting. These intermediate settings correspond to resonances in the robot's mechanics, where the physical mechanics conspire to make the load on the motors appear greater than it really is.

You could try tracking down the various interplays in the mechanical elements, tightening here or loosening there, in an effort to remove (actually, change) the resonances. A simpler approach involves starting the pulses at a slow rate, then quickly accelerating the pulses until they reach the desired rate. This technique, known as ramping, can prove very effective in overcoming mechanical resonances. And since the solution lies in software, you can fine-tune your solution easily.

Note that you don't need a very complicated ramping algorithm to overcome even serious mechanical resonances. I once worked on a table-top device that used several steppers to move paper through a mechanism. The mechanism had some very nasty resonance problems, and the first-order solution, throwing in a bigger motor, wasn't available due to physical constraints.

But experimentation proved that a very simple ramping technique solved the problem. We ended up issuing four pulses at a slow frequency, four pulses at a faster rate, and issued the remaining pulses at the fastest, target rate. Those first eight steps, ramping from a slower rate up to the target rate, helped the steppers overcome the mechanical resonances and yielded very reliable performance.

You can find many books and papers dealing with ramping algorithms, and some of the ramping strategies get quite complex. But try starting out with a simpler two- or three-speed ramp such as I described above; you'll likely find it clears up any problems you might have with stalling steppers on startup.

I covered the details of issuing stepper pulses in "Stepper Motor Basics," so I won't go over that ground again. I will, however, spend some time on the concept of a motion table for describing exactly how the robot moves about. Unlike previous robots such as Huey, which reacts to its environment, this machine blindly moves in a prescribed pattern. Each element of this pattern contains at least three pieces of information for the right and left motors. For each motor, you have to provide the delay between steps, the number of steps in the element, and the direction of motion. In SBasic, you can use a DATA table to house this information (See Table 1).

This table entry describes a motion element that causes the robot to spin clockwise about its center point for a total of 2200 steps on either motor. Each step in the motion requires 5000 time tics on the 68hc11, which at 8 MHz and default timer prescaler works out to 2.5 msecs per step, or 400 steps per second.

The next two entries contain the number of steps to issue for each motor. Note that this motion table contains no information regarding half- or full-stepping; the step type must be set elsewhere. The entry shown will move each motor exactly 2200 steps, but obviously a half-stepping robot will only travel half as far as a full-stepping robot. The last two entries contain previously-defined constants that control the direction of motion.

```
DATA        5000,  5000              ' left  and  right  step  delays
DATA        2200,  2200              ' left  and  right  step  counts
DATA        STEP_FWD,  STEP_REV      ' left  and  right  step  direction
```

Table 1. DATA table.

Despite its simplicity, this table scheme can prove very flexible. For example, you can use it to incorporate ramping by including a couple of entries with only three or four steps each, but with decreasing step delays. I have written an SBasic library for stepper motors built around this table concept, and made it available on my web site. Feel free to build on my original code, adding features you think necessary.

One change you might include would add the option of disabling the motors for a fixed delay. Disabling the motors removes the current from the coils, dramatically reducing current drain from your battery. Such a feature might be handy if your robot needed to execute several sequences with long delays between them.

You will need to alter my step count values, based on the diameter of the wheels you put on your robot. The math for calculating the basic move and turn functions is straightforward, though you might want to throw in a little ramping to get your robot up to speed. Remember to allow for the steps used in ramping in your overall step count. I know it's only a few steps, but not taking them into account would certainly knock you out of an SRS competition.

Adding Wrinkles

After you have your DR robot running, you can think about changing the event slightly to add complexity and crowd appeal. The DR event basically requires a robot that can work in a Cartesian grid system, a feature you can exploit in highly cool wrinkles to the basic event.

For example, you could require that a robot leave the starting point and map out a more complex pattern before reaching the ending point. A pair of marked points, perhaps about a meter apart, could

serve as pylons. Robots could be required to navigate either an ellipse or a figure-8 around these pylons before heading for the ending point. Obviously, the coordinates of the two pylons would have to be announced prior to the event.

Another option would involve a tour of several different marked points, visiting each point in a predetermined order. The robot wouldn't necessarily have to hit each point exactly, but might have to pause and announce, using lights or beeper, that it "knows" it has reached a waypoint before proceeding on. As above, the coordinates of each waypoint and its order in the visit sequence would be announced prior to competition.

Or you could offer a variation geared for the more artistic robot builders. A robot would be placed at the starting point, then given up to two minutes to go anywhere on the arena, in whatever path it chooses, before returning to the starting point. A panel of judges would assign a numeric score to the artistic value and the difficulty of the "floor exercise" performed. The average of this score, added to an accuracy value based on how precisely the robot returned to the starting point, would determine the robot's overall score.

To jazz up this last version, a robot could use a single R/C servo motor to lower a pen onto the arena's surface, which would now be covered with a large sheet of paper. This essentially turns the robot into a movable plotter head, allowing your software to trace out complex designs. This event would place a premium on careful design and artistic creation, and should be a real crowd-pleaser.

Carried one step further (sorry about that), two or more robots could work together as a team, tracing out intricate patterns around each other on the arena floor. After a specified amount of time, each would have to find its way back to the proper ending point, with the judges scoring on both artistic merit and precision.

The dead-reckoning robot, with its premium on precision mechanics and software, opens up new possibilities and makes a great showcase for your skills. While I'm sure you enjoy building behavior-based robots as much as I do, take the time to build one of these stepper-motor machines. I think you'll like the new opportunities it presents.

Wowser Web Site

Kevin Ross, president of the Seattle Robotics Society, has set up a first-rate Robot Builder's Page. This site, attached to his home page, is actually an extensive HTML article on building a small robopet using a BOTBoard and one of Marvin Green's plastic, two-tiered bases. Links within this document describe vital subelements, such as how to modify an R/C servo so it goes round and round instead of just back and forth. Kevin also has included many beautiful, high-quality photos of his construction techniques. The photos in the servo modification section are super, and really add to the textual instructions.

Kevin's RoboGuide, as he calls it, is an expanding work and will doubtless mutate as his time and inclinations allow. Check in now and then on what I know will become a much-visited robotics site.

Hercules, My Smallest Robot

My December 1996 column for *Nuts & Volts* highlighted a little robot that you can build in a weekend or so, provided you have a well-stocked junk box. Hercules consists of a four AA-cell battery holder, small motors, a BOTBoard, and some other goodies. I built Hercules just to see how small a robot I could make easily. Given more machining tools and more motor skills, I'm sure I could do better, but Hercules is pretty cute, and not bad for a first effort.

Next, I look into building some high-current output ports. The 68hc11 output port lines can handle a few mA of current, but sometimes you need an Amp or so of current, so you need to look for an alternative. I describe how to wire several 74hc595 serial latches in a chain, then add ULN2803 high-current driver chips to their outputs, giving you two dozen high-current outputs that you can modify using the 68hc11's SPI. This technique could come in handy on the robotic pinball project...

Finally, I look at an interesting subgroup of MIT students, the Cyborgs. These guys have built wearable computers, PCs so small and low-power that they fit in backpacks and run off batteries. Some of this technology will transfer across directly to robotics, and I spent some time tracking down low-power PCs that could serve as robot brains. Take a look at the Cyborg web pages for a glimpse at truly low-power computing.

It's the end of October here in the Pacific Northwest, and time for fall cleanup. Those with outdoor interests rake the leaves, clean the gutters, and prepare the garden for the long winter months. I, however, close out those long-standing robot projects that have piled up on my desk, workbench, and floor.

Hercules

I've seen some tiny robots in my day. The smallest by far is Goliath, a robot made by MIT students in the Robot Lab. Goliath measures less than a cubic inch, yet includes a 68hc11 with sound, light, and bump sensors. The little machine uses surplus pager motors for drive power, and the circuit boards and MCU form the mechanical frame. It is a miniature work of art, and started me thinking about making a small robot of my own.

I don't have ready access to the machining tools used to fabricate Goliath, but I was curious about just how small a robot I could make, given nothing more sophisticated than workbench tools and a Radio Shack. You can see the results in the accompanying photos of Hercules.

I built Hercules from a 68hc811e2 BOTBoard and a Radio Shack 4-cell AA battery holder. The underside of the battery holder forms the frame for

Figure 1. Hercules stands less than three inches tall and measures only four inches across. Here you can see the BOTBoard strapped to the four-cell battery holder that serves as power source and mechanical frame.

mounting the motors and front skid. The BOTBoard rests on top of the battery pack, protected from contacting the batteries with a piece of thin cardboard cut from a business card. I used an elastic restraining tool (rubber band) to hold the BOTBoard in place.

The front skid is a metal furniture foot, sold in packs of four at most hardware stores. I shopped around until I found a style that included threaded studs on the feet, so that the skid's height is adjustable. Mounting the skid to the bottom of the battery holder becomes a simple, two-step process. First, hot-glue the nylon sleeve of the skid to the battery holder in the proper location, then let the glue set. Next, screw the metal part of the skid into the sleeve and adjust to the proper height. When you've got the right height, coat the junction of the nylon sleeve and the skid's metal threads with hot glue and let it set. That's it!

I pulled the motors from my junk box of spare robot parts. They are beautiful ball-bearing gearhead motors that I picked up long ago from some surplus store, so you are unfortunately on your own here.

My motors measure slightly over 1/2 inch in diameter, with an overall length of 1-3/8 inches, not including the shaft. These dimensions made them a perfect fit for the 2-1/2 inch- wide battery holder.

I could have simply hot-glued the motors to the battery holder, but it's hard to get the alignment right when you're mounting a cylinder to a flat surface, and you don't get a lot of surface contact area. The latter can lead to a weak mechanical connection. I found the perfect item in my junk box: a pair of C-shaped metal snap-in clamps, used to hold large electrolytic capacitors. I simply snapped a motor into a clamp, then hot-glued the clamp to the battery holder. As a bonus, the clamp lets you reposition the motor laterally, so you have a little flexibility in alignment, even with the clamp firmly glued in place.

Choosing suitable wheels and mounting them on the motor shafts presented a problem. The motors have very thin shafts, slightly less than 0.1 inch in diameter. My usual wheels, the Dave Brown Lite Flite items, were right out; far too large. I rooted through my junk box again and came up with some old Red Pepper racers I had purchased at a Radio Shack long ago. I ripped a couple of wheels off their shafts and checked the fit. The axle bore was still too large, but only by 1/16 inch or so.

Figure 2. Another view of Hercules, showing the circuitry I added to the BOTBoard's prototyping area. The rubber band runs alongside one of the TC4427 motor driver ICs; two photocells look out from either side of the front.

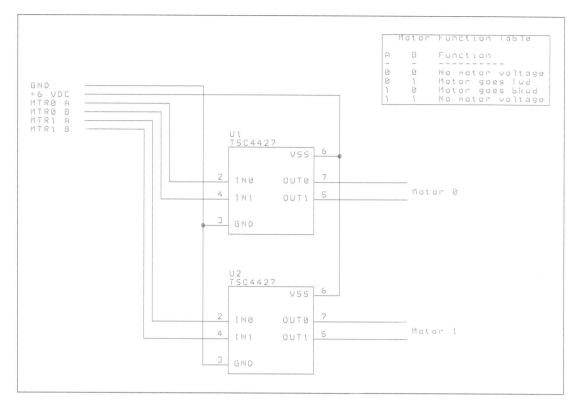

Figure 3. Schematic of Hercules' motor driver board.

To make a wheel fit firmly on the motor shaft, I first cut a piece of 3/32-inch heatshrink to a length slightly longer than the motor shaft. I slipped the heatshrink over the motor shaft, then shrank it down. This left a soft plastic cover firmly bonded to the motor shaft, and also expanded the motor shaft considerably. Next, I pressed a wheel onto the shaft; it fit like a glove. Finally, I pulled the wheel back off, filled the bore with hot glue, then pressed the wheel back onto the shaft. When the glue set, the wheel was firmly (and I mean firmly) seated on the motor shaft.

The major challenge in building such a small 'bot lies with the motor drivers. For Hercules, I used drivers built from a pair of TelCom Semiconductors TC4427 power MOSFET drivers. Each 8-pin DIP device contains a full MOSFET H-bridge rated at 1.5 Amps (peak), which gives me plenty of current for

my tiny 6 VDC gearhead motors. I've mentioned this chip before; its small size, digital inputs, and ability to use a separate, higher motor voltage make it a natural for small machines. And the accompanying schematic shows that the hookup is a no-brainer (See Figure 3). The two motor drivers fit comfortably in the compact prototyping area on the BOTBoard, and at $1.79 in singles from Digi-Key, the 4427 is a cheap low-power H-bridge.

I wired my motor drivers to bits PB2 through PB5 on the BOTBoard; I also have a two-color LED hooked to PB0 and PB1. To change motor direction and LED color, I simply write a new value to port B. Since the gearhead motors move so slowly, I didn't need to bother with pulse-width modulation (PWM) speed control. The following chunk of SBasic code shows how easy it is to control the two motors:

```
'    Declare motor and LED constants
'    for port B.

const   FWD   = %010111
const   LEFT  = %011001
const   RIGHT = %100110
const   STOP  = %000000


'    Move forward


pokeb   portb, FWD
```

Hercules, like many of my robot projects, is evolving. For now, it serves as a base for behavior experiments. I've added two photocells for sensing ambient light, and have lots of port lines left for adding other sensors and actuators. Right now, the biggest obstacle is finding room on the BOTBoard or frame for adding more electronic circuitry.

But Hercules shows how quickly you can built a bantam 'bot; I have about one weekend in Herc. If you built just one a week, you could have a fleet of four going in a month. Add a little software for herd or swarm behavior, and you'll knock 'em dead at the next club meeting.

Figure 4. Roll over, Hercules! Here you can see the small gearhead motors, mounted to the battery holder using capacitor clamps. The flat furniture foot at the front makes an excellent skid.

Lots o' Ports

One recent 68hc11 robot project required 24 output-only I/O lines, each capable of handling at least 250 mA. This seemed like a natural for the 68hc11 Serial Peripheral Interface (SPI), Motorola's high-speed synchronous serial bus. I've already used the 74hc595 serial-to-parallel converter chip on other projects, such as my serial LCD driver. It works very well with the SPI, and its serial output line lets you chain together as many 74hc595s as you need. Refer to the accompanying schematic (Figure 5).

Unfortunately, the 74hc595 output lines can't supply a lot of current. To get the required oomph, I followed each 74hc595 with a ULN2803 high-current driver. Many companies make this octal driver IC; my prototype circuit uses Motorola parts, but devices by SGS-Thompson and others will work as well. This chip contains eight power Darlington drivers with built-in snubbing diodes, so you can use the outputs to directly switch relays or other inductive devices, without fear that back EMF could fry the driver chip.

My schematic uses three separate IDC connectors, one for each of the 8-bit output ports. On the connectors, I've matched each output line with a ground line, keeping the grounds on even-numbered pins. This arrangement lends itself to hacking. You can mount male IDC headers in the connector layouts, then use a two-pin female connector to hook up just one output line, if you want to try a quicky experiment.

I could have used 16-pin IDC connectors for J3 through J5, but I intentionally used a 20-pin layout so I could take advantage of the cheap XT floppy disc cables available at local surplus stores. For a buck or so, I can buy a two-foot cable complete with 20-pin connector, which gives me a cheap, high-quality solution to my cabling problem.

Wired as shown, the three 74hc595 ICs form a large, 24-bit SPI power output latch. I brought the necessary SPI signals and logic power line out on a six-pin male Molex KK-style connector. Using the SPI to control this super latch is much like using a

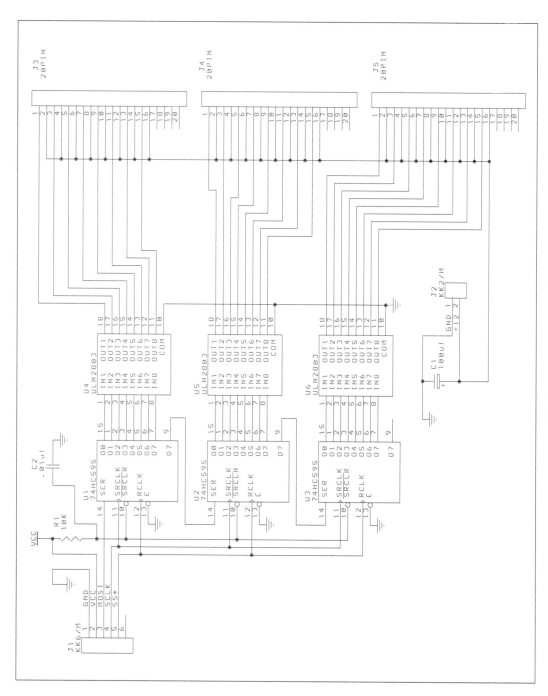

Figure 5. Schematic of 24-line output board.

```
`
`    SendSPI()        send 3 bytes to SPI
`
`    Enter with the three bytes pushed onto the data stack.
`    Remember that the bytes will go out right to left!
`

SendSPI:
pokeb portd, peekb(portd) and $df    ` pull SS low
pokeb spdr, pull()                           ` send without checking
waituntil spsr, $80                          ` wait until transfer
done
pokeb spdr, pull()                           ` send second byte
waituntil spsr, $80                          ` wait until transfer
done
pokeb spdr, pull()                           ` send third byte
waituntil spsr, $80                          ` wait until transfer
done
pokeb portd, peekb(portd) or $20    ` pull SS high
return



`
`    Send three bytes to SPI super-latch.
`

gosub SendSPI, $33, $22, $11            ` $11 goes out first!
```

Figure 6. Listing of code that sends three bytes to SPI.

single 74hc595 serial latch. My 68hc11 software must pull the slave-select (*SS) line low, then set the master-out-slave-in (MOSI) line to the desired level and pulse the clock (SCLK) line. After sending out 24 pulses with the proper data level on MOSI, my software must pull *SS back high. This last change causes all 74hc595s to transfer the newly-received data simultaneously to their respective output lines.

This is only slightly more complicated than the usual SPI transfer (Figure 6). The following chunk of SBasic code shows how easily you can update all 24 outputs. Remember also that this transfer takes place at 2 megabits per second, since the 68hc11 SPI subsystem handles the transfers in hardware.

The circuit is very simple, and you shouldn't have any trouble wiring one up using the Radio Shack experimenter boards or wirewrap.

Embedded PCs

The first show of the new season for PBS' television program *Scientific American Frontiers* featured several ongoing projects at MIT's Media Lab. This show aired in October of 1996, and Alan Alda, the show's host, demonstrated some way cool technology, including plenty of virtual-reality projects. But the segment that really got me charged dealt with the Cyborgs, a group of MIT students who build and use wearable computers.

Briefly, a wearable computer contains a full DOS, Windows, or Linux machine, complete with video display, keyboard, hard drive, and other PC essentials, housed in a backpack or beltpack. The user puts on this computer, complete with battery supply, and walks out the door. The five Cyborgs shown on the MIT campus used the Private Eye video display system, a small LED-based graphics device that clips to a pair of glasses or headband. For the keyboard, the universal choice was the Twiddler, a one-hand device containing a chording keyboard with built-in mouse. (Chording means you press two or more keys simultaneously to generate another, single key.)

Some of the Cyborgs wore systems that included RF modems, letting them become mobile links within the school's intranet, which in turn made such a wearable computer a moving node on the Internet. Someone sitting at a desktop PC in the Media Lab could connect to a modem-equipped Cyborg and exchange email or chat directly. And if the Cyborg system included a live video camera, as one system did, the chair-bound viewer could see what the Cyborg saw, using a small MPEG window.

My robotics interest in all this stems from the PC used to drive the Cyborg systems. You can't grab some motherboard, hook it to a battery, and expect any decent results. First off, many of the motherboards require +/- 12 VDC supplies to function. Secondly, since the designers knew in advance you would have a 230-watt supply sitting in your case, they took no pains to reduce power consumption. And I won't even mention the difficulties with physically mounting such a large board, or main-

taining good mechanical and electrical connections with all the necessary plug-in peripheral boards.

No, for either a wearable computer or a PC-based robot platform you need a small, low-power, fully-contained PC single-board computer (SBC). For robotics, I would extend this to include solid-state bootable RAM or flash EPROM drives, to cut down even further on power consumption and mechanical weight and fragility.

So I began a search for the perfect PC SBC. I found several that got close, but weren't exactly right for one reason or another. Usually this was power draw; I really wanted an SBC that could run on less than 3 watts, which is only 600 mA at 5 VDC. And I dismissed any SBC that required any voltages other than 5 VDC, even if it only needed a few milliamps.

Finally, I found the WinSystems 386SX PC-104 module. This board, dubbed the PCM-SX, measures only 3.6 inches by 3.8 inches, weighs just 2.5 ounces, and draws a mere 375 mA from its lone 5 VDC supply. You get a 386SX CPU running at 40 MHz, complete with two serial ports, one parallel port, an IDE hard-drive controller, floppy disc controller, and a standard PC-104 bus connector for expansion. The board can boot off of a single 32-pin memory device, which currently gives you the options of a 512K SRAM or flash, a 1M EPROM, or the 12M DiskOnChip device available from MSystems.

As I write this, the WinSystems board is targeted for production in January, priced at $395 in singles. From my research, this is a smoking deal, and I hope to have one in my hands by the time you read this. I'll give you an update later. If you'd like a look at the WinSystems' 386SX board, or any of the company's large array of products, stop by their excellent web site.

My Marble Maze Machine

This column contains an entire robotic project, from initial design to final implementation. Appearing in the February 1997 issue of *Nuts & Volts*, my marble-maze robot was a hit at both the Seattle Robotics Society meeting and at the regional Mensa conference, where Kevin Ross and I gave a presentation on hobby robotics.

The marble maze-runner consists of a two-servo joint mounted to the underside of a flat platform. Powered by a 68hc11 BOTBoard, the platform can tilt in all directions, causing a marble to roll through a small cardboard maze built into the top of the platform. The software I wrote for the 68hc11 allows three different modes of operation. You can have the 68hc11 run the marble through the maze, using a fixed set of tilt instructions and delays, or you can use keys on a small RS-232 terminal keypad to control the platform yourself. The third mode lets you level the platform prior to playing, so the marble stays at the starting point before you begin.

This was a neat little project, and could make an excellent beginning for larger, more complex Science Fair projects or robotics exhibits. I know it was cool, because the crowds at Mensa and the SRS couldn't leave the maze-runner alone.

Not all robots need wheels, nor do they have to run around on the floor. This month's machine rolls a marble through a tabletop cardboard maze, and can serve as the starting point for kinetic sculpture or other robotic art.

The Mensa organization asked some of us from the Seattle Robotics Society to give an informal talk on hobby robotics. This seemed like a good reason to build another 'bot, so I started sketching ideas for my newest creation. Rather than build another wheeled platform, I set out to design a tabletop unit that could do "something cool." Gradually, the disorganized wish list sorted itself out, and I determined to build a robotic device that could tilt a platform in two degrees of freedom, so as to move a marble through a small maze.

Having decided on the problem I wanted to solve, I then began sketching out possible solutions. I figured I could build this easily, using a 68hc11 BOTBoard and a couple of unmodified Futaba S-148 hobby R/C servos. Longtime readers of this column are no doubt familiar with how I hook these components together, and past columns are filled with info on using them.

My first attempt at the maze robot involved a platform resting on four 2-inch legs, each leg fastened to one corner of the platform. Each leg would be hooked to a clevis, in turn hooked to the control horn of a servo motor. In theory, each servo motor could then raise or lower its corner of the platform. With coordinated movements of two servos, I could then tilt the platform in any direction.

Unfortunately, this scheme proved unworkable for me. I did not have the necessary in-line ball joint, needed to mount each leg onto a corner of the platform. Unless these joints can pivot through 360 degrees, the table isn't free to tilt. And while such items undoubtedly exist, I couldn't find anything suitable in my price range.

So I took the easier course and rethought my design. This time, I came up with a pair of servo motors, fastened together to form a powered joint with two degrees of freedom. The body of the upper servo is firmly fastened to the control horn of the lower motor. This lets the lower servo swing its partner from side to side. I then fastened a shelf, made of thin Sintra shaped into an L-bracket, to the upper servo's control horn. This arrangement lets me roll and pitch the shelf in any direction. See the accompanying diagrams (Figures 1 and 2).

As you can see, the mechanism is quite simple, and can be extended by adding extra servos for more degrees of freedom. Note, however, that actually putting these together can get a little tricky. First of all, hook each servo you intend to use to some type of test rig, such as a BOTBoard with suitable software, and verify that the motor actually works. Also verify that the motor will give you at least 90 degrees of motion. Most will, but sometimes a clunker sneaks through QA, and you're better off finding it now. Once you bolt this rig together, you won't want to take it apart again!

Next, fashion the L-bracket shelf from some light, sturdy material. I found some 1/8-inch Sintra scrap at the local plastics store, already formed in the necessary right angle; a piece eight inches long worked for me. You could also use very light aluminum or ABS. Note that you must leave clearance between the underside of the shelf and the top of the upper servo when the assembly is put together. Therefore, you will need a flange at least 2 inches wide where you can attach the servo's control horn.

Cut this L-bracket material as shown, then drill a 1/8-inch hole where you will put the center of the servo's control horn. Next, take a spare control horn and align its center hole with the hole you just drilled. Select three evenly-spaced holes on the perimeter of the control horn and mark through these holes onto the L-bracket material. Drill these marked holes, using a 3/16" drill bit.

Now you should be ready to mount the shelf to the control horn. Note you must FIRST insert the control horn's mounting screw into the bore of the horn! With the screw in place, use double-sided foam tape to stick the control horn to the shelf. Take care to align the three screw holes in the shelf's flange with the selected holes in the control horn. Run three small sheet metal screws, shipped with the servo hardware package, through the holes in the shelf flange and into the control horn. Use heavy nippers or horseshoe nail clippers to cut off the protruding ends of the screws.

Temporarily fit the shelf and control horn to a servo to verify alignment. You should be able to move the shelf through at least 90 degrees without hitting any part of the servo's case. If the mounting ears on the servo case interfere, just cut them off with heavy wire cutters, then sand or file the case smooth. After you are satisfied with the alignment, heat up the hot glue gun and run a bead of glue around the perimeter of the control horn to solidly fasten it to the shelf flange.

Next up, you need to fasten the upper servo to the control horn for the lower servo. In this case, you must first bolt the control horn to the lower servo before proceeding. Put a couple of strips of foam tape on the control horn surface, then lightly press the upper servo to the tape to check alignment. With the shelf fastened to the upper servo, you should be able to roll and pitch the shelf flange through 360 degrees without interference. If necessary, you can also cut the mounting ears off the lower servo.

When you are satisfied with the performance, firmly press the upper servo into place, then run a bead of hot glue around the perimeter of the lower

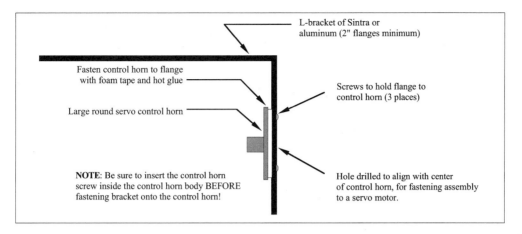

Figure 1. Side view of the L-bracket.

servo's control horn. When cool, this glue will bond the assembly into a single powered joint with two degrees of freedom. You should be able to hook these two servos to a BOTBoard and move the platform as needed.

Next up, you have to mount the lower servo to some type of base. I used a base made from a sheet of 1/4-inch white Sintra, cut to 11 inches by 11 inches. The dimensions were intentional; I wanted the whole unit to sit comfortably in an empty Xerox paper box, making it easy to transport. I placed the two-servo assembly on the base so that the L-bracket's pivot point sat at the base's center. I then marked two dots on the base, one on either side of the lower servo at the center of the longer side. After drilling two 1/8-inch holes at these dots, I used foam tape to stick the lower servo in place, ran a nylon wire tie through the holes, and cinched it down to clamp the motor. A

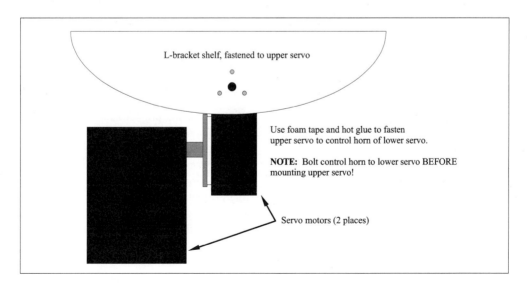

Figure 2. Front view of the L-bracket.

bead of hot glue around the servo at the base finished the mounting.

With the mechanical mounting complete, I was able to grasp the upper shelf of the L-bracket and slowly move the bracket right, left, forward, and backward from stop to stop on both servos. Be sure that all mechanical connections are firm and solid, since any slop will cause the maze platform to wobble during use. Note that some slop is inherent, due to the design of the servo motors. You can expect a few degrees of play around both servo shafts. But do what you can to make all the other mechanical connections solid.

Next up, I added the electronics. This consisted of a 68hc11 BOTBoard loaded with a 68hc811e2 microcontroller (MCU) and two battery packs. The larger pack holds four C cells, used to provide power for the two servo motors. The four AA cells in the smaller pack run the MCU. You almost have to use two supplies for this machine, since the servos will draw so much current as they change position that the voltage will sag below the MCU's reset threshold, resetting the processor.

Mounting the electronics took nothing more than some foam tape on the underside of the battery packs and a dab of hot glue at each corner of the BOTBoard. I also wired two SPST switches, one in series with each battery pack, so I could control the voltage from either pack individually. This came in very handy when I was testing the software, since I didn't want the maze whipping back and forth during my trials.

Everything mechanical was done now except the fun part: designing a maze to run a marble through. I started by cutting a piece of 1/4-inch Gator board to a 10 inch by 10 inch square. Gator board is a tough, plastic-coated version of foam-core, sometimes available at art and industrial art stores. You could probably substitute heavy corrugated cardboard as well. I decided against Sintra for weight reasons; I wanted to keep the mass at the end of the servo's arm as small as possible.

I then doodled around on a piece of quadrile paper, trying different maze arrangements. I only used segment lengths that were multiples of 1 inch, to make the final construction as easy as possible. After a few trials, I came up with the maze layout shown here (Figure 3). When I was satisfied with the layout, it was time to build it up.

Since the Gator board had a white surface, I hunted up some white 1/16-inch cardboard, similar to shirt cardboard. Using an Exacto knife, I cut out many one-inch wide strips of this cardboard. I then cut these into an assortment of shorter strips, in 1-inch, 2-inch, 3-inch, and 4-inch lengths. I also cut a few 5-inch and 6-inch lengths as well.

I began the maze construction by building a one-inch-high wall completely around the outside of the maze base. I held this in place with masking tape on the lower outside edge and a bead of hot glue on the lower inside edge. I also used hot glue on the four inner corners.

Working from my layout page, I then started at one corner of the maze base and carefully placed each maze wall. A quick dab of hot glue at the seam between wall and floor held the wall in place. Since every wall touches at least one other wall at a right angle, I ran a bead of hot glue along that vertical edge, providing more support. After the glue cooled, I was ready to fasten my maze to the top shelf on the L-bracket. Just a couple of pieces of foam tape and the project was nearly complete; I only needed a little bit of software.

The Maze Software

I envisioned this project as an interactive design, one in which a user could try to guide the marble through the maze. I also wanted the robot to be able to run the marble itself, to show how easy the task is. Finally, the software had to support a means of leveling the platform. I could have tried shimming the base every time I set up the maze, but that's kind of lame. Using the software to adjust the resting position offered a more elegant solution.

All of the above features require some type of control for the human to use, and I had a lot of options available. I could have used some slide

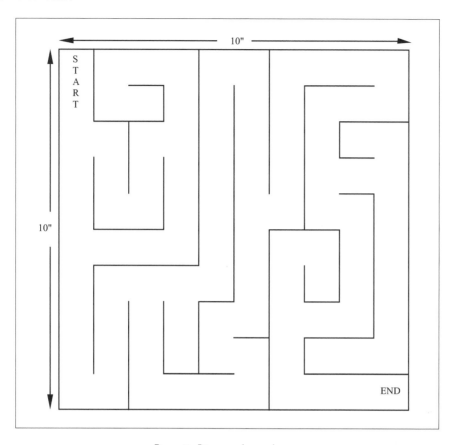

Figure 3. Diagram of maze layout.

switches to select what mode to use (adjust, robot play, or human play), two frob knobs for tweaking the maze's pitch and roll during leveling, and a joystick for actually controlling the maze.

But a little luck came my way at exactly this point in the design. Vetco, a local surplus store, set out a box of small, single-board RS-232 terminals. Each handheld terminal sported a two-line by 24 character LCD display and a 12-key keypad that included all ten digits, an Enter key, and (oddly enough) a blue wheelchair symbol. The price was right, at under $20, so I snapped one up.

When I got it home, I removed the backlighting, which really sucked the juice, and I shorted out the voltage regulator. This left me with a terminal that only needed +5 VDC to run. I then wired a four-conductor cable so I could hook the terminal to my BOTBoard. I tapped into the receive and transmit signals on the MCU side of the terminal's RS-232 level shifter, so both signals were logic-level as needed by the BOTBoard.

This terminal added some real capabilities. Now I could have the robot software report its pitch and roll settings, so I could trim the maze quickly. I could even build in some simple menuing, allowing me to switch between the three different modes with a single keypress.

The final software breaks down into three major components, with the active component selected by the operator through a menu system. The first element, dubbed GAME, lets the operator control the position of the platform using the major nine keys on the keypad. Thus, the 5-key centers the maze, while the 2-key pitches the maze platform forward and the 9-key simultaneously pitches it backward and rolls it right.

The second element, which I call PLAY, causes the robot to execute a fixed series of pitches and rolls, moving the marble on a preset course through the maze. This sequence assumes that the marble is already in the starting position at the upper-left corner of the maze. Since the robot has no sensors for feedback, this element is a simple open-loop set of commands.

Finally, the third element (ADJ) lets me adjust the initial position of the maze platform, using the main nine keys as described in GAME above. Each press of one of the eight outer keys tilts the maze slightly in that direction; pressing the 5-key returns the maze to the currently defined center position, so I can start over if I choose. Once I have the maze centered, I can press the 0-key to lock in the position as the new center position. This adjustment is written to RAM and stays in place so long as the MCU has power.

I began the software by building the ADJ element, so I could determine the servo values to use to center the maze on my work table. These two servo values became the starting position following power-up. With ADJ running, I then added the menuing system so I could toggle between ADJ and GAME, which was the next step.

The GAME element let me actually try running the marble myself, and I spent about half an hour in "research." Though the maze is pretty simple on paper, getting the timing and motions down proved quite a challenge.

Using the GAME element, I mapped out the series of moves needed to run the marble from its starting position to the end. The move sequence involved time duration as well as pitch and roll instructions. For each move, the robot had to tilt the platform, then wait long enough for the marble to respond. It only took a few minutes to build up a table of tilt instructions and durations, using the keypad to provide steering. These instructions formed the core to the final element, PLAY.

I converted my list of moves into a data table and added a subroutine that would step through the table, first doing the required tilt motion, then waiting the prescribed delay. When the code was finished, I could use the keypad to select PLAY, then watch my robot move the marble completely through the maze on its own.

Needless to say, this robot was a hit at the Mensa gathering, and the guys at the Seattle Robotics Society got a kick out of it as well. Randy Sargent, one of the Newton Labs team that has been making such a splash in the world-wide robotics contests recently, got all excited about hanging a camera over the maze, then having a robot run the maze on its own, without a table of instructions. Sounds really cool, and I'll keep you posted if the team finishes the project.

For now, I've got a cool show-bot that is both interactive and entertaining. You can put one together in a weekend or so, using little more than a BOTBoard, a couple of servos, cardboard, and hot glue. I'll move the code to my web site so you can build from it. You won't have the small terminal like I did, but you could hook the maze robot to your PC and use the PC's keyboard instead. Or you could rewrite the software to use switches and joysticks. Regardless of your approach, I'm sure you'll have a blast with this machine.

Tackle-bot, a Backyard Explorer

My *Nuts & Volts* column from August of 1997 focused on the Mars Sojourner robot. In particular, I wanted to develop a robotics project that would give at least a glimpse of the challenges faced by the scientists and engineers responsible for the Pathfinder project.

Although it's a far cry from a Mars explorer, Tackle-bot is at least easy to build and rugged enough for use outdoors. I provide you with details on selecting a suitable housing, and guidelines for adding motors, control electronics, and wheels.

In keeping with the remote-control aspect of Mars exploration, I also describe a contest that Kevin Ross of the Seattle Robotics Society dreamed up. This event would be great for most robot clubs and, given suitable terrain, could present a serious design challenge.

I close with a description of Robo-Screen, a PC-based utility that I wrote so that you can use a graphics interface to control your robot, without having to write a graphics program of your own. On the PC side, you use a text file to define the controls and displays you will use. All you need to do on the robot's side is write some simple code and hook up a serial cable or RF link; now you have a working graphical interface for your 'bot.

For my generation, it was landing a man on the moon. That glorious series of flights fired the imagination of millions and led many a young student to a career in science. The graphic proof that impossibly tough problems could be solved with brains and teamwork fixed society on its present high-tech track, ultimately giving us Silicon Valley, the Macintosh, and the Information Age.

The space race spawned industries, fortunes, technologies, and attitudes that changed our world, both for good and forever. After that one giant leap, we rolled across the moon's surface, sent emissaries to most of the planets, and watched in awe as two of our machines traveled beyond our own system.

But somehow the magic seems gone, or at least waning. Today, the Hubble space telescope and the orbital shuttle are largely taken for granted, though their technology is orders of magnitude greater than that used to visit the solar system years ago. Perhaps these technical miracles suffer at the hands of a jaded public, whose desktops hum with computing power unthinkable when the first Pioneer lifted off.

But can a society able to visit any web site on Earth still get cranked up? After all, you can surf to Zimbabwe or Camelot with no physical or mental effort. How far must "far" be to inspire the young scientists of today? How difficult a challenge is needed to fuel the public's imagination? How about Mars?

On the Fourth of July, 1997, NASA's Mars Pathfinder landed on the Red Planet's surface. As I write this in early July, the Sojourner rover has rolled off the Mars lander and the mission is on track. By the time you read these lines, the world will have spent several weeks feasting on images from the surface of a planet long the staple of science fiction novels.

Speaking of the web, stop by any of the various NASA mirror sites to see what's what with Mars Pathfinder. These sites will overwhelm you with up-to-date information at every technical level, complete with pictures, charts, and graphs.

Back on Earth...

Sure, it would be great to have a hand in the mission to Mars. And hitting the NASA web pages every day can be a real kick. But face it, I won't be involved in a project nearly as grand as a robotic mission to Mars. Still, this hobby is great for spawning dreams and for using your imagination. If Mars is out of reach, how about something a little closer to home, say the backyard or a nearby green belt?

One day, I was out bumming around with Kevin Ross, president of the Seattle Robotics Society, and he mentioned the Mars Pathfinder mission and lamented the fact that no one in the club would be involved. Then he described his ideas for a Mars exploration contest that the SRS could run. He saw it as an open-ended event for remotely guided machines. Using a club-provided RF data link, each team would run their robot around an outside area, gathering as much data as possible about the terrain and beaming it back to the operator. The operator, using only the telemetry, would guide the machine, recognize problems, and overcome any obstacles.

The contest design deals with two main requirements; a need for realism and the hobbyist's budget. The real Pathfinder mission is an enormous technical gamble. Our best engineers have tried to stack the deck in favor of success as much as possible, but the whole mission is a high-stakes crap shoot. Launch a small robotic probe millions of miles into space, land it on an unknown surface, and hope the mission survives every possible fatal problem, from busted solar panels to an off-balance landing.

Once the lander successfully deploys, the data stream back to Earth has to contend with the ten-minute one-way light time and the low data rate. The latter, on the order of 9600 baud, means that sending a single 256x256 pixel frame of 8-bit video will take over 68 minutes without compression. So you can forget about the traditional type of joy-stick interactive control used in Earth-bound robots; a one-hour delay isn't interactive by my definition.

Making this kind of bandwidth and delay work in a club contest isn't feasible; the crowd would fall asleep before anything happened. But Kevin figured we could add a challenge for the robot builder and still keep the contest interesting for the audience. He suggested limiting all telemetry and control information to a single 9600 baud RF link. Your robot design can use whatever data and information you can cram through that pipe. If you want video, fine; just shove it through at 9600 baud. Want to include sonar mapping? OK, but send the data back at 9600 baud. As for the delay, he described inserting a PC into the link, so all exchanges are intercepted, delayed, then retransmitted. This would let him introduce any type of delay he wanted, though ten seconds seems about right.

The 9600 baud RF link has a lot of appeal, especially since it will be easy for the builder to simulate. Just hook a serial cable between the robot and the operator's console and you're done. So even if the builder doesn't have a pair of RF modems, she can still test her robot and get the kinks worked out of the software before showing up at the contest. Once the contest starts, the SRS will provide a pair of RF modems to replace the cable, and the robot should function identically.

This is more of an event than a contest, since scoring is going to be a little difficult. The idea is to return as much useful data as possible. Remember that the operator and her team will not be able to watch the robot or see the terrain it is exploring, so getting accurate and usable data back is paramount. Guiding a robot you can't see over unknown terrain will place a premium on good sensor design and robust software.

If your robot can report back accurate information about wind speed, temperature, light levels, soil content, and other parameters, you'll get bonus points for the additional, non-navigational data. Retrieving a sample or two will probably be worth extra credit as well, even though the real Mars Pathfinder won't have any way to retrieve or return samples.

Kevin added a few more restrictions and guidelines, some based on actual Pathfinder requirements and others to tailor the contest more towards hobbyist robots. To summarize, the major design goals are:

1) The robot must fit within an 18-inch cube,
2) It must be self-powered, battery-operated, and RF tethered,
3) All data must flow through a single 9600 baud link,
4) The operator must guide the robot and collect as much useful data as possible without being able to see either the robot or the terrain.

Tackle-bot

I like a challenge as much as the next robo-nerd, so I began dreaming up a Mars explorer robot. I started with the basic design for BYRD, my backyard research drone from a couple of years ago (see "Build BYRD, a Back Yard Research Drone"). BYRD was the start of a good explorer design, but it was way too big. Still, I could reuse the 2L011 Grainger motors and the two 6 VDC, 12 AHr batteries. And the BotBoard-2 would make a good beginning controller, though I knew I would want to drop in a 68hc12 board as soon as possible. But I still needed a frame small enough to meet the 18-inch cube size restriction.

After several days of thought, I settled on a 16-inch plastic tackle box with three internal, staircasing trays. I chose the Flambeau 1832 box, which I purchased at an outrageous price from a local drug store. You can also hit the sporting goods stores, large general stores, and, if you aren't too fussy about smell, garage sales.

I must admit, the Flambeau tackle box makes a super robot frame. It measures 16 inches long, well

Figure 1. Tackle-bot Built from a fishing tackle-box, this robot measures 16" long by 14" high and is less than 12" wide. It uses two lawnmower tires in back and a single furniture caster up front.

within the size limit, and the three trays are made of thick ABS-type plastic with well-made hinges and supports. The top tray contains a large center compartment that can hold a good-sized computer board, and each tray is tall enough that you can use its front wall as a panel for mounting hardware such as switches and fuse holders.

I also like the carry handle and the two large front latches. The entire robot frame closes down into a compact, easy to carry shape. No more lugging a large box around just to carry the robot in. And the trays contain so many handy compartments that you could fill the robot up with any tools and spare parts you might need for on-site maintenance.

But the real payoff comes when you begin installing your robot electronics. The three trays and the box's floor give you four horizontal mounting surfaces. You can reach all four surfaces very easily

by simply lifting the three trays up and out of the way on their staircasing hinges. Thus, though the floor of the box only measures 16 inches by 7 inches, the cantilever tray design gives you an effective floor space of almost 16 inches by 28 inches.

I started building my Tackle-bot by bolting on the two drive motors. This involved a plea to Dan Mauch, the club's resident CNC guru, for a little help with the heavy metal. I needed a flat piece of 1/4-inch aluminum cut to 3 inches by 7.5 inches, to use as a mounting surface for the two motor L-brackets. Mounting the two 3-inch motor brackets directly to the plastic body of the tackle-box simply wouldn't work. The box floor is so pliable that the weight and lever arm of the motor/wheel combination would twist the box out of shape.

Dan got me fixed right up, and it only took a few minutes to drill the proper holes into the floor of the

Figure 2. Under the hood With the top open and the trays extended, Tackle-bot has lots of room for circuitry. Here I've added the MCU board (top), the motor control board (lower left), and the power switches.

Figure 3. The motors This view shows the underside of Tackle-bot, with the two gearhead motors each mounted to its own L-bracket. The two brackets are then bolted to a flat piece of aluminum plate that spans the underside of the box, adding support.

box, bolt the mounting bracket and motors in place, and add on the 3/4-inch shaft adapters that Dan had made for BYRD earlier. For those who haven't been following this column, I'll just mention that these shaft adapters consist of a 3/4-inch bolt 2-1/4 inches long with a 3/8-inch bore drilled about an inch deep from the head end. One shoulder of the bolt is then drilled and tapped for a 4-40 set-screw. This combination lets us fit the adapter onto the shaft of a typical industrial gearhead motor such as the Grainger 2L011, then bolt on a standard lawn mower wheel using a 3/4-inch Nylok nut.

The front caster presented a greater challenge. I wanted to use one of the hollow, wide plastic wheels off of BYRD, which started life as a pull-around beer cooler. But the caster was not a good match for the 7-inch rear wheels and would have lifted the front end of the robot too high off the ground. I never did find the ultimate caster, but I settled for a 4-inch industrial caster from the local hardware store. I had to add a 1/2-inch shim made from some Sintra plastic to

get the robot level; this also made a strong mounting plate for bolting the caster to the underside of the frame.

Next, I measured and drilled mounting holes for the BotBoard-2 computer board and for a homebrew two-channel motor driver board. The BB2 fits perfectly in the top center tray on the tackle-box, but I had to do a little surgery on one of the tray walls to make room for the motor driver board. I used heavy tin snips to make two vertical cuts in the tray wall at the desired locations, then fired up the Dremel and used a standard cutting wheel to buzz away the plastic between the two cuts.

I'll include my standard Dremel warning: Always, **always** wear eye protection whenever you use a Dremel tool! The business end of that hummer really spins, and if the cutting tool shatters, it will spray the area with shards that could easily take out an eye. Play it safe and always wear your goggles or glasses.

To add the two 6 VDC batteries, I first used foam-tape to secure two pieces of corrugated cardboard

Figure 4. The motor board | cut out part of a tray wall to make room for my usual relay-based motor control board, built on a Radio Shack experimenter's perfboard. Refer to the article for mounting details.

to the bottom of the box where I intended to fit the batteries. The tape and cardboard shimmed the batteries up away from the screw heads for the front caster. I then added a couple of small pieces of foam tape to the top of the cardboard pieces, so the batteries wouldn't slide around. Finally, I drilled two 1/8-inch holes in the box's floor, about two inches apart, between the pieces of cardboard. I used these holes to mount a pair of 4-40 threaded aluminum spacers, each 1-1/4 inches long.

The spacers stick up from the box's floor, between the two batteries. With the batteries stuck horizontally to the cardboard shims, I then placed a second piece of 1/4-inch Sintra plastic, suitably drilled, over the batteries and ran two 4-40 screws into the vertical spacers. This makes a simple clamp to hold the batteries flat against the box's floor.

When I built BYRD, I ran wires to the main power switch and the motor power switch, housed in a separate aluminum box and mounted to the outside of the frame. For Tackle-bot, I decided to mount both of these switches inside the box, using the vertical wall on one of the lower trays as a panel. This makes a clean, easy to use arrangement. Throwing the main power switch applies power to the 68hc11 microcontroller, but the motors won't turn unless you also throw the motor power switch. For safety's sake, I also need to add a second main power switch, accessible from the outside of the box, but I haven't done that yet.

After I took the accompanying photos, I redid the wiring to make it a little neater. First, I tore out all the wiring you see here. Then I drilled holes in the bottoms of the appropriate trays, next to each switch and circuit board. Finally, I ran the wiring along the underside of each tray, bringing the wire ends up through the holes for the connections. This leaves me with a cleaner wiring harness that is much easier to service.

I then installed a small switching power supply, picked up some months ago when National Semiconductor was running their "Simple Switcher" give-away. The PCB for this supply uses the LM2595 five-pin device and the board is smaller than one square inch, yet it delivers up to one amp of current at 5 VDC from an input of 12 VDC. If you want to build a switcher of your own using this design, hit the National Semiconductor web page. This is a dynamite source of info on any National part; the layout is easy to follow and you can quickly get to data sheets and applications notes. It even provides budgetary pricing and availability of samples. Since you asked, free samples of the LM2595 in a TO-220 package are available; see the National web site for details.

A little wiring, a little drilling, a little bolting and gluing ... I was ready for the first test of my new robot frame. I didn't even change the code in the old BB2; the firmware left over from BYRD would serve for my first tests. I hooked up a small LCD-type terminal to Tackle-bot's serial port and hit the power switch. The display showed BYRD's power-on prompt, so I was ready to go. Pressing the proper keys on the terminal ran the motors through their paces, though the speed is a wee bit too fast for indoor use. Still, it was a success; my new Tackle-bot is a going concern, with plenty of torque and carrying capacity.

What About Software?

Controlling an autonomous rover and pulling data back from it are serious challenges, requiring some sophisticated software. Here, I'm using the term "software" in the PC-based sense, rather than in the embedded firmware sense. After all, a small robot armed with even a few sensors can spit out an overwhelming amount of data. You need some kind of PC (or Mac or ...) software that can control that data flow and present it to you in a meaningful way.

I've done a few projects in Visual Basic, so I figured I could crank out a simple program for col-

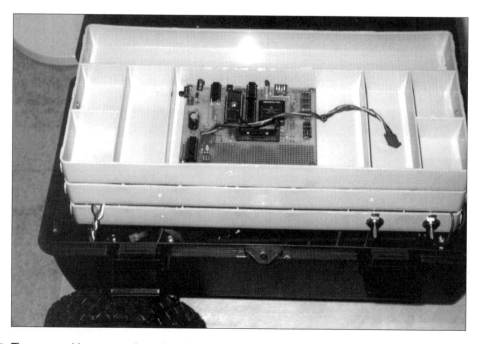

Figure 5. The top tray Here you see the 68hc11-based BotBoard-2 that runs the robot. I bolted the PCB into the large open area in the top tray. On the right side of the lowest tray, you can see the main power switch (right) and the motor switch.

lecting data, controlling the robot, and displaying the data. But then I started thinking about other robot builders. After all, not everyone has access to software for developing graphical programs, or what's more, has the ability or inclination to write such programs. Perhaps, if I structured my program properly, I could make it easy for others to run their robot using my software. This would spare them the grief of coding their own PC-based user interface.

With that thought, I started work on a program I call roboscrn, for Robot Screen. The core of the program is written in VB3 and I run it under Windows 95. Please note that this program is far from finished, and I'm only describing the very first phase of its development. But I hope you'll get a feel for what the finished software might provide.

I designed the program to support a serial connection to a small robot. The program issues commands as two-letter strings followed by zero or more arguments. My design assumes a fairly simple feature set on the target robot; eight digital inputs, eight digital outputs, and eight 8-bit analog inputs.

The display consists of three groups of buttons, boxes, and controls for each major type of I/O subsystem. For example, data from the analog ports are displayed in a set of eight text boxes, complete with a label for each analog channel and a units string that describes what the analog channel measures.

Rather than simply display the 8-bit hex or decimal value for each channel returned by the target robot, I allow you to define a mathematical function for each analog channel. The program reads an analog channel, applies the appropriate function, then displays the result as a floating point number in the proper text box. This means that the program displays the value for the channel "Battery Voltage" as the number of volts DC, in floating point, as you would expect to read it.

You define all the characteristics for your data displays with a text file and a simple description language. Using a text editor such as Notepad, you enter the proper commands for defining your data display. When you later execute the roboscrn program, it opens this file, processes the commands,

and tailors the display accordingly. Commands in this text file can determine the rate at which the host PC polls the target robot for data, what COM port it uses, and what mathematical functions are applied to each data channel.

Consider an analog channel that returns a data value from a thermal sensor of some kind. To convert the sensor's 8-bit value to degrees Centigrade, you must first apply the function:

$$Tc = 40.1 * V**2 - 0.5 * V + 37.8$$

(The above is purely hypothetical; I don't know of or have such a sensor.) Further suppose that your robot will return the 8-bit value for this sensor as data channel 2, that you want the display labeled "Temperature" and you want the units to be "Degrees C." The following commands would set up the roboscrn display properly:

> **ad_label 2 "Temperature"**
> **ad_units 2 "Degrees C"**
> **ad_function 2 37.8 -0.5 40.1**

As you can see, I use a straightforward format. The first line contains the name of the command (**ad_label**), the channel involved (2), and the exact text to use as a label for the display ("Temperature"). The second line uses the **ad_units** command to define the text string ("Degrees C") associated with channel two. Finally, the last line gives the conversion function for the value returned by channel two. Here, the first floating point number is the zeroth-order coefficient of the conversion polynomial, the second floating point number is the first-order value, and the last argument holds the second-order value. You can readily see how the **ad_function** command line maps into the above conversion polynomial. As I write this, my preliminary version of roboscrn supports up to second-order conversion polynomials, though that could change on a whim.

A similar set of commands controls the group of digital outputs. These appear on the roboscrn display as a vertical line of eight radio buttons. Clicking a button on or off causes the software to issue a command to the target robot to set the corresponding digital output to logic 1 or logic 0. What actually happens on the target robot is up to your firmware. In fact, your firmware can use one of these digital output signals as a command to trigger a more complex array of actions, if you choose. All that really matters is that clicking a digital output button on the screen causes roboscrn to issue a known sequence to the target robot.

For example, you might decide that digital output 0 will control power to the on-board video camera. Setting this up requires just one command line in the text file:

digi_out 0 "Camera"

This line uses the **digi_out** command to assign the label "Camera" to digital output 0. When you run the roboscrn program, you will see a radio button labeled "Camera" in the first position in the group of digital outputs. Clicking that button will cause roboscrn to issue a command to the target robot instructing it to turn digital output 0 on or off, as appropriate.

The key to making this program work with robots that I didn't design lies in the interface between roboscrn and the target robot. So long as your robot's firmware follows the rules laid down by my roboscrn design, your robot will work fine with roboscrn. For example, when roboscrn wants your target robot to send eight channels of analog data, it issues the sequence:

D*

Your robot firmware should echo each character it receives from roboscrn. Upon detecting the above sequence, your firmware should immediately respond with:

D<0><1><2><3><4><5><6><7>

where each field of the form **<n>** is a single binary byte containing the analog data for the corresponding channel. Your firmware should terminate this line with a carriage-return (CR) code.

When roboscrn receives this 10-character sequence, it will parse out each byte of data, apply the proper conversion function, and show the converted value on the display in the proper location.

The same type of interface rules apply to the other I/O channels. Thus, by following a few simple guidelines and tailoring your firmware to match, you can have a polished interface to your roving robot up and running in a matter of minutes.

Please note that roboscrn is in its very earliest stages, and the program's design and interface rules will likely mutate over the coming weeks and months. However, I'll stick to these general guidelines as much as possible. Check my web page for the latest version. For now, I intend to use VB3, so you will need some flavor of Windows to run the final application.

I hope that roboscrn will help you get your own rover robot up and running. Ideally, this discussion will spur others to developing their own versions of roboscrn; if so, I hope you'll let others share and enjoy your efforts. This kind of tool, written once, will benefit countless other hobbyists who might not have the time or talents to build such an app.

That takes care of this month. I hope Sojourner is still motoring around up there on Mars and still sending back pictures. That little robot has certainly spurred a lot of interest in science and in the field of robotics. And even though you can't go to Mars (at least, not yet), you can still build your own rover robot. Just grab a tackle box and some software, and go to it. See you on Mars!

```
*
*    A  sample  command  file  for  roboscrn.
*
*    This  file  defines  three  analog  channels,
*    eight  digital  outputs,  and  sets  the
*    data  query  rate  at  once  every  two
*    seconds,  using  COM2.
*

ad_label  0  "Battery"
ad_label  1  "Temperature"
ad_label  2  "Light Level"

ad_units  0  "Volts"
ad_units  1  "Degrees  F"
ad_units  2  "Lux"

ad_function  0  2.2  1.0  1.5
ad_function  1  50  0.4  0
ad_function  2  0  1  0

ad_rate  2

port  2

digi_out  0  "Camera"
digi_out  1  "Forward  lights"
digi_out  2  "Aft  lights"
digi_out  3  "Siren"
digi_out  4  "Flamethrower"
digi_out  5  "Shredder"
digi_out  6  "Whirling  blades"
digi_out  7  "Tear  gas"
```

Figure 6. A sample roboscrn command file.

Try Your Hand at a Mini-Sumo Robot

This column from September of 1997 tried to explain how to start a robotics club like the Seattle Robotics Society. Judging from my email, the need is out there, all across the country. But putting together a group of hackers like the SRS isn't easy, and I certainly don't have a sure-fire formula. So I offered here some guidelines, a bit of history, and a large dose of encouragement.

The Mini-Sumo robotics event, a pint-sized version of the Sumo robotics contest spawned in Japan some years ago, has been popular in the SRS for quite a while. This column provided information so you could build the necessary Sumo ring to hold your own Mini-Sumo events and build a suitably sized robot. I also discussed a weekend hack that Marvin Green and I did, to try and get a contest robot running in just an evening. I even included a condensed version of my SBasic code for the little machine, so you could get started on your own software.

I get a lot of email from readers, and some of it would just break your heart. It's usually from some poor soul stuck out in a robotics hinterland, trying to piece together a robopet or a controller board or just anything, but he (and sometimes she) has no one to talk to, no one to bounce ideas off of. Finally, they send off an electron to me asking for help.

I usually get back to them with either a workable solution or a pointer to help them take the next step. But once in a while, a reader will ask the larger question, the solution to which will actually solve all the other, smaller problems: How do I start a robotics club like the Seattle Robotics Society?

Tough question, that. Actually, that is two tough questions. The first question really is: How do I start a robotics club? Since I've never actually started a club myself, I can only offer what I think are sensible suggestions. Start by communicating with as many other people of the same persuasion as you can. If you're enrolled in a junior college class on microcontrollers, talk to the others in your class, to your instructor, and to those in similar classes, to see who is interested in robotics. If you're hanging out at the library Internet terminals, leave a card on the bulletin board and see who responds. Post a notice at the local surplus electronics store or computer store. If your city or town has a web page dedicated to local activities, see if you can post a notice to it.

The point here is to get your interest in robotics across to others similarly inclined. You can almost approach this as if it were a job search. Call all of your friends and ask if they are interested in building robots with you. If they are not (and sadly, many will not be), always ask them for the names of at least two other people that you can contact. The goal isn't to find a lot of people interested in robotics. You are trying to get in touch with as many people as possible, period. In the course of your efforts, you will eventually find the few who want to do 'bots. So when you start calling, be sure to call ALL your friends and relatives, even if you know (or think) they won't do robots.

Be sure to contact those you know in hobbies or activities that use similar thinking skills. For instance, if you play Magic or chess, talk to others who play. These hobbies use much the same grey matter as designing robots, and should provide you with a good starting point in your search.

If your area has a local high-tech tourist attraction, hang out with the people who work there. I did an article some time ago about MBARI, the Monterey Bay Aquarium Research Institute (see "Deep-sea Submersible Robots.") Talk about a bunch of ready-made robo-nerds! Any two or three of the MBARI staff could have made an excellent core to a robotics club. I'm sure if I had just hung around the aquarium and asked enough questions, I could have connected with some of the engineers.

Then there is the Internet. You can find hundreds (thousands?) of web sites devoted to homebrew robotics, university and government research projects, and robotics designs from private industry. If you visit enough of these and use email appropriately, I'm sure you can contact others who share your interests. You might even find people close enough to you geographically that you can visit each other.

And remember to investigate Internet sites in related fields. Rather than focus on just robotics,

look into geotechnical, astronomy, oceanography, model rocketry, remote control, homebrew electronics, embedded control, model railroading, and ham radio, to name just a few. I'm sure you'll find many people with interests in several of these fields, and you can probably interest them in robotics as well. Again, you might find people who live within visiting range.

I mentioned earlier that there were really two questions involved here. The second question is: How do I make a club like the Seattle Robotics Society? The short answer is time, luck, and hard work. The long answer deals with people and synergy.

The SRS has been around for a long time; I attended my first meeting in late 1989. The club still has members who recall the very early days of 1981 and 1982, when the SRS first formed. Back then, the club members spent most of their time on mechanics and simple electronics. Microcontrollers in today's form simply didn't exist, and the computers of that day, such as the Apple, were too expensive and too power-hungry to bolt onto a robot frame. This marked the "heavy-metal" phase of the SRS.

Meetings took place each month at a surplus electronics store in downtown Seattle. The store carried lots of heavy surplus metal and motors, and the SRS members were always a sure sale. The club had a mascot robot, a very large metal frame on very large casters and carrying a very large battery supply. Since almost no one had enough work space in which to tinker with this machine, it usually spent its days in the front of the electronics store, where it served as a nerd-lure and sales gimmick.

By the time I arrived in 1989, regular attendance was on the decline, with the average meeting drawing just ten or so. Membership continued to drop, and at its lowest ebb, I recall attending a meeting with only about four other people. But the faithful hung in there, the membership slowly rebounded, and today's meetings routinely pull fifty or more attendees.

I think sheer enthusiasm kept the club from dying out altogether. Even when only five of us showed up, we always brought our works in progress, always talked about the robots we would build, and always shared our experiences. To this day, the most important element of each meeting is the time spent describing what we've done and how we did it.

This usually occupies at least an hour of each meeting, and is open to all who wish to talk. To make sure everyone gets a chance to speak, Kevin Ross, the club's president, polls each and every attendee at the beginning of the meeting. If someone wants to talk, Kevin notes the name and topic on the blackboard. After working the room, Kevin goes back to the blackboard and gives each person who signed up their five or ten minutes of speaking time.

I get a chuckle out of the first-timers; they are always shocked that they would be asked if they have an item for the agenda. But that's part of what makes this club work; it doesn't matter if you're a first-timer or a founder, everyone gets a chance to contribute.

And asking questions can be just as important as describing your project. Some of our most interesting discussions have started with someone wanting to know if someone else had solved a certain problem. One such question, regarding a power supply that didn't poop out when the battery dropped below six volts, spawned a discussion that turned into a design session, ultimately leading to the Junk Box Switcher, the topic of an article a year or so ago (see "Try This Junk-Box Switcher Supply.")

The years since I arrived have seen a huge change in the type and number of robots in the club. The early "large-iron" designs have given way to small, light machines built of plastic, wood, brass rod, and PCB stock. Where robot builders used to describe batteries in terms of Amp-hours or pounds, they now speak of C cells or solar cells. And the new machines don't rely on the heavy motors of yesteryear's designs; today's builders go with pager motors and R/C servos.

Since the materials are so light and easy to work, the robots are small and easy to build. The club's periodic nerd-fest, Robothon, will routinely draw 100 attendees or more. Many of the SRS members show up with not just one but several working robots, and the arena is usually thick with little robotic critters.

But the biggest change of all in the time I've been a member concerns the microcontrollers used in the club's robots. In the early days, you could choose from maybe two or three viable microcontrollers, but there weren't a lot of development tools around and few people knew how to write software well enough to get the robot going. Sure, you could build in a 68hc11 board, but that required point-to-point wiring on a perfboard. So making your first microcontroller was a major effort, and building the second one required just as much work. But by then the excitement of getting the first one running had worn off, so making the second one became a chore more than anything else.

Even when you did get a microcontroller built, it was still a major uphill battle to get any software written for it. Everything had to be done in assembly language, assuming you could even find a suitable assembler. If you got stuck on a hard problem, such as setting up interrupt vectors, you had almost no one to talk to about it. And after you got the microcontroller to do anything at all, you then had to write and test the real robot code. The sheer enormity of the struggle stopped a lot of people.

What a difference a few years make! Today, the club sports groups of people with serious design experience on microcontrollers as diverse as the 68hc11, 68hc12, 68332, 80c188, 8051, and others. Anyone in the club who wants to start a design simply needs to raise a hand and she'll find SRS members who already have etched PCBs, written software tools, and built libraries of working code. Take, for example, the 68hc11. Besides the stock Motorola tools such as asmhc11 and pcbug11, SRS members have developed power tools such as Interactive-C and SBasic, with large libraries of tested code. Any member can draw on these resources to help build their own machine, saving a bundle of time and work.

But all of the above doesn't really answer the question of how to make a club like the Seattle Robotics Society. The SRS became an extension of the talents and energy of its members. The club became what it is because so many people wanted so much to spend time with other robot builders, to learn from

Figure 1. Top view of Marvin's Mini Sumo frame. You get IR object detection, prototyping area, and four servo controllers, in less than a 10 cm by 10 cm square. Not bad!

others, and to help others. In all my years in the SRS, I can't recall hearing a single cross word or seeing a heated argument of any kind. It isn't that we have rules against such, only that they just don't happen.

This is as close as I can get to telling you how to build a club like the SRS; always contribute more than you take.

Mini-Sumo

Some years ago, a robot competition in Japan introduced the world to robot Sumo. Modeled after the traditional Japanese wrestling sport, two small machines square off against each other in a black-surfaced ring. To win, one robot must push the other off the ring. Size and mass restrictions apply to keep the robots within sensible limits.

The Seattle Robotics Society has run Sumo competitions for a few years now, and Bill Harrison, one of the club's stalwarts, has kept the club's gear in primo condition; Bill also runs the Sumo event during the club's periodic Robothons.

But building a Sumo contender can prove difficult. The 20 cm (about 7.87 inches) footprint and 3 kg mass limit translates into a hefty robot packed into a small space. This places a premium on machining skills and tools. To open up the Sumo event to a larger audience, Bill developed the Mini-Sumo contest.

Mini-Sumo robots mass just 500 g and fit into a 10 cm footprint; the contest takes place on a ring half the size of the larger 'bots, only 77 cm (about 2.5 feet) in diameter. The smaller size and weight limits let you build a Mini robot using lighter materials, which makes the robot easier to build. Of course, you now have the challenge of getting all the electronics, motors, wheels, and batteries to fit in the smaller size and mass limits. But the club has far more Mini robots than full-fledged Sumos, so I think Bill's move to the smaller machines was a good one.

To date, most of the machines are blind pushers. They patrol the Sumo ring in a pattern, hoping their bumper switches will contact something so they can turn and push it off the ring. Think of these designs as first-order solutions, little more than brainless, sightless machines. Naturally, many in the club are pushing for stronger, smarter solutions.

During a recent visit to Seattle, Marvin Green brought along his first version of a Mini robot. Built in a 10 cm square base, it uses hobby R/C servos for motors and has a front skid or scoop for sliding under the opposing machine. Marvin and Bill collaborated on the mechanics, and the result is a clean, simple mechanical solution. Marvin then added a custom-designed BOTBoard-style computer to run the robot. He designed a new PCB for this 68hc11 microcontroller, adding features long missing from his original BOTBoard product, such as on-board 40 kHz IR object detection and 32K of RAM. The final PCB is smaller than the 10 cm square space restriction for the Mini robot, making it an ideal general-purpose 68hc11 robot controller.

Marvin and I cleared out a working space in my very cluttered robot room, then settled down to some serious hacking. It was the Friday night before the SRS' monthly meeting, Marvin hadn't even built the first robot from his new design, and there wasn't any software for running his new machine. So while he heated up my soldering iron and unpacked his robot parts, I fired up my laptop and began designing the first cut at the Mini Sumo robot code.

Just four hours later, with a short break for sugar and fat, we had a working Mini Sumo robot, complete with IR object detection and tracking software. The fact that we could get so much done in such a short time speaks well for the tool set we used. Marvin's electronics design worked beautifully with no surprises, and my large library of SBasic code let me focus on navigation and strategy without bogging down on the low-level problems such as making a motor move. It was quite a thrill that night, watching our little warrior chase a box around on the carpet, knowing that shortly after dinner it had been sitting in various plastic bags in Marvin's luggage.

Figure 2. Front view of the Mini Sumo. This shows the front scoop, the rubber-rimmed wheels, and the new BOTBoard PCB. The silver cube in the center of the PCB's front edge is the IR detector.

Figure 3. On the underside of the Mini Sumo, you can see the four penlight cells for power. The vertical bar on the left side is the front scoop with its down-looking photodetector for sensing the white line at the ring's rim.

We strode into the SRS meeting next morning, ready to show off our new creation. Naturally, Murphy was in fine shape and managed to throw a small spanner into our demo. My robot room uses incandescent lighting, which seldom presents a problem with IR object detection systems. The meeting room, however, has flourescent lighting, which is almost guaranteed to screw up an IR system. Needless to say, we had neglected to add a filter to our IR detector and the little robot saw enemies everywhere. So the demo didn't go as well as we would have liked, but the guys understood what we had done and were appreciative.

By the way, if you've ever run into this lighting problem with IR detectors, just cut a piece of exposed film from the tail end of a roll of color print negatives and glue it over the front of your IR detector. The film is transparent to IR but nearly opaque to regular light, and makes a great filter.

I have included in this article a partial listing of the minisumo.bas file, so you can see how my code translates the IR object-detect signals into a plan of action (Figure 4, located at the end of the article).

The code also looks at the floor to make sure the robot isn't sitting on the white line that marks the ring's boundary.

Look at the code in the GetInput routine. It first uses the IR LEDs and IR detector to look for an object. If it finds an object nearby, the code sets a variable to show the desired motor action, then exits immediately, without waiting to look at the floor. This means that the code gives priority to chasing an object over staying away from the edge of the ring. This bit of subsumption gets around a common tactic in Sumo robots, that of deploying a white skirt around a robot so an opponent thinks it's near the edge and breaks off an attack.

Obviously, this code isn't going to compile, as large chunks are missing. But you should get a feel for what is involved in writing software for a simple-minded attacker. Build on this code and you can create a killer Mini of your own. For more details on Sumo in general, see the SRS web page. Be sure to check out Marvin Green's web site for the latest on his 68hc11 and 68hc12 projects. And you can send email to Bill Harrison about his Sumo efforts.

Robot Ideas

My wife, Linda, suggested I add a little section here on how I decide what kind of robot to build next. I'll admit to a blank stare at this concept, since I seldom think about how I think. In response, she pointed out several techniques I use to get over the hurdle and start on the next robot.

For example, she mentioned that I usually stare vacantly out of the window. Well, yes, window staring is important to robot design. In particular, the dark clouds and gray skies of Seattle seem to help spawn robot ideas. There's some aspect of rain clouds scudding across the horizon and sleet pattering on the window that helps me focus on the motor mounts for Tackle-bot or the walking mechanism for my little servo robot. And windows are absolutely the best thing for writing software. I don't know how many times I've turned away from the screen in frustration, only to see the solution in the swaying branches of a fir tree or the manic blossoms on a nearby rhodie.

But windows provide another resource. I've looked outside at spiders in their webs and thought of a rope-climbing robot, then spent much time dreaming up possible mechanisms for scaling a rope or string. Watching ants troop across a window sill helps me get a perspective on walking robots. The rhythm of an ant's legs doesn't translate well into a hobby robot unless you've got a serious junk box, but thinking is free and easy, and you never know where it will lead.

Speaking of junk boxes, Linda points out that I'll spend long hours digging through my tons o' junk, moving parts and pieces around, just playing with how things fit together. Guilty again. I especially like laying out brass rod, battery holders, and BotBoards, then seeing how I can fit them together

into a robot design. The brass rod lends itself to designs of kinetic sculpture, and the BotBoards are small enough to fit just about anywhere, so they don't detract from the effect you're trying to create.

And doing this effectively requires LOTS of junk boxes. I've got cardboard boxes stacked five feet high across one wall of my robot room, and the closet of my robot room, intended to hold clothes and shoes, is packed with books, boxes, half-finished robots, and electronics. I have boxes filled with nothing but motors of various kinds, other boxes crammed full of wheels, and a bin or two packed with batteries. Just as Linda fondles fabric to get ideas for her next quilt design, I juggle junk while searching for that next robot.

But the most important element of designing robots, according to my wife, is laughing. She claims that I'll often get stuck on some part of a robot, so I'll just drop what I'm doing for a while. Instead, I'll pick up a book or watch the squirrels outside our window, and before long I'll be chuckling about something. Shortly thereafter, she'll find me back at the workbench, the problem solved and I'm on to the next phase.

I'll have to take her word on this one, since I don't pay that much attention to my work habits. But I do laugh a lot, and find humor in just about everything, so she's probably right. I know I've got more Dave Barry books than I do robot books, but that could just be that Dave Barry is more prolific than all the robot authors combined. (OK, maybe not.)

But it's worth a try. Next time you're stuck on a robot design, grab a little Dave Barry or Jay Leno, tune in the Comedy Channel, or head for the theater. Spend some quality time yukking it up and see what happens. Maybe it'll shake a few ideas loose and get you over the next hump. If not, you can always try my sure-fire cure for robot-block: choclate eclairs.

```
'
'     minisumo.bas
'
'     Program to control a mini Sumo robot, based on Marvin Green's
'     mini-Sumo platform.
'


main:
gosub  Initialize          ' set everything up
do                         ' do forever...
      gosub   GetInput     ' look around
      gosub   Process      ' do something about it
      gosub   Delay, 40    ' wait a bit
loop


'
'    Look around and return result in Action.
'
'    Note that chasing an object has priority over
'    avoiding the ring's perimeter line.
'
GetInput:
temp = 0                             ' start with nothing
gosub  IRLEDOn, RIGHT_IR                ' turn on right IR
gosub  Delay, 2                      ' wait a bit
if peekb(porta) and $80 =   0  ' active low
      temp = 1                       ' found something
endif
gosub IRLEDOn, LEFT_IR               ' turn on left IR
gosub  Delay, 2                      ' wait a bit
if peekb(porta) and $80 = 0    ' active low
      temp = temp + 2                ' found something
endif
gosub  IRLEDOff                      ' turn off IR LEDs
select  temp                            ' see what we found
      case  0                        ' nothing out there
      Action = IDLE                      ' forget it
      endcase

      case  1                        ' something to the right
```

Figure 4. Partial listing of minisumo.bas.

```
        Action = TURN_RIGHT            ' head for it
        print "R ";
        return                         ' leave now
        endcase

        case   2                ' something to the left
        Action = TURN_LEFT             ' head for it
        print "L ";
        return                         ' leave now
        endcase

        case   3                ' something in front
        Action = FORWARD        ' head for it
        print "F ";
        return                         ' leave now
        endcase
endselect
gosub   CheckLine               ' need to check for line
print peekb(adr4);
if peekb(adr4) > LineThreshold        ' if line is there...
        Action = REVERSE        ' need to back up
        print "B ";
        return                         ' leave now
endif
print "I ";
return




'
'   Do something, based on value in Action. If
'   Action = IDLE, do a little dance while we look
'   for a target.
'
Process:
if Action = IDLE                ' nothing out there...
        IdleCount = IdleCount + 1
        if IdleCount < 8        ' sit for a while
             return                    ' leave now
        endif
```

Figure 4 continued.

```
          if IdleCount < 16        ' go forward for a while
               Action = FORWARD
          endif
          if IdleCount < 24        ' turn right for a while
               Action = TURN_RIGHT
          endif
          if IdleCount = 32        ' turn left for a while
               IdleCount = 0          ' reset counter
               Action = TURN_LEFT
          endif
     endif
     select  Action                ' based on Action...
          case   FORWARD                 ' go forward
          poke   RTOC, RIGHT_FWD
          poke   LTOC, LEFT_FWD
          endcase

          case   TURN_RIGHT        ' turn right
          poke   RTOC, RIGHT_REV
          poke   LTOC, LEFT_FWD
          endcase

          case   TURN_LEFT         ' turn left
          poke   RTOC, RIGHT_FWD
          poke   LTOC, LEFT_REV
          endcase

          case   REVERSE                 ' back up
          poke   RTOC, RIGHT_REV
          poke   LTOC, LEFT_REV
          endcase
     endselect
     return

     end
```

Figure 4 continued.

I Start on a Fire-fighting Robot

This column, which appeared in the November 1997 issue of *Nuts & Volts*, marked the start of my ill-fated attempt to design and build a robot for competition in the Trinity College Fire-Fighting Home Robot Contest. I had the best of intentions, you know, but other robot ideas kept diverting me, and I eventually missed the deadline completely.

Still, the columns describing my efforts were very well received, and even though I finally abandoned the task, those columns contained some valuable robotics information. This article, for example, discusses my search for a suitable software tool, which ultimately led me to Software Development Systems' demo package. This excellent tool set includes a 68000 C compiler, an assembler... well, just read the column and find out for yourself.

The 1998 Trinity College Fire-fighting Home Robot Contest takes place April 18 and 19, 1998, on the Trinity College campus in Hartford, Connecticut. The contest's home page claims "this is the largest, public, true Robotics competition held in the U.S. that is open to entrants of any age, ability or experience from anywhere in the world." Be that as it may, the contest offers a serious robotics challenge, so I figured I'd give it a shot.

Briefly stated, your robot must move through a mock floor-plan of a house, locate a fire (actually, a lit candle), and extinguish the flame. Bonuses and penalties are applied, as necessary, to arrive at an operating score for each of three runs. The lowest operating score on the two best runs serves as the robot's overall score, and the lowest overall score wins the contest.

Note, however, that it is not enough just to put the robot on the floor and hope for the best. Your robot must find the candle and put out the flame in at least two runs just to qualify for a prize; any run in which your robot fails to put out the flame isn't counted for scoring.

The contest rules include several penalties that will further separate the very best machines from the rest of the herd. Touching a wall, with either the robot body or a feeler, gets your machine an immediate 5-point penalty for each occurence. Run a feeler along the wall... that'll cost your machine 1 point per inch traveled, plus the contact penalty. And touching the candle means a fast 50 extra points slapped on your robot's time score.

As for positive reinforcement, the contest format rewards those robot designers who put in the extra effort. If your robot starts its run based on a sound signal supplied by the referee, it gets 5% off of its time score. Putting out the flame, then returning to the starting point nets a 10% reduction. If your robot does its thing with furniture (actually, yellow steel cylinders) in each room, it nets a 30% bonus. And any robot that puts out the candle on all three runs gets an additional 10% off of its best score.

This is not an exhaustive list of the penalties and bonuses; get a copy of the rules or check the web site for the official information. But these details show the care the organizers have put into designing the event. The format gives you a wide range of options in your strategy. For example, you might emphasize speed and the ability to return to the starting point, hoping to make a fast enough run to outclass another builder whose robot runs more slowly but in "furniture-mode."

One interesting aspect of this event concerns the floor plan. It is known, within limits, well before the event starts. In fact, I've included a copy of the approximate floor plan for you to review (Figure 1). Note, however, that this is only an approximation. The actual contest floor plan may vary by as much as one inch in any dimension from those shown.

The designers have made a few aspects of the contest robot-friendly. The walls are all painted flat latex white, the floor is flat black, and the starting point is a 12-inch white circle painted on the floor. Even the candle is easier to find than it might be; the designers have mounted it on a 3-inch by 3-inch yellow wooden base. Other navigational aids, such as 1-inch wide white lines to mark the doorways, will make your job as robot designer somewhat easier, but only somewhat. This is still a challenging contest.

Jake Mendelssohn has been involved with this contest for years, and serves as a contact point. Jake has arranged for VHS video tapes of recent Trinity events; check the Fire-Fighting web site for ordering details.

The fire-fighting contest runs in two divisions. The Junior section is limited to those in high school, while the Senior section covers everyone else. Talk about a perfect contest for that Science Fair project! What's more, first place in each division garners a cool $1000 cash first prize, with additional awards for other top finishers. Again, check the official rules and web site for details.

OK, so this is a way cool robot event. But I'm stuck in the rain and drizzle of the Puget Sound, so why am I pumped about a contest being run clear across the country? Because the Seattle Robotics Society is hosting the Northwest Regional for the Fire Fighting Home Robot Contest, on March 21, 1998.

Last year's event in Hartford drew over 80 robots, and the running time became so long that it basically filled an entire weekend. To try and contain the chaos, Trinity College decided to hold several regional events, with winners moving on to the finals in Hartford, and the SRS is happy to be involved. Current plans call for first prize in both divisions to include some subsidy from Trinity College on airfare to Hartford, so the winners can participate in April. Those of you who live elsewhere in the robotic hinterlands should check the contest's web page for details on the other regional events.

Several SRS members have already swung into action, committing to build the necessary frames and venues needed to host the event, and to build robots that can compete. You can find out more information on the SRS' involvement by hitting our web page. Also watch this column for more details as they become available.

Getting Started

I began my first-order solution to the robot design by stretching out on the couch with a beer nearby. After much mulling, I decided to begin with a round base and direct-drive stepper motor drive system. The contest rules limit the robot to a 12-inch cube and the running time shouldn't be too great, so I

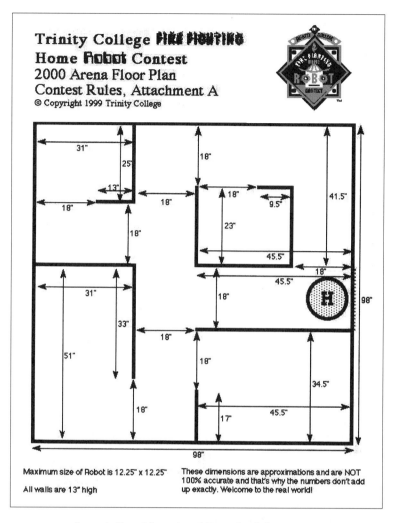

Figure 1. Typical floor plan of Trinity Fire-Fighting contest.

should be able to keep the weight low enough so the steppers will work. I'm expecting the stepper motors to help with another problem, that of navigation.

The robot's computer will already have a workable map of the house, but knowing how to get from point A to point B means you have to move reliably. One method of moving precisely involves encoders attached to wheel shafts or to a rolling floor contact. Another technique uses very precise motion increments, such as those supplied by steppers.

Another advantage with going the stepper route concerns borrowing technology. I can draw on Bill Bailey's stepper driver electronics and the wealth of experience other club members have gathered over the years, to make my task easier. Add a gel-cell battery of the proper size and weight, a little bit of wiring, and the easy part should be done.

Now for the hard part. This robot will need much more than a BOTBoard for brains. At first I considered a 68hc12 computer, since I really like the newest microcontroller family from Motorola. Even though I eventually chose another chip, I suggest you check the Motorola web site for details on these new devices. Fast, small, low-power, and source-code compatible with the 68hc11, they provide plenty of oomph for your robot designs.

But I decided instead to go with a 68332 MCU. I've used this 68000-like device in the past, but never in a robot design. The 32-bit architecture, powerful addressing modes, and lots of on-board I/O and timer subsystems make it an ideal choice for upper-level robot designs. Plus, several SRS members, including the Newton Labs gang that keeps winning all of those world-wide robot contests, use and like the 68332.

So I started looking across the Internet for a suitable computer. Note that I could have purchased a blank board from at least one SRS member and built up my own 68332 system from scratch. But I opted instead for a commercial board, and finally settled on the New Micros NMIX-0332 unit.

I've used New Micros' products many times in the past, both for my job and my hobby, and always been happy with their designs and their quality. They sell a large variety of single-board computers and peripheral boards, and I suggest you consider their products anytime you're looking for an off-the-shelf board. They can even provide boards in OEM configurations, if you want to build a commercial product around their electronics.

My $239 netted me a populated and tested 68332 SBC, complete with four JEDEC 32-pin sockets for my mix of memory chips. As shipped, the board contains 256K bytes of static RAM and 128K bytes of EPROM, both arranged on a 16-bit data bus. The board contains provisions for battery-backup of the RAM, and you can restrap the electronics to support a wide variety of different chip sizes. The 4 inch by 6 inch PCB includes a useful-sized prototyping area and pad layouts for I/O signals. It also sports an expansion connector that mates with any of the New Micros peripheral boards, should I need to add more robot horsepower quickly.

The SBC arrived with two strong pieces of firmware already installed. The pair of EPROMs contain CPU32BUG, Motorola's 32-bit monitor/debugger system, and Max-FORTH, New Micros' workhorse Forth compiler. I've used the 8-bit version of Max-FORTH on several projects before, mostly with the 68hc11 SBCs, so I like having that option available when it comes time to write code. And CPU32BUG provides some real power for debugging and running any assembly language programs I may need to try out.

As with other New Micros' products, I went through my usual likes and dislikes with this board. The design is clean and well-done, and the board is well-made and has lots of features on it that I like. But the documentation is very haphazard, and could serve as a barrier to those not well versed in details of microcontroller design. For example, the docs include two pages of schematics, both D-size and reduced to single 8-1/2 inch by 11 inch sheets, then photocopied, so you can imagine how difficult they are to read. The page labeled "Memory Map" contains a single paragraph of vaguely relevant information, but no memory map. There is no well-designed manual that takes you step-by-step through the board and its capabilities, just 31 pages of printed material on assorted topics. I also got a floppy disc containing additional software, such as a public-domain assembler and a public-domain C compiler, and text files related to Max-FORTH and the NMIX-0332.

To their credit, New Micros does include a spiral-bound manual on CPU32BUG and a copy of the MC68332 User's Manual, both vital tools, but they are actually printed and distributed by Motorola. The bottom line: I love New Micros' hardware and firmware, but if this is your first microcontroller, you might have more success going with another vendor.

The Software

Next up, I needed some form of compiler and assembler for developing my robot code. I briefly considered modifying SBasic so it could support the 68332, but discarded the idea for this project. That doesn't mean I've given up on the notion entirely; you may yet see a 68000 SBasic. But it won't likely be done for the Fire-Fighting contest.

No, this kind of project needs a higher-horse-power solution than SBasic. Assembly language for the whole project is right out; I'm not that big of a masochist. But elements of the project will require some low-level work, so an assembler is necessary. This still leaves me with a decision on the high-level tool.

After much thought, I ended up opting for C. I actually prefer doing robotics in Forth, but face it, C is more widely known, and explaining C functions to someone without a lot of programming experience is going to be easier than explaining those same functions written in Forth. (Been there, done that.)

So having chosen a language, I now needed to find the toolset. My requirements list for this assembler/compiler package was short but firm. One, it had to be free. Not cheap, free. I figure several of you will try to duplicate or improve on my efforts as these columns go by, and I'm not going to start off with a $2,000 compiler suite and expect you to follow suit. What's more, I'm just as cheap as the next cash-strapped hobbyist.

Second, I wanted an assembler/compiler system that was well-regarded by other users, sported a good set of features, and didn't take a Master's in Computer Science to use or maintain. Note that wanting a commercial-grade package doesn't necessarily conflict with item one above; it just means I have to look a little harder and be a little more discriminating.

Finally, the compiler had to generate good, compact code. I realize that after working on a 2K platform for several projects, 128K of code space seems like oceans of room. But bloated code builds up fast (can you say "Windows?"), and at this point I can't see the end of the project, so I'll err on the side of caution.

With this list in mind, I headed for my favorite freeware shopping mall, the Internet. After a few hours with AltaVista, I uncovered some candidates. The Free Software Foundation maintains a suite of tools based around the Gnu C compiler. Doing a web-search for gcc turned up a raft of various Gnu spinoffs. Many of these sites contain the full source code and build files for making gcc compilers for nearly any chip, including the 68000. As an aside, you can even find a gcc for the 68hc11, should you want to try this compiler on the 8-bitter. Oli Kraus' web site contains a full 68hc11 implementation of the gcc cross-compiler.

Those hardcore software hackers out there should note, however, that this is an executable only, no source. Also note that despite repeated efforts, I was unable to get a working set of va_arg macros running, so I could not create functions with variable-length argument lists, such as printf(). But I really like this compiler, even with these problems, and you may yet see a 68hc11 or 68hc12 design based on this tool.

But the Gnu tools really require more software expertise to use and maintain than I was willing to accept for my project and my readers, so I continued looking. Another possibility lies in the public-domain C compiler provided by Motorola and shipped on the New Micros floppy disc. The disc contains the executable for both the C compiler and the matching assembler. It also contains what looks like the full source code for the C compiler, though I haven't tried rebuilding the compiler yet.

But the .doc file on the compiler indicated a few problems that left it short of my requirements list. Again, however, those software hackers out there could have a lot of fun with this compiler. Wally Brandt, the original developer of this compiler, did a terrific service to the 68000 comunity when he released his work, and I'd like to see others carry onward from there. Given the long expected life of the 68000 family and the ever-growing number of hobbyists on the prowl for cheap tools, an upgrade to this compiler would find a lot of grateful users. Think about it.

Meanwhile, back at the ranch, I was running out of options. I had found two good compilers that were more of a software project than I was willing to tackle just to get a robo-tool running, and I still needed a compiler and assembler. Then, in a fairly typical Zen thunderbolt, the answer called me up one morning at work.

Louis Meadows is a sales rep for a company called Software Development Systems, which makes a suite of very powerful software tools for the 68000 and other high-end chips. Lou had sent me a couple of demo systems of SDS' 68000 compiler tools some time ago, in conjunction with a project investigation my company was pursuing. The project didn't pan out and I ended up putting the unopened demo packages on the shelf.

So I had to disappoint Lou when he made his follow-up call. But we spent some time talking about related matters, and I mentioned my search for cheap 68000 C tools. He replied that the demo packages he had sent might actually solve my problem, as the demos are widely used in the university environment for teaching C on microcontrollers. I had originally dismissed this outright, since most such demo packages are crippled and can't produce very much code. Lou said he thought the limit on the demo compiler was about 100K bytes of code. He answered my shocked question by saying that you can't really do any serious projects in less than 100K of code.

As soon as we hung up, I ripped open one of the demo packages and installed it. The more I read and tested, the more convinced I became that this was the end of my software search. The seven-floppy package contains SDS' C/C++ Starter Kit for the 68K, version 7.03, which runs under Windows 3.1, Win95, and WinNT. The main product is the SingleStep 68000 debugger and simulator, a powerful Windows-based program for testing your final object files. SingleStep is crammed with powerful features such as source-level debugging and breakpoints, an editor, and built-in downloading capabilities.

The suite also includes a command-line assembler called as68000 and a matching command-line compiler, cc68000. These tools are easy to use from regular DOS batch files, or you can grab a public-domain Make system off the web and build up make files for automatically compiling and assembling your projects. The assembler supports macros and relocation, and the package includes a linker (but no librarian) for creating a large project out of several smaller library files. In short, this demo kit is a godsend for the hobbyist or student on a budget.

One feature of this kit that I really like is the ability to generate absolute listings from the relocatable output files. If you've used any of the older relocating compiler suites, you know how irritating it can be to try and use a monitor to debug your code. All of the listings contain labels addressed at 0, since the linker will fix them up later, so you never know where a label or variable lies in memory. The SDS kit, however, includes a program called abs, which merges the linker information with the relocatable listings, giving you printouts with real addresses for all those variables. Mighty handy, and way cool for a free compiler.

There are a few restrictions on this system worth noting. As I said, the linker does not include a librarian, so you cannot alter the two main library modules supplied with the package, nor can you create your own object library modules. Also, each invocation of the linker supports a maximum of three different object files plus the library files, so you end up having to create large source files to stay within the three-file limit. Finally, the compiler and assembler apparently are restricted to using available memory for holding their working tables, which puts an effective but unknown limit on the size of files you can assemble or compile. Still, if Lou's approximation is on-target, I'll write a lot of robot code before I hit that wall.

The bottom line: Run, don't walk, to your web browser, telephone, or fax, and order a copy of the SDS Starter Kit for the 68000. Check their web site for full details. I have the floppy disc package, but Lou mentioned a CD-ROM version as well. If the CD-ROM version includes the user manuals, it is a must-have.

Making It Work

Now comes the tricky bit. I have this first-class compiler and a first-class microcontroller; I just have to make the compiler create code that the board can run right out of reset. Those of you who have created Windows programs using tools such as Visual Basic have been able to skip over this next problem, as you have an operating system shielding you from the low-level tasks. But there are no such shields in place in robotics, and your software design has to cover everything that happens from the very first CPU clock cycle.

The reset startup task requires that you, as the system engineer, know all you can about the microcontroller your code will run on and the run-time module your compiler will graft onto your object file. The 68332 MCU contains a variety of extra I/O subsystems, each more complex than its 68hc11 counterpart, if there is such. Still, the general startup flow can be (grossly) simplified to the following steps.

First, initialize such system-wide elements as the MCU clock frequency and the CPU's stack pointer. Next, set up the MCU's chip-select lines so that the external memory is remapped from its reset state to its final working configuration. Finally, begin executing the top-level program by jumping to the address of the function main().

The key element behind this startup task is an assembler source file called start.s. The SDS demo kit includes a good start.s file to show you the basics, but I modified it extensively during my experimentation. I've included a copy of my current version of start.s, so you can see what is involved in setting up the 68332 after reset (Figure 2, at end of article). Note that this is a preliminary version and will change rapidly and often as my development continues. For now, however, this file will give you a good start. Keep an eye on my web site for ongoing updates to this and other 68332-related files.

Immediately after reset, the 68332 fetches the 32-bit value at address $0 and uses it as the stack pointer (register A7). It also fetches the 32-bit value at address $4 and uses it as the program counter (PC). The CPU sets up its internal registers and flags, then passes control to the program at the address currently in PC. Given the start.s file below, that starting point is the label START. Browse through the file, checking both the code at START and the code in the vector table, so you understand how this reset mechanism works.

Notice that the source at START follows an assembler SECTION directive that references an area called code. This SECTION directive tells the linker that subsequent object code should be placed in an area of memory named code. Later, when you run the linker to build the final executable, your linker commands must assign an address to this code section. The linker in turn will pull together all the various code sections, regardless of what object file they appear in, and stack them up at the designated code address. The linker will perform this same task for all the other SECTION directives it encounters. As it does so, the linker will automatically adjust the addresses of any referenced variables and labels, so the final executable file contains the correct addresses for the memory areas you have defined.

This operation is known as relocation, and lets you write a program without knowing exactly where in memory it will ultimately execute, with one small caveat. The SDS compiler will always generate code that can be successfully relinked, since that is how it was designed. The assembler, however, relies on you to make sure your code is relocatable. The linker will trip up if you try to make it relocate something that isn't relocatable, but the back-and-forth edit, assemble, and link operations can get to be a drag. For this reason, it's usually handy to write in C rather than assembler.

At START, the code sets up some internal registers, then begins a large loop that sets up the hardware registers. The 68332 has about a billion I/O registers, some of which are vital to your program and many of which are irrelevant. But the vital ones must be set up before your program will run successfully, and which ones you need to alter will vary depending on your design. So the register setup loop uses a table that you can customize quickly without getting bogged down in a bunch of nasty looping code and special conditions. Simply check the code further down in the file, at label config. This is a typical configuration table and should provide all the details you need for building your own table.

Next, the program prepares two areas of memory reserved by the C compiler. One area, in the section named ram, contains all uninitialized C variables. By convention, these variables must all contain 0 following reset, and the short loop around label ZLP makes sure that this task gets done. Immediately after this loop lies a second loop with a similar function. The code around the label ILP takes care of those variables that must be filled with preassigned values following reset. Notice how the code copies values from the section named DATA, which lies in ROM, to the section named data, which lies in RAM. Again, the command options you specify to the linker assigns these addresses so the linker gets everything sorted out properly.

Finally, everything is set up and the CPU is ready to run your top-level program. It does this by executing a JSR instruction to the address _main, which is where the compiler put the start of your main() function. Now your top-level code takes over, causing your robot to roll forward, find the candle, and claim first prize. By convention, control should never leave the main() function in an embedded control program such as this. If it should, however, the tight loop at label _exit will snag the CPU, preventing it from running away and executing trash.

Every C program you write must include the object file created by assembling the file start.s or something similar. Therefore, your work with the SDS compiler package is immediately limited to just two other files, as the third and final file you get will be start.o, the object file for start.s. Note that I have included a pair of character I/O routines in my start.s. These provide the low-level code for exchanging characters via the 68332's SCI. In this version of start.s, I've used a TRAP instruction that invokes the CPU32BUG's character I/O subfunctions. In later versions of start.s, which will run when the CPU32BUG monitor has been removed, I will include code that performs these I/O operations directly.

That's a Wrap

This completes the first phase of my robot design. I realize there isn't much in the way of hardware or electronics here, but I foresee software as the biggest hurdle, and I want to get a quick start on it. I will keep you informed in upcoming articles as to my progress on this project. Bear in mind that this design is evolving as I write these articles, so elements of the design will change from month to month. Given the time lag between my writing and your reading, I will always be about six weeks ahead of what you see here. And you won't even see the final project before the contest runs. But it should prove interesting, and I hope you'll stay tuned.

```
;
;    Typical 68332 startup file, based on original provided by
;    Software Development Systems in their SDS Startup Kit.
;


;
;    Use the Motorola equate file to declare the 68332 I/O regis-
ters.
;

#include equ332.asm

;
;    Declare some important externals and publics.
;

     XDEF   START,__brkp,__brksz,__exit
     XREF   STKTOP,DATA,_main

;
;    I removed the C++ initializers. See the original start.s file
;    for their declaration.
;


;
;    Define variables to track memory allocations for mbrk().
;
     SECTION     ram
__brkp
     DS.L  1                   ; point to available memory
__brksz
     DS.L  1                   ; number of available bytes



;
;    The reset vector will point to this code (START).
;    By the time we get here, A7 will have been initialized.
;
     SECTION     code
     .STK  "none"                    ; terminate call chain for -OG
```

Figure 2. The startup file (start.s) for my robot.

```
START
      MOVE.L       #STKTOP,A7  ; set the stack pointer
      MOVE.L       #0,A6       ; terminate call chain for -Og

;
;   Perform low-level configuration of system hardware. Configura-
tion
;   relies on the table at label config below.
;

      move.w       #$2700,SR          ; setup the status register
      move.l       #config,a0         ; point to start of config tbl
tbl1:
      move.w       (a0)+,d0               ; get size word
      move.l       (a0)+,a1               ; get destination addr
      move.l       (a0)+,d1               ; get data to move
      cmp.w #0,d0            ; write a byte?
      bne   tbl2            ; branch if not
      move.b       d1,(a1)               ; write byte to addr
      bra   tbl1            ; do next
tbl2:
      cmp.w #1,d0            ; write a word?
      bne   tbl3            ; branch if not
      move.w       d1,(a1)               ; write word to addr
      bra   tbl1            ; do next
tbl3:
      cmp.w #2,d0            ; write a long?
      bne   tblx            ; if not, all done
      move.l       d1,(a1)               ; write long to addr
      bra   tbl1            ; do next
tblx:


;
;   Zero out uninitialized RAM.
;
      MOVE.L       #`BASE(ram),A1   ; A1 = base of region ram
      MOVE.L       #`SIZE(ram),D0   ; D0 = size of region ram
      LSR.L #2,D0           ; compute size in longs
      BRA   ZDBF            ; enter a fast loop
```

Figure 2 continued.

```
ZLP:
      CLR.L (A1)+                ; clear four bytes at a time
ZDBF:
      DBF    D0,ZLP             ; up to 256K in inner loop
      SUB.L #$10000,D0          ; rest in outer loop
      BHS    ZLP


;
;   Initialize other RAM from ROM.
;
      MOVE.L       #DATA,A0      ; A0 = ROM base of region data
      MOVE.L       #`BASE(data),A1   ; A1 = RAM base of region data
      MOVE.L       #`SIZE(data),D0   ; D0 = size of region data
      LSR.L #2,D0                ; compute size in longs
      BRA    IDBF                ; enter a fast loop
ILP:
      MOVE.L       (A0)+,(A1)+ ; move four bytes at a time
IDBF:
      DBF    D0,ILP             ; up to 256K in inner loop
      SUB.L #$10000,D0          ; rest in outer loop
      BHS    ILP


;
;   Initialize memory allocator.
;
;     MOVE.L       #`BASE(malloc),D0 ; address of malloc region
;     MOVE.L       #`SIZE(malloc),D1 ; size of malloc region
;     .IF "ptrd"?"2"
;     SWAP  D0                  ; handle 2-byte C "pointers"
;     .ENDIF
;     .IF  "long"?"2"
;     SWAP  D1                  ; handle 2-byte C "longs"
;     .ENDIF
;     MOVE.L       D0,__brkp    ; vars referenced by mbrk()
;     MOVE.L       D1,__brksz


;
;   Invoke main() with no arguments.
;
      JSR    _main               ; "int" return value in D0
```

Figure 2 continued.

```
;
;    Control  reaches  this  point  if  control  leaves  main().
;

__exit
DONE
      BRA    DONE                  ; loop  if  main  ever  returns



;
;   If  your  program  needs  standard  character  I/O  routines,  this
makes
;   a  good  place  to  stash  them.  The  following  routines  assume  that
;   CPU32  is  available,  and  rely  on  the  monitor's  TRAP  mechanism.
;


;
;   outch        output  char  in  D0  to  active  port
;

      XDEF  _outch
      .FDEF _outch,4

_outch
      move.b      d0,-(a7)                  ; save  char  on  stack
      trap #15               ; use CPU32
      dc.w $0020             ; output  a  char
      rts                    ; and  leave



;
;   inch        get  char  from  active  port,  return  in  D0
;

      XDEF  _inch

_inch
      subq.l      #2,a7                ; make  room  on  stack
      trap #15               ; use CPU32
      dc.w $0000             ; input  a  char
      move.b      (a7)+,d0                ; put  char  in  D0
```

Figure 2 continued.

```
        rts                     ; and leave



;
;    Define the system configuration table.
;
;    The following table works for the New Micros 68332 board.
Change
;    this table as needed for your hardware.
;
;    Each entry in the configuration table specifies the
;    size of a value to write to memory, the address affected, and
;    the data to write to it. The config implementation is based
;    on the TINIT routine used by New Micros in their MAXForth/332
;    firmware. I liked their idea so much I had to use it in my
;    system.
;
;    Each entry in the table consists of a word and two longs, in
the
;    following order:
;
;       dc.w  0, 1, or 2        ; 0=byte, 1=word, 2=long
;       dc.l  addr              ; address to change
;       dc.l  data              ; value to write to addr
;
;    For example, to write the word $0038 to address SCCR0 (sets
;    baud rate to 9600 baud), you would use the following table
entry:
;
;       dc.w  1                 ; SCCR0 (word)
;       dc.l  SCCR0             ; set baud rate
;       dc.l  $0038             ; to 9600
;
;    Mark the end of the table with a size field other than 0, 1, or
2.
;
;    Note that the configuration values are written to the hardware
;    registers in the order of appearance.
;
```

Figure 2 continued.

```
config:
;       dc.w        1               ;16.667MHz
;       dc.l        SYNCR
;       dc.l        $7F00

;       dc.w        0               ;BME, NO WATCHDOG TIMER
;       dc.l        SYPCR
;       dc.l        4

;       dc.w        1               ;1st rom, 128K at 0000
;       dc.l        CSBARBT
;       dc.l        $0004

;       dc.w        1               ;
;       dc.l        CSORBT
;       dc.l        $7B70

;       dc.w        1               ;2nd rom, 16k at 4000
;       dc.l        CSBAR0
;       dc.l        $0042

;       dc.w        1               ;ram is mapped via cs1
;       dc.l        CSBAR1
;       dc.l        $0007

;       dc.w        1               ;and cs2 - 1024k each
;       dc.l        CSBAR2
;       dc.l        $0107

;       dc.w        1               ;io space, 128k at f00000
;       dc.l        CSBAR3
;       dc.l        $F004

;       dc.w        1               ;wait states for CS
;       dc.l        CSOR0
;       dc.l        $7B70

;       dc.w        1               ;
;       dc.l        CSOR1
```

Figure 2 continued.

```
;       dc.l        $7B70

;       dc.w        1                   ;
;       dc.l        CSOR2
;       dc.l        $7B70

;       dc.w        1                   ;
;       dc.l        CSOR3
;       dc.l        $7B70

;       dc.w        1                   ;sci=9600 baud
;       dc.l        SCCR0
;       dc.l        $0038

;       dc.w        1                   ;enable rcv and xmt
;       dc.l        SCCR1
;       dc.l        $000C

;       dc.w        1                   ;setup internal ram
;       dc.l        TRAMBAR
;       dc.l        $FFE8

        dc.w    $ffff               ; marks end of table

;
;   Declare the reset vector, stored in supervisor space at address
0.
;   Address 0 holds the 32-bit address to write to A7, the stack
pointer.
;   Address 4 holds the 32-bit address to write to PC.
;
        SECTION     reset
        DC.L STKTOP             ; initial stack pointer
        DC.L START              ; initial execution address

;
;       OTHER EXCEPTION VECTORS: to supervisor data space at address
8,
;       or 8 bytes beyond where the vector base register will point.
;       This table is commented out because no actual interrupt rou-
```

Figure 2 continued.

```
;       tines  are  provided.
;*******************************************************************
;       SECTION     vects
;       DC.L  BUSERROR,ADRERROR              ;  0x08
;       DC.L  ILLEGAL,ZERODIV,CHK,TRAPV          ;  0x10
;       DC.L  PRIVILEGE,TRACE,EMULA,EMULF         ;  0x20
;       DC.L  RESVD,PROTO,FORMAT,UNINIT           ;  0x30
;       DCB.L 8,RESVD                         ;  0x40
;       DC.L  SPURIOUS,AUTO1,AUTO2,AUTO3          ;  0x60
;       DC.L  AUTO4,AUTO5,AUTO6,AUTO7             ;  0x70
;       DC.L  TRAP0,TRAP1,TRAP2,TRAP3             ;  0x80
;       DC.L  TRAP4,TRAP5,TRAP6,TRAP7             ;  0x90
;       DC.L  TRAP8,TRAP9,TRAPA,TRAPB             ;  0xa0
;       DC.L  TRAPC,TRAPD,TRAPE,TRAPF             ;  0xb0
;       DC.L  FUNORD,FNEXACT,FZERODIV,FUNFLOW     ;  0xc0
;       DC.L  FOPND,FOVFLOW,FSNAN,RESVD           ;  0xd0
;       DCB.L 8,RESVD                         ;  0xe0
;       DCB.L 192,USER                        ;  0x100
```

Figure 2 continued.

Part VI

Adventures in Hacking

Some of my most popular articles dealt with hacking a piece of electronics or a mechanical toy into something completely different. These columns all focus on finding a product suitable for robotics, then redesigning it. In one column, Marvin Green and I spent an evening deciphering the bit stream from a surplus TV remote control, then added electronics to a robot so we could use the remote control to guide the machine around my lab. A similar hack of a nifty toy truck, the Ready-Set-Go, resulted in a great little robot frame for very few robo-bucks.

I devoted several columns to hacking a couple of Practical Peripherals' external modems, turning them into high-quality development boards for the underlying microcontrollers. One modem board contains an 8051 processor with lots of memory, which I reconfigured to make it easier for development use. Another board uses the Motorola 68302, a high-end communications controller with lots of horsepower for robot

projects. These two projects resulted in development boards that cost far less than comparable commercial products, and show how a little judicious hacking can pay off.

The surplus market offers many hackable products, such as the Proxim RDA-100 RF modems featured in another article. It was easy to get these running, and having a long-range RF link between a robot and my lab PC opens up lots of possibilities. I can upload data, send commands, and even download complete new programs, all without having to see or touch the robot.

This section concludes with an article on hacking the Nintendo GameBoy® into a robot controller. I describe the inner workings of the GB, and provide pointers to Web sites filled with tools and information. Even the older GB, featured in this column, offers plenty of capability for robot control, and you sure can't beat the price.

Decoding a TV Remote Control

This is one of my favorite columns, from the November 1994 issue of *Nuts & Volts*. The first part gives you a graphic, almost play-by-play description of a way cool hack that Marvin Green and I performed on a surplus TV remote control. Doing the hack was great fun, more so because I got to work with a very bright and dedicated robo-nerd. And Marvin likes chocolate eclairs as much as I do!

The article also describes one of Marvin's designs, a simple but elegant frame that could help anyone get started in robotics. His BBOT product still shows up at SRS meetings, and it still has one of the best bumper-skirts I've ever seen. I also spend a little space talking about the glories of double-sided foam tape, a "must-have" addition to any robot builder's toolbox.

Finally, I close with a quick review of the Rug Warrior computer kit and its companion Interactive-C software. This kit grew out of the *Mobile Robots* book, written by Joe Jones and Anita Flynn; see my "Inspiration to Implementation" article for a review of *Mobile Robots*.

"OK," I said, "let's hit it." Marvin Green and I trooped upstairs to my workroom, two Men on a Mission. We intended to do some serious robot hacking that night, and it was time to get started. We had selected a nontrivial project. Months ago, I had bought several TV IR remote controls for 99 cents each at a local surplus shop. Tonight, Marvin and I were determined to deduce the coding scheme used by the controls, then use the handhelds to control Max, my newest robot.

Marvin sat down at my electronics workbench, grabbed a small screwdriver, and started opening the control. He said, "First, I want to use your 'scope to look at the pulses you get when you press a button on this thing."

"No problem," I responded, and gave him a quick rundown on my old Tektronix 454 oscilloscope. I watched Marvin clip the 'scope's input lead to the control's IR LED, at the junction between the LED and its current limiting resistor. When he pressed a button on the remote, a pattern blinked briefly on the 'scope's screen. Some buttons gave a repetitive pattern, making them easier to examine.

"I set the horizontal timing way down, and you can clearly see this remote is sending 40 kHz bursts for each pulse," he noted. "This thing should directly drive one of those Sharp IR detector modules that Electronic Goldmine sells."

"Great!" I had hoped this was the case. "Some of the older controls used other frequencies; I'll take this as a good omen." I left him with my engineering notebook and pencil, staring at the oscilloscope and making notes. I wheeled my chair up to the computer desk and hit the power switch on my 386.

Our plan of attack was simple. Marvin would glean what he could about the pulse stream by observing traces on the 'scope. I would use that information to write a simple program to monitor the output of an IR detector module. Then, we would aim the remote at the detector module, press some buttons, and the program would report the incoming pulse stream. Based on this new information, I would then write a second program that (hopefully) would let us run Max using the remote control.

The SRS Meeting

Marvin had driven up from Portland the day before, to attend that month's meeting of the Seattle Robotics Society. As we worked, we discussed the meeting.

"The guys really liked your little robot frame kit, Marvin," I said, referring to his latest creation, the BBOT frame. "You did a beautiful job on the design and the workmanship."

"Yeah," he replied, "I found a plastics shop that would make them for me just like I want them. You think $29.95 is a reasonable price for a two-level robot frame kit?"

"Of course!" I was kind of surprised at the question. "That black plastic frame, clear dome, and plastic skirt look first-rate. How many did you sell at today's meeting?"

"Oh, I sold ten or so. They went fast; I just wish I had had a few more ready for today."

"Well, think about what you are offering for $30. Someone can take your frame kit, add a couple of servo motors, a BOTBoard, and some batteries, and create a working robot in just an evening or two." I glared for a moment at my code that was not working out, then went back to the conversation.

"Your bumper skirt is ingenious, so simple and easy to build. Plus, you get 100% coverage on the robot's perimeter. Guys in the club have spent years trying to come up with something even half as good, and failed."

"Well, I just modified some other designs I saw, eliminated things that looked too difficult to make or buy, and replaced them with rubber bands and common hardware," he said. "No big deal."

I had noticed this before about Marvin's designs. They were always the simplest, most effective solutions to some nagging problem. He never added anything that wasn't necessary, and he always used pieces that you could easily buy or make.

"Well, as soon as word gets out about your frame kit," I said, "you're going to get a flood of orders. Someone could use it for a Science Fair, a class project, or almost any kind of robot experiment. Heck, the frames are so cheap, you could even build a herd of robots and play with interaction between robots."

Marvin changed the subject. "Well, I saw a lot of other really neat robots today. The guy with the two robots using tank treads did a great job building his machines."

"That was Bob Dain," I replied. "Yeah, those were nice 'bots. He said he got the tank tread at a hobby store, then just strung it around the wheels he had built onto his frames. And boy, that one machine was fast!"

"And Tom Dickens had a couple of nice little servo-driven robots," said Marvin. "I thought his front bumper switch was pretty clever."

"Yeah," I agreed, "I'm glad Tom is working with BOTBoards now. He is really sharp, and I can't wait to see what other kinds of robots he comes up with."

"Tom is using your BOTBoards as the core of the microcontroller classes he teaches at Cogswell College in Bellevue. I always thought that would be a great use for BOTBoards."

"Do your meetings usually include a tech session like that talk you gave on Forth?" Marvin asked. "I thought you did a good job on that."

"Well, we always try to have something technical after the general meeting," I answered. "That way, people have a chance to learn from a club expert in

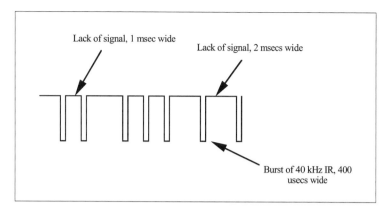

Lack of signal, 1 msec wide

Lack of signal, 2 msecs wide

Burst of 40 kHz IR, 400 usecs wide

Figure 1. Timing of signals from IR remote control.

some matter. Everyone in the club is experienced in some aspect of robotics, everyone has something to contribute. The technical sessions are where that happens."

Back to the Hack

We had been working while all this talking was going on, but Marvin was making better progress than I was. "OK, I've got a handle on the pulse stream. Here, look at my diagram on the timing of the IR signal..."

He had wired an IR detector up to my oscilloscope input, connected a set of four AA cell alkalines, and examined the waveforms he got from some button presses.

"You can see here that each burst of 40 kHz IR gives a low-going output pulse about 400 μsecs long," said Marvin, pointing to one of the narrow pulses on his diagram (Figure 1).

"The time between each pulse is always either 1 msec or 2 msecs, which I assume indicates a zero or a one," he continued. "And each key generates 13 IR pulses, which means you get 12 bits of information."

I picked up the thread. "That means we just need to wire an IR detector to an input on the BOTBoard, then have code time the duration between each two pulses. Anything less than 1.5 msecs is a zero, any-

thing greater is a one. Then the code mashes all the ones and zeros together to make a character. Since it is easier to work in eight bits instead of 12, and since the remote only has 18 keys on it, I'll make a lookup table that translates each 12-bit pattern into some ASCII character."

Marvin then asked, "But how are you going to know when a character starts?"

"Simple," I replied. "We'll connect the IR detector's output line to PA7 on the BOTBoard's 68hc11, then configure that pin to generate a pulse-accumulator input (PAI) interrupt when it sees a low-going edge. Pressing a key on the remote generates the first burst of IR to mark the start of the character. That first burst will make the IR detector's output go low, causing a PAI interrupt.

"Once control reaches the PAI interrupt service routine, or ISR, software can start timing the pulses using a simple counting loop. When all the pulses have been tallied, code rearms the PAI interrupt and leaves the ISR. Then the 68hc11 will be ready for the next keypress."

"OK," said Marvin, "I'll wire an IR detector to Max's computer board. How do you want the connector hooked up?"

"Just use this drawing as a guide," I replied, handing him a diagram I had scribbled out (Figure 2). "Remember to connect the IR module's ground lead to its metal case."

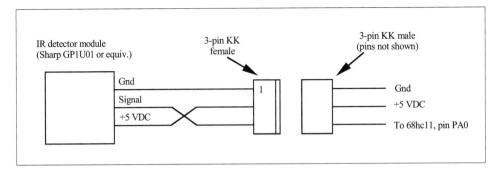

Figure 2. Diagram for wiring an IR detector module to the 68hc11.

"Yeah, I know all about that," he said. "If you don't make that connection, all you get is noise on the output. Then nothing seems to work."

Software, Round One

I started off by writing a small tiny4th program that did some initialization, then sat in a loop waiting for a variable's value to change. The real work in this program was done inside the ISR, written in 68hc11 assembly language. The ISR's code timed the incoming pulses, marking them as ones or zeros based on the timing Marvin had determined.

Next, the code stashed the values in an array, previously declared in the tiny4th portion of the program. The ISR code then wrote a special value into a global variable, as a signal that a keypress had been detected. Finally, the ISR code rearmed the interrupt and returned, leaving everything ready for the next incoming character.

The tiny4th program, upon seeing that the flag variable had changed, displayed a stream of ones and zeros on the terminal to show the incoming data. It then cleared the flag variable and went back into the waiting loop.

Simple stuff, but it wasn't working. I double-checked the ISR code, rethought the tiny4th code, but no luck. I even hooked an IR detector up to a spare BOTBoard (I've always got a few of those lying around) and hit it with a few blasts from the IR controller. No dice.

"That's it, Marvin," I said in disgust. "I don't know what this stuff is, but it isn't code. Let's take a break."

"Suits me," he responded. "Let's go have some brain food."

This referred to the giant chocolate eclairs we had picked up earlier, while we were getting the pizza and beer for dinner. These were the second-largest eclairs I had ever seen, and each had about a month's worth of sugar and fat in them. Programmer food!

After polishing off an eclair and a glass of milk each (we know the importance of a balanced meal), it was time to tackle the remote control again. By now, it was after 8:00 PM, and we wanted to wrap this thing up tonight.

Marvin went back to the electronics bench to tinker with the 'scope some more, while I headed for the PC. During the break, I had figured out that the problem lay in the initialization of the PAI interrupt. Somehow, I wasn't doing something exactly right. I had already gone through the Motorola 68hc11 Reference Manual several times, rereading the section on the Pulse Accumulator Input pin. One more scan wouldn't hurt.

The PAI (pin PA7) is normally used in one of two modes. The gated-time mode counts ticks derived from the MCU's internal clock for so long as the PA7 input remains at a selected state. This lets your software tell how long the PAI input stays at a specific level.

The event-counting mode increments a register, called the accumulator, each time the PA7 input detects the selected edge. This lets your software count the number of times the PAI input goes to a specific level.

Note that either mode can be configured to generate a PAI interrupt when the PAI input changes to the specified state. The PAI subsystem can generate interrupts on selected events, and whenever the 8-bit accumulator reaches $ff. As with all 68hc11 subsystems, local masks can be used to enable or disable any combination of these interrupts.

I didn't really want to use either mode; I just wanted an interrupt when the PAI input saw a low-going edge. Therefore, I wrote my initialization code to disable the PAI subsystem, though I did enable the edge-detect interrupt. After careful reading, I decided instead that I needed to enable the PAI system and disable all unnecessary PAI interrupts. The initialization code then looked like this:

```
: init-pai
pactl c@
40 or
pactl c!
tmsk2 c@
10 or
tmsk2 c!
10 tflg2 c!
;
```

The first three lines of INIT-PAI fetch the current value for PACTL, set the PAEN bit to enable the PAI subsystem, then write the new value back to PACTL. The next three lines perform a similar function for the TMSK2 register, setting the PAII bit to allow interrupts on the selected input edge at PA7. By default, this interrupt responds to falling edges on PA7. Finally, the code writes a logic 1 to the PAIF bit in the TFLG2 register, resetting the PAI interrupt flag so the next edge will generate an interrupt.

I had already written the PAI ISR code, so I compiled the program, used pcbug11 to download the object code into my test BOTBoard, reset it,

and pressed a key on my remote. The program dutifully displayed a stream of ones and zeros across the screen!

Marvin heard my excited "Yes!" and came over to look, notebook in hand. "Watch this, Marvin," I said, as I played with some of the buttons. "Now we can see what the codes really look like."

"Great! See how they compare with my notes here." He had carefully written out the pulse pattern for each of the 18 keys, using Ss for short gaps between pulses and Ls for long gaps. A quick check showed that my code wasn't matching up properly with his notes. I found and fixed a couple of small bugs, and everything synced up beautifully. Now the BOTBoard software would help us plan the next step.

Software, Round Two

"What's next?" Marvin asked, after we had satisfied ourselves that we had doped out the remote's transmission scheme.

"Well, I need to write the next version of this software, so that it translates keypresses into specific ASCII characters and puts the characters into a queue for later use. How about if you build a foot-long, three-wire cable so we can stick an IR detector on a piece of brass rod, fix the rod so the detector is pointing straight up, then attach the rod to Max's top platform?"

Marvin started rooting around in my wire bin for the appropriate connectors and cable, leaving me to tackle the software. But this program would prove easier than the first, since the hardest part was already working.

I started with the prototype software we had just tested. The bit patterns we had logged actually consisted of a three-bit prefix, a five-bit data section, and a four-bit suffix. The prefix and suffix portions were identical for all 18 keys, so I decided to use this fact to help weed out noise and bogus signals. I added code to the existing ISR module that tested for the proper prefix and suffix patterns. Other code verified that each pulse was no more than the prescribed length, and that no gap between pulses exceeded arbitrarily-defined minima or maxima.

This added lots of code to the ISR, but I felt sure that the extra security was worthwhile. After all, Max weighs quite a bit, and I wanted to avoid having it start up at full speed because of a false signal.

I built in a conversion table, so that each button on the remote control was translated into a specific uppercase ASCII letter. I also added a queue mechanism, so incoming characters would be saved by the ISR into a small buffer. They would sit there until the main program fetched them from the queue, or until the queue wrapped around and clobbered them.

I felt the queue was necessary since some of the remote's keys send a stream of characters for so long as the key is held down, rather than only as you press it. The queue would help make sure the program saw all the incoming characters.

Finally, I wrote code to connect the character queue to tiny4th's **KEY** and **?TERMINAL** words, normally used to handle characters input via the serial port. With this mechanism in place, my tiny4th code only needed to invoke **?TERMINAL** to check the incoming queue, or invoke **KEY** to fetch the next available character.

"OK, Karl, what do you think of this?" Marvin asked, holding up the finished sensor. "I used your 100 watt Weller soldering iron to solder the IR detector module to one end of this 8-inch brass rod, then hooked on the cable."

"Looks good," I said, "but how do you plan to mount it onto Max's top platform?"

"Easy," he replied, "I've bent the last inch of the rod into a horizontal triangle, and I'll just blob some hot glue onto it and stick it onto that open spot on Max's board. Should hold just fine."

"Well, I've got the first cut of this control software written, so set Max up here on my desk and I'll try downloading the software into him." I hooked the serial cable into the connector on Max's computer board, then started the transfer using pcbug11.

"Here's the control," said Marvin, offering me the unit. "What keys do what?"

"I picked out these four keys to do forward, backward, right, and left," I said, pointing to a diamond-shaped pattern in the center of the unit. "And this repeating key in the upper left corner is stop."

The software download finished, so I disconnected the serial cable, reset Max, and put it on the floor. You could almost hear the drum roll. I pointed the control at Max, with its newly-acquired IR detector facing straight up. I pressed the key I had assigned to be forward, and Max obediently lurched forward. I was so shocked that the code actually worked that I forgot which key was stop, and Max slammed into the wall.

"This is great!" we both said at the same time. We took turns driving Max all over the upper floor of my house, using the remote control. After about five minutes of enjoying our success, we both got a little more critical. "It moves way too fast to control easily with this thing," noted Marvin. "Yeah," I replied, "and we lose a lot of range because we have to use a bank shot off the ceiling to get the IR to its sensor."

I solved the first problem by editing the software to use different PWM values during turns and backing up. This made Max much easier to maneuver, but still left him with plenty of speed for the straightaways. I wouldn't solve the second problem until after Marvin left for Portland. I used four IR detector modules, each wired to one input of a 74hc20 dual quad-input NAND gate. See the accompanying schematic (Figure 3).

The output of this first NAND gate is inverted by the second NAND gate. Its output is then hooked to the connector originally used by the first IR detector module. This circuit generates an low-going pulse if any one or more IR detectors sees a signal. Essentially, I have built an IR detector with 360 degree vision.

A refinement, as yet untested, is to bring the output of each IR detector to an input port. The ISR software could then check this input port to determine exactly which of the four IR detectors saw the signal. This information, coupled with the character received, would enable Max to run some very sophisticated beacon contests, or to locate

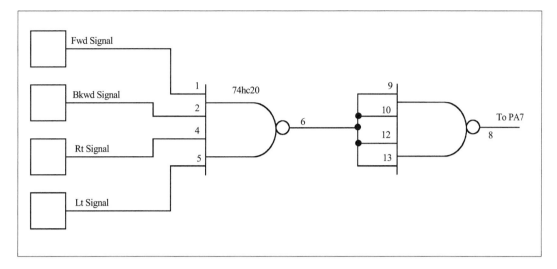

Figure 3. Wiring four IR detector modules to produce a single output.

any object marked with an IR control such as these surplus units.

The BBOT Frame Kit

Marvin's BBOT robot frame kit, mentioned above, is the latest addition to his line of simple, inexpensive robot projects. He is selling two versions of the kit, one with two platforms stacked vertically, and another, smaller version with just a single platform.

Marvin calls the double-decker version the BBOT frame kit and sells it for $29.95. The smaller unit, dubbed the BBOT jr. kit, costs $19.95. Be sure to contact Marvin about shipping costs and availability.

The BBOT frame kits have quite a bit in common. I have a BBOT frame to review, so I'll describe that. Remember that the jr. version only has one platform.

The BBOT frame consists of two circles of 1/8th-inch thick black plastic, precut to 5.5-inch diameters and drilled for assembly. The two circles bolt together using supplied 4-40 hardware and threaded spacers to form a two-tier platform about 1.25 inches high. The lower platform has shaped cutouts, one on each side, to accomodate hobby-store wheels, that you supply, up to about 2.5 inches in diameter.

Marvin designed the frame kit to use two modified hobby R/C servo motors (not included), similar to the Futaba S148 units featured in earlier Amateur Robotics columns. First, modify the motors so they turn 'round and 'round, then fasten them to the lower platform using double-sided foam tape or cable ties from the local hardware store.

Then, thread the servos' cables through the holes provided in the two platforms until the cables reach the top level. From here, you can plug the cables into the proper connectors on a BOTBoard (you supply), and you're just about ready to go.

The frame contains enough room for the BOTBoard and user-supplied batteries and sensors. It comes with a clear plastic dome for the top and with a six-inch diameter plastic ring to serve as a bumper skirt. Marvin designed his bumper skirt to hang from two rubber bands (included) and rest against three tiny microswitches (also included).

The mechanics for holding this skirt in place are simple and elegant, and the skirt provides full coverage of the robot's perimeter. Properly adjusted, it can detect inward contact on any point on the outside of the ring's one-inch tall surface. The kit also includes a black furniture knob and mounting hardware; the knob serves as a front skid for the robot.

Note that you will need to do quite a bit of drilling on this frame. You need to add holes in the top platform to mount your computer board, holes in the plastic dome to mount it to the supplied angle brackets, and holes in the bumper skirt for threading the rubber bands.

I used double-sided foam tape to mount my sensors, battery holder, and the supplied bumper switches. If you don't use double-sided foam tape, plan on drilling even more holes.

I made one change to Marvin's intended layout. I replaced his single large black drawer knob with two smaller red knobs, mounted front and back. The single knob would let the robot tip over backwards, even though the battery was mounted as far forward as possible. I could have turned everything around, and put the large knob in back, but sooner or later the robot would have had to back up, and then it would have tipped over forwards.

The finished robot, named Arnold, has a very polished, professional look to it. The round dome, seamless bumper skirt, and smooth black plastic platforms give Arnold a touch of class. And Marvin's design is very well suited to cruising through the pitfalls and dangers in my living room. This frame kit would make a great starting point for a Science Fair project, or for experiments in robo-herd behavior.

No Holes!

I recently picked up a roll of double-sided foam tape from Electronic Goldmine. This tape (#3905) is one inch wide, wound into a roll about 10 inches across; cost was $3 in their #222 catalog. Marvin has used this tape for nearly all his robots. He mounts battery holders to frames with it. He fastens servo motors down with it. Bumper switches. You name it.

I never believed that the tape would really work, though, until I bought some of my own and tried it. The stuff is just shy of magic! I can stick a four-cell AA battery holder to a piece of plastic and the holder WILL NOT come off.

I'm sold; I have already started using double-sided tape for all of my projects. Be sure to get the foam-type tape, such as Electronic Goldmine sells.

Mobile Robot Kit

Speaking of kits, I also got an early look at the Mobile Robot kit, offered by A K Peters, Ltd., the publishing house that sells the book *Mobile Robots*. I reviewed the book several months ago, and recommend it without reservation to anyone interested in building small robots (see "Inspiration to Implementation.")

I want to make something clear from the first. This is not a kit for building a robot; this is a kit for building the electronics of a robot. Some of the earlier advertisements were not clear on this point, and several people in the Seattle Robotics Society misunderstood what was being offered. So I will describe in detail what arrived in my kit.

I received a partially assembled 68hc11a1 microcontroller board, complete with 32K of battery-backed static RAM. As it comes out of the box, this board supports RS-232 communications to a PC or Mac; you must specify which host you are using when you order your kit. My review kit did not include a serial cable for connecting to my 386.

The kit's instructions described how to build such a cable, though, and I had it done without too much hassle.

The kit came with a PC version of Interactive C (IC), a program development system created at MIT and used in, among other things, the 6.270 Robot Design class.

I plugged in the serial cable, hooked up 6 VDC to the board, and turned on the power. I was immediately able to download the required portions of IC. I could then download and run several of the provided test programs, including a cute routine for playing a tune on the board's piezo speaker.

As of this writing, I have wired about 50% of the board, using the supplied sockets and parts. This kit was designed by several people involved with the original *Mobile Robots* book and its Rug Warrior robot, so the design is well done and the board goes together easily.

One nice feature is the on-board LCD. My kit was supposed to have the 16 character by two line display, but instead had a note indicating that the display would be shipped separately. So I grabbed a spare LCD from my junk drawer and plugged it into the on-board connector. With the LCD in place, IC displays its initial sign-on text, as well as a little heart-shaped character that "beats" to show the 68hc11 is alive.

The kit comes with an assembly guide that describes how to use IC, as well as providing details on debugging, operation, and assembly instructions. When fully constructed, this board provides support for bumper switches, an IR object detector circuit, a cadmium-sulfide photocell for light-level detection, and a microphone for checking ambient noise level.

The board also features a 754410 dual H-bridge motor driver IC, for running two DC motors at up to 1 Amp current drain. This is a bipolar driver chip, so the motors you hook up will receive about two volts less than the motor battery voltage, owing to the voltage drop inside the IC.

Though the assembly guide is fairly complete, you will really want a copy of *Mobile Robots* nearby to get the most from this kit. The supplied assembly guide is more of a technical reference manual; the book will help you expand your kit and learn how it works inside.

I haven't had the kit running long enough to really wail on IC. I do know that this is a very mature system, having been through several years of testing "under fire" in the MIT robot classes. IC offers a multi-tasking system, floating-point support, and an interactive environment for writing robot software right on your machine. I am really looking forward to using it.

A K Peters, Ltd., is selling the Mobile Robot electronics kit for $359; you can contact the company directly for details. I consider this a good price for what you get. Sure, you can buy general-purpose 68hc11 single-board computers (SBCs) for less, but an SBC does not a robot make. Besides the com-

puter and memory, this electronics kit includes tested designs for motor control and sensor circuitry: design elements that have broken the spirit of more than one robo-hacker.

The board is compact, measuring only 4.5 by 3.5 inches. It is well laid out, complete with silk-screened legends for mounting components and labeling sockets. And the array of connectors means you can easily add or rearrange external devices.

The IC system, customized for this board, creates a powerful combination. Software already designed to work with well-crafted hardware means you can focus on writing your robo-code, not figuring out why the two elements don't work together properly.

Naturally, I found a few nits to pick about this kit. I would like to see a pre-built serial cable made available, perhaps at a slight additional cost. And I really, **really** wish the authors' would include pin numbers on all of their schematics! Not that I expect to have to trace anything out, but if I did, the supplied schematics are almost useless for debugging.

[Author's update: The current version of the kit, called "The Brawn," contains a base, motors, wheels, and hardware for mounting your finished board. The kit called "The Brains" includes a disk of robot programs from the second edition of *Mobile Robots*; the disk also contains a good library of selftest programs.]

Some SRS members thought $359 was too much to pay for a kit that didn't include a motorized base. On the one hand, I can see their point. If you are building this for a child's Science Fair project, or if you are on a student's budget, the cost may be prohibitive.

On the other hand, a scaled-down version of Max, my latest robot base described in the last several columns, should only cost between $50 and $100, depending on how fancy you want to get. And you can probably trim that down quite a bit if your junk box is well-stocked.

The gripping hand is that the board's designers have solved nearly all the tough robot design problems for you, and packaged it up in one small board and one floppy disc. Add in the *Mobile Robots* book, filled with closely related information, and you have an unbeatable combination.

As of now, this kit is by far the best method of building the electronics for your own medium-sized custom robot that I have seen. The design team for this kit includes Joe Jones and Anita Flynn, authors of the *Mobile Robots* book. It also includes Randy Sargent and Fred Martin, who helped originate the 6.270 Robot Design contest and the IC programming environment.

Wiring Up an RF Modem Link

This article from the *Nuts & Volts* of December 1994 showcased a terrific surplus find, the Proxim RDA-100 RF modem boards. As the start of this article describes, writing about surplus bargains is a mixed blessing. It's great to find a really powerful piece of gear at a cheap price, but it can take so long to get an article into print that sometimes the device is no longer available by the time the readers see the article. In this case, the surplus dealer, Timeline, had enough units on hand to warrant running the article. By the time you read this, of course, they have long since sold out.

Still, it was a fascinating project, and the end result was so impressive that I decided to include the article. You will always be able to find similar deals in the surplus market, and this column can help you turn your precious bargain into robo-gold. Had I not done the extra digging and work you see here, I would not have had access to a robust wireless connection for my robots.

The lesson here is to keep your ear tuned to the surplus market and to the Internet, be creative, and be prepared to pounce on the right item when it comes by. You can save a bundle of cash and have a lot of fun at the same time.

Few surplus bargains make good subjects for this column. Often, I find the item at a local store; great for others in the Seattle Robotics Society, but useless for readers scattered across the country (or world). Or, if I did find it in a mail-order house, the item is too expensive, too bizarre, or in too short of supply to fit well for an article I have to write nearly two months in advance. But once in a while something turns up that is simply too cool not to feature in this column. Case in point: Timeline's $99 19.2 Kbaud spread-spectrum RF modems.

On the Net

I first heard of these modems through posts that began appearing in the Internet comp.robotics newsgroup. One reader mentioned seeing the ad in an issue of Circuit Cellar INK; another followed up by saying he had already ordered a pair. Later posts gave more information on the units. Readers described voltage requirements, operating frequency, physical size, and other details. Eventually, a reader posted that he had actually gotten one of the modems to talk to his PC's serial port at 9600 baud. He gave details on the menu-driven software embedded in the modem and used for configuration.

By now, I was interested enough to plunk down my $198 plus shipping to get a pair of modems. I called Timeline and ordered my toys.

A First Look

This is the OEM version of the Proxim, Inc., RDA-100 RF modem. The OEM version is a pair of boards, each 2-3/8 inches by 3-3/4 inches, plugged together to form a stack only 1/2 inches thick. The two boards make heavy use of surface-mount technology (SMT), yielding a compact assembly.

Note the words "OEM version." You get wired and tested boards only. No case, no power supply, no connectors beyond those soldered onto the board, not even an antenna. You wire up a power supply, mount the boards in an enclosure of your design, and add the circuitry so your computer can talk to the modem. Happily, this is easy to do, and the resulting high-speed serial link can add a slew of capabilities to your robotics projects.

As shipped, each modem comes bolted together in its stacked configuration, with 4-40 studs at three of the corners. This simplifies mounting the board set to a chassis; drill three properly located holes, add the right hardware, and you're done. The modem uses 5 VDC at up to 230 mA. It exchanges data with other units using frequency-hopping in the 902 to 928 MHz band. The 100 mW RF output into a 50 ohm antenna allows operation without FCC license.

The Proxim literature accompanying the RF modem lists a nominal range of 500 feet with throughput to the receiving system of up to 19.2 Kbaud. Based on posts to comp.robotics, you will more likely see 9.6 Kbaud throughput; the higher rate is under ideal conditions, and few in the newsgroup had reported sustained 19.2 Kbaud speeds.

These units are some of the highest quality electronics I've ever encountered in the surplus market. The boards are beautifully etched, silk-screened, and solder-masked, stuffed with SMT components and a Rockwell 6502 CPU as the main chip.

Proxim is discontinuing the RDA-100 in favor of an upgraded model. According to reports from comp.robotics, commercial packaged modems using this board set sell for about $600 each. To save $500 each for a 9600 baud RF modem, I'll gladly drill a few holes and wire a few connectors.

My Plans

I see this modem as a major part in my long-term plan: building a tele-operated robot. As I envision it, my robot would have a video camera and transmitter mounted onboard, so I can see what lies around my machine. The robot would also carry a bidirectional serial link, so I could exchange commands and data, using my home PC or my Toshiba 3100 SX laptop.

This technology would let me put my robot on the back deck, then sit down in front of the living room TV and drive the robot around my back yard from the sofa. I could guide the machine out into the wetlands at the end of yard, then park it until nightfall. I would love to use the charge-coupled device (CCD) video camera's night-vision capabilities to check out the after-hours animal life in the wetlands. I know the bushes and grasses are home to mice and other small critters, since our cat drags in a carcass from time to time. It would be fun to watch the creatures in their "wild" habitat (minus the cat, of course). And the robot's motors would let me guide it to the action.

Most video RF links, such as the Gemini Rabbit, provide hookups for audio as well. This would let me add a microphone to the robot, for eavesdropping on the nightlife and locating where I want to visit next. I already have the Gemini system and a Chinon video camera; more on this in a later column. But the serial data link has been a major problem for some time. I have considered, and discarded, options such as IR links and modified walkie-talkies.

The Proxim modems fill this gap perfectly. They hook directly to the PC or the robot's 68hc11 serial port, handle all the low-level details of packetizing the data, and act exactly as if the two computers had a wire running between them.

The Base Unit

I started by converting one of my RDA-100 boards into a base unit. By this I mean a modem, mounted in a chassis, that I can connect to an RS-232 cable. The chassis would also take a power cable from a wall-wart. The host PC would act as if it were connected to either a modem or a null-modem cable, depending on my needs.

I chose a Radio Shack aluminum project box (270-238) as my chassis. This box measures about 5 by 3 by 2 inches, a perfect match for the RDA-100 electronics. I marked mounting holes on the top (larger half) of the box's shell, leaving about 3/4 inch of clearance at either end of the chassis. I then drilled the three required mounting holes, using an 1/8-inch drill bit. This mounting arrangement will let the board hang down from the top panel of the chassis.

The RDA-100 boards have 4-40 studs on the mounting posts, but the studs aren't long enough to clear the pins of a 16-pin header already installed on the circuit boards for making electrical connections. So I added a threaded 9/16-inch spacer to each 4-40 stud. This left clearance between the pins and the chassis when the board is bolted into place.

I wanted some room between the board and the top of the chassis for another reason. I wanted to put a chassis-mount TNC female connector in the center of the top panel, and I needed to move the board far enough below the chassis top to provide clearance.

With the mounting holes drilled, but the board not installed, I then drilled holes in the two end panels of the top half for the other assemblies I wanted to add. Refer to the accompanying diagram of the base unit panel layouts (Figure 1).

I mounted a red LED on the front panel so I can tell when the unit has power. Just about anything will work; I used a diffused LED rather than a clear one, so I could see the light from an angle. Note that the Proxim literature supplied with each modem discusses a power LED that blinks with a distinctive pattern under certain conditions. The LED in my schematic does not blink. If you want to add the "blinking LED" option to your modem, you will have to add extra circuitry. Timeline's documentation includes a partial schematic for adding this feature.

The non-maskable interrupt (NMI) pushbutton can be any normally-open momentary contact switch. I used one from my junkbox, but the Radio Shack 275-1571 or 275-1547 units should work as well. I'll describe the NMI pushbutton in a moment.

I added a DB9 female solder-cup connector (276-1538) to the back panel, for the RS-232 connection to a host machine such as my 386. The 2.1mm power jack (274-1567) lets me plug a Radio Shack 9 VDC wall-wart (273-1651) into the base unit, for unlimited power.

Note that I'm not through drilling holes in the base unit chassis yet. I'll mention the remaining holes as they come by. The holes for the DB9 and power jack are oddly shaped, and were a royal pain to make, given my low-tech mechanical tools and matching skills. I made the job a little easier by using TopDraw, my favorite Windows drawing package.

I measured the mounting outline of each component, then drew that shape, with screw holes, using TopDraw. Then I printed the shapes on my laser printer, cut them out, and taped them into place on the chassis panel using clear packing tape. Now I

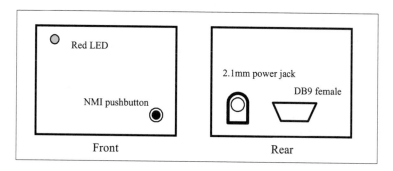

Figure 1. Base unit panel layouts.

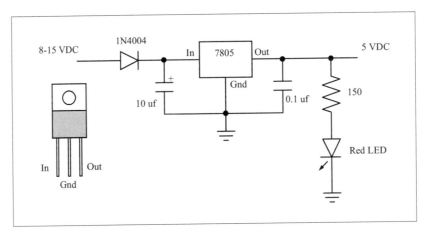

Figure 2. Schematic of base unit power supply circuitry.

could drill and file to my heart's content, staying within the lines of the drawing. I then pulled up the tape and the holes were finished.

Finally, I drilled a hole in the exact center of the top panel, to hold the chassis-mount TNC connector. I used a top-mount configuration to gain as much ground-plane as possible from the chassis. You can just as easily mount a right-angle connector to the rear of the chassis if you want.

The next step involved wiring up the base unit's power supply. This uses a 7805 regulator to drop the supplied 9 VDC down to the 5 VDC needed by the RDA-100 electronics. Refer to the accompanying schematic of the power supply circuitry (Figure 2).

This circuit is your garden-variety 7805 regulator. I've added the red LED (mounted on the front panel) as a power indicator. I've also added a 1n4004 silicon rectifier in series with the input connector. This diode eases my paranoia. There are too many wall-warts running around with output plugs wired opposite from what this base unit requires. The diode serves as protection against plugging in the wrong polarity wall-wart.

Note that the recommended Radio Shack wall-wart provides the positive connection to the plug's shell and the negative connection to the plug's center. Always verify the polarity of your wall-wart, and

the connections to your panel-mounted jack, before applying power to anything you build into a chassis.

The 7805 circuitry is so compact that I didn't bother using a project board. I just drilled a 1/8-inch hole in the rear panel, above the power jack hole. Then I smeared the 7805 regulator with heat-transfer silicon grease, and bolted it to the chassis with 4-40 hardware. Finally, I wired the two capacitors directly to the proper 7805 pins. I made the remaining connections using lengths of 22 AWG stranded wire.

I verified my wiring, then plugged in the wall-wart and checked that the LED lighted and that my 7805 provided the expected +5 VDC. Be sure you take this step before ever plugging anything critical into a home-built power supply.

Next up is the RS-232 level shifter. The RDA-100 exchanges serial data using TTL levels, 0 to 5 VDC. The PC uses RS-232 levels for its serial communications. I had to provide some type of circuitry to shift the serial data between these two levels.

I could have used my typical 4-pin serial connection, described in previous columns. As you recall, it uses a MAX232 chip built into a plastic DB25 connector shell. The MAX232 gets its power from the 4-pin Molex KK-style connector, normally plugged into a robot computer board. I decided instead to wire up a dedicated level shifter and mount it inside the chassis. I chose this method for one big reason.

I intend to use this modem with machines other than my main 386. The Toshiba, for example, has a DB9 connection on its serial port. I didn't want to have to carry around the DB25 level shifter AND a DB25-to-DB9 adapter, just to use the modem with my laptop.

So I grabbed half of a Radio Shack 276-148 project board, a low-profile 16-pin solder-tail socket, and four axial 10 µf capacitors, and wired up a very thin MAX232 level shifter board. Refer to the accompanying schematic of the MAX232 level shifter (Figure 3).

I took great pains to make this board as thin as possible. I needed to mount it to the top panel, between the RDA-100 electronics and the chassis top.

There is only about 5/8-inch clearance between these two surfaces, and the level shifter electronics, with mounting hardware, has to fit in this space.

I located and drilled two holes to hold this small circuit board to the top panel, then covered the chassis immediately underneath the mounting location with foam double-sided tape. This insulated the board's wiring from the chassis panel, and helped hold the board in place. Finally, I bolted the level shifter board into place using two 4-40 screws and very short spacers.

Now I was ready to wire up the electronics. The RDA-100 has a row of 16 pins mounted on the hardware side of the assembly. These pins fit a special single-inline connector with 2mm spacing between pins. Rather than try to find such a device, I chose to wire directly to the connector's pins. Several of the

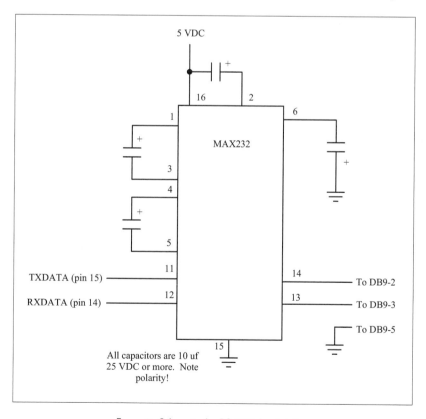

Figure 3. Schematic for RS-232 level shifter.

accompanying schematics call out pin numbers in parentheses; these numbers refer to the 16-pin connector on the RDA-100 board.

I used 30 AWG wirewrap wire to tie all the RDA-100 +5 VDC pins together. I then used stranded 22 AWG wire to connect between my added components and the appropriate RDA-100 pins. I added a piece of heat-shrink over each connection, both for insulation and for strain relief.

Connecting the antenna required a short length of RG-174A 50-ohm coax and some care. The RDA-100 boards use a very tiny SMA connector, not usually found in your local Radio Shack store. Again, I opted for expediency. I soldered the coax's braid to the connector's shell at the circuit board. I then carefully soldered the coax's center conductor to the circuit pad wired to the connector's center pin.

Be EXTREMELY careful making this second connection! There is an SMT chip capacitor soldered onto this pad. Be sure you connect your coax lead to the proper end of this chip, and be sure you do not damage or disconnect the capacitor in the process. I then wired the other end of the coax to a panel-mount TNC female connector bolted to the chassis' top panel.

I hooked a panel-mounted pushbutton to the RDA-100 pin for the NMI reset (Figure 4). This signal provides the user with a means to access the on-chip software for configuring a modem. The user can change network configuration, handshaking protocol, baud rates, and other important characteristics, all through the modem's serial connection.

This completed the hookup. Then I mounted the RDA-100 electronics to the panel, verified all my wiring, tightened down all the hardware, and bolted the chassis together. I will mention here that the Radio Shack chassis are not finished very well. In par-

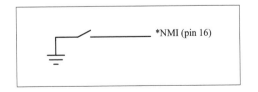

Figure 4. Schematic for NMI pushbutton.

ticular, my boxes had very sharp edges on them, sharp enough to cut me if handled improperly. If you encounter the same situation, take a moment to run a trimming tool or file over the edges and round the corners off.

Time for the big test. I plugged the serial cable from my PC into the DB9 connector, plugged in the wall-wart's power cable (no on-off switch here!), and started up my PC's comm package. When I pressed the NMI pushbutton, I was rewarded with the Proxim Radio Module User Interface menu. The PC was communicating with the modem!

The Remote Unit

The second unit, intended for use on a robot frame, proved much simpler to build. For one thing, I dispensed with the RS-232 level shifter. I was going to drive it directly with the 68hc11's serial communications interface (SCI) port, which is already using TTL levels.

I also left out the 7805 regulator and associated circuitry. I would use a four-pin cable with a Molex KK connector on it to tie directly to the SCI port on my robot's electronics. This connection includes ground and 5 VDC, so I wouldn't need the 7805 (Figure 5).

I did, however, include the NMI pushbutton and the red LED. I also added a special two-pin Molex KK connector, so I could connect a separate battery supply to the modem if necessary. Refer to the accompanying diagram for the remote unit panel layout and recessed controls layout (Figure 6).

Since I wanted to protect the controls and electrical connections from accidental contact, I cut a 1/2 inch by 1-1/2 inch slit in one end panel of the top half of a Radio Shack aluminum project box (270-238). I then wired the connectors, pushbutton, and LED to one half of a Radio Shack 276-148 project board, then drilled suitable holes and mounted the board behind the slit. Set back from the panel on 4-40 spacers, the controls cannot be disturbed accidentally, yet are easy enough to reach.

I used an Omron board-mounted pushbutton for this unit, rather than a panel-mount switch. Elec-

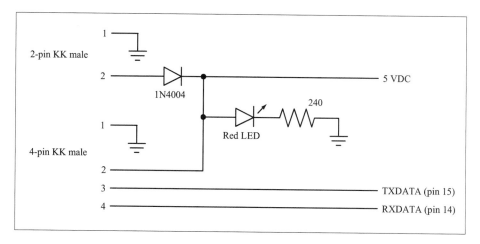

Figure 5. Schematic for remote unit controls.

tronic Goldmine sells this tiny switch, measuring only 1/4 inch on a side, that is ideal for this project. Cost was 5 for $1 in the #222 catalog, item number G2340. I use these switches on most of my projects, and recommend them highly.

I then wired up the antenna connection and mounted the electronics, as described above for the base unit. All that remained was making a suitable antenna for my two units.

Proxim recommends a quarter-wave antenna which, by their documentation, should be four inches long. Frank Haymes, the club's RF expert, expressed concern about the accuracy of this number. Consulting my trusty 1989 ARRL Handbook yields the following formula for a half-wave antenna at frequencies above 30 mHz:

$$\textbf{Antenna length (in inches)} = (5904 * K) / f$$

where K is a multiplying factor based on the antenna conductor's width, and f is the frequency in mHz. Using an arbitrarily selected K of 0.97 from the book's table produces:

$$\textbf{Antenna length} = 5727 / 910 = 6.3 \textbf{ inches}$$

Since this is a half-wave antenna, a quarter-wave element would be 3.2 inches long.

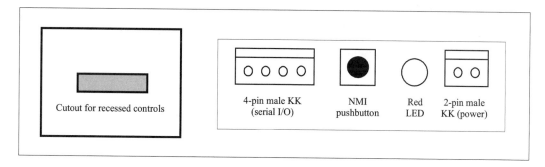

Figure 6. Remote unit panel layout and recessed controls layout.

Actually making the antenna is simple. I just clipped a piece of 22 AWG solid wire to a length of six inches or more, then stripped 1/4 inch of insulation from one end.

Next, I fed the trimmed end of the wire into the barrel of a TNC male connector, and carefully pressed it into the jaws of the center pin at the opposite end of the connector. I could feel the wire sinking into the pin for the full length of the stripped wire. Then, I filled the open end of the connector's barrel with a plug of hot glue, to provide strain relief and to hold the wire in place.

Finally, I measured and trimmed the wire to length. Note that you must measure the wire from the end of the connector's barrel, not from the connector's pin. The barrel is electrically connected to ground when it is screwed into the chassis-mounted connector.

Checkout

This completed construction of both units. All that remained was to try communications. I hooked up the base unit to my PC. I then plugged a set of four Renewal alkaline C cells into a battery holder and hooked it to the remote unit's two-pin Molex KK connector, to act as a power source.

Then I added the necessary adapters to hook my Toshiba laptop to the remote unit's four-pin Molex KK serial connector. I started comm packages on both machines, then used the embedded menu software to configure both units for a point-to-point network.

When you reach this point, the menu will prompt you to remotely configure the opposite unit. Answer "yes" to this question. It provides a quick operational test of both units, and ensures that they both get programmed properly. Note that you must have the opposite modem in this point-to-point setup turned on, or the remote configuration will fail.

Finally, I used the menu's Launch command to switch both devices into modem mode. When I typed characters on my laptop's keyboard, they echoed on the PC's screen. Characters also transferred properly going the other way. Both units worked perfectly!

Just for grins, I downloaded a text file from one machine to the other. I can't vouch for 9600 baud throughput, but the text scrolled by at a good clip.

Next, I took my laptop down into the garage and set it up on the hood of my truck. The file transfer worked flawlessly, through the length of the house and down one story. Granted, this wouldn't push a modem designed to provide 500 feet range, but I enjoyed seeing the two units work over a connection I couldn't have possibly made using wire.

A later demonstration at the SRS meeting showed a top range considerably below 500 feet. However, this was through thick concrete walls, next to a high-voltage panel, using batteries that had been drained pretty hard the night before. I will continue testing these units for range.

Both units initially power up in their saved configurations, in modem mode. This means that you just have to switch the units on, and they are automatically ready to exchange serial data. You only need to use the NMI button and resulting menu whenever you need to reconfigure your modems.

Wind-up

I have talked to Ron at Timeline, regarding his expected inventory of these modems. At the time of our conversation (early October), he still had 800 units left from an original order of about 1000. This was after a fairly spirited discussion in the comp.robotics newsgroups. Based on this conversation, I am fairly confident that *Nuts & Volts* readers will be able to order some of these modems by the time this article comes out. And if you did purchase these units and have had difficulty getting them to work, perhaps the answer to your problems appeared here.

I will use these modems in later projects, especially in my developing tele-operations robot. And I will continue my investigation into other suitable methods for exchanging data between a computer and a remote robot. For now, however, I am very happy with my new toys. Time to go do some more research...

A Dirt-Cheap 8051 Development System

This column from March of 1995 marks the first of several columns on hacking commercial modems. It proved to be a popular series, spawning plenty of email from readers who had suggestions and comments. For the robot hacker, the older external (and even internal) modems are a true gold mine. Development boards for the high-end MCUs in these modems are ridiculously expensive, yet a modem, which contains basically the same chips and peripherals, is dirt cheap. Taking a few evenings to modify one of these modems can net you a valuable development platform for under $100.

I also used this column to finish up a simple robotic eye project, suitable for use in a line-following robot. I built my eye using surplus phototransistors, and stripped the project way down to keep it simple. You can scale up what I've done here and make a more powerful or sophisticated sensor with little effort.

I was talking about simple vision systems with my friend, Marvin Green. Regular readers of this column will remember Marvin as the designer of the BOTBoard, a single-chip 68hc11 circuit board that I mention often in these pages. I had described to Marvin my trip to MIT last year, to cover the 6.270 Robot Design competition (see "A Visit to the MIT Campus.") While at MIT, I toured various labs and at one point ended up in the Artificial Intelligence library.

There, I saw an impressive demo of an autonomous robot equipped with a stereo video vision system. The robot stood about three feet tall, with a cylindrical base and a tall, slender body topped with a pair of miniature video cameras.

This baby could move! It raced around the carpeted library floor, at speeds faster than I could comfortably walk. It dodged tables, people, books, furniture, and even changes of flooring. And the only sensor it used was the vision system.

The robot was not tethered or restricted physically; it was free to roam wherever it wanted. But the control software specifically caused the robot to avoid elements its vision system tagged as obstacles. One definition of an obstacle was a dark horizontal line in the cameras' fields of view. The library carpet was bordered by dark rubber molding, which the robot saw as something to avoid. It would tear across the floor, see the dark strip of molding, then veer off in another direction.

After explaining all of this to Marvin, I began wondering aloud if there was a simple way to build similar abilities into a small, servo-based robot such as Huey. Marvin replied that he had saved an old *Byte* magazine article that described exactly that.

Two days later, I received in the mail a folder containing some pages Marvin had trimmed out of the February and March 1979 (!) issues of *Byte*. The articles, entitled *Designing a Robot from Nature*, were written by Andrew Filo. The first article covers the biological components of a frog's vision system. Filo describes the method frogs use to detect and track fast-moving prey, such as flies. Briefly, frogs determine the speed and angle of an object that crosses their field of view using a net convexity detector. They also use a sustained contrast detector to sense background light levels.

The net convexity detector is actually a group of two types of receptors on the eye's retina. The vast majority of receptors are low-sensitivity types, providing a small voltage when triggered by the presence of a visual image. The second kind of receptor provides a significantly higher voltage in response to a visual stimulus. These high-sensitivity receptors are relatively few in number, and are scattered across the retina. Note, however, that the high-sensitivity receptors never occur at the outer edge or in the center of the retina. Instead, they appear at various off-center sites.

This placement, coupled with the types of signals each receptor delivers, enables the frog's brain to deduce the speed and direction of travel of any small object in its field of view.

The sustained contrast detector works more like a charge-coupled device (CCD) camera. This detector contains a matrix of receptors that monitor the general light level of the image seen by the frog. Each receptor provides a signal if its portion of the image is above a certain threshold of brightness.

The frog's brain transforms this matrix of information into edges of optical contrast.

I'll not go into greater detail on the frog's eye's operation, but instead refer you to the *Byte* article and to the bibliography in that article. Of more interest to me is the second article. Filo gives details of his ingenious designs of two robotic eyes, patterned after the frog's eye. He created a sustained contrast detector by arranging 20 phototransistors in a 4x5 matrix. Light, directed onto the matrix by a suitable lens, is converted into discrete ones and zeros by simple electronics. The computer samples the resulting data and calculates a crude image in software.

The net convexity detector also consists of a 4x5 matrix of phototransistors, placed in the focal plane of a lens (Figure 1). Again, each phototransistor converts detected light into an electronic signal which is then sampled by the computer. Here, however, the computer ignores the output of a phototransistor unless the device signals a sudden change in light. This means that the computer does not process cells seeing a constant light level, whether that light is bright or dim.

When a cell registers a sudden change in light level, the computer generally accepts this change as a "one". The only exception is the single phototransistor designated as the high-sensitivity receptor. The computer translates its signal as a higher value, say, two. This arrangment means that any object crossing the robot's eye will generate a string of ones with perhaps a "two" thrown in. The placement of the high-sensitivity receptor lets the software quickly determine the exact angle at which an object crosses the lens' field of view.

Filo's articles go into considerable depth on how his software translates the signals created by these two eyes into useful data for his robot system. He provides detailed drawings and explanations that you can easily adapt to your own use.

I hope you get a chance to read through Filo's articles. Besides the intriguing descriptions of the frog's eye, he used some very clever engineering to overcome the low level of technology available 15 years ago.

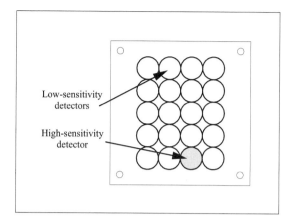

Figure 1. 20 phototransistors arranged as a 4x5 net convexity detector.

Remember When...

It's hard to believe how little was available to robotics experimenters in the late 1970s and early 1980s. Today, we often take for granted our 486 or Pentium processors, with their attendant half-gig hard drives and 32M of on-board RAM. The computer hobby was considerably different back then.

The pages of these articles included advertisements for circuit boards for adding memory, in the form of static RAM, to an S-100 system. Assembled and tested boards containing 32K (yes, kilobytes) went for $649 each, but serious experimenters could save $50 by soldering a kit together.

And robot builders didn't have 68hc11 BOT-Boards lying around, ready to drop into a frame and go. Filo's robot project consisted of a homemade mechanical hand, wired to and controlled by an 8008 microcomputer. Anyone out there remember the 8008?

Your Own Eye

I spent an evening putting together an eye of my own. Refer to the accompanying diagram of a simple robot eye (Figures 2 and 3). This eye, or light sensor, consists of a plastic lens fitted to a heavy cardboard tube. At the focal plane of the lens, I mounted one-half of a Radio Shack experimenter board (276-148) containing three phototransitors and associated resistors. I then added a cable for hooking the electronics to a small 68hc11 system.

I bought the phototransistors and lens from Electronic Goldmine. The lens (G3068) sells for $2 for a

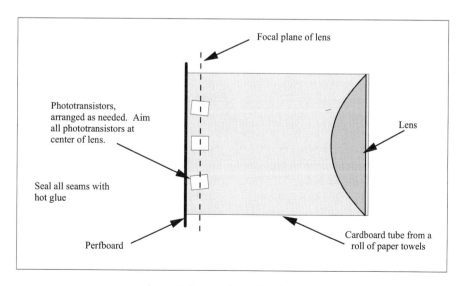

Figure 2. Design of a simple robot eye.

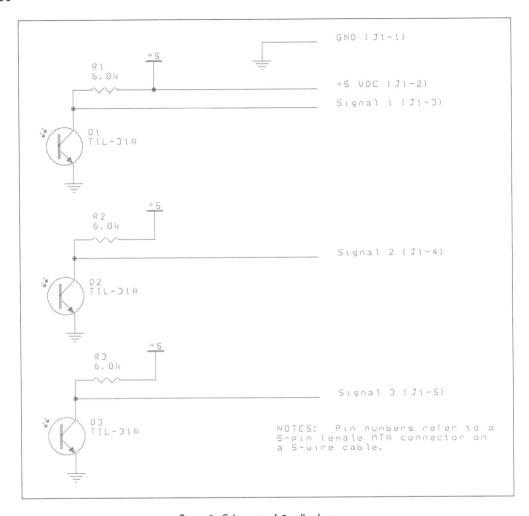

Figure 3. Schematic of 3-cell robot eye.

pair, and measures 1-1/2 inches diameter, with a focal length of about 1-1/2 inches. I scavenged three phototransistors from a surplus assembly, also offered by EG (G911). This assembly contains four TIL-31A photodetectors and four TI-305 IR LEDs, plus connectors, wire, and a red LED: cost is $1 per assembly.

The TIL-31A photodetectors make ideal components for a robotic eye. They are housed in a small metal can that mounts vertically on a board, so that the light-sensitive element faces away from the board. These detectors include their own high-qual-

ity lens built into the case, they are sensitive, and they cost very little.

I used only three photodetectors in my robot eye, as I was building an eye for a specific purpose. You can certainly add more detectors if you like; the techniques I used will work in general. I began by cutting a piece of cardboard tube from an empty roll of paper towels. This tube is about 1/4 inch larger in diameter than the lens, but it is easy to get and cheap. I cut the tube to a length of two inches, using an Exacto knife. I then slit the tube lengthwise, using a single straight cut.

Next, I put one edge of the cut inside of the second edge, so I could roll the tube slightly smaller. I fitted the lens into one end of the tube, rolled the tube tightly against the lens, then temporarily taped the tube in place. As shown in Figure 2, I positioned the lens with the convex (outward bulging) side to the inside of the tube. I also took care to place the lens as close to the end of the tube as possible, making sure it was set perpendicular to the length of the tube.

When I had the lens properly positioned, I held it in place with three small dots of hot glue, applied at the joint formed by the lens and the tube wall. After that glue cooled, I then added a bead of glue to the seam formed by the overlapping edges where I had earlier cut the tube. When that glue cooled, I then removed the temporary tape and sealed that section of the seam with hot glue as well. I now had a small lens and holder, ready for the addition of my robotic eye electronics.

The accompanying schematic shows the eye's electronics are quite simple (Figure 3). The simplicity arises because all of the A/D conversion and sampling occurs in the robot's 68hc11 microcontroller. I started by choosing the arrangement of phototransistors. I used a triangle design because I was building an eye that my robot could use for following a dark line on a white surface. Refer to the accompanying diagram of a 3-cell robotic eye (Figure 4).

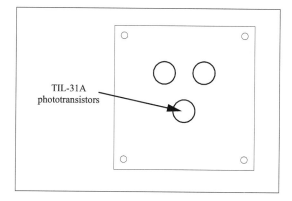

Figure 4. Diagram of 3-cell robot eye.

TIL-31A
phototransistors

After choosing the site for each phototransistor, I soldered them in place. I took care to mount the detectors about 1/8-inch above the circuit board. This small gap is critical because it allows the following sensor alignment. Once soldered in place, I then carefully pushed each phototransistor slightly off-center, aiming it at the center of the lens. You can make this step easier if you place each phototransistor on the board so that its leads are properly aligned for bending in the desired direction.

I then completed the wiring of the sensor board, adding the resistors and cabling. My sensor board needs five connections to the host computer: +5VDC, ground, and three analog signals. I terminated the cable with a female 5-pin Molex KK-style connector. While placing the remaining components, I always checked to make sure nothing interfered with the planned placement of the lens tube. The lens tube mounts directly to the circuit board, in exactly one location, so I checked often to make sure I wasn't putting a part in a wrong place.

After soldering all the components, I positioned the lens tube onto the circuit board. While holding it in place, I visually checked alignment. When I was satisfied that I had the lens where I wanted it, I used three small dabs of hot glue to fix the lens tube in place. After making one last check that the lens was properly positioned, I finished construction by adding a bead of hot glue along the entire seam formed by the tube and the circuit board.

The completed robotic eye measures about two inches long, with four mounting holes on the circuit board. It provides an analog signal from each detector, suitable for connection to the analog inputs of a 68hc11. I mounted my new robotic eye to a small stand on my work bench, hooked it to a BOTBoard, and ran some experiments to determine its field of view.

I used a solid white piece of paper as a background. For a target, I attached a dime-sized piece of black paper to a length of white 22 AWG wire. I could move the target around on the background, without having shadows or my finger disturb the readings. I lit the background using my incandescent desk lamp. I then wrote a small program that would sample all three signals and display the values read on my laptop's screen.

I placed the eye about six inches from the background, started the program, and began moving the target around, noting the changes in the displayed readings. Despite the target's size, only one detector ever saw it at any one time. The readings from a detector changed by about $40, or 64 decimal, between light and dark. This wide range of values will ease the job of writing the final line-following software.

You can use this same technique to create a net convexity detector, as shown in Figure 1. Remember that this design uses 20 identical photodetectors. The single detector marked as high-sensitivity is actually no more sensitive than its companions; the software you write simply gives this detector's output more value. And remember that you might need to alter the base of the lens-holding tube slightly to accomodate the larger array of sensors.

Cheap Development Systems

I'm always on the lookout for inexpensive development systems for microcontrollers. My favorite chip is the 68hc11, but I'll check out nearly anything. Recently, I had to pull the modem card out of my main PC. While I was holding the card, I found a QA sticker on the main chip and, by reflex, scraped off the sticker. I wasn't actually expecting to find a number I recognized; most modem boards use house-numbered MCUs.

Lo and behold, this board used a stock Motorola 68302. At this point I had to sit down for a moment. I was holding in my hand a small circuit board containing a 16 MHz 68302, 256K of socketed EPROM, 64K of static RAM, and 2K of serial EEPROM. I had paid $89 for this board over a year ago.

Those of you who have looked into the 68302 and other recent Motorola MCUs already know the major reason why experimenters don't use the chips. The leads are so tightly spaced that the average hobbyist can't solder them. And that assumes a hobbyist could even make or buy an etched board with such fine detail. Sure, you can hunt around and

find commercial development systems for the 68302 and other high-end parts. But I guarantee you the price will be way beyond $89.

But the laws of supply and demand enable modem manufacturers to build and sell these high-tech systems for less than $100. I got on the phone to the modem's maker, Practical Peripherals, and talked to their technical service desk. What follows is the result of several conversations and some independent digging of my own. Practical Peripherals sells two classes of modem, each in two different forms. The regular class modem contains an 80c31 variation, with 32K of socketed EPROM and 32K of static RAM. The processor runs at over 22 MHz, and is a house-numbered Siemens part.

The Pro class modem contains a 68302, which is a CMOS version of the 68000 with three on-chip high-speed serial ports, four controllable chip selects, 32 bi-directional I/O lines, and other goodies. This modem carries 256K of socketed EPROM (in two chips) and 64K of static RAM. Practical Peripherals sells both classes of modem in either internal or external format. The latter is by far the more interesting to robotics hobbyists, as it includes an on-board serial port, eight LEDs, and a small wall-wart power supply.

The circuit boards in either class of external modem measure about 6-3/4 inches by 4-1/2 inches and make extensive use of surface mount technology (SMT). Note that the above refers to the Practical Peripherals 14.4 Kbaud modems. I haven't had a chance to explore the newer, 28.8 Kbaud units yet. Besides, the price of the slower modems is dropping fast; as of this writing, mail-order prices for the Pro class 14.4 Kbaud external modems is $129. Check the pages of Computer Shopper for suitable suppliers.

Other modem makers will likely use different chips, so if you want to try other MCUs, look around. I know that the older Supra internal modems used a Rockwell MCU with two socketed EPROMs on-board, which could point to a 16-bit 6502 variant. And some of the newer AT&T modems sport DSPs; anyone want to try hacking robot code with a DSP?

I have already turned a regular class Practical Peripherals external modem into a nice little 8051 development system. I started by heading down to the local Computer City and picking up a new modem for $99. I opened the case, took the modem board out, and spent some time looking over the electronics. I noticed a 62256 SMT 32K static RAM chip already soldered in place, with space next to it for a second chip. I suspect you could add an extra 32K of RAM by simply soldering in another chip.

But I was headed in another direction. I wanted to know where the ROM and RAM appeared in the 8031's I/O map. And the only way to learn that was to look inside the EPROM.

I want to stress something here. I have absolutely no intention of ever again using this board as a modem. I couldn't care less how Practical Peripheral's code works, how they designed their code, or how their code could be altered to change the modem's performance. All I care about is locating the various I/O resources in the MCU's address map.

I pulled the EPROM out of its socket, put it in the socket of my EPROM programmer, and downloaded its contents to my PC as an Intel HEX file. Next, I needed an 8051 disassembler. I also realized that, eventually, I'd need a good 8051 assembler. Where better to look for such tools than the Internet?

A little bit of prowling through the comp.robotics Frequently Asked Questions (FAQ) file turned up pointers to several valuable FTP sites. The Intel FTP site contains several tools and files supplied by Intel. There is lots of stuff for high-end MCUs, but not too much for the 8051. Still, it's an interesting site, and worth a look.

The two winners were the bode site and the pppl site. I picked up some very strong tools from both of them, but I don't have notes on exactly what software came from which system. I encourage you to try both of them. Note also that the bode site carries copies of nearly everything on the Motorola FREEWARE web site. From bode and pppl, I grabbed a very nice assembler from MetaLink. asm51 appears to be a complete MCS-51 assembler, and it supports include files, macros, and segmentation. It is very fast, with good error descriptions, and templates for several different processors beyond the 8051.

I also picked up the dis8051 disassembler from DataSync. This is an excellent disassembler, capable of translating a .HEX file into legal-looking 8051 assembler source. I say "looking," because I never actually tried to assemble the disassembled source. But the output is certainly good enough to skim through and understand.

I turned dis8051 loose on the .HEX file from the Practical Peripherals EPROM, and took a look at the output source file. I used the information gleaned from my examination to start writing an 8051 monitor of my own.

Actually, I went back to the Internet and grabbed a copy of mon31-11.asm from one of the above FTP sites. This file is the source code for a simple but useable 8051 ROM monitor. Using this source as a starting point, I began rewriting large sections and adding the features I wanted to see in my new monitor.

The result, which I call mon31-20.asm, for version 2.0, serves my purposes well. I can now use it to explore other parts of the Practical Peripherals modem's electronics. The next mods to make to the modem electronics involved rearranging the locations of the I/O resources. I needed to do this to make the board more useful for robotics.

As purchased, the EPROM resides in code space from $0 to $7fff, with a second image from $8000 to $ffff. The 32K static RAM runs from $0 to $7fff in data space. This arrangement is nearly useless for robotics. The only memory in code space is the EPROM. This means that you have to burn your program into EPROM, try it, change it, burn it again, etc.

I need to change this layout somewhat. I wanted the EPROM to appear only from $0 to $7fff in code space. I also wanted the static RAM to reside at addresses $8000 to $ffff in BOTH data and code space. This last feature is most important. If the RAM exists in both spaces, I can use the monitor to download data into data space RAM, then jump to an address in code space RAM. This lets me use the RAM as a ROM emulator, speeding up code development.

This conversion is actually quite easy, but be warned that you will need to do some careful work. The mods involve unsoldering one or two SMT pins

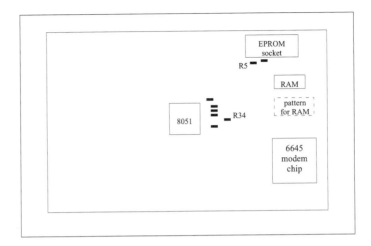

Figure 5. Layout of Practical Peripherals 14.4 Kbaud card.

and devices, and must be done with care. You will need a fine-point soldering iron with a very clean tip, and some fine solder. You will also need a magnifying glass, fine-tipped tools such as tweezers and a screwdriver, and some means of removing excess solder, such as a solder pump.

The following instructions deal with board type A1033600 (Figure 5). You can find the board type on a small barcode sticker on the solder-side of the board. In fact, Practical Peripherals even puts a barcode sticker, complete with the board number, on the outside of all of their boxed modems. This lets you verify that you are buying the above board without even opening the box.

If you already have a Practical Peripherals modem, but it doesn't have this board type, read through the instructions anyway. You should be able to locate comparable junctions for your board.

Making the Mods

My board contains a Siemens MCU, house-number SAB-C501-L20N; this is a square SMT chip in the center of the board. I spot-checked a few pins of this IC and verified that it seems to match the pinout of an Intel 8051 in a PLCC 44-pin package.

This board also contains a 32-pin socket with a 28-pin EPROM plugged into it. The ROM socket sits in the upper-right corner of the board. Note that in the following paragraphs, if I refer to pins on the EPROM, I mean counting from pin 1 of the chip. If I refer to pins on the socket, I mean counting from pin 1 of the socket.

The EPROM, a 27c256, is installed in the socket "bottom-aligned," so chip pin 14 fits into socket pin 16. Therefore, pin 15 of the EPROM is the same as pin 17 of the socket.

Similarly, the SMT static RAM IC is bottom-aligned in a 32-pad pattern at U6. Here, I will refer to device pins, numbered from 1 to 28, or to pattern pads, numbered from 1 to 32.

Begin by locating R5, an SMT resistor marked 470 (47 ohms), mounted on the component side of the board, near pin 15 of the EPROM. Remove this resistor by unsoldering it. Take care not to damage any of the tiny adjacent traces. Cut a 2-inch length of 30 AWG wirewrap wire and strip 1/16 inch of insulation from either end. Solder one end of this wire to the empty pad at R5, nearest pin 17 of the EPROM.

Locate R34, a 47 ohm SMT resistor near the MCU. Solder the other end of this wire to the pad on R34 closest to the MCU. This mod removes the copy

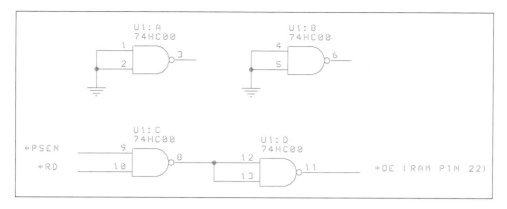

Figure 6. Address decoder schematic.

of the EPROM that appeared from $8000 to $ffff in code space. Carefully unsolder and lift pin 20 of the static RAM chip at U6. Be extremely careful doing this. You only need to lift the pin enough to clear the solder pad below it. Use a magnifying glass and a clean soldering iron, and take your time.

If you bend the pin up too far making this mod, DO NOT try to bend the pin back down nearer to the board. Doing so will surely fatigue and break the lead, ruining your board. Instead, flow a small amount of solder onto the pin so the solder acts to strengthen the lead. When you are done, inspect your work to ensure that no filaments of solder short the pin to any of the adjacent pins or pads.

Now perform the same operation on pin 22 of this chip. Wire up a socketed 74hc00 per the accompanying address decode schematic (Figure 6). Do not make any of the connections to the board yet; simply wire from pin to pin on the socket, using 30 AWG wirewrap wire and a solder-tail socket.

When finished, you should have a socketed "dead-bug" 74hc00 with five 2-inch wires coming off of pins 14, 7, 9, 10, and 11. The first two of these pins will carry +5 VDC and ground to the IC. Strip 1/16 inch of insulation from the ends of these five wires.

Carefully solder the wire from pin 14 of the 74hc00 to pad 32 of the blank SMT IC pattern adjacent to U6, the static RAM chip. As mentioned, this pattern contains 32 pads, and U6 is bottom-aligned on a

similar pattern. Pad 32 provides +5 VDC to the static RAM chip. Next, solder the wire from pin 7 of the 74hc00 to pad 16 of the same blank SMT IC pattern. This provides the ground connection to the static RAM chip.

Now solder the wire from pin 11 of the 74hc00 to the lead at pin 22 on the static RAM IC at U6. Note that you are soldering this wire to the chip's lead, NOT to the pad of the U6 pattern!

Make this connection by laying the wire along the lead, then heating the connection. DO NOT try to wrap the wire around the lead! This wire provides the *OE signal for the static RAM chip.

Solder the wire from pin 10 of the 74hc00 to pad 24 of the blank pattern at U12. This pad would correspond to pin 22 of a static RAM chip, if it were soldered onto the pattern bottom-aligned. This pad provides the *RD signal from the MCU. Solder the wire from pin 9 of the 74hc00 to the empty pad next to the legend C14, near pin 14 of the static RAM at U6. This pad provides the *PSEN signal from the MCU.

Finally, prepare a 2-inch length of wirewrap wire, trimming 1/16 inch of insulation from both ends. Solder one end of this wire to pad 22 of the blank pattern at U12. This would be pin 20 if a static RAM chip were soldered into place. Solder the other end of this wire to the lifted pin 20 of the static RAM chip at U6. Take care to connect only to pin 20, not to any adjacent pin, pad, or trace.

This completes modification of the 8051 modem board. You now have a development system with 32K of socketed EPROM and 32K of ROM-emulation RAM. You will need to mount the 74hc00 IC and its socket someplace safe on the board, so its exposed pins don't short out something. I cut a small piece of double-sided foam tape to size, then stuck the 74hc00 upside down on top of the small SMT IC just to the left of the pattern for U12.

The next obvious step would be adding battery-backup to the 32K of static RAM. This would let you develop your robot code on your desktop, then bolt the whole board onto your robot frame and restart the computer, running your code out of RAM. I haven't installed such a battery-backup scheme yet, but I'll keep you posted.

The modified Practical Peripherals 8051 board seems ideally suited for robot hacking. Many of the 8051 signals have an SMT 47 ohm resistor soldered in series. I'm sure Practical Peripherals did this so they could add or subtract features by simply adding or removing resistors. But this works to our advantage as well. Now you can easily isolate port pins and solder wires to the fat SMT pads, if necessary.

This board contains a number of interesting chips. The Rockwell 6645 chip drives the modem circuitry, which is fairly useless for robotics. But it might have additional features that a robot builder can exploit.

Of more interest to me is the Atmel 24c16 2K serial EEPROM chip, located at U2, just to the left of the EPROM socket. Be sure to contact Atmel and ask for a copy of the data sheets for this chip. It uses a stream of pulses to store and retrieve bytes of data from EEPROM, and could make a great place to stash power-up features, a small bootstrap, or other data.

I also noticed two Motorola 74hc595 8-bit serial-in, parallel-out converters. These are bizarre chips to have laying around inside a modem. I have a feeling Practical Peripherals used them as a poor-man's 8-bit output port, since running the 8051 in expanded mode uses up nearly all of its I/O lines.

Now you will need some software. My monitor isn't quite finished as of this writing, though it certainly will be done by the time you read this. I will make arrangements for storing this on an FTP site somewhere so you can retrieve it.

Those of you more adventurous can use the MON31-11.ASM source file as a starting point for your own efforts. Or you can roll your own monitor or robot code, using C, BAS051, or other language. Note that you can find the source for BASIC52, a BASIC interpreter for the 8052 chip, at the Intel FTP site. With suitable modification on your part, this might prove a welcome addition to your development system. After all, you do have 32K of EPROM to fill up.

A Dirt-Cheap 8051 Development System, Part Two

This column, written for the April 1995 issue of *Nuts & Volts*, continues on the hack of the Practical Peripheral's PP144 external modem. I go into some details on using the board's 74hc595 ICs as parallel output latches, for driving goodies such as LEDs and motors. I also explore the software needed to pull all of this together, and show how the serial bus talks to the parallel latches.

The SRS members are always looking for ways to move data between a robot and a host PC, so I took some time to explore the use of IR transmission. The results were a little disappointing, but it was still ground well worth covering. This column describes some simple steps to try in setting up your own IR serial link, and suggests ways you might get around the problems I found.

Hacking the PP144 Modem

As I discussed in last month's column, you can turn a Practical Peripherals 14.4 Kbaud external modem into a super 8051 development system with just a little work (see "A Dirt-cheap 8051 Development System.") These modems sell new for $99 at large computer stores. Refer to last month's article for details on how to rearrange the board's memory map.

Note that I am referring to the lower-end Practical Peripherals 14.4 Kbaud modem here, not the ProClass version. The ProClass modems use Motorola 68302 microcontrollers (MCUs), and I have not yet hacked one of those boards. Hereafter, I will abbreviate Practical Peripherals 14.4 Kbaud modem to PP144.

The PP144 board contains a pair of 74hc595 ICs. These are serial-in/parallel-out shift registers, used to create two 8-bit latches for driving the front-panel LEDs and other electronics within the modem. Refer to the accompanying schematic of the PP144 74hc595 parallel ports (Figure 1).

I have discussed the 74hc595 before, in conjunction with the 68hc11 MCU (see "A Look at the SPI.") Therefore, I won't go into a lot of detail on the

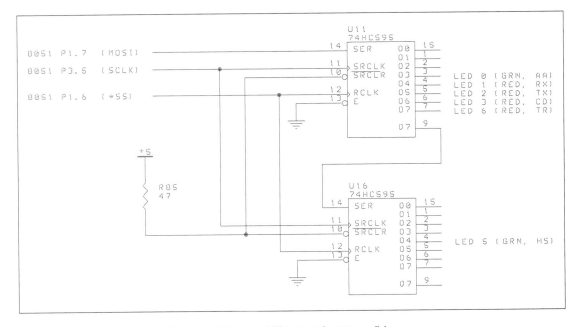

Figure 1. Schematic of PP144 74hc595 parallel ports.

74hc595 here; you can check out the June 1994 column for more information. The PP144 uses three MCU port lines to create a Motorola-style serial bus. This bus, dubbed the SPI, allows eight-bit transfers at speeds of up to 2 Mbits per second.

You need just three signals to drive a 74hc595 on such a bus. MOSI stands for Master Out Slave In, and carries data from the master device, the 8051 in this case, to the slave device. SCLK provides a clock signal for moving each bit of data into the slave device.

Finally, the slave device uses *SS as a slave-select signal to mark the start and end of a data transfer.

The Motorola 68hc11 generates these signals in hardware, which lets your code perform SPI transfers without software intervention. You just have your code write a data byte into a special I/O register, and the SPI transfer takes place automatically. The 8051, however, has to do all the signal gyrations in software. The task isn't difficult, but it does limit the top transfer speed to much less than a 68hc11 can provide. Still, if all you want your code to do is

light LEDs or move servo motors, these serial-parallel ports will do the job. And the code is quite small and instructive.

The PP144 code uses bit 7 of the 8051's port 1 (P1.7) as the MOSI line, P3.5 for the SCLK line, and P1.6 for the *SS line. Note the connection between U11-9 and U16-14. This causes the data clocked out of U11 to appear at the input to U16. Thus, these two chips form a 16-bit serial latch. Your code cannot simply change U11 or U16; it must rewrite all 16 bits in the latches in one transfer.

Additionally, these latches are write-only. This means that setting or clearing a single bit on one of the latches takes a little work. Your code has to maintain a RAM copy of both latches, then alter the copy and write the new values out to the latches.

It only takes a few bytes of code to support the serial latches. Refer to the accompanying listing of 8051 code to send data to the PP144 latches (Figure 2). The routine SETLATCH transfers the two bytes, passed in R4 and R5, to the 74hc595 latches. This code handles the entire transfer; it sets up the con-

```
;
;      The following equates define pins used to
;      control the PP144 74hc595 serial latches.
;      See also the code in setlatch below.
;

SS      equ     P1.6                    ;slave select
SCLK    equ     P3.5                    ;shift clock
MOSI    equ     P1.7                    ;data, MCU -> latches


;
;      setlatch            send 2 bytes to the 74hc595
;                              parallel latches
;
;      To use, store the MSB of the data bytes in
;      R4 and the LSB in R5. This routine alters
;      ACC and B.
;

setlatch:
        clr   SCLK                      ;drop the shift clk
        clr   SS                        ;signal the slave
        mov   a,r4                      ;get MSB of data
        call  _setlatch                 ;send to latch
        mov   a,r5                      ;get LSB of data
        call  _setlatch                 ;send to latch
        setb  SS                        ;xfer data to outputs
        ret


_setlatch:
        mov   b,#8                      ;get a counter
set10:
        rlc   a                         ;move a bit into C
        mov   MOSI,c                     ;move bit to latch
        setb  SCLK                      ;raise clock
        clr   SCLK                      ;drop clock
        djnz  b,set10                   ;count this bit
        ret
```

Figure 2. 8051 code to send data to PP144 latches.

trol lines, toggles the clock line, and finally releases the 74hc595 latches so the transferred data appears on the chips' output lines.

The code shows quite clearly the sequence involved in mimicing a 68hc11's SPI. The transfer begins by forcing the SCLK signal low, although it should have been low already as a result of the previous transfer.

Next, the code pulls SS low, signaling the slave device that a transfer is about to occur. This action is followed by two invocations of _SETLATCH, a smaller routine that actually sends the byte in ACC out the serial port. After both bytes are sent, this code pulls SS high; this action actually latches the transferred data into the 74hc595s' output lines, so it can appear on the chips' outputs.

_SETLATCH sequentially shifts each bit in ACC into the carry flag, and from there to the MOSI line. The code then pulses the SCLK line so the slave device can read the data on MOSI. _SETLATCH returns control to the calling routine after sending all eight bits in ACC.

You can add these routines to your own monitor code for setting LEDs. Just make sure your code keeps a copy of the last value written to the latches, and it uses that value in setting up the next bytes to send. The 74hc595 can supply about 20 mA of current from each output line, and the PP144 LEDs draw about 1 mA each. This means there is plenty of current capacity left over for driving other peripherals.

For example, you could hook the control lead of a hobby servo motor to one of the port lines and use the 74hc595 to control the motor. Note that you cannot drive high-current devices such as regular motors or relays with the 74hc595; you will have to add an external buffer or MOSFET for that. Also, be aware that the LEDs are active-low. This means they light if your code writes a zero to the corresponding output line. Writing a one to an LED line will turn the LED off.

I do have a few other mods to make to my PP144 development system, as time permits. I want to cut a square opening in the LED panel, so I can route a cable from the board to the outside world. This will let me bring servo control signals and other I/O lines out.

I also intend to add a Motorola 145041 serial A/D. Since the serial bus has already been implemented on this board, it would be a simple matter to find one more select line and another line to use for receiving the serial data from the A/D chip. This would give me 11 channels of 8-bit A/D to go with my 16 output lines.

Finally, I need to remove the on-board RS-232 level shifters, and wire up one of my standard 4-pin serial connectors instead. This would make the PP144 robot-ready, able to work as either an AC-powered development board or a robot brain.

This concludes my hacking on the PP144, at least for now. I am very impressed with the quality of the Practical Peripherals design. I know the company set out to make a modem, but they also designed a dynamite 8051 development system. Have fun with this one!

IR Communications

The February meeting of the Seattle Robotics Society, as usual, created a lot of discussions and project ideas. One project that surfaces often deals with exchanging serial data between a robot and a remote PC. Adding even a low-speed data link between your 'bot and a computer can open up whole new areas of experimentation. Your machine can send back telemetry data, such as temperature or light levels. Your PC can provide mapping data or new program updates to your robot. You can even use the serial link for a simple tele-operation control.

Several people in the club pressed for some form of IR link between the robot and a PC. In theory, this is simple. Just gate some type of pulse generator with your serial data, then hook an IR LED to the generator's output. You also need a receiver that decodes the gated pulse train into matching ones and zeros.

In practice, however, this simple design has some shortcomings. IR is notoriously difficult to work with. We tend to treat IR like regular light, except that you can't see it. In reality, IR has its own characteristics, many of which have to be taken into account when designing such a communication system. For example, IR is absorbed by, among other things, shiny black metal objects such as filing cabinets. Other materials, such as some plastics, are opaque to regular light and nearly transparent to IR.

Thus, you cannot always predict how IR signals will behave in the real world. But you can build an IR serial system very easily, and experimenting with it can prove fun and instructive.

The biggest obstacle to a reliable IR serial system lies with the receiver. Since it must detect short bursts of pulses at the selected frequency, its design can prove difficult.

One easy solution to this problem, though, is to use a stock receiver module such as the Radio Shack IR detector (276-137). They're cheap, easy to get, and quite reliable.

Bear in mind, however, that this module's 40 kHz detection frequency and lock characteristics will limit the baud rate of the data you can send. In general, 1200 baud is about the top end for this system. If you want to send data at a faster rate, you will need to design and build a receiver with a correspondingly higher detection frequency.

For our purposes, however, 1200 baud will prove sufficient. Using the IR detector module really simplifies the receiver side of the system. It is nothing more than a detector module, +5 VDC, and an RS-232 level shifter such as a MAX233 chip.

Note that the detector module puts out an active-low signal; that is, its output goes to logic 0 when it detects a 40 kHz IR signal. This signal polarity is exactly what the MAX233 chip needs for its input. That only leaves the design of the transmitter side. Refer to the accompanying schematic for the IR serial data transmitter (Figure 3).

The IR transmitter uses a single 74hc00 IC to generate the 40 kHz carrier signal, gate the carrier with the serial pulse stream, and drive the IR LED. You can easily fit this circuit onto one half of a Radio Shack 276-148 experimenter's board. Note that R2 controls the frequency of the carrier signal, which

you want to be as close as possible to 40 kHz. You might need to change R2's value slightly to get the desired frequency. If you want to be really precise, replace R2 with a 4.7K resistor and a 10K trimpot in series. Then monitor the signal as you tweak the pot to get exactly 40 kHz.

Be sure to include C2 and C3, especially if you run a transmitter and receiver circuit from the same battery. The IR LED can draw a lot of current as it fires, and this current spike can feed the transmitted signal directly into the IR detector, causing it to falsely echo characters.

In fact, those of you building a transmitter and receiver pair on the same board should perform a little test to make sure your receiver isn't picking up false signals on the power lines. Just temporarily replace D1 with a green LED, then try sending characters. If you have noise on the power lines, your receiver will still "echo" the transmitted characters. If your power lines are clean, the receiver will not echo the characters with a green LED, but will echo them with an IR LED.

The IR LED isn't critical; just about any type will do, though some will give you a better signal than others. Try experimenting with different units.

Note also that R4's value provides more current than some LEDs can handle. As shown, the transmitter will supply bursts of about 40 mA of current to the LED. Even though this is beyond the sustained current that a 74hc00 is specified to handle, the short duration pulses don't seem to present a problem.

Some LEDs, on the other hand, may be too fragile to take the relatively large amount of current. You can spot LED failures due to excessive current during sustained transmission streams. After some time, the receiver will begin returning garbled and trashed characters. If you turn the transmission off for a while, then try again, the stream will start out just fine. This is an indication that you have stressed the LED and need to replace it (and reduce the LED drive current).

I built up a board consisting of a transmitter and receiver section, then hooked up the unit to my stan-

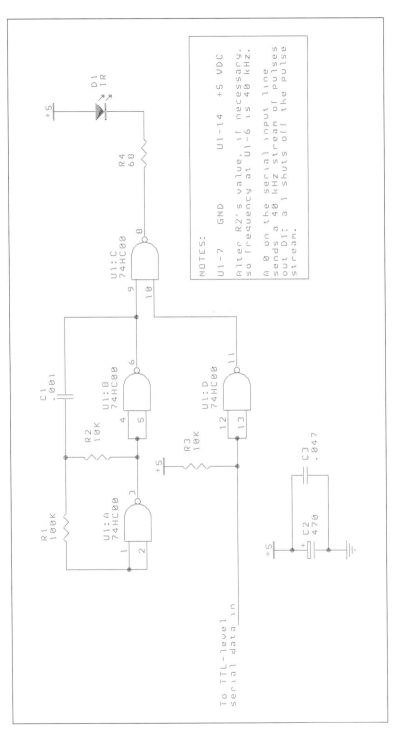

Figure 3. Schematic of IR serial data transmitter.

dard RS-232 robot connector. As I've mentioned before, this connector has an embedded MAX232 chip in it, so it converts the RS-232 signals from my PC into TTL-level signals for my robots.

I stuck the LED to the table top, about four inches from the IR receiver's case, using a small piece of double-sided foam tape. I aimed the LED and the receiver's photodetector at a white flat surface about five feet away. I fired up the Toshiba laptop, hooked up my serial cable, and started my communications program at 1200 baud. Whenever I typed characters on the keyboard, the IR transmitter/receiver faithfully echoed them back to the laptop.

Next, I did an ASCII file transmission and watched the quality of the character stream being echoed on the screen. At 1200 baud, with no delay between characters, the reception quality was disappointing. After a short burst of perfectly-received characters, the stream would degrade to just noise. Often, the comm program would lock up, as if it had received a control character sequence to halt transmission.

I could improve the quality of reception by adding a delay between each character. However, to get nearly 100% accuracy, I had to introduce so much delay between characters that the effective throughput dropped to about 300 baud.

Interestingly, slowing the actual baud rate down to 300 baud severely degraded performance. This leads me to suspect that the IR detector's output may be remaining true (low) too long after the actual signal disappears. This in turn confuses the comm software. Inserting a delay between the 1200 baud characters gives the IR detector's output time to recover before the next character arrives.

The bottom line: This system could be what you need for short range serial communications between host and robot. Granted, you will want to try different IR LEDs, and probably vary the characteristics of your data flow. But it certainly makes a good starting point.

By the way, don't be disappointed because my test was against a target surface just five feet away. This is a reflected signal, not a direct signal. Past comparisons between direct and reflected signals have shown a four-fold increase in distance by moving to a direct signal. Thus, I would expect this system to yield about 20 feet range going direct.

Note that you will need to keep the robot and the host lined up optically. The host side doesn't present much of a problem; just stick the host's LED/detector pair on a pole or up in a corner of the room. The robot's sensor system can prove more challenging. I suggest you wire three or four detectors in parallel, using the 74hc20 circuit I described in a previous column "Decoding a TV Remote Control." By arranging the multiple sensors in a circle, you can give your robot a 360 degree field of view.

For the IR LED, you can try driving a 2n2222 with the output of the 74hc00, then using the transistor in turn to drive a group of LEDs, arranged in a circle. This will allow your robot to "talk" in all directions.

All that remains is adding a little software. The exact code you add will depend greatly on how you intend to use the IR link. For example, if you will have only your robot in the arena, perhaps giving a tele-operations demo, than your robot can assume any characters it sees are intended for it. This will drastically simplify your code. Just send the characters and wait for a response.

If, however, you expect to have two (or more) robots roaming around, each using this type of IR link, you will have to add some type of addressing scheme. It could be as simple as sending a unique character at the front of each message, identifying the target robot.

Regardless of your transmission scheme, I would suggest some type of verification method, to minimize problems due to lost or mangled characters. You could have the robot echo each character back before the host sends the next; if the echo doesn't appear, the host re-sends until it gives up or the robot responds.

These schemes will all slow down the throughput, but are needed to insure good data flow. As you can see, this IR link isn't high speed, but it might be just the solution to your current problem. So where can you go with this concept? Higher speeds will require a higher-frequency carrier and a redesigned IR receiver. The IR LEDs are good for frequencies much higher than 40 kHz, and multiplying the carrier frequency by 10 would yield a corresponding increase in baud rate.

As for redesigning the IR receiver, I'll leave that to others in the SRS more capable in hardware design than I. Perhaps after I show this IR link at the next meeting, Keith Payea or Bob Nansel may run with the idea. You may yet see a 20 Kbaud IR link in *Nuts & Volts*, care of the Seattle Robotics Society.

Hacking a 68302 Modem Board

Back in June of 1995, when *Nuts & Volts* ran this column, Web access was a little dicey, using FTP to copy files took a little fiddling around, and the Motorola FREEWARE BBS was still in use. A lot has changed since then, so this column really looks dated; think of this as an Internet time capsule. But it does cover some improvements to SBasic, my recently released 68hc11 Basic compiler, including support for interrupt service routines.

Of more relevance, this column marks the beginning of a multi-article treatment of the Practical Peripheral's ProClass modem hack. Unlike the standard PP modems, the ProClass units contain an onboard 68302 MCU and a bucket of RAM and EPROM. Development boards for the 68302 device were going for more than $800 back then, so paying $129 for a modem and modifying it seemed like a reasonable path to take.

In this first column on the 68302 hack, I walk you through the board's layout and describe how the MCU comes out of reset and configures the board's memory devices. I also describe my first program, a little routine that (naturally) blinks an LED.

Last month, I described my latest major software effort, a small Basic compiler for the 68hc11 (see "A First Look at SBasic.") I'll continue that discussion this month, highlighting some of the changes I've made recently.

By the time you read this, I should have a solid release of 68hc11 SBasic waiting for you. I have already posted the latest release of tiny4th to my web site, so those of you interested in trying out a new language can help yourselves.

SBasic

Since last month, I have made several improvements to SBasic. One important upgrade concerns variables. The early tiny Basics usually provided 26 different variables, named with the letters of the alphabet. It was easy to write a compiler or interpreter using this technique, but Basic programs written with single-letter variables could prove difficult to maintain. After all, B doesn't describe a variable's purpose nearly as well as LeftMotorSpeed does.

The current release of SBasic (beta.z5a) no longer supports the single-letter variable names of earlier versions. Instead, SBasic now provides the **DECLARE** statement for defining variables.

By the way, all of my examples will show SBasic statements in uppercase. SBasic is actually case-insensitive with regard to commands, statements, and variables; FOO, foo, and FoO are all the same to SBasic.

You must now **DECLARE** all variables your program will use, before you refer to them. Syntax for the **DECLARE** statement is:

DECLARE foo

This statement creates a 16-bit integer variable named FOO, and assigns it an address in SBasic's variable space. You may have as many variables as you like, subject to the amount of target system RAM.

I also added the **INTERRUPTS** statement, used to enable and disable system-wide interrupts. The **INTERRUPTS** statement requires a single argument, either **ON** to enable system interrupts, or **OFF** to disable them. For 68hc11 SBasic, **INTERRUPTS ON** compiles to a CLI instruction, while **INTERRUPTS OFF** becomes an SEI instruction.

Note that the **INTERRUPTS** statement only affects system-wide interrupts. If your program needs to enable or disable interrupts from subsystems, such as the Serial Communications Interface (SCI), you must include **POKEB** statements for specifically controlling those interrupts.

Because SBasic compiles down to assembly language, it runs far faster than tiny4th, which compiles down to interpreted pcode. In fact, it is more than fast enough to allow you to write interrupt service routines (ISRs) in SBasic. Those of you familiar with compilers will probably be disappointed in the quality of code generated by SBasic. Currently, SBasic does no code optimization across source lines, so it does not, for example, compress register usage or variable accesses.

But the SBasic compiler has sparked a lot of interest among Seattle Robotics Society (SRS) members, and plans are already underway to add optimization at a later date. For now, though, the code created is slightly bloated but runs much faster than anything tiny4th could create.

I have provided a sample program that shows off several important features of SBasic, and is worth studying. Refer to the accompanying listing for counters.bas (Figure 1, at end of article).

The **INCLUDE** statement allows you to include existing files of SBasic source code in your source file. In this case, the SBasic compiler will open the file regs11.lib, compile all the SBasic source lines it finds in that file, then close it and resume compiling counters.bas. regs11.lib contains SBasic **CONST** statements that assign labels to all of the 68hc11 I/O registers. The labels used correspond with the common Motorola register names. For example, regs11.lib contains a **CONST** statement that defines PORTA as $1000. The **INCLUDE** feature lets you maintain libraries of working SBasic source code, and reuse them when appropriate. It also helps keep your main program file down to a manageable size.

The two **DECLARE** statements create the variables TIMER1 and TIMER2. The **INTERRUPT** statement defines an ISR, associated with the interrupt vector at $fff0. The 68hc11 automatically passes control through $fff0 whenever a real-time interrupt (RTI) occurs. The end result is that the code following the **INTERRUPT** statement will run each time an RTI occurs.

I prefer this method of handling interrupts on the target system, over the more traditional ON ... used by other Basics. I find this ISR technique separates the interrupt handling from the main code, reducing confusion over which section of code does what.

The actual work done by the ISR occurs between the **INTERRUPT** and the **END** statements. In this case, both timer variables are tested to see if either has reached zero; if not, that variable is decremented. The ISR then rearms the RTI subsystem by writing a one to the proper bit in the 68hc11's TFLG2 register. Finally, the ISR executes an **END** statement, which SBasic will compile into a ReTurn from Interrupt (RTI) instruction.

The second half of this example, starting with the label MAIN, contains the main-line code for the program. The label MAIN is required; every SBasic program must contain this label or the compiler will generate an error message. MAIN marks the location in your SBasic program where the 68hc11 will start execution following reset. You can place the label MAIN anywhere you like. You might prefer to have MAIN at the top of the file, with the ISRs and subroutines below. Or, you could place MAIN at the bottom of the file, with all the supporting routines above it.

The two **POKEB** statements following MAIN set up the SCI for 9600 baud serial operation. Note that the value written to SCCR2 only turns on the transmitter; this program only sends data, it doesn't receive it.

The next two lines initialize the TIMER1 and TIMER2 variables. Unlike traditional Basics, SBasic does not preset any variables. Your variables will contain whatever garbage was in those RAM locations when the 68hc11 powered up.

The next two **POKEB** statements turn on the real-time interrupt subsystem and clear any pending RTI interrupts. All that is left is to enable system-wide interrupts, which is done by the following **INTERRUPTS** statement.

The **DO-LOOP** at the bottom of MAIN forms an endless loop structure. The code monitors the state of TIMER1 and TIMER2, checking to see if either has reached zero. If so, the code uses a **PRINT** statement to inform the user, then reloads the timer variable.

The end effect of this program is to display periodically either "timer1" or "timer2" on the serial terminal. The values loaded into TIMER1 and TIMER2 correspond to roughly three and five seconds each, and set the delays between each display update.

The SBasic release includes library files that support **PRINT** statements using the SCI. Although this example **PRINT**s only fixed strings, SBasic also supports printing the results of algebraic expressions. It also handles empty **PRINT** statements, terminating semi-colons, and terminating commas.

FREEWARE and the Web

You can now access Motorola's FREEWARE files directly from the Internet. I've always been impressed by Motorola's microcontroller support efforts. The company provides free samples, documentation, and software for all of their MCUs. The latter has always been distributed through the FREEWARE BBS.

But the BBS isn't as easy to use as one might want. The phone call is long-distance, and the user interface mutates often, sometimes taking on irritating or frustrating features. This morning, for example, the BBS file search system would let me search for any keyword in the files' descriptions, but would not let me simply browse all of the file descriptions. In fact, the search prompt specifically precluded using wildcard searches!

But none of that matters to me now, because Motorola's Internet connection makes the FREEWARE files available to Web browsers and FTP transfers. Take a few minutes to try out this newly updated Internet source.

A Digression

I use the Enhanced NCSA Mosaic for Windows to do my browsing; I really like the clean feel of this program. You can get a copy of the Enhanced Mosaic browser by buying *The Mosaic Handbook for Microsoft Windows*, an excellent guide to this powerful Internet tool.

The Mosaic Handbook is published by O'Reilly & Associates, Inc., and is one of the company's popular "Nutshell" handbooks. If you've ever grazed a large technical bookstore, you've undoubtedly seen many of the Nutshell books. These are paperbacks with a colored spine and a white cover; the cover usually features a pen drawing of an animal or scene loosely tied to the book's theme. For example, the cover of *The Mosaic Handbook* shows a lookout, aloft in a sailing-ship's crow's-nest, bundled against the wind and looking out to sea through an old telescope. The drawing supports the book's theme of navigating the Internet.

The book gives accurate, well-written information on setting up Enhanced Mosaic, using various Internet tools from Mosaic, and building up your own Web home page. It also goes into considerable detail on HTML, or HyperText Markup Language, used to create the various headlines, lists, and graphics links that are the hallmark of the Web pages.

If you have, or can get, a suitable SLIP or PPP connection for your PC, I suggest you check out this book. I've always found the Nutshell books to be concise, on-target, and accurate; *The Mosaic Handbook* continues that tradition. Recommended.

Where Was I?

Oh yes, Motorola's BBS files. I originally went searching for a Motorola 68000 assembler. I started by using Mosaic to access the FREEWARE home page at the above http: address. The home page is nothing fancy, just a text file describing the underlying directories and giving some background information on the FREEWARE BBS and files.

The page is associated with the root directory of the Motorola files system. The bottom of the home page contains a list of directory names: no descriptions, no helpful text, just one directory name per line. Clicking on one of the directory names moves you to that directory and shows the underlying files and subdirectories. Mosaic maintains a "backward" button and prompt to move back up the the previous level. This helps in the navigation, since the Motorola page provides no help whatsoever.

You will get the most use out of this system if you have any experience using the real FREEWARE BBS; the directory layout is identical. I clicked on [pub] to access the subdirectories for each MCU. From there, I went to [m68k] and looked around. I also checked out [pub/mcu302], since I needed information on using the Motorola 68302 MCU. I also spent time scanning [pub/ibm], searching for PC-based development software.

Clicking on a specific file within a directory gets you appropriate access to that file. For text files, this means you get to see the entire file, displayed on the Mosaic screen. For binaries, such as .exe files, this means that Mosaic opens a command box so you can start transferring the file onto your PC.

The text file viewing is especially nice. I can review the text in a file before saving a copy to my system, a feature that the BBS couldn't provide. Unfortunately, Mosaic doesn't have a one-button method for moving a text file onto your PC. You can look at a text file easily, but you cannot directly save it: odd. Mosaic's Save As command, under the File menu, doesn't do the trick; it reports an error complaining that this is not an HTML file.

But there is a way, none the less. With the text file displayed in your Mosaic window, press Ctrl-A to select all of the text. Then use the Edit menu to Copy the text to Window's clipboard. Next, bring up the Write text editor; this is usually available in the Main program group. Starting with an empty Write document, use the Edit menu to Paste the clipboard into the Write window. Now you can use the File menu to Save As a file anywhere you want.

Some of the directories under the Motorola root directory are empty. If you click on an empty directory, such as [bin], you get a message that Mosaic cannot load the item. After playing with Mosaic for a while, I decided to try FTP access to the Motorola site. I use the Chameleon Sampler Internet tool package for my non-Mosaic netsurfing, so I clicked on the FTP tool to start it up.

The Motorola FTP site supports anonymous FTP login. Use your userid as your password; this is usually the default password in most FTP tools for any sites marked as using anonymous login.

The FTP view of the Motorola site looks just as you would expect, a tree of subdirectories anchored at the root, named [/]. The [bin] directory is empty; it has only the parent ([..]) in its directory list. Clicking on any of the files in a directory's list starts the familiar FTP transfer.

The Web and FTP access to the Motorola FREEWARE files makes finding and fetching the assorted software simple, even fun. Right now, though, the HTML pages are really barren, with nothing like the pointers and hints usually found on polished pages.

Of course, I'm perfectly happy with the lack of graphics; fancy pictures and talking displays really slow down performance over a 19.2K baud link. But Motorola could do more in the way of guidance through the directory structure of this vital 'net resource.

Hacking the ProClass Modem

I started work on hacking the Practical Peripherals ProClass 14.4 Kbaud external modem. As mentioned in earlier columns, the ProClass series of modems use Motorola 68302 MCUs, and contain 64K of static RAM, 2K of data EEPROM, and 256K of socketed EPROM (see "A Dirt-cheap 8051 Development System.") Cost for a new modem: $129 and falling. I call that a terrific price for a 68302 development system.

Of course, this is a modem, not a development system; at least, not yet. I just have to do a little modification to the board. Please note that I have not had a lot of time to play with this modem yet. This article covers the beginnings of my hacking efforts; I'll try to finish the project by next article.

I'll begin with a word of caution. The 68302 is WAY more complex than a 68hc11. It is based on a CMOS 68000 core, so anyone with 68000 experience will feel at home, at least with the assembly language. But the 68302 was designed to handle very complex serial communication tasks. Its three on-chip serial processors provide background support for protocols such as HDLC, and they can sustain very high data rates with little 68000 intervention.

But this increased power means a corresponding increase in setup complexity. All serial communication is managed through on-chip buffers and interrupts. Even a task as simple as sending a single character out a serial port can require up to 40 lines of setup code.

The bottom line: Call your nearest Motorola field support office or distributor, and get a copy of the *MC68302 Integrated Multiprotocol Processor User's Manual* (MC68302UM/AD Rev 2 or higher).

Also cruise through the FREEWARE FTP site and pick up all the 68302 example code you can find. And if you don't have any experience with the 68000 yet, be sure to pick up some Motorola literature on

using that chip. You'll also find lots of 68000 source code and utilities on the FREEWARE site.

I might as well include a plug for my favorite 68000 book. *68000 Assembly Language*, by Donald Krantz and James Stanley, is as entertaining and readable a book on such a complex subject as you could want. The authors cover 68000 software systems design, of which assembly language is only a small part. The book includes designs for LARGE 68000 projects, such as a multi-screen RAM-based text editor and a graphics package. The coverage of these projects includes full printed source listings.

I know that at least the editor source files assemble and work properly, because I typed every single line of them in by hand, to generate a text editor for a 68000 system I was developing. The resulting editor ran fast, smooth, and was very useful.

Note that this book is slightly dated, as it first came out in 1986, and my copy was purchased shortly after that. But I still refer to it often, and it covers subject areas, such as disassembly and op-code layout, that other works just don't touch. And with all the information packed into this book, the authors display a great sense of humor. I can sympathize with software hackers who feel that an important part of code debug is to let the MCU know "you've killed before."

I hope Addison-Wesley has produced later editions of this book. Krantz and Stanley's wonderful text got me through some tough times with earlier 68000 projects; I'd like to see others enjoy this book as much as I did. Recommended.

Where Was I?

Why spend the time to work with a chip as complex as the 68302, when the 68hc11 is simple, cheap, and somewhat easy to get? Because the 68302 has several advantages over its 8-bit cousin. First is raw speed. The 68302 on my PP144PRO modem board runs a 64K-iteration empty loop in about a tenth of a second, depending on the number of wait-states selected for the EPROM device. A 68hc11 is considerably slower.

Secondly, the 68302's serial channels support direct-memory access (DMA) transfers. This means your code can set up a serial channel to receive data in a particular format, then go off and do other tasks. When the data arrives, it will be automatically written into the specified memory locations, without 68000 intervention. Later, your code can get around to analyzing the data.

This feature meshes well with the new CCD chips coming out of Scotland. These chips provide a serial data stream video image. The video image can be large (on the order of 64K) and it can come in at a high baud rate. This means you need a fast processor that can grab a large data stream quickly.

Third is the large memory space. A 68hc11 robopet project only needs about 2K of code space, but video processing and more sophisticated robotics functions will push the 64K boundary of a 68hc11. The 68302's 24-bit address space leaves room for lots of RAM and EPROM for your future projects.

The Modem Board

The board contains eight LEDs along the "front" edge, seven of which are under software control. Refer to the accompanying diagram of the port assignments for the PP144-PRO panel LEDs (Figure 2). The EPROM and RAM chips use no traditional chip-select logic. Instead, their chip select lines are tied to two of the 68302's four programmable chip-select outputs. These outputs allow the software to dynamically place the EPROM and RAM anywhere in the 68302's address space desired.

When the 68302 comes out of a full reset, the EPROM, which is tied to chip-select 0 (*CS0), is automatically located at address $0 and is sized at 8K

bytes. This places the EPROM at the start of the 68000's vector area, so the processor can find its reset vector at $0 through $4.

The initial code in this EPROM should then alter the memory map to move the EPROM up out of the lower memory addresses, and instead move the RAM into that area. Having RAM here lets your code change the contents of the 68000's vector table whenever necessary, to support new interrupt service routines (ISRs).

As of this writing, I have not yet located a suitable PC-based 68000 disassembler. I was forced to hand-disassemble the startup code in the PP144PRO EPROMs, in order to duplicate the sequence used to rearrange the memory space.

I used the ASM 68000 assembler, available from the Motorola FREEWARE site, to assemble my source file. I then erased the original 27c1001 EPROMs and burned my program into them, using a Needham PB10 EPROM program I picked up from Digi-Key.

I had included a small routine in my first program that simply turned an LED on and off while running through repeated 64K-iteration empty loops. This makes a great visual signal that you have code running, and provides a little "instant gratification."

The fact that the LED blinked after power-up means that my code successfully switched RAM and EPROM, then properly jumped to the correct continuation point. It seems so simple, but this startup is vital and hacking it into a system you didn't design isn't always easy.

I won't provide source listings for this part of the code. It is a large chunk of code that performs a trivial task. The important part of the code, the 68302 initialization ritual, will appear in later examples that I will make available on my web site.

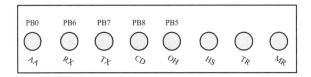

Figure 2. Port assignments for PP144-PRO panel LEDs.

I have a lot more ground to cover on this modem, but no more time before this article has to go out. By next month, I intend to add several useful features to support robotics use.

I plan to add battery-backup capability to the static RAM, as well as a monitor program that lets me download S19 object records from a serial port. These two features would allow me to load a program into RAM, turn off power, install the board in its host robot, then switch on the power and begin running the downloaded program.

I also intend to strip all circuitry off the board that does not run on +5 volts. This includes the 1488 and 1489 RS-232 drivers, located just behind the DB-25 connector. It also includes most of the power supply circuitry. I will use a 4-pin Molex KK-style connector to bring the CMOS-level serial data signals out of the board, so I can plug in my usual serial connector; see previous articles on building a MAX232 converter circuit for details.

I might even try adding an additional 64K of RAM. It would be easy to do; just piggyback a second chip on each of the two static RAM chips already supplied. I'll have to run wires from the new RAM chips to one of the remaining *CS signals, to provide chip select, but that shouldn't be tough. And I'll end up with 128K of RAM, for those really BIG projects.

Well, I can't wait to get back on the modem hack. Guess I'll wrap this up now. See you next month.

```
'

'     counters.bas

'

'     Creating down-counting timers with SBasic.

'

'     This program sets up two variables as down-
'     counting timers. It also sets up an
'     interrupt service routine (ISR) to actually
'     perform the down-counting function.

'

'     Finally, this program uses the SCI at 9600
'     baud to alert the user whenever a timer falls
'     to 0.

'


include    "regs11.lib"        ' set 68hc11 regs

declare    timer1              ' first timer
declare    timer2              ' second timer

interrupt  $fff0               ' RTI vector
if timer1 <> 0                 ' if not yet 0...
    timer1 = timer1 - 1        ' decrement timer1
endif
if timer2 <> 0                 ' if not yet 0...
```

Figure 1. Listing of counters.bas.

```
        timer2 = timer2 - 1          ' decrement timer2
endif
pokeb tflg2, $40                     ' rearm RTI intr
end                                  ' return from ISR


main:                                ' marks main code

pokeb baud, $30                      ' use 9600 baud
pokeb sccr2, $08                     ' turn on xmtr
timer1 = 3*256                       ' preload timer1
timer2 = 5*256                       ' preload timer2
pokeb tmsk2, $40                     ' enable RTI intrs
pokeb tflg2, $40                     ' arm RTI intr
interrupts   on                      ' enable all intrs

do                                   ' start a loop
     if timer1 = 0                   ' if hit 0...
          print "timer1"             ' tell user
          timer1 = 3*256             ' reload timer1
     endif
     if timer2 = 0                   ' if hit 0...
          print "timer2"             ' tell user
          timer2 = 5*256             ' reload timer2
     endif
loop                                 ' loop forever

end
```

Figure 1 continued.

Hacking a 68302 Modem Board, Part Two

My July 1995 column in *Nuts & Volts* continues the hack of the Practical Peripherals' PP144 Pro-Class modem. This 14.4 kBaud modem contains a Motorola 68302 microcontroller and enough assorted peripherals to make an excellent robot brain. In this column, I show you how to rewire the board so it can run on four C cell batteries instead of using the wall-wart power supply. I also describe how to mod the board so it uses my customary 4-pin serial cable, rather than the original DB-25 connector.

This month, I'll take you well down the path to getting the Practical Peripherals 14.4 KBaud ProClass external modem (PP144Pro) robot-ready. Be warned, this requires some very delicate work, so make sure you have a fine-tipped soldering iron, an Exacto knife, and a magnifying glass.

As mentioned in previous columns, Practical Peripherals makes two versions of an external modem. The regular class uses an embedded 8051 variant, and provides an excellent set of resources for robot hackers looking to develop with that chip (see "A Dirt-cheap 8051 Development System.")

The ProClass modems, however, use an embedded Motorola 68302 MCU. Working with this chip is normally beyond the reach of hobbyists, due to the extremely fine leads and close pin spacing.

But the PP144Pro modem design offers a unique opportunity to bolt a powerful MCU, loaded with high-speed I/O, into your robot. You just have to buy the modem, make some cuts and jumpers, and write your own robot code. This article will take care of the second item, and get you started on the third.

The Tool Set

This project calls for some unusual tools. The list starts with an EPROM programmer that supports the 32-pin 27c1001 EPROMs. I have been using the Needham's PB10 programmer, available from Digi-Key and other mail order houses. The PB10 consists of a PCB that plugs into a slot inside your PC, and a 40-pin zero-insertion force (ZIF) socket on a frame, that sits outside your PC and connects to the card via a supplied ribbon cable.

The PB10 comes with software that provides a crucial function in 68302 development. The PP144Pro uses two EPROMs, one containing the odd bytes of the executable program, the other containing the even bytes. To develop code successfully for the 68302, you must have some means of splitting your executable files into these two halves.

For the PB10, the O-option lets you split an object file into two or more sections, then program selected sections into each EPROM. Note that this splitting only occurs during EPROM programming. The PB10 does not generate separate, split files.

I have mixed feelings about the PB10. It certainly does everything I need to do regarding EPROM programming, it is very fast, and it handles some large devices. But I find the standard display colors awful, and the mouse cursor is a large blinking block that annoys me no end. I wouldn't mind a block that doesn't blink, but the flashing cursor is too much. Unfortunately, Needham's couldn't provide me with a way to turn off the cursor or stop its blinking.

The PB10 supports macros, and you can customize the initialization file, if necessary. You can also buy separate adapters to go with your PB10, so it can handle programmable MCUs and other devices. And the price, about $140 from Digi-Key, is quite reasonable. In all, this is a solid tool for any development requiring EPROM support.

You will also need a 68000 assembler or compiler. The 68302 uses a 68000 as its core CPU, so you can use nearly anything that generates 68000 executable files. I found a few public-domain 68000 assemblers on the Internet, and for now have settled on the program ASM.

If you use a C (or other) compiler for your development work, you must be able to insert a run-time initialization file, so your code can set up the 68302 before running your main program. I'll get to the reasons for this shortly.

You will also need an EPROM eraser and a supply of 27c1001 EPROMs. I use the Walling MOD-D II UV eraser, available from Digi-Key for about $40. Walling makes two models, one with an adjustable timer and one without the timer. I strongly prefer the unit that lacks the timer. Invariably, the timer, even at its maximum setting, doesn't completely erase whatever EPROM I'm using. Then, the timer goes off and shuts down the eraser. So besides having the irritating timer sound, I still have to run the eraser one more time. So I just use the cheaper model and let it go. Sometimes, I forget and erase the EPROMs for a couple of hours, but haven't damaged anything that way, at least not yet.

As to the EPROMs, I bought six chips from Prime Electronic Components, who advertise regularly in *Nuts & Volts*. They shipped quickly, sent confirming paperwork, and took plastic over the phone. They have good prices, too.

Last Month...

I started this hack last month; please review "Hacking a 68302 Modem Board" for background details. The 68302 contains three very sophisticated serial communications controllers (SCCs), which provide serial I/O. Though vaguely related to the 68hc11's serial communications interface (SCI), the SCCs provide far more complex features, and are naturally far more difficult to set up.

The 68302 SCC contains hardware (actually, RISC microcode) support for serial formats such as HDLC, BiSync, DDCMP, and V.110 protocols. Each of the three separate SCCs can handle up to eight transmit and eight receive DMA buffers, and can use either internal or external clocks. Clock rates using the 68302's internal clock generator can hit 5.56

MHz, which translates into 347 kBaud. With some work, you can even set up an SCC to act as a 9600 baud UART.

I will never again complain that the 68hc11 SCI is complicated or hard to use. Not after struggling through the ritual needed to set up one of the 68302's SCCs. I started with code contained in the Motorola *MC68302 Integrated Multiprotocol Processor User's Manual*, available from the Motorola literature center or through your local Motorola field office. The literature number for this book is MC68302UM/AD. Don't even think about using the 68302 without a copy of this book. If you want to hack the 68302 but don't have a copy already, stop reading this article right now and go order one.

The sample code contained in the User's Manual sets up SCC3 as a plain-vanilla UART, using two DMA buffers for both transmit and receive. All buffers are only one-byte long, so this code creates a double-buffered UART. The code was designed to run on some type of commercial 68302 development platform, so by the time this code actually started running, the 68302 would have already been fully initialized. Unfortunately, I needed to add custom code to initialize the PP144Pro before I could run the UART code.

The PP144Pro, like most 68000 systems, uses a two-step configuration process immediately following power-up. Initially, the EPROM resides at address $0, where the 68000 expects to find its reset and interrupt vectors. Eventually, however, the application program will want RAM in the $0 address space. This lets the program modify the interrupt vectors in low memory. Thus, the startup code must begin running in the $0 area, then somehow move itself to a higher address range and move RAM down into the $0 area.

The 68302 makes this process fairly easy, as it has four software-configurable chip select lines and 1152 bytes of on-chip RAM. The upper half of this RAM is assigned to support the SCCs; these 576 bytes are called parameter RAM. The lower 576 bytes can be used for any purpose, such as stack space or data storage. Motorola calls this section of on-chip memory system RAM.

The technique for shifting memory devices then becomes straightforward. The application program writes a few bytes of executable code into an area of system RAM, then jumps to the code. This code shifts the memory devices as needed, then jumps back to the correct address in the newly-moved ROM.

Note that your code must first position this block of on-chip RAM somewhere in memory. Since it can reside anywhere in the 68302's 24-bit address space, my code puts it in the same place the sample code did, at $700000. When your code fixes the on-chip RAM in place, it also sets the addresses of the 68302's I/O registers. These registers sit just above the on-chip RAM, and serve functions similar to the 68hc11's I/O registers.

I won't go into any more detail on initializing the 68302. Instead, I will refer you to the file checkout.asm, available from my web site.

The code in checkout.asm not only initializes the 68302 on the PP144Pro board, it also sets up SCC3 to echo any characters it receives via the board's RS-232 connector. Unfortunately, the stock PP144Pro modem board isn't wired properly to do this. Therefore, you will have to modify the board before you can use its serial port for your own programs.

Hardware Mods

I'm going to describe three different mods, the first two involving the RS-232 serial connector. The first mod lets you use the original RS-232 connector for your serial hookup. The second mod disables the original connector, replacing it with my standard 4-pin Molex KK-style male connector. The third mod makes your PP144Pro modem board completely battery powered, ready for installation in a robot frame.

If you want, you can perform just the first mod, and continue to use the RS-232 connector for communication. You can always perform the second and third mods later, if necessary. If, however, you intend to use the PP144Pro as the heart of a robot, you will eventually have to perform the last two mods. Note that the checkout.asm program will work with all mods.

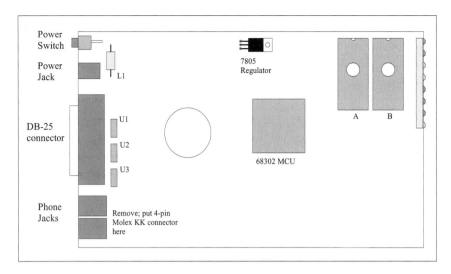

Figure 1. Diagram of PP144 board layout.

First, a comment about orientation. Whenever I describe the location of landmarks on the PP144Pro PCB, I am viewing the board with the DB-25 connector to the left and the LEDs in the upper-right corner. Refer to the accompanying layout diagram (Figure 1).

Mod #1 is straightforward. You must unsolder and lift pins U3-3 and U3-6 from their respective pads. Then you need to run a wire from the lifted pin 6 of U3 to pin 3 of the DB-25 connector. This is easy to say, but can be difficult to do. Begin by locating U3, a surface-mount (SMT) chip mounted just to the right of the DB-25 connector. This is a 1488 level shifter, though the chip on my board is numbered 5788.

Use a fine-tipped soldering iron and a piece of high-grade copper braid to wick the solder from pins 3 and 6 of this device. Using the soldering iron and an Exacto blade, carefully heat one of these leads and pry it up far enough to clear the board. Do the same on the other lead.

Next, cut a piece of 30 AWG wire-wrap wire about two inches long, and trim 1/16 inch of insulation from one end. Carefully solder this trimmed end to the lifted pin 6. Use a magnifying lens to check your work. Make sure you don't accidentally create a solder bridge between the lifted pin and the pad beneath it. Trim insulation from the other end of this

wire, route it around the right edge of the board, and solder it to pin 3 of the DB-25 connector on the underside of the board.

This completes mod #1. You can now burn the checkout program into two EPROMs and install them in the sockets. Note that the EPROM holding the even addresses goes into socket B (U16). With the checkout program installed, connect a serial cable to the PP144Pro board and start a comm package such as ProComm. Set the connection for 9600 baud, eight data bits, no parity, and one stop bit.

Plug the PP144Pro wall-wart into the power connector and turn on the board's switch. Three LEDs should light on the front panel; the green LED nearest the board's center will blink. If you press a key on the PC, you should see the character echoed on the screen.

Congratulations! You now have access to an extremely powerful robot brain. Now would be a good time to look through the 68302 User's Manual, to get a feel for all the other resources on this chip besides what I've discussed here. To give you an idea of the processor's speed, the blinking LED changes state after running through a polling loop $ffff times. Those of you used to the 68hc11's speed are in for a pleasant surprise.

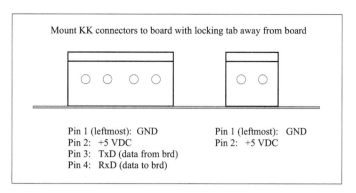

Figure 2. Diagram showing wiring and mounting of KK connectors.

Mod #2 configures the PP144Pro board to use my standard serial connector. This mod is more involved than mod #1, and requires some careful trace cutting. Begin by cutting four pieces of 30 AWG wire-wrap wire, each about 2 1/2 inches long. Solder one end of each wire to a pin on a 4-pin Molex KK-style male connector. Refer to the accompanying diagram for pin identification (Figure 2).

The next step requires clearing some space on the PCB for mounting the Molex connector. Since I no longer need this board to work as a modem, pulling off all the modem components makes sense. If you hold the PCB up in front of a strong light, you will clearly see where the opaque digital ground plane ends and the translucent modem area begins.

I sat down at my work bench and begin removing components from the modem area of the PP144Pro board. The first items I removed were the two phone jacks, J3 and J4, located just below the DB-25 connector. This clears the space for mounting the Molex serial connector later. I also pulled off C22, C23, and U5. This frees up some important solder pads that you will use in a moment.

Do not remove any components, other than C22, C23, and U5, that are not mounted wholly within this modem area! I ended up leaving the small SMT parts in place, because it was too much hassle to remove them.

Now you can finish wiring up the Molex serial connector. Solder the wire from Molex pin 1 to the upper pad of C23. This is the ground connection for the serial port. Solder the wire from Molex pin 2 to pad 5 of U5. This is the +5 connection for the serial port.

Solder the wire from Molex pin 3 to the via (plated-through hole) between pads 10 and 11 of U3. This via is located just above the "R4" silk-screened legend, and is barely large enough to take a 30 AWG wire. Note that you will probably not be able to force the wire completely through the via. Simply hold it in place, then add a touch of solder. Now carefully cut the trace that runs diagonally from this via to the underside of U3. Be sure you make your cut on the diagonal length of the trace, to the left of the via. Do not damage any of the adjacent traces or connections, and verify that you cut completely through the trace.

Locate the trace running to the via that is inside the "C" of the silkscreen for C19. You must connect the wire from Molex pin 4 to this trace. You can try inserting the wire into the via and soldering it. You can also do what I did, which was to carefully scrape some of the solder mask off of this trace, immediately to the right of the "C19" legend, then solder the wire to the exposed copper.

Now turn the PP144Pro PCB over and orient it so that the DB-25 connector lies to the right and the modem area to the lower right. Locate the trace that comes from the via labeled T131 and runs to the right, passing just below resistor R60. This via is electrically connected to the trace now wired to

Molex pin 4. You can verify this with an ohmeter. Cut this trace, using an Exacto knife or other sharp blade.

This completes mod #2. The DB-25 connector is no longer available; all serial communications must now go through the Molex serial connector. Remember that the Molex serial connection is digital, not RS-232. You must provide some external means of translating the digital signals at the Molex connector to the RS-232 levels needed by your PC. I suggest you use one of my MAX232 adapters, described in "Your First 86hc11 Microcontroller."

You should be able to run the CHECKOUT program as before, now using the Molex connector. The CHECKOUT program serves as a good verification tool. If you make other mods to the PP144Pro PCB, you can always plug the CHECKOUT EPROMs into the board and verify that everything still works. If the board no longer echoes characters, stop and figure out what you changed, and fix it before continuing.

Robot-Ready

Unfortunately, you cannot use the PP144Pro PCB on a robot in its present state. The 1488 and 1489 RS-232 drivers require +12 volts and the +12 VDC power supply circuitry is still on-board. The following mod fixes that. Note that you cannot proceed until you have installed and tested the second mod described above.

The first step towards making the PP144Pro robot-ready involves removing the 1488 and 1489 drivers. This is a simple, though brutal, task. Using a pair of sharp, fine-tipped diagonal cutters, carefully cut each lead on the right-hand side of U3, U2, and U1. Cut each lead as close to the body of the chip as you can. Make sure you don't accidentally damage any of the nearby traces or components.

After these leads have been cut, carefully lift the right-side edge of one chip, hinging it up on the left-side leads. Work the chip body up and down through about a 45-degree arc, fatiguing the left-side leads until they all break. Do NOT try to pull the chip free of the left-side traces; you will destroy the board. Simply hinge the chip back and forth until the

leads all break from fatigue. Repeat this procedure on the other two chips.

Now use a fine-tipped soldering iron to carefully clean all of the IC pads of metal chips and solder. Make sure you don't accidently leave a solder bridge between pads or to adjacent traces.

This removes nearly all of the on-board ICs that use +12 VDC. I say "nearly," because the LM386 amplifier chip (U4), used to drive the speaker, remains. I decided for now to leave the chip in place. I may later add enough voltage to power the LM386 back up, and it would be fun to have an amplifier already available.

Next up, you need to remove the on-board power supply so you can run your 68302 system from batteries. Begin by removing the power jack, located between the power switch and the DB-25 connector. Next, lift only the lower lead of ferrite bead inductor L1, mounted next to the power switch. This isolates one of the switch leads, which we will later use as a power supply point.

Now clip all three leads from the 7805 voltage regulator. This is the flat 3-pin device riveted to the PCB, about midway across the top of the board. I don't recommend trying to drill out the rivet; just cut all the leads off and unsolder the clipped leads from the board.

Clip the cathode lead of a 1n4001 diode to about 1 inch long, then solder this lead to the center lug of the power switch. Twist a pair of wires, one red and one black, into a two-conductor cable about three inches long. Be sure to use stranded wire; 28 to 22 AWG should be fine.

Strip about 3/8 inch of insulation from all ends of the wires. Slide a short length of thin heat-shrink over one end of the red wire. Solder this wire to the anode end of the 1n4001, then shrink the insulation in place using the tip of your soldering iron. Solder the matching end of the black wire to the negative terminal of the large electrolytic capacitor, located near the now-defunct 7805 voltage regulator. Solder a two-pin male Molex KK-style connector to the other ends of this power cable. Watch the polarity of the hookup; use Figure 2 for reference.

Finally, cut a piece of stranded wire about three inches long. Solder one end to the free lead of ferrite bead inductor L1. Dress this connection so that it does not contact any other circuitry. If you like, you can slide a piece of heat-shrink over the connection and shrink it down. Solder the remaining end of the this wire to the lowest of the three cleared pads to the left of the 7805 regulator. This connection provides power to the board when the power switch is closed.

This finishes the power supply circuitry. The board will work just fine with a power supply made of four alkaline C-cells. The diode provides polarity protection, so you don't burn up your board by accidently installing the batteries backwards.

The diode also provides about 0.7 VDC drop. The board contains a couple of 74ls parts, which require a power supply more nearly +5 VDC than the 'hc parts. Note that four alkalines provide roughly +6 VDC, but four NiCds only give about +4.8 VDC. If you know you will only use NiCds in your battery pack, you can try using a Schottky diode, such as the 1n5817, in place of the 1n4001. A Schottky diode only gives about 0.2 VDC drop, which could leave enough juice for the board. I haven't tried this, though, so check your voltage measurements before actually turning on the board.

All that remains is a little cleanup. Use some 5-minute epoxy or hot-glue to glue the 4-pin Molex serial connector to the board, in the same location previously used by the phone jack nearest the DB-25 connector. Similarly, glue the 2-pin power connector in place, in the same location previously used by the power jack.

While the glue sets, find the plastic case originally used to house the modem board. (You did save the case, didn't you?) Using a small file, rough out the power jack hole in the case until it is large enough to provide access to your power connector through the case wall. When the glue has cured completely, carefully slide the PCB back into its case. Bolt the board into place using the original 4-40 hardware.

Put two 4-inch strips of double-sided foam tape on the top of the case, over the slots originally used as a speaker baffle. Press a plastic 4-cell C-size battery holder onto the foam tape strips. Add a suitable power connector, hook up your power cable, and throw the switch. If you made all of your mods correctly, the LEDs should blink in their familiar pattern.

Wrapping Up

This completes the major effort in turning a PP144Pro modem into a robot brain. You can now begin the job of writing your robot code, using the CHECKOUT program as a starting point.

Next month, I will show you how to convert the now-defunct DB-25 connector into an I/O port for experimentation. This conversion will let you hook a ribbon cable, terminated in a DB-25 pin connector, into the modem's electronics. The cable in turn brings selected I/O pins out to a header for easy access.

For now, feel free to hack away. Refer to the accompanying chart of LED connections for details on a few of the I/O lines (Figure 3). You can tie into these lines temporarily, to try some quick experiments. Remember that the LEDs are active low; you must write a zero to the I/O port pin in order to turn on the LED.

Brain Exercise

I often worry that my creativity doesn't get enough workout. Yes, I design robots and robot sub-

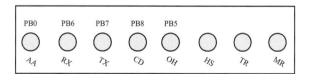

Figure 3. Port assignments for PP144-PRO panel LEDs.

systems, but that just exercises the same brain areas that I usually exercise. What about those brain cells that I don't use so often? How do I keep them in shape?

A couple we know recently introduced Linda and me to Tangoes. Those of you with small children probably already know about these little plastic geometric puzzles. For the rest of you, Tangoes consists of seven flat shapes, usually made of brightly colored plastic. You can arrange the plastic pieces in seemingly endless patterns, creating everything from geometric designs to stylized pictures of animals or other objects.

To provide both challenge and instruction, you can buy separate decks of Tangoes cards. Each card contains two large (usually black) designs for you to try duplicating, one design on each side of the card. Each side also contains the solution to the design pictured opposite, with the proper arrangment outlined in white.

Note that you can get Tangoes card decks in three different difficulty levels. Linda and I started with level 2. My first attempt was pretty discouraging. It was as if my brain simply lacked some key element, as if I had a sponge where there should have been neurons. I spent several minutes just pushing the pieces around on the table, hoping that my efforts would create a solution out of simple Brownian motion.

Eventually, though, I began to "see" differently. The pictured shapes I was trying to make would take on internal lines that matched the available Tangoes pieces. After enough practice, I could imagine the Tangoes pieces moving about on the design, rearranging themselves, until the solution appeared.

Those of you accustomed to working and thinking in graphics images may find this pretty small potatoes. But I've always been a text-based life form, and graphics has never held much fascination for me. Consequently, I've let that section of my brain grow dormant; at least that's how it seems to me.

Doing the Tangoes puzzle forces me to exercise that little-used section of my brain, and I could "feel" those neurons rousing themselves from their long naps, stretching themselves, and starting to move about. And who knows; perhaps this creative stimulation will result later in a cool robot design or mondo hack that I might have missed otherwise.

Regardless, I'll continue to work the Tangoes. If you want to try your hand at this ancient Chinese puzzle, hit a nearby toy store. Four sets of pieces and a deck of cards will only set you back about $12. Consider it an investment in your brain's good health.

The Ready-Set-Go Toy Truck

I like getting Christmas presents as much as the next kid, and by now my wife, Linda, has accepted the fact that I usually dismember my toys after a few hours, to find out what makes them tick. But the Christmas toy described in this column from the August 1996 issue of *Nuts & Volts*, is one of my favorites. It took me a while to get around to rewiring it with my own microcontroller, but I had great fun doing it, and ended up with one of the classier looking robots in the club.

The Ready-Set-Go, or RSG, is a brightly colored plastic toy truck with a pretty sophisticated on-board microcontroller. I ended up cutting out the microcontroller, leaving its circuit board intact, then wiring up my own 68hc11 BOTBoard as the new controller. This article details the techniques I used for tracking down the different functions and figuring out how to replace them with my own chip.

This article concludes with a look at the pulse-width modulation (PWM) software, written in SBasic, that runs on the BOTBoard and provides speed control of the RSG motors. Similar software appears in other columns, and the technique used is very instructive. Understanding how this code generates the required pulse trains can help you set up your own motor driver software.

Sometimes you can find a toy that makes a perfect robot platform. With just a little effort on your part, you can convert a battery-powered car or tank into a great experimental base for your robot projects. I found just such a toy in the Ready-Set-Go (RSG) computer-controlled electric truck, which my wife, Linda, gave me last Christmas. (Who says 40-something guys can't get toy trucks for Christmas, anyway?)

The accompanying pictures, being black and white, can't do justice to the color scheme on this foot-long jacked-up truck. The bright red plastic body, white bumpers fore and aft, and yellow undercarriage make this baby really stand out. What would be the truck bed contains an array of nine large, brightly labeled pushbuttons. You use these keys to enter commands into the truck's computer memory. When finished, you just press the GO button, and the truck executes the commands you've entered in sequence.

Even cooler, the RSG has a very high quality sound system on it. Each time you press a key, the truck speaks the desired action, such as "Left" or "Reverse." Additionally, the truck has a very realistic engine noise that runs while the truck is sitting still, as if it were idling. The RSG even has a "wheelie" button, which causes it to go real fast and rear back on special struts as it races forward, accompanied by the appropriate tire squeals.

And I haven't mentioned the two bright white lights built into the truck's roll bar. They add a nice touch as the truck takes off across the floor, lights flashing to get the cat out of the way. All told, the RSG is a perfect toy for the robot hacker in your family, and I couldn't wait to tear mine apart.

Inside the RSG

The RSG innards contain several excellent robotics subassemblies, already designed and tested for you. The motor assembly consists of two inexpensive Mabuchi-style motors, built into a dual gear train to provide independent drive on both rear wheels. The gear system is open, which means you have physical access to all gears in the system. This lets you add encoders to each drive train, by positioning a small IR emitter-detector near the surface of a gear and adding a black ink mark on the gear's surface.

The keypad consists of a separate printed circuit board (PCB) laid out in a matrix arrangement. Wires connect this PCB with the main board, which contains the 28-pin microcontroller (MCU), motor drive electronics, sound chip and amplifier. This board in turn is wired to the motor assembly and the two battery compartments through a rat's nest of wires.

The two battery compartments are a nice touch. The smaller compartment holds three AA batteries, providing 4.5 VDC to the computer board, assuming you use alkaline batteries. The larger section carries three D cells for the motors. This two-supply design really keeps the motor noise out of the MCU circuitry, and it also protects the MCU from voltage sags when the motors kick in from a standing stop.

I played with my RSG for a while, to determine the system's external design elements, then reached for my screwdriver. A handful of self-tapping screws hold the plastic shell together. I removed the cab seat, which contains a cute little plastic driver, complete with blue sunglasses and backwards-facing baseball cap. With the seat gone, the interior space of the cab opens up, yielding extra room for batteries and electronics.

I spent some time trying to trace out the motor control circuitry on the MCU's PCB, but finally gave up. It uses several discrete transistors, arranged in some bizarre bridge system, to provide PWM drive in both directions for each motor. I finally decided that I really didn't have to know what the circuit looked like, only how it functions.

I traced out the wiring to determine the four MCU pins responsible for driving each element of the motor circuitry. Actually, this PCB uses two I/O lines, wired in parallel, to drive a single motor control line. Apparently, the MCU, which is house-numbered and therefore unknown, has pretty weak drive so the designer had to double up the I/O lines.

I hooked my voltmeter to one of the I/O pairs, then made the RSG move forward. The voltmeter showed a reading of 2.0 VDC. Since I knew from the circuitry that the I/O lines were pulled high to 4.5 VDC, this meant the MCU was using some form of PWM to control the motor's speed. Additional probing and motor motions produced a complete table of control signals (Tables 1 and 2). As you can see, the MCU uses separate lines for forward and backward drive on each motor.

Next, I traced out the control lines for the lights. No surprises here; the MCU uses one I/O line to control the right light and one to control the left. Bringing a light control line low turns off that light, while driving it to 4.5 VDC turns that light on.

The sound system proved much more sophisticated. A small amount of on-board electronics drives a second, tiny PCB mounted vertically on the main board. This tiny board looks like the guts of one of those talking birthday cards, the kind you can use to record your own greeting. The sound board connects to the main board with seven solder pads, which provide both electrical hookup and mechanical mounting. Three of the pads connected to I/O lines on the MCU. I probed these three lines with my oscilloscope while I pushed various buttons to make the truck talk.

Two of the lines have obvious functions. Pin 1 of the voice chip is a serial data line; the patterns on it vary as a different button is pushed. Pin 2 carries

Cable Pin	BOTBoard pin	MCU pin	Comments
1	PE7	N/C	
2	PE3	N/C	
3	PE6	N/C	
4	PE2	N/C	
5	PE5	N/C	
6	PE1	N/C	
7	PE4	N/C	
8	PE0	N/C	
9	PB0	24, 25	1 = right motor forward, 0 = off (must use PWM)
10	PB1	22, 23	1 = right motor reverse, 0 = off (must use PWM)
11	PB2	20, 21	1 = left motor forward, 0 = off (must use PWM)
12	PB3	18, 19	1 = left motor reverse, 0 = off (must use PWM)
13	PB4	10	1 = right light on, 0 = off
14	PB5	11	1 = left light on, 0 = off
15	PB6	N/C	
16	PB7	N/C	
17	PA0	N/C	
18	PA1	N/C	
19	PA2	N/C	
20	PA7	N/C	
21	+5 VDC	28	MCU Power
22	+5 VDC	28	MCU Power
23	GND	12, 14	GND
24	GND	12,14	GND

Table 1. Connections between BOTBoard and RSG MCU pad layout, 24-pin cable.

Cable pin	BOTBoard pin	MCU pin	Comments
1	+5 VDC	N/C	
2	GND	N/C	
3	PC0	1	keybrd, column 1 (input)
4	PC1	27	keybrd, column 2 (input)
5	PC2	26	keybrd, column 3 (input)
6	PC3	6	keybrd, row 1 (output)
7	PC4	5	keybrd, row 2 (output)
8	PC5	4	keybrd, row 3 (output)
9	PC6	8	voice board, serial data
10	PC7	9	voice board, serial clock

Table 2. Connections between BOTBoard and RSG MCU pad layout, 10-pin cable.

eight clock pulses, used to clock in the data from the MCU via pin 1. Pin 3 shows a complex burst of data, lasting much longer than the clock stream. Ultimately, this last line stymied me. As of this writing, I do not yet have the voice system running. If I get it figured out, I'll pass the solution along in a later column.

Only the keypad remained. The nine keys on the pad form a 3x3 matrix, with one exception. The ignition key, in the lower-left corner, acts as a MCU reset switch, and is brought out separately. The remaining eight switches are arranged in a standard scanned matrix. I probed each MCU pin hooked to the keypad and noted the signal changes as I pressed each key. It only took a few minutes to develop the wiring layout for the keypad. I finished up by noting items such as the power and ground hookups, and jotted down all the details in my robotics notebook.

Now came the serious step. Using a set of narrow-jawed wire cutters, I carefully cut each pin of the MCU as close to the chip's body as possible, then removed and discarded the chip. Next, I carefully turned the board over and removed all of the residual pins, leaving empty solder pads for all 28 pins. Obviously, once I've taken this step, I can no longer go back to verify any operation, so my notes had better be accurate.

My goal was to replace the MCU with a BOTBoard running a 68hc811e2 in single-chip mode. The 2K of code space on the 'e2 would hold enough code for most simple applications, and I could always rewire a larger board in place if necessary. I soldered a 24-pin dual-row male header in the larger connector pattern on the BOTBoard, to give me access to port E, port B, and parts of port A. I also added a 10-pin male header to the smaller connector layout, so I could use port C.

NOTE: I used a 24-pin dual-row header for the larger connector, even though the BOTBoard layout has 26 holes. I mounted my header so pins 1 and 2 of the original layout are empty. Thus, pin 1 of my large ribbon connecter actually hooks to pin 3 of the BOTBoard layout. All references to pin numbers in

Figure 1. My Ready-Set-Go truck, complete with BOTBoard mounted on the cab's roof.
I removed the yellow plastic cab and little plastic driver to make room for the cabling.

this article are to my 24-pin connector, NOT to the BOTBoard's 26-pin layout.

I changed the MCU power supply wiring slightly, to accomodate the BOTBoard. I removed the AA-battery connection from the MCU board and wired it to a small DPDT slide switch that I installed in the truck's roof. I then wired from the center terminal of that switch to the BOTBoard's power connecter. Finally, I hooked the two +5 VDC wires in the BOTBoard's 24-pin ribbon cable to pad 28 on the MCU layout. This arrangement means that the BOTBoard gets its power from the truck's AA batteries, and the MCU board gets its power through the large ribbon cable.

If you intend to use the voice chip, I suggest you not exceed the 4.5 VDC voltage of the original MCU power supply. All of the greeting cards I've seen that use a similar chip have used 4.5 VDC, so I have a feeling that the voice chip isn't very tolerant of excessive voltages.

I added a second DPDT slide switch, also mounted in the truck's roof, to control the voltage to the motors. I also wired this switch in series with the remaining half of the first switch. This means that I cannot have voltage on the motors unless the MCU is active, so there is no chance that the motors will accidently see the full 4.5 VDC battery supply. Perhaps this is excessive caution, but the RSG designer took great care to never let the motors see the full battery voltage, so I figured I'd better do the same.

I then hooked up the remaining wires on the two ribbon cables to the appropriate pads on the MCU board. This gives me full control of the RSG subassemblies from my BOTBoard. Refer to Tables 1 and 2 for the RSG connections.

Please note that support for the keypad requires that you add three pullup resistors on the BOTBoard. You must pull lines PC0, PC1, and PC2 to Vcc or your code will not be able to read the keypad properly.

Figure 2. I mounted the BOTBoard to the roof using 4-40 hardware, then ran the cables through the cab window. Here you can see the large keypad available for input to your programs. The truck is slightly over one foot long.

The Software

The 2K of EEPROM in the 68hc811e2 lets me write my software in SBasic, which sure beats using assembly language. I've included source code for the two most important functions, motor control and keypad scan (Figures 4 and 5, at end of article). Both functions turn up often in amateur robotics, and the RSG design makes the motor control code different enough to warrant taking a look.

I'll start by describing the real-time interrupt (RTI) interrupt service routine (ISR) code. This module generates the PWM signals for both motors, and is executed once every 4.2 msecs. The mainline code sets up the 68hc11 RTI subsystem, then enables interrupts. When an RTI interrupt occurs, control automatically transfers into this routine.

First, the ISR rearms the RTI subsystem, so another interrupt can occur 4.2 msecs later. Next, the code handles a couple of variables, WAIT and KEYWAIT, treating them as down-counting timers. If a variable is not yet zero, this code decrements the variable. This action causes the variables to decrement "magically" every 4.2 msecs, until they eventually reach zero, where they stay until changed later. Thus, code can generate a 1-second delay by loading a variable such as WAIT with 250, then periodically testing WAIT. When WAIT hits zero, the 1-second delay is finished.

Next, the ISR code must build up a PWM value, used to control the two motors. Recall that the motor control lines are tied to PB0 through PB3, the low four bits of port B. I'll describe the technique used to control the right motor; the left motor code is similar.

The ISR code first tests the low bit of the variable RPWM. If the low bit equals one, the motor should receive voltage, so the ISR code stores the current right motor control mask in variable T. If the low bit equals zero, however, the motor should not receive voltage for this time slice, so the right motor control mask is not added to T. After going through this same process for the left motor, the ISR code modifies the low four bits of port B to hold the new control mask in T. The code then rotates the RPWM value one bit position to the right, updating the low bit for the next time slice in 4.2 msecs.

The interaction between RPWM and RMTR make this motor control scheme work. RPWM holds a 16-bit value that serves as a PWM mask. Each bit in RPWM is one if the motor should receive voltage, or zero if it should not. The pattern stored in RPWM gets rotated one bit position on each RTI interrupt, so the low bit is constantly updated. This means that the motor will see a series of voltage pulses matching the pattern of one-bits in RPWM.

For example, a value of $5555 in RPWM means that the right motor will see alternating bursts of 4.5 VDC and 0 VDC, with each burst lasting 4.2 msecs. This works out to a 50% duty cycle. Using a value of $aaaa in RPWM results in the same 50% duty cycle, while a value of $eeee generates a 75% duty cycle.

The variable RMTR controls the direction and motion of the right motor. Whenever the low bit of RPWM is one, the ISR adds the value in RMTR to the motor mask being built. Thus, if RMTR contains a one in the bit position for controlling the forward motor driver, the right motor will see a burst of voltage on the forward control line. If, however, the RMTR variable holds the mask for the reverse motor driver, the motor will see a burst of voltage on the reverse driver. RMTR is also used to stop the right motor. If code stores a value of zero in RMTR, neither control line will be set to one, even if the low bit in RPWM is one at that moment. Thus, your code can uses RPWM to control a motor's speed, while using RMTR to control direction or stopping.

The code for reading the RSG's keypad is much more complex. The keypad looks like a 3x3 switch matrix, with columns 1 through 3 wired to PC0 through PC2, and rows 1 through 3 wired to PC3 through PC5. Note that the code sets up PC0 through PC2 as inputs, with the other three lines configured as outputs.

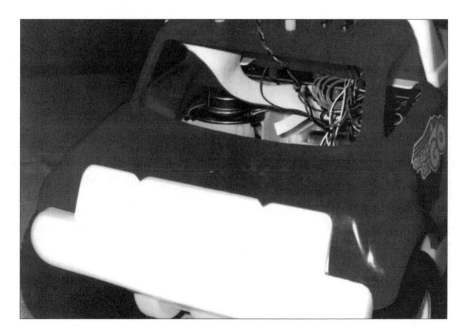

Figure 3. Looking through the front window at the rat's nest of wiring that goes to the MCU board. The speaker for the voice card sits to the left. Work carefully when moving the MCU board around, so you don't break any vital connections.

The scan operation is straightforward. The code in ScanKeys steps across each of the three rows, momentarily pulling a row line low. It then tests the three column lines to see if one of them went low also. If so, then the row and column lines uniquely define the pressed key. If no column line goes low, ScanKeys moves to the next row and tries again. If no column line goes low for any row, then no key was pressed.

If ScanKeys detects a pressed key, it converts the column lines into a number from 1 to 7, for later use as an index into an array. This explains the XOR operation at the end of ScanKeys, used to invert the column value into an index. The math operation on N converts the row and column values for a pressed key into the final array index, which is returned to the calling routine.

The code in GetKey calls ScanKeys to check for a pressed key. If no key is pressed, GetKey returns a zero to the calling routine. If a key is detected, however, GetKey sets up variable KEYWAIT as a debounce timer. After a suitable delay, GetKey again calls ScanKeys, to see if the pressed key is still down.

If the same key is down after the debounce delay, GetKey compares this new key value (in NEWKEY) to the last known key (in OLDKEY). If the two match, then the pressed key is simply the original key, still being held down. In this case, GetKey ignores the pressed key by putting a zero in NEWKEY.

Finally, GetKey uses NEWKEY as an index into an array of key codes in KEYTABLE. If NEWKEY holds a zero, GetKey returns the value in the zeroth element of KEYTABLE, which is a zero. Otherwise, GetKey returns an ASCII character corresponding to the pressed key. I arbitrarily mapped keys that made sense to me; thus, pressing the right arrow key returns an ASCII R, while pressing the GO key gets a value of G.

You can modify the code in GetKey to produce a typematic effect, if you want. This involves recording how many scans elapse with the same key held down, then issuing a second keypress for the same key after enough successful scans. I chose the simpler method of only returning a keypress when the key is first pressed, then ignoring all subsequent scans for that same key until I get a scan with no key pressed.

Conclusion

This article has described converting a popular, well-designed toy into a useful robotics platform. As you can see from the charts of connections between the 68hc11 and the RSG, a number of I/O lines remain available for your own use. Particularly, the SPI lines on port D and all of the A/D inputs on port E remain unused. You should have plenty of space for adding IR detectors, bumper switches, and other goodies.

I'm very pleased with the care shown in the RSG's design. The separate power supplies, open gear drive system, and cool packaging can make this a showcase robot. The interior, once you remove the little plastic driver and cab seat, can hold extra circuitry or a larger battery. And having a large keypad always available lets you do on-the-fly control of your robot.

I will post the complete code for this project on my Web page. I will also include a few color images of the RSG robot, so you can take a look at how it is evolving. And if any of you have modified other toys for robotics, drop me an electron and let me know what you did.

```
const    RMTR_FWD  = $01
const    RMTR_REV  = $02
const    LMTR_FWD  = $04
const    LMTR_REV  = $08
const    RMTR_STOP = 0
const    LMTR_STOP = 0


'
'    RTI  interrupt  service  routine
'
'    This ISR supports down-counting timers, such as WAIT. It also
'    supports PWM of both motors, through the RPWM and LPWM pattern
'    variables and the two motor control variables, RMTR and LMTR.
'
'    This routine assumes port b controls the motors. See the
'    constant declarations above for pin assignments.
'

interrupt   $fff0                      ' RTI ISR
pokeb tflg2, $40                       ' rearm rti
if wait <> 0
      wait = wait - 1
endif
if keywait <> 0
      keywait = keywait - 1
endif
t = 0
if rpwm and 1 = 1
      t = rmtr
endif
if lpwm and 1 = 1
      t = t or lmtr
endif
pokeb portb, ((peekb(portb) and $f0) or t)
rpwm = rroll(rpwm)
lpwm = rroll(lpwm)
end
```

Figure 4. Listing of real-time interrupt ISR code.

```
const   DEBOUNCE = 10

keytable:
datab 0, 'W', 'F', 0, 'H', 'L', 'B', 0, 'R', '?', 'G', 0, 'X'



ScanKeys:
keyrow = $30                                         ' mask to pull one row
low
for n=0 to 2                                         ' for all three rows...
      pokeb  portc, ((peekb(portc) and $c0) or keyrow)        ' pull
a
' row low
      scnkey = peekb(portc) and $07 ' read the cols
      if scnkey <> 7                                ' if one col is low...
            exit                                    ' leave the loop now
      endif
      keyrow = lshft(keyrow) + $08 ' move to next row
      keyrow = keyrow and $38       ' keep within proper 3 bits
next
pokeb  portc, (peekb(portc) or $38) ' leave with all rows high
scnkey = scnkey xor 7                               ' invert result bits
if scnkey = 0                                       ' if 0, no key was
pressed
      n = 0                                         ' show as return value
else
      n = (n * 4) + scnkey                          ' convert to unique
value
endif
return    n




GetKey:
newkey = usr(ScanKeys)              ' check the keys
if newkey = 0                        ' if nothing pressed...
      oldkey = 0                     ' mark this as a released key
      return  0                      ' show it and leave
endif
keywait = DEBOUNCE                   ' get debounce delay
```

Figure 5. Listing of keyboard scanning code.

```
do
loop until keywait = 0              ' wait it out
if usr(ScanKeys) = newkey               ' if the same key is down...
       if newkey <> oldkey              ' and different from last
                                        ' time...
              oldkey = newkey           ' record the new key
       else                             ' this is same key as last
                                        ' time...
              newkey = 0                ' only show key on first press
       endif
else                                    ' key changed between scans;
                                ' noise
       newkey = 0                       ' show nothing pressed
       oldkey = 0                       ' erase memory also
endif
return       peekb(addr(keytable) + newkey) ' convert to ASCII
```

Figure 5 continued.

Reworking the GameBoy®

In this column from March of 1998, I described how to wire up and talk to a Vector-2X electronic compass board using a 68hc11 BOTBoard. This neat little compass from Precision Navigation lets your robot orient itself to magnetic North, and played a part in an SRS entry in the 1998 Trinity College Fire-Fighting Robot contest.

I also used this column to describe one of the cooler hacks I've seen, turning the Nintendo GameBoy (GB) into a compact robot-brain. I received a lot of email about this idea after the column appeared, and I was glad to introduce readers to some of the fine work done by people like Jeff Frohwein and Pascal Felber. There are so many tools and so much information available on the Internet for hacking the GB that you owe yourself a look at this project.

A vision problem has hampered progress on my fire-fighting robot; I can't stay focused. It isn't that I don't want to work on the robot I'm due to enter in the Northwest Regional of the Trinity College Fire-Fighting Robot contest. It's just that too many interesting diversions keep appearing.

For example, I recently received email from Christine Sherer of Precision Navigation, Inc., makers of the Vector-2X magneto-inductive compass boards. She wanted to know if I would be interested in trying out one of the low-end boards. We exchanged some email to the effect of, sure, I'd love to try one, and a few days later the package arrived on my doorstep.

The V2X compass board measures 1.3 by 1.5 inches and contains several surface-mount (SMT) parts, including what looks like a small microcontroller (MCU). The board also sports two slim coils of wire, oriented at right angles to each other. Seventeen pins on 0.1-inch centers along the board's perimeter provide electronic connections to the outside world.

Specifications for this board put it in the high-end of the hobbyist domain. For the retail price of $50, you get a small PCB that can resolve headings from due North to within 1 degree, draws less than six mA running and 100 µA when idle, and interfaces cleanly to many of the popular MCUs.

The module's designers evidently put a lot of thought into making the board as flexible and powerful as possible, while keeping to a minimum the fuss of wiring it up. The V2X uses a serial bus, compatible with Motorola's SPI, for sending data to the host computer. Since the board doesn't accept incoming commands, you can get by with just the clock, outgoing data, and select lines. You also need to provide a polling signal to trigger a reading, and you need to monitor an end-of-conversion (EOC) signal to tell when the V2X has a valid reading available.

These five signals, plus +5 VDC and ground, constitute a minimal V2X configuration, capable of providing a compass reading at least twice a second. The remaining pins on the V2X board allow you to activate other options, such as performing a calibration for surrounding metal objects, selecting high or low resolution mode, selecting output data formats, or operating the V2X mounted upside down.

The package I received also include a copy of the Vector Electronic Modules' Application Notes, version 1.06. The app notes go into great detail on how to wire up and use the V2X module, and was invaluable in getting my board running. The notes discuss the many options available to you when designing a V2X into your robot. For example, the V2X can output data to the host MCU in any of three different formats: binary, BCD, or raw. Each format has its strengths, and a quick rundown of each will show the care this board's designers took in their engineering.

If you strap the V2X for binary operation (its default mode), your MCU gets two bytes of heading information when it takes a reading. These two bytes, concatenated together, form a 16-bit binary number from 0 to 359. This is the integer value of the compass heading, where 0 is North and 90 is West, assuming the V2X is sitting with its pins pointed down. Binary mode makes the most sense when the MCU will perform some kind of computation on the returned value.

In BCD mode, the MCU also gets two bytes of heading data, but now the concatenated 16-bit number holds the compass reading as a three-digit BCD number. This format comes in handy when the V2X is driving a set of LED displays. The data can be clocked serially through a set of shift registers, then latched into the display, all without requiring an external MCU.

Raw mode may be the most powerful format of all. In this case, the V2X clocks out four bytes of data, arranged as two 16-bit signed integers. The first integer contains the orientation of the X-axis, in a range from -32,000 to 32,000, while the second integer handles the Y-axis. This format gives your MCU increased resolution of the readings, at a penalty of having to do some serious trig to compute an angular heading.

Where Am I?

Enough of this esoterica, it's time to wire this hummer up. I started with a BOTBoard fitted with a 68hc811e2. After looking over the manual, I opted to wire my V2X in the minimal configuration. I soldered a 10-pin IDC male connector to the BOTBoard's prototyping area, then made the connections shown in the accompanying wiring diagram (Figure 1). I chose PE0 for my EOC signal, since all port E pins work as both analog and digital inputs, without having to monkey with a data-direction register. The same holds for PB1, an output line used to start a conversion. (I would have used PB0, but it's already wired to an LED on this BOTBoard.)

I rooted through my junk box and found a foot-long 10-wire ribbon cable, complete with a female connector on one end; those long-ago trips to the surplus stores always pay off. I stripped insulation from the appropriate wires on the other end of the ribbon, tinned the leads, and carefully soldered them to the proper pins on the V2X board. Take care during this step and any time you handle the leads of the V2X. The pins look like IC leads, but actually are quite soft and could break with rough handling.

As you can see, the wiring required isn't much; all the effort lies in the software. Since this didn't look like a tough problem, I fired up my SBasic compiler and hacked out the code you see here (Figure 2,

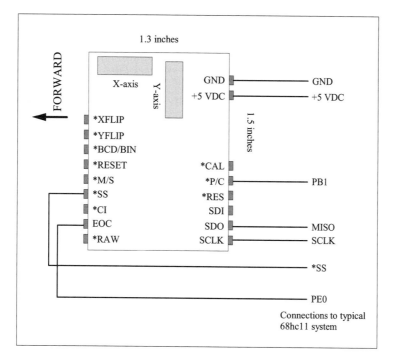

Figure 1. A minimal Vector-2X configuration using a 68hc11.

at end of article). Though the program isn't very long, it takes a bit of line-twiddling to get a reading out of the V2X, and a walk-through should prove helpful. Keep in mind that the V2X is strapped by default for binary mode, so my program will be reading two bytes of binary data that yield a 16-bit heading from 0 to 359.

As usual, I start by including the file regs11.lib, which contains SBasic declarations of all the 68hc11 I/O registers. This lets me refer to port B as PORTB, not as $1004; makes the program easier to follow. Next, I declare two variables, WAIT and VALUE. WAIT is a general-purpose variable that will act as a down-counting timer, while VALUE will store the heading returned by the V2X. Next, I define an interrupt service routine (ISR) at address $fff0, the interrupt vector reserved for the real-time interrupt (RTI). The code for this ISR checks WAIT to see if it is already zero; if not, this code decrements WAIT. The code then rearms the RTI so it can generate another

interrupt at the next interval. Since this program uses the default value for the RTI intervals, WAIT decrements at the rate of one count every 4.1 msecs. Setting WAIT to 250 means that it will take just about one second for WAIT to reach zero.

The first section of code after the label MAIN sets up the SCI so I can monitor my program using the PC's serial port. This section also sets up the RTI subsystem, clears WAIT to zero, then turns on interrupts. At this point, the RTI ISR can begin fielding interrupts and checking WAIT.

The next section of code sets up the SPI for communication with the V2X board. First, my code forces *SS high, then makes *SS (actually, PD5) an output. Doing these steps in this order insures that the V2X doesn't get a false slave-select signal during setup. Next, my code configures the SPI for the proper SCLK polarity, phase, and frequency. The V2X wants an SCLK signal that idles high (CPOL=1) and it will provide valid data on each rising edge of SCLK

(CPHA=1). The frequency isn't that critical, since the V2X can handle clock rates of up to 1 MHz, so my value of 62.5 kHz is downright leisurely. My code then reads both SPSR and SPDR to clear the SPI registers, then pulls PB1 high to put the V2X in sleep mode. Finally, my code prints a line of text so I can see that control at least got this far.

The rest of the program is a large **DO-LOOP** that triggers whenever WAIT holds a zero, indicating that it is time to read the V2X. Taking a reading begins by pulling P/C low for at least 10 msecs, which forces the V2X to wake up and sample its sensors. When the sampling starts, the V2X pulls EOC (wired to PE0) low. After my code releases P/C, the V2X continues its conversion; when the conversion is complete, the V2X pulls EOC high. My code then adds an additional 10 msec delay before beginning the transfer of data from the V2X.

Data transfer is pretty straightforward; it starts when my code brings *SS low. Following a five msec delay required by the V2X, my code writes a zero to the SPDR register, effectively sending a zero out the SPI bus and fetching a byte of data from the SPI bus. The outgoing byte doesn't matter, as nothing is wired to the MOSI line; what really counts is the byte sent back by the V2X. This byte, the more-significant half of the 16-bit data value, is multiplied by 256 and stored in variable VALUE. My code then sends a second zero out the SPI, collects the second byte from the V2X, adds that to VALUE, and pulls *SS high. All that remains is to print the value in VALUE to the screen so I can see where my compass points. Finally, this block of code stores 250 in WAIT, setting up a one-second pause before the next reading.

Getting my code to work proved quite simple, thanks to the well-done manual provided by Precision Navigation. I will point out one subtle element, however, that must be addressed or you won't be able to make the V2X give you data. I had to insert a five msec delay after bringing *SS low, but before I started an SPI transfer. Without this delay, the electronics behaved as if it was working, but I never got meaningful data. This requirement is mentioned in the app notes' text, but does not appear in any of the

timing diagrams. I think PNI should make this requirement more obvious in their graphics. Refer to the accompanying timing diagram, which I've redrawn to show the needed delay (Figure 3).

That said, I'll conclude by saying that this compass module rates as a "must-have" for anyone looking to add navigation to a robot. The low cost, ease of use, flexibility of design, small size, and other features make this little board a tool well worth adding to your collection.

The GameBoy®

A couple of years ago I did a series of columns on hacking a pair of Practical Peripherals external modems, turning them into excellent 8051 and 68302 development systems (see "A Dirt-cheap 8051 Development System" and "Hacking a 68302 Modem Board.") Judging by the mail I received, these were some of my more popular columns. I never got around to building a complete robot using either board, but the hacks were great fun and I've always been on the lookout for similar projects.

But hacking modems into development systems has some disadvantages. Sure, you can save about 90% over the cost of a comparable development system for the same MCU, and nearly all of the hairy design work has already been done for you. But modem technology evolves at breakneck speed, with companies introducing a terrific product, then canning it after a few months in the market, in favor of the next terrific product. This leaves the hobbyist with a very small window of opportunity. By the time you buy a modem, hack it, document what you've done, and tell your friends, the market may have evolved to the point that no one else can buy the same modem and duplicate your hack.

So in these last couple of years, I've been on the lookout for another technology, some other device that, properly hacked, could serve as a robot platform or brain. I'd be willing to forego the latest high-speed chips in favor of guaranteed availability and a strong software toolset. Hopefully, if the device has been around long enough, the market's

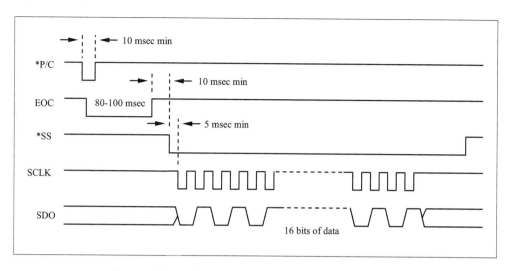

Figure 3. Timing diagram for exchanging data with the Vector-2X.

economies of scale could kick in and assure widespread availability for a low price.

Enter the Nintendo GameBoy. These devices have been around at least ten years, an eternity in today's fast-mutating consumer electronics market. At a retail price of about $50, you get a black-and-white sprite-addressable screen with 160x144 pixel resolution, stereo sound, battery operation, ROM/RAM cartridge support, and a minimalist keyboard design, all in a handheld unit. If you hit some of the larger game stores, you can usually find clean, used GameBoys for $25 or less.

But I didn't really feel up to the challenge of buying a GB, opening it up, and doing the serious hacks to make it work, because I didn't even know enough about it to get started. My major concern centered around tools for writing GB software. After all, the electronics on something this old would probably be straightforward, but I didn't even know what MCU was inside, and I certainly didn't know where I could find software development tools.

And then I stumbled across the mother lode while surfing the Web. I was looking for something totally unrelated and, as usually happens on the Web, took a wrong turn on Alta Vista. Jeff Frohwein's web page contains a TON of valuable inside information

on the GB. Jeff's site covers both hardware and software, and includes several schematics for GameBoy-related hacks and projects. I learned that the GB uses a modified Z80 microcomputer running at 4.91 mHz, the ROM cartridge supports a variety of banked ROM and RAM schemes, and the GameBoy sports an external bidirectional serial port very similar to the Motorola SPI, though much slower.

Jeff has created several circuits for adding functionality to the GB, and his web site includes schematics, in the form of graphics files, for many of these projects. The most ambitious project is probably his Carbon Copy Card (C3), used to program a specially modified GameBoy cartridge with software of your own development. To use the C3 adapter, you first need to buy a GB cartridge that contains ROM, RAM, and a backup battery for the RAM. You then replace the surface-mount (SMT) ROM with an Atmel 29f040 flash EPROM of your own purchase. Once modified, you can download any GB program you care to write into the modified cartridge, using the C3 adapter, then play that game in any GameBoy machine. Jeff's site contains considerable details on which cartridges can be successfully modified, how to modify the cartridge, and how to build your own C3 adapter.

But hardware projects are only the beginning. On Jeff's site and other sites linked from there, I found scads of GameBoy-related software. Perhaps the most important development tool you can grab is one of the many GameBoy emulators, such as the Virtual GameBoy (VGB). I'm using VGB-DOS version 0.86, released by Hans de Goede and based on the original work by Marat Fayzullin. Installation was a breeze, requiring little more than unzipping the VGB-DOS.ZIP file. The emulator (VGB) turns your PC into a very faithful rendition of a GameBoy, complete with stereo sound effects. You use the PC's arrow keys, the Shift key, and the Alt key as the six keys on the GB keypad. The emulator supports many other options during runtime, all well designed and smoothly functioning. The emulation is so good, in fact, that the program will execute ROM images of about 85% of the GB games on the market. This means that if you write a program targeted for the GameBoy, you can probably test your program on the emulator before transferring it to a cartridge.

Which leads me to my last concern, namely software development tools specific to the GameBoy. The GB uses a microcomputer similar to the Z80, but the differences are great enough that Z80 assemblers and compilers won't work for software development. Not a problem, I just grabbed a copy of GBDK 2.0, the freeware GameBoy Developer's Kit distributed by Pascal Felber. I'm using version 2.0b11, updated 24 November 1997, so it looks like Pascal stays on top of bugs and design changes. You can get the latest GBDK by following the links on Jeff's web page.

The GBDK includes an ANSI C compiler with source code for several common library files, an assembler for the GB microcomputer, and a linker. You can generate fully compatible GameBoy ROM images, and the library routines included with the distribution handle all the low-level chores for you. For example, the GBDK supports **puts()**, the standard C library routine for console text output. Since the GB has no built-in alphabetic or numeric symbols, the GBDK library functions automatically prepare the GB sprite tables for use as characters. Once your program is loaded into the GB emulator or the GameBoy itself, an invocation of **puts()** causes the specified string to appear on the GB graphics display as text.

The above elements comprise a full development package for the GameBoy. You write your software in either C or assembly language, using Pascal's GBDK. The resulting object file is a complete GB ROM image, in a PC file with a .gb extension. You then invoke the Virtual GameBoy emulator to test your program, then edit and retest as necessary. Finally, you can use Jeff's Carbon Copy Card to move your new GameBoy program into the 29f040 flash memory of a modified GB ROM cartridge, insert the cartridge into an unmodified GB, and away you go.

The GameBoy play unit has many, but not all, of the elements of a decent robot controller. You've got your ROM, RAM, and battery-backed RAM (in most game cartridges) for program and data storage. You've got your six keys on the keypad for rudimentary user input. You've got your low-res black-and-white graphics display with four shades of gray for pretty nice user output. About the only item lacking for a really good robot control is a wad of I/O, but the GameBoy's design gives you two ways to solve this lack.

First, you can attach memory-mapped devices to the GameBoy by wiring them into a game cartridge and using the address and data busses available on the cartridge connector. The information on Jeff's web page contains enough detail so that you should be able to graft on devices such as output ports and A/D converters without much trouble. This should be even easier when you take advantage of the GameBoy's memory banking scheme. Though it's a huge waste of addressing space, you could assign a single A/D chip, for example, to RAM bank 1, then switch to that bank whenever you needed to talk to the A/D.

I should point out here that the GameBoy microcomputer, unlike the Z80, does not support INP and OUTP opcodes, normally used on Intel-style chips for accessing a dedicated set of I/O device addresses. This actually works to the hacker's advantage here, since it simplifies the addition of I/O devices; you don't need access to the I/O signals to get the job done.

The second way to add I/O to the GameBoy involves the link port. This port appears on the GB as a six-pin connector; it supports a serial bus very similar to the Motorola SPI. Like the SPI, this bus is bi-directional, with data going out on one pin and coming in on another, synchronized with a data clock provided on a third pin. There is even a chip-select line that goes low whenever the GB accesses the link port. About the only real difference between the GB's link port and the Motorola SPI is speed; the SPI can hit 1 Mb per second transfer rate, while the GB is fixed at 120 μsec per bit time, or 8.3 Kb.

Transfer speed aside, you can still do all the standard SPI tricks with the GB's link port. If you need an 8-bit output port, just connect a 74hc595 to the port. If you need more output lines, daisy-chain a second or third 'hc595 to the port. You can add a serial digital-to-analog converter to the port so your GB can generate analog voltages. For a really sophisticated project, check the schematic on Jeff's page for adding two 8-bit output ports and two 8-bit input ports.

With all of this capability, the GB can indeed serve as the core to a powerful little robot. In fact, Jeff has photos of a small robot wired up to a GameBoy. His machine, a LynxMotion robotic gripper, looks especially cool with its little GameBoy controller alongside. I can imagine the GB's graphics display would make an excellent tool for showing gripper targeting, orientation, and force feedback information.

As if all this isn't enough, Jeff has created a floating-point (!) GB Basic interpreter. This interpreter executes out of GB ROM, accepts your program from any of several sources, stores it in RAM as tokens, then executes the tokens when you enter RUN or following power-up. I haven't tried his Basic yet,

but the command summary and docs remind me of the old line-numbered Basics. Jeff has included commands in his Basic interpreter for drawing simple graphics images and for controlling R/C servo motors from the GB.

According to the GameBoy Basic (GBB) FAQ, you need to modify a GB cartridge such as Donkey Kong so it holds an EPROM, burn an image of GBB into the EPROM, and install this chip in the cartridge. You can enter your GBB programs into the GameBoy in a number of ways; perhaps the easiest involves a printer-cable-to-GameBoy hack called the GB Terminal for DOS. Once you have saved your program into the battery-backed RAM on the GB as a file of tokens, you can have the GB execute your program automatically on powerup. Check Jeff's web page for details on installing the Basic, wiring up the keyboard, and developing and downloading Basic programs.

Speaking of FAQs, be sure to grab a copy of the GameBoy FAQ, originally compiled by Marat; you can find links to copies all over the web. Last updated in 1995, it still answers a lot of questions, and contains excellent detail on some of the sophisticated video, sound, and I/O functions provided by the GB CPU.

I've always viewed the GameBoy and similar game machines as simple, single-function toys; boy, was I wrong! Other machines, such as the Super Nintendo Entertainment System, are also candidates for this kind of hacking, but people like Jeff and Pascal have provided such powerful tools for the GB that you could cover a lot of robo-ground simply by sticking with the GameBoy. Next time you need a nice-looking control center for your robot, give a thought to the GameBoy. All the tools and information already exist on the web; you just need to do a little design and some typing. Have fun!

```
'
'    vector2.bas            test file for the Vector-2X compass module
'
'    This program uses a BotBoard (with a 68hc811e2) to talk to a
'    Vector-2X compass module via the SPI. This is a bare-bones
'    configuration, based on Appendix B of the Application Notes
'    (Ver. 1.06) supplied by Precision Navigation, Inc.
'
'    This program expects the following connections:
'
'    Vector-2X              68hc11
'    ------     ---
'    +5                     +5
'    GND                    GND
'    EOC                    PE0
'    P/C                    PB1
'    *SS                    *SS
'    SCLK                   SCLK
'    SDO                    MISO
'
'    The program will output the compass heading as an integer
'    from 0 to 359 to the SCI; use the TERM command in pcbug11
'    to see the information. The heading is updated once per
'    second.
'

include    "regs11.lib"


'
'   Declare variables used in this program.
'

declare    wait      ' delay counter
declare    value     ' holds returned value



'
'   Declare the RTI ISR, used to generate timed delays.
'

interrupt  $fff0           ' RTI ISR
```

Figure 2. Listing of vector2.bas.

```
if wait <> 0              ' if not done yet...
     wait = wait - 1      ' count this tick
endif
pokeb tflg2, %01000000 ' clear RTI flag
end


'
'   The main program
'


main:
pokeb baud, $30                    ' 9600 baud
pokeb sccr2, $0c                   ' turn on SCI
wait = 0
pokeb tmsk2, peekb(tmsk2) or $40    ' permit RTI interrupts
pokeb tflg2, %01000000             ' clear RTI flag
interrupts on                      ' allow interrupts

pokeb portd, peekb(portd) or $20    ' force *SS high
pokeb ddrd, %00111010              ' *SS=output
pokeb spcr, %01011111              ' CPOL=1, CPHA=1, E/32
value = peekb(spsr)               ' dummy read to clear SPI
value = peekb(spdr)               ' dummy read to clear SPI
pokeb portb, peekb(portb) or $02    ' force P/C high

print "vector2.bas"                ' announce our presence

do
          if wait = 0                 ' if time to read...
          pokeb portb, peekb(portb) and $fd ' force P/C low
          wait = 4                          ' 10 ms delay (or
so)
          do loop until wait = 0            ' time it out
          pokeb portb, peekb(portb) or $02  ' force P/C high
          waituntil porte, $01              ' wait until PE0
is high
          wait = 4                          ' 10 ms delay (or
so)
          do loop until wait = 0            ' time it out
          pokeb portd, peekb(portd) and $df ' force SS low
```

Figure 2 continued.

```
            wait = 3                    ' 5 ms delay (or
so)
            do loop until wait = 0      ' time it out
            pokeb spdr, 0               ' send a 0 to
start xfer
            waituntil spsr, $80         ' wait until xfer
ends
            value = peekb(spdr) * 256   ' get msb of data
            pokeb spdr, 0               ' send a 0 to
start xfer
            waituntil spsr, $80         ' wait until xfer
ends
            value = value + peekb(spdr) ' get lsb of data
            pokeb portd, peekb(portd) or $20  ' force SS high
            print "Value = "; value; "    ";
            outch $0d                   ' make it pretty
            wait = 250                  ' set up a 1-sec
delay
      endif
loop

end
```

Figure 2 continued.

Part VII
The 68hc11

Even though many of the articles deal with the 68hc11 microcontroller in some manner, the two articles in this section go into great detail on key subsystems of this popular chip. The first column covers the Serial Peripheral Interface (SPI), which you can use to add devices to your 68hc11 design without the burden of memory expansion. I take you through the basics, show you how to wire up a popular IC for use as a cheap output port, and conclude with plans for controlling a liquid-crystal display (LCD) with the SPI.

When you must add external memory, it helps to have some guidelines. The second article covers a variety of techniques for adding different sizes and types of memory to a basic 68hc11 design. I cover topics such as battery-backed RAM, RAM/ROM chip selects, and simple address decode circuitry. If you want to try your hand at designing an expanded-mode 68hc11, or if you want to understand how an existing design works, this chapter should be a big help.

A Look At the SPI

This column, printed in June of 1994, marked the first of many looks at the 68hc11's Synchronous Peripheral Interface, or SPI. This high-speed serial bus has long been a powerful element of the Motorola family of microcontrollers, and it appears in many of my robot designs. Figuring out how to use the SPI can stop a lot of beginners, since the Motorola documentation is more reference than instructional. In this column, I explain how the bus works, and I also provide a real-world, useful example of an SPI device.

The 74hc595 IC, when tied to the SPI, gives you a low-cost 8-bit output latch using only a few serial lines. Most robot designs need lots of output lines that only need to change occasionally, and this serial latch technique is a powerful tool to add to your toolbox.

Finally, this article shows you how to wire up one of the popular liquid-crystal displays (LCDs) to the serial latch, and how to control the LCD so your software can display text. This makes a dynamite debugging tool to add to your robot, letting you monitor your robot's decisions in real time.

Most Motorola microcontrollers (MCUs) include a high-speed serial bus, called the Serial Peripheral Interface, or SPI. You can use the SPI to move bytes of data at up to 1 Mbits per second between a 68hc11 and a compatible chip or circuit. The SPI can come in handy when you need to exchange data with another circuit, but your design has already used up most of the 68hc11's parallel I/O lines.

Additionally, the SPI can cut down on your cabling needs, since you can move eight bits of data using only three or four data wires, rather than the eight normally required. Data are exchanged across the SPI bus between two devices, known as the master and the slave. The master device provides the clock signal for timing the data exchange. The slave device accepts incoming data and supplies outgoing data, using the clock supplied by the master.

On the 68hc11, the SPI appears as four dedicated pins on port D. Port D pin 2 (PD2) carries data to the master from the slave. The signal on this pin is known as Master In Slave Out, or MISO. PD3 carries data from the master to the slave. This signal is known as Master Out Slave In, or MOSI. PD4 supplies the clock signal from the master, and carries the clock signal to the slave. A rising or falling edge on this line strobes the data bits being exchanged between the master and slave. This signal is known as SCK. PD5 is an SPI-specific select line. The master may use this line to indicate a data exchange is started. The slave uses this line to determine when an SPI transfer is started. This active-low signal is known as Slave Select, or *SS.

In general, an SPI transfer involves the following steps. First, the master drops its *SS line, informing the slave that a transfer is started. The master then issues eight pulses on the SCK line. At each pulse, the master sets the state of MOSI to reflect one bit of the outgoing data byte. It simultaneously samples the state of MISO to get one bit of the incoming data byte. At the same time, the slave sets the state of MISO to reflect one bit of the outgoing data byte, and also samples the state of MOSI to get a bit of the incoming data byte. After the eighth SCK pulse, the master raises *SS, signaling the end of the SPI transfer.

As you can see, an SPI transfer is bidirectional, with the master and slave simultaneously exchanging a data byte. Note, however, that all this bit-banging and line-twiddling occurs without any software overhead. Your program only needs to write a couple of bytes to specific I/O registers to send data, then read a register to get the incoming data.

The speed of the transfer is set by the master, since it controls the SCK line used to strobe data bits in and out. On the 68hc11, the master can pulse SCK at up to one-half the internal clock rate. For a 68hc11 using an 8 MHz crystal, this works out to one-half of 2 MHz, or 1 μsec per data bit. Note that a 68hc11 configured as a slave can receive SPI data at twice this rate, or 2 Mbits per second.

The SPI Registers

As with most 68hc11 subsystems, software uses memory-mapped I/O registers to access and control elements of the SPI. The following paragraphs describe the four I/O registers involved in the SPI. All addresses given below are the defaults used by the 68hc11 immediately after reset. Refer to the accompanying table of SPI registers (Figure 1).

The SPI control register (SPCR) appears at address $1028. The MCU uses bits in this register to

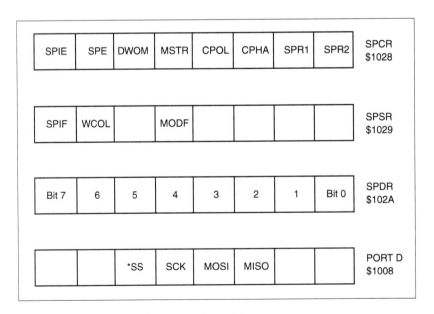

Figure 1. 68hc11 SPI registers.

enable the SPI subsystem, enable SPI interrupts, select master or slave mode, and set the SCK clock rate, phase, and polarity.

SPSR ($1029) is the SPI status register. Bits in this register indicate that an SPI transfer is complete, and whether an error occurred during the transfer. The MCU uses this register to determine if a transfer has finished, and to clear any SPI transfer errors.

The SPI data register (SPDR) appears at address $102a. The MCU writes a data byte to SPDR for transfer, and reads SPDR to get an incoming data byte. Finally, the port D data direction register (DDRD, at $1009) controls the direction of the four SPI data lines. Software must properly initialize the bits in DDRD before SPI transfers can occur.

The DDRD Bits

Setting up the SPI begins by configuring the four port D lines involved in SPI transfers. If the MCU is to be an SPI master, software must set MOSI and SCK as outputs, and MISO as an input. Additionally, the software should set *SS as an output if the MCU uses this signal as a slave select line. Writing $38 (0011 1000) to DDRD will perform these functions.

If the MCU is to function as a slave, these bits must be configured a little differently. Software must set MISO as in output, so the slave can send outgoing data. All other port D bits will be set automatically when the MCU puts itself into slave mode using the SPCR (see below). Writing $04 (0000 0100) to DDRD will perform this setup. Once configured, your software will not normally alter any of these SPI-related bits.

The SPCR Register

Next, the MCU's software must select the proper mode (slave or master) and configure the SCK signal. These functions are handled by writing to the SPCR. The SPCR bits provide great flexibility in using the SPI. I will describe the more commonly used features. For details on all the SPCR options available, consult the *Motorola 68HC11 Reference Manual*.

Setting SPCR bit 6 (SPE) enables the SPI subsystem. If this bit is zero (default following reset), no SPI transfers can occur. Setting SPCR bit 7 (SPIE) enables interrupts from the SPI subsystem. If this bit is set and if interrupts are enabled, the MCU's program will receive an SPI interrupt whenever an SPI transfer finishes or an SPI error is detected. If the SPIE bit is zero, the MCU's software must poll the SPI status register to determine if a transfer is finished or if an SPI error occurred.

The MCU software sets SPCR bit 4 (MSTR) if the MCU is to function as a master. Clearing this bit configures the MCU as a slave. Note that enabling the SPI subsystem and making the MCU a slave (SPE = 1 and MSTR = 0) automatically configures the port D pins MOSI, *SS, and SCK.

Software uses SPCR bit 3 (CPOL) to determine the idle state, high or low, of the SCK signal. Setting CPOL means that SCK idles high; clearing CPOL means that SCK idles low.

SCPR bit 2 (CPHA) selects which edge of SCK will serve as the data strobe. Setting CPHA means the MCU will use the trailing edge of SCK as the data strobe. Clearing CPHA means that the MCU will use the leading edge instead.

The interplay between CPOL and CPHA lets you create data strobes for almost any type of synchronous serial device. Just read the technical literature for the device you want, determine the clock's idle state and strobe phase, and set CPOL and CPHA accordingly. If you are designing an SPI system around two 68hc11s, you must use identical CPOL and CPHA settings on both units.

Finally, SCPR bit 1 (SPR1) and bit 0 (SPR0) determine the frequency used for SCK by the master. The frequency is expressed as the internal clock (E) divided by a scalar. The four available scalars are:

SPR1	SPR0	Scalar
0	0	2
0	1	4
1	0	16
1	1	32

Thus, if the MCU uses an 8 MHz clock and software clears both SPR1 and SPR0, the SCK frequency will be 2 MHz divided by 2, or 1 MHz. Note that the slave does not generate an SCK signal. The states of SPR0 and SPR1 are meaningless on MCUs configured as SPI slaves. As with the DDRD SPI-related bits, your software will normally write to the SPCR just once, prior to the first SPI transfer.

The SPSR Register

The software uses bit 7 (SPIF) of the SPSR to test for a finished transfer. This bit is set by the MCU hardware whenever a transfer completes. This bit is automatically cleared whenever the incoming data byte is read from the SPDR.

Therefore, software in the master MCU exchanges data via the SPI by first starting a transfer, then polling the SPSR until SPIF goes high. Reading the incoming byte from the SPDR clears the SPIF, preparing the SPI for the next exchange.

In programs using the SPI interrupt, the master simply starts a data transfer, then continues with normal processing. When the transfer completes, the MCU triggers an SPI interrupt and control jumps to the address stored in the SPI interrupt vector at addresses $ffd8-$ffd9. The software responsible for servicing an SPI interrupt must read the incoming byte from SPDR to clear the SPIF, save the byte somewhere useful, then return from the interrupt.

Note that slave MCUs running with the SPI interrupt enabled don't even have to start a transfer. The incoming data byte will trigger an SPI interrupt automatically. The slave's interrupt service software must fetch the incoming byte and save it, then return from the interrupt.

A Closer Look

An example might help to clarify some of this. Suppose we want to configure an MCU as a polled master, using an SPI clock rate of E/2, with data strobe occurring on a high-going trailing edge of SCK.

Based on the above, we would write the value $5c (0101 1100) to the SPCR. Refer to the accompanying SPI transfer timing diagram (Figure 2). Software in the master MCU signals the start of an SPI transfer by writing a zero to port D, bit 5, bringing *SS low. The software then writes the outgoing data byte to the SPDR.

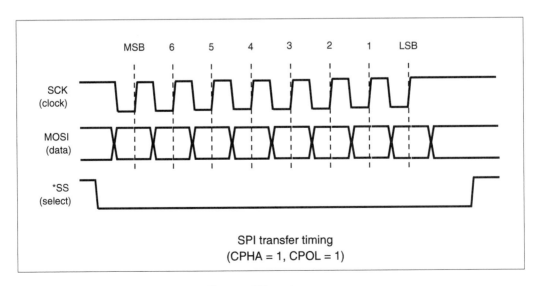

Figure 2. SPI transfer timing.

Now the 68hc11 hardware takes over, transferring each bit of data out the MOSI line and simultaneously reading each bit in from the MISO line. Each bit exchange takes one μsec, using the E/2 clock rate; the entire exchange requires only eight μsecs.

After the MCU has sent its outgoing byte and fetched the incoming byte, the hardware sets the SPIF bit in the SPSR register. Software running in both MCUs polls the SPSR register, looking for a one in the SPIF bit. When the one appears, the transfer is complete and the software can fetch the incoming byte by reading the SPDR register. Software in the master MCU then raises *SS, indicating the end of the transfer.

The diagram shows the relationships between the SCK, MOSI, and *SS signals for the master MCU in the example case. Sending each bit of data to the slave occurs in the following sequence.

The MCU's hardware first drops the SCK line, then sets MOSI high or low, depending on the value of the outgoing bit. The hardware then raises the SCK line and latches the value of the incoming MISO line. The value on this line at the instant SCK rises is the level (zero or one) of the incoming bit in the slave's data byte. The slave MCU uses the SCK pulses provided by the master MCU to time incoming data bits. The slave sends data out on MISO and brings data in on MOSI.

I realize some of this is pretty heavy-duty, especially compared to the much simpler parallel I/O normally used. Perhaps a real-world example of using the SPI will help.

Serial In, Parallel Out

The Motorola 74hc595 is a 16-pin IC that converts a serial data stream into eight digital output lines. It works well with the SPI, though it will work with other serial data busses also. Refer to the accompanying diagram of the 74hc595 for details (Figure 3).

This chip contains an eight-bit shift register that receives serial data on its input pin. The eight outputs of the shift register are internally connected to an eight-input data latch. The latch's eight outputs appear at the 74hc595's output pins.

The 74hc595 receives the incoming serial clock on pin 11, and the incoming data on pin 14. Data are transferred to the 74hc595's output pins on a low-to-high transition on pin 12.

I have supplied corresponding SPI labels to these pins for reference. Note that the 74hc595 also provides a serial data out line, which could be tied to the master's MISO pin. This line serially carries the states of the eight shift register outputs, and could be used to determine the last data transferred to the 74hc595.

You can use the 74hc595 as an SPI-driven 8-bit output port by providing just a few connections to the 68hc11. You must supply MOSI, *SS, and SCK, as well as power and ground. You may optionally connect the 74hc595's serial data out to MISO, though I haven't found a need for this yet. Additionally, you must configure your 68hc11 master MCU to use CPOL equal to one and CPHA equal to one. This provides the active high, trailing-edge strobe needed by the 74hc595.

This creates an 8-bit latched port capable of supplying 10 to 20 mA of current on each output pin, with a Vcc of up to 6 VDC. The chip is fully compatible with the 68hc11 in this configuration.

To get even more latched output lines from the SPI, you can cascade two 74hc595s. Simply connect pin 9 of the first chip to pin 14 of the second. Also tie both pin 11s together and both pin 12s together. In this arrangement, the 68hc11 MCU performs two SPI transfers, leaving *SS low throughout. After the second transfer finishes, the MCU raises *SS, latching all 16 data bits simultaneously.

You can repeat this cascading layout, using as many 74hc595s as you like. Just remember that the master MCU should not raise *SS until all SPI transfers have finished. Otherwise, the output lines of all 74hc595s will momentarily carry data intended for other devices.

Liquid Crystal Displays

LCDs make great output devices for robots. You can display data and debug information as your program

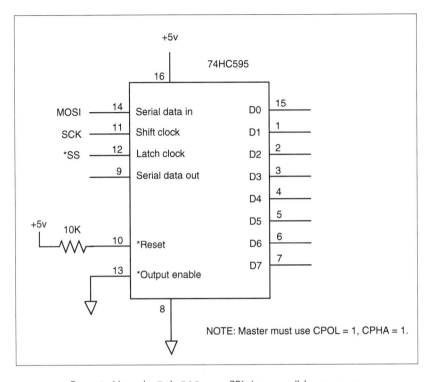

Figure 3. Using the 74hc595 as an SPI-driven parallel output port.

evolves. You can also provide information and requests to the humans who interface with your 'bot. Mail order surplus firms carry these displays in various formats and sizes. I find the two line by 20 character displays most useful. These are usually available with characters nearly 0.5 inches tall, which are very easy to see from moderate distances.

You can also get displays in 2 by 40, 2 by 16, 1 by 16, and many other formats. Prices range from $10 to $40, depending on where you shop. Try to stay with the larger character size, if possible.

All of these LCDs use a simple, 8-bit parallel interface. The signals are usually brought out on a 14-pin connector block along the bottom edge of the display. Refer to the accompanying diagram (Figure 4) and pin functions list (Figure 5) for details.

LCD pins 1 and 2 connect to ground and +5 VDC, respectively. A voltage on pin 3 can be used to control the display contrast. Normally, you can con-

nect a 10K or 20K trim-pot between +5 VDC and ground, then connect the pot's wiper to pin 3. This will provide adequate display contrast. If you don't want to bother with the pot, you can tie pin 3 directly to pin 1. This will give a fixed contrast that might be acceptable.

Pin 4 (RS) selects the LCD's data register if high, or the command register if low. Pin 5 (R/*W) indicates a read of the LCD's data bus if high, or a write to the LCD's data bus if low. Pin 6 (E) is the enable line. It is normally low, but goes high when the controlling MCU wants to exchange data with the LCD. Note that the states of pins 4 and 5 dictate the type of function performed as E changes state. Data on these pins must be valid on the rising edge of E. Pins 7 through 14 transfer eight bits of parallel data between the MCU and the LCD. For a write operation, data from the controlling MCU must be valid on the falling edge of E.

Figure 4. A typical LCD.

The sequence for writing a byte of data to the LCD is straightforward. The MCU sets RS high to write data or low to write a command. It also sets R/*W low to perform a write operation. Next, the MCU brings the E line high. It puts the proper data out on the data bus, then brings E low. Note that if your application only involves writing data to the LCD, you can leave the R/*W line connected to ground.

The controlling MCU can write to the LCD command register to set characteristics such as cursor visibility and format, number of display lines, font style, and cursor location. Any data written by the controlling MCU to the LCD data register will normally appear on the LCD's screen. Note that it is possible to write data to locations not visible on the LCD's screen. Consult data sheets from any of the larger LCD makers for full details on available LCD commands and their effects. Companies such as Optrex and Hitachi make these displays, and their distributors can supply data sheets. You can sometimes get copies of the appropriate sheets from larger mail order houses.

Hooking Up the LCD

You normally connect an LCD to a MCU one of two ways. You can hook the LCD directly to the MCU's address and data busses, provided the MCU's clock is slow enough. Most of these LCDs need E to remain high a minimum of 450 nsecs, a pulse width that can prove unacceptably slow for some MCUs.

You can also drive the LCD with a latched output port, such as an 82c55. With this technique, you need to simulate RS, R/*W, and E, using selected output lines. This will create cycle times many microseconds long, but the LCD has no maximum cycle time, so the long delays are not a problem.

Simple addition indicates that this technique requires at least 10 output lines, which is very inconvenient. Fortunately, the LCD designers planned ahead, and provided a way to run the LCD using a 4-bit data bus. You can use the smaller bus with a latched output port by providing data to the LCD using DB4 through DB7; the lower four data bits aren't used. You then use two of the remaining output port lines to simulate E and RS; remember that R/*W is tied low for write-only applications.

This arrangement lets you drive the LCD from a single 8-bit port, using only six output lines. You would have to use a seventh line and a bidirectional port if you wanted to read data from the LCD.

Pin	Function
1	Vss (ground)
2	Vcc (+5 VDC)
3	Vee (display contrast)
4	RS (register select)
5	R/*W (read/write)
6	E (enable)
7	DB0
8	DB1
9	DB2
10	DB3
11	DB4
12	DB5
13	DB6
14	DB7

Figure 5. LCD pin functions.

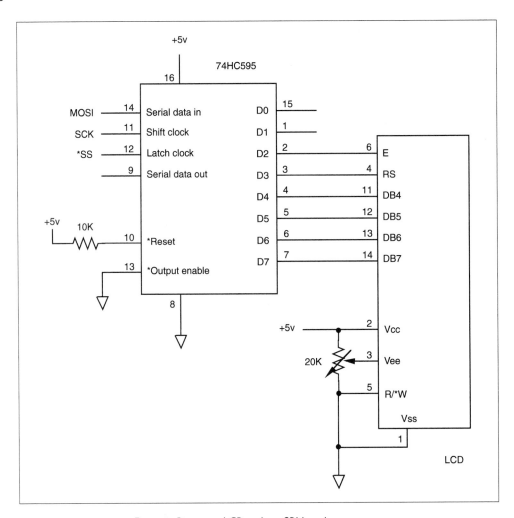

Figure 6. Driving an LCD with an SPI-based output port.

Putting It All Together

Since we only need an 8-bit output port to control the LCD, we can hook it up to the 74hc595 and create an SPI-driven liquid-crystal display. Refer to the accompanying schematic (Figure 6).

Here, I have hooked the high four data bits of the LCD's data bus to the high four output bits of the 74hc595. I am also using D3 of the 74hc595 to supply the LCD's RS signal, and D2 to supply the E signal. The low four bits of the LCD's data bus are not used.

Bits D0 and D1 of the 74hc595 are available for other functions, such as driving an LED or beeper.

Data are sent to the LCD in two groups of four bits each. The first transfer writes the high four bits, or nybble, to the LCD. The second transfer writes the low nybble. Each nybble transfer requires several SPI operations, to simulate the actions of E, RS, and the data lines. Refer to the accompanying chart of SPI nybble operations (Figure 7).

Note that transferring a full byte requires two nybble sequences. Also, the software must set up

D7	D6	D5	D4	D3	D2		
0	0	0	0	0	0	RS = 0, E = 0	Writing a command nybble to the LCD
0	0	0	0	0	1	RS = 0, E = 1	
D	D	D	D	0	1	Data, RS = 0, E = 1	
D	D	D	D	0	0	Data, RS = 0, E = 0	
0	0	0	0	1	0	RS = 1, E = 0	Writing a data nybble to the LCD
0	0	0	0	1	1	RS = 1, E = 1	
D	D	D	D	1	1	Data, RS = 1, E = 1	
D	D	D	D	1	0	Data, RS = 1, E = 0	

Figure 7. SPI transfers for sending nybbles to the LCD.

the second transfer properly by moving the low nybble up to the high nybble of the data byte.

The only tricky bit remaining is the LCD's power-up initialization ritual. Most manuals do not explain this very well, and the ritual shown in the accompanying diagram is based on several previous projects (Figure 8). Note that the first four transfers are nybbles, not bytes. All data in the initialization ritual are written to the LCD's command register, not the data register.

After your software completes the initialization ritual, the LCD will be in 4-bit mode, 2-line format, using 5 by 7 dot characters. The LCD screen will be cleared, the cursor turned off, and the cursor positioned at character location 0.

Characters written to the LCD's data register will now appear on the top line of the LCD's screen. As each character is written to the LCD's data register, the cursor location will advance one position. To begin writing data to the second line of

Wait at least 15 msecs after power-up, then send a nybble of $03

Wait at least 4 msecs, then send a nybble of $03

Wait at least 100 usecs, then send a nybble of $03

Wait at least 4 msecs, then send a nybble of $02

Wait at least 1 msecs, then send a byte of $28

Wait at least 40 usecs, then send a byte of $08

Wait at least 40 usecs, then send a byte of $01

Wait at least 1 msecs, then send a byte of $06

Wait at least 40 usecs, then send a byte of $0c

Figure 8. Power-up initialization ritual for the LCD.

the LCD's screen, you must send the byte $c0 to the LCD's command register. This positions the LCD cursor at character location $40, which is column one of the second line. To clear the display and return to character location 0, you must write the byte $01 to the LCD's command register, then wait at least 2 msecs.

I have described only some of the formats and functions of these common display devices. For more details, consult the supplier's literature.

Winding Up

I have developed a library of tiny4th routines that simplify controlling the LCD via the SPI. The source for these routines is available in the current release of tiny4th, version 2.1. For those of you who prefer C to Forth, the tiny4th source files are well commented, and you shouldn't have any trouble converting them.

A quick note on the Pacific Science Center (PSC) robots project, discussed in "Designing an Interactive Robot Display." If you want to build the beacon receiver circuit for use in your own robot, be sure to add a .01 µf capacitor between the phototransistor and the LM380's input. This blocks the DC voltage generated by the phototran-sistor in high ambient light. Also, add a 0.1 µf capacitor between Vcc and GND at the LM380, to bypass any electrical noise on the power supply.

We ended up replacing most of the servo motors on the PSC robots; the heavy use wears them out after a few weeks. We are now using the bearing replacement kits, available from Tower Hobbies, to upgrade our Futaba S148 servos. Preliminary results look excellent.

Kevin Ross rewrote the tiny4th code that drives the robots, essentially replacing the beacon electronics with software. This was necessary because the robots were getting too many false triggers by confusing the beacon signal with the object detect signals.

The new system works quite well, and Kevin still has several bytes of code space left in the 2K EEPROM of the 68hc811e2s we used. Eventually, I'll write something about the final PSC code, as Kevin's software solution looks both elegant and robust.

68hc11 Memory Expansion

The basics of adding memory and I/O devices to a 68hc11 system can prove frustrating. On the surface, the technical books make it seem simple, but actual examples aren't easy to come by. In this column, published in July of 1994, I go through the design elements of such expansions. Whether you want to add an 8K EPROM, a pair of 32K RAM chips, or an 82c55 parallel port, you'll find enough info here to get the job done.

This column also touches on related elements, such as adding battery-backup to static RAM chips, and using PCBUG11, Motorola's must-have 68hc11 development tool, to program 8K EEPROM devices.

Finally, I describe an 8x751 development kit offered by Philips Semiconductor. Priced at less than $50 at the time, it was a cheap and well-designed tool for starting out on the Intel family of chips.

The 2K of EEPROM in a Motorola 68hc811e2 microcontroller (MCU) can hold enough code for fairly complex robots, but sometimes you just need more code space. You can gain that extra space many ways.

Adding external RAM or EPROM is a popular choice, for several reasons. First of all, you can add up to 64K bytes of memory, for those really BIG projects. 32K chips, both EPROM and static RAM, are cheap and readily available, helping to keep system cost down. And the denser chips mean you can fill the MCU's address space with just two 28-pin devices, keeping your board space down.

Expanded Mode

In the old days, running a 68hc11 in expanded mode took a lot of board space and chips. You needed an 8-bit latch, such as a 74hc573, to demultiplex the address and data busses. You also needed some glue logic to properly combine the R/*W and E signals, as well as logic for creating chip select signals for your memory and I/O chips. Then, you had to throw in the crystal and its passives, the RS-232 level conversion for your serial port, reset circuitry, pullups, and other odds and ends. And that didn't even count the memory and I/O chips!

Your finished design could easily reach six or seven chips. Even simple expanded-mode 68hc11 systems were so tedious to build that making one by hand was a pain, and I needed a very serious reason to do a second one.

Today, however, CGN's 1101 system board carries the 68hc11 socket and all the expansion circuitry on a single PCB about 2 inches by 3 inches in size. Since it uses SMT components, it costs less and takes up less space than any comparable circuit you could likely develop.

You can think of the CGN 1101 as a super-component. It is a working, tested 68hc11 board, lacking only the MCU chip, which you supply. All you have to do is wire up power and your memory or I/O devices. Its wirewrap connections make installation on any perfboard a breeze. I reviewed two of CGN's boards in "The Rapid Deployment Maze" and "Quick and Easy 68hc11 Expansion." If you haven't heard of this company's products before, contact CGN directly or check out my previous articles.

One weekend, I decided I needed a new computer board for my next robot project. I grabbed a CGN 1101, a large Radio Shack experimenter board (276-147), and some wirewrap sockets. By Sunday night, my expanded mode 68hc11 system was up and running.

Design Elements

Anytime you design an expanded-mode system, you have to deal with chip select signals. These signals divide the MCU's address space into blocks, assigning each block of addresses to a specific memory or I/O device.

You need to combine the 68hc11 E signal with your chip select logic. E serves as the 68hc11's master system clock. This combination ensures that your chip select signal is valid at the proper point in the MCU's clock cycle. Also, you must design circuitry to properly condition the 68hc11's R/*W signal. This usually means combining R/*W with E.

Adding One 32K Device

Perhaps the simplest expanded-mode memory system involves a single 32K memory device, such as a 27c256 EPROM. Since the 68hc11 wants to find its reset vector at $fffe, wiring the 27c256 to appear in the top 32K of memory is a natural choice.

Part A of the included circuit is about as simple as it gets (Figure 1). It generates an active-low chip select (*ROM) for any address in the upper 32K of the 68hc11's memory space. It also creates properly conditioned read-enable (*RE) and write-enable (*WE) signals.

Most memory devices that you might add to your 68hc11 system will use *WE, but not *RE. I included the logic for *RE for those rare times when you might need to use an Intel-style I/O chip. These chips require separate active-low signals to read and write data.

Part B performs the same functions, using signals found on the CGN 1101 board (Figure 1). Note that you do not need the R/*W conditioning circuitry. The 1101 pin labeled WE is a properly conditioned, active-low write signal. The 1101 pin labeld RE serves as a read-enable signal. Remember to follow good CMOS practice and tie all unused 74hc00 inputs high or low, to prevent excessive current drain. *ROM and *WE are the two signals that require the most circuitry to create and, especially for the CGN hookup, that isn't much. The rest of the wiring for a 27c256 involves connecting the appropriate pin of the IC to the proper CGN 1101 pin.

The accompanying table gives the connections for wiring 32K, 8K, and 2K JEDEC pinout devices (Figure 2). The 27c256 32K EPROM is such a JEDEC device, as are most 28-pin static RAMs. The 6116 2K static RAM is also a JEDEC device. Note, however, that these older devices are 24-pin DIP ICs.

I normally don't bother to draw up a schematic for adding memory to a homebrew 68hc11 system. I just design the write-enable and chip-select logic. The address and data bus circuitry follows the wiring table shown in Figure 2.

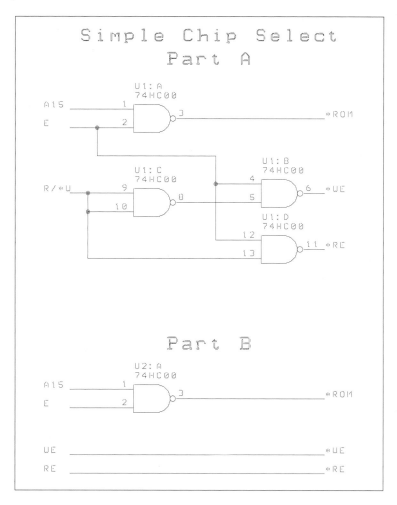

Figure 1. Some simple chip select circuits for a single 32K device.

Adding Two 32K Devices

Adding two 32K memory chips to a 68hc11 requires more complex circuitry. Here, you need to create chip-select signals for all addresses in the upper half and the lower half of the address space.

Trying to build the two chip-select signals and the write-enable signal out of NAND gates would require two ICs. The 74hc139 provides a one-chip solution. Refer to the accompanying schematic of RAM/ROM selects (Figure 3).

The 74hc139 is a dual 1-of-4 demultiplexer (demux). I'll describe the function of the device's first half (pins 1 through 7). The second half functions identically. This chip uses its two inputs, A and B, to decode one of four possible outputs. These outputs, labeled Y0 through Y3, correspond to the four arrangements of one and zero on the two inputs. Input A (pin 2) is the less-significant input and B is the more-significant input. Thus, a one on A and a zero on B forms a binary 01. This in turn selects output Y1.

Signal	32K 28-pin	8K 28-pin	2k 24-pin	Notes
A0	10	10	8	
A1	9	9	7	
A2	8	8	6	
A3	7	7	5	
A4	6	6	4	
A5	5	5	3	
A6	4	4	2	
A7	3	3	1	
A8	25	25	23	
A9	24	24	22	
A10	21	21	19	
A11	23	23		
A12	2	2		
A13	26			
A14	1			
D0	11	11	9	
D1	12	12	10	
D2	13	13	11	
D3	15	15	13	
D4	16	16	14	
D5	17	17	15	
D6	18	18	16	
D7	19	19	17	
R/*W	27	27	21	WE on CGN 1101
*CS	20	20	18	Must include E
	22	22	20	*OE; see text
Vdd	28	28	24	+5 VDC
Vss	14	14	12	GND
		26		CS1; pull to +5
		1		No connection

Figure 2. Table of pinouts and connections for JEDEC standard memory devices.

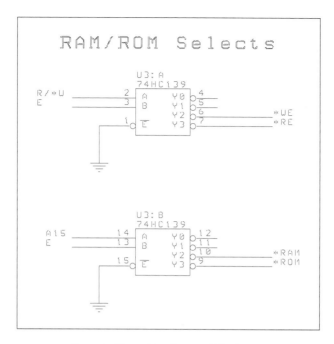

Figure 3. Chip selects for two 32K devices.

The selected output goes low, while all other outputs remain high. In the above example, pin 5 will go low while pins 4, 6, and 7 remain high. This chip lets me easily build the needed chip-select and write-enable logic. I simply use R/*W and E as inputs, then use the appropriate outputs as the *RE and *WE signals.

I arbitrarily wired R/*W to the A input and E to the B input. Since I want *WE to go low whenever E is high and R/*W is low, I need to use the output that corresponds to a binary 10 on the two inputs. This output is Y2, or pin 6.

Similarly, *RE must go low whenever E is high and R/*W is high. This creates a binary 11 on the inputs, which causes Y3 (pin 7) to go low. I use the second half of this chip in a similar manner to create my two chip selects. The *ROM signal must go low when A15 is high and E is high. This forms a binary 11 on the inputs, and selects Y3, or pin 9.

The *RAM signal must go low when A15 is low and E is high. This combination creates a binary 10 on the inputs, which selects Y2, or pin 10.

This means that *ROM goes low for all addresses in the range $8000 through $ffff. Similarly, *RAM goes low for all addresses in the range $0 through $7fff. As before, if you use the CGN 1101's *WE signal, you don't need the write-enable part of this logic. In this case, remember to pull the A and B inputs high, to prevent excessive current drain in the unused half of the 74hc139.

Adding 8K Devices

Adding memory in 32K chunks is easy, but not always the best choice. Depending on your robot's requirements, you might be better off adding 8K chips instead. Suppose your robot only needs a small amount of additional RAM, and your code will fit comfortably in 8K (or 16K) of EPROM. More importantly, you decide you must add a parallel I/O chip, such as an 82c55, to your system.

Adding the 82c55 to a 68hc11 system that already has 64K of memory can prove awkward. You

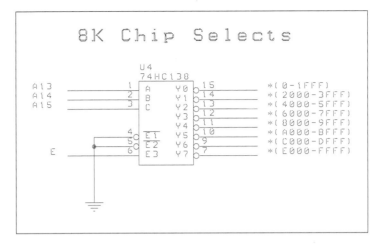

Figure 4. Chip select logic for 8K devices.

will need to carve out a chunk of address space from one of the memory chips, to make room for the I/O addresses. This can make the design of your chip-select logic more complex.

I find that going to 8K memory devices simplifies this design. My latest 68hc11 system currently supports 16K of memory and an 82c55. The I/O appears in its own address space, without conflicting with any of the memory devices. Again, I needed only one IC to create the chip-select signals. Refer to the accompanying schematic of 8K chip selects (Figure 4).

Here, I am using a 74hc138 demux to convert the top three address lines into one of eight possible select lines. I connect E to pin 6, which is a positive enable line, and ground pins 4 and 5, the two negative enable lines. I also wire the top three address lines to the three inputs. Note that address line A13 goes to the least-significant input (pin 1) and A15 goes to the most-significant input (pin 3).

The 74hc138 provides eight output lines, Y0 through Y7. If the three enable lines have the proper signals on them, then one of the outputs will go low, based on the levels of the three inputs. Consider the results if E is high, A13 and A14 are low, and A15 is high.

The enable lines all have the proper signals, so the 74hc138 treats its three inputs as a 3-bit binary

number. The example levels generate a binary 100 on the input lines. The 74hc138 will then pull output Y4 (pin 11) low and leave the other outputs high. Note that if the enable pins do not have the proper signals, all outputs go high.

This circuit creates eight chip-select lines, one for each 8K of memory space. The schematic lists the addresses assigned to each output. Thus, output Y7 (pin 7) goes low whenever the MCU accesses an address in the range $e000 through $ffff.

Note that I didn't create a write-enable signal this time. The CGN 1101 provides that signal for you. If you aren't using the 1101, you can use one of the earlier write-enable circuits.

I find this 74hc138 circuit very useful. I can quickly add I/O devices, by simply tying the device's chip-select line to one of the 74hc138's outputs. The I/O device then automatically appears at the selected address range.

More I/O

The accompanying schematic shows how easy it is to add an 82c55 parallel I/O chip to a 68hc11 system (Figure 5). By wiring in just 14 connections, you end up with three bidirectional parallel ports hooked directly onto the 'hc11's bus.

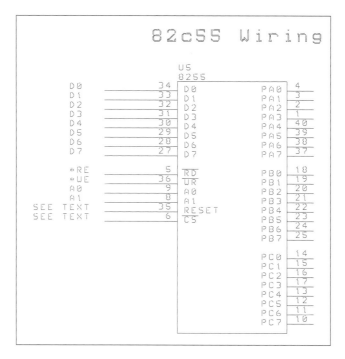

Figure 5. Schematic for wiring an 82c55.

The 82c55 has some limitations to it that you need to keep in mind. It can only handle about 2.5 mA per I/O pin, compared to the 20+ mA of a 68hc11 I/O pin. And you cannot arbitrarily assign the direction of each and every I/O line, as you can with some of the 68hc11 ports. Still, at $5 each, it is a cheap, simple way to add 24 I/O lines to your system.

Most of the connections between the 82c55 and the 68hc11 are straightforward. Lines D0 through D7 form the data bus, and are direct connections. Lines A0 and A1 are the two lowest address lines from the 68hc11 bus. The 82c55 uses these to access any of its four internal registers. *RE and *WE act as read-enable and write-enable signals. You can use any of the earlier circuits to generate these signals. Note that the CGN 1101 board provides these signals already, on pins labeled RE and WE.

You will normally keep the RESET pin low, and bring it high momentarily to reset the 82c55. The

CGN 1101 board provides an active-high reset signal, labeled RST, on one of its pins. Or you can invert the 68hc11's reset signal using a 74hc04 or 74hc00. As a last resort, you could just ground the RESET pin. This means your system software will have to set up the 82c55 fairly promptly, as it may power up in an unpredictable state.

The *CS signal is an active-low input for chip select. You can use any of the outputs from the 74hc138 circuit to assign a block of four addresses to the 82c55. Note that some choices are better than others, primarily because the chip-select logic "partially decodes" the 82c55's addresses.

This phrase means that the 82c55 registers appear at more than just one set of four addresses. In fact, if you use the 74hc138 circuitry, the 82c55 registers will repeat throughout an 8K address space. This multiple appearance happens because the addressing logic only uses five address lines to determine the 82c55's location. Only A15, A14, A13, A1, and

A0 have any effect on address selection; all other address lines are "don't care."

This means that using the 74hc138's Y7 output as the 82c55 chip select would be a bad idea. Any access of your EPROM would cause a bus contention with the 82c55, as both devices would activate at the same time.

Y0 also makes a bad choice as the 82c55 chip select, for a different reason. This line goes low when the software accesses any addresses in the $0 to $1fff range. The 68hc11's on-chip RAM and I/O registers usually occupy this area of the address map.

When I used Y0 as the 82c55 chip-select, writes to 68hc11 RAM or I/O registers corrupted data previously written to the 82c55. Apparently, the 68hc11 drives the external busses when accessing these addresses. Even though it only uses its internal busses for reading data from its RAM or I/O registers, it still generates the required addresses externally. This causes the 82c55 to respond, reading garbage data from the data bus during write operations.

Note that the 68hc11 reference manual states "The internal address decode circuitry automatically protects against conflicts... between an internal and external resource." Based on my experience, this protection applies only to the 68hc11 processor. External resources in a conflicting address range have no protection against writes to internal resources.

Therefore, you should assign an unused 74hc138 output, such as Y1, to be your 82c55 chip-select line. This seems wasteful, since you are mapping four I/O registers throughout an entire 8K address space, but you have little choice with the simple decoding circuitry shown. As noted, Y1 corresponds to the address range $2000 through $3fff.

The 82c55 is simple to use. It contains four registers, appearing sequentially at its assigned address. In the above example, the 82c55 first appears at address $2000, so the following discussion will use that as the 82c55's base address.

The first register ($2000) reads and writes data to port A, an 8-bit parallel port. The second register ($2001) reads and writes data to port B, an 8-bit par-

allel port. The third register ($2002) reads and writes data to port C, two 4-bit parallel ports. The upper and lower halves of port C may be independently configured as input or output ports.

The fourth register ($2003) is an 8-bit command register. Data written to this register controls the 82c55's configuration. Ports configured for output latch values written to them by the software. Reading these output ports returns the latched value, regardless of the loads present on the output pins.

This device supports three modes of operation. The most useful mode (at least to me) is mode 0. Mode 0 configures the 82c55 as three parallel I/O ports. Your software can place the 82c55 in mode 0 at any time, by writing the proper data to the command register. Refer to the accompanying table of 82c55 mode 0 options (Figure 6). For example, writing a $90 to the command register would configure port A as an input port, port B as an output port, and both halves of port C as output ports. I have intentionally omitted descriptions of the other 82c55 modes. For information on all available modes, refer to the Intel (or other vendor) literature.

Non-Volatile Memory

EPROMs are very handy devices to add to an expanded-mode 68hc11. They are non-volatile, in that they retain their data almost forever, even if you cycle power repeatedly. The major drawback with EPROM, of course, is the long delay between each programming cycle. It can take up to 15 minutes to erase a chip, then up to a minute or more to reprogram it.

Bit #	Function
7	1 = set mode option
6-5	00 = mode 0
4	0 = port A is output, 1 = input
3	0 = upper half of port C is output, 1 = input
2	0 = mode 0
1	0 = port B is output, 1 = input
0	0 = lower half of port C is output, 1 = input

Figure 6. Table of 82c55 mode 0 options.

Which brings up the expense and hassle of an EPROM programmer. Most units capable of programming the 27c256 chips will run you about $120 to $170. They usually require installation in your PC, which means you have to burn up a card slot. (Those of you trying to do robotics on a Mac have probably already discovered the expense of serial-based EPROM programmers.)

Another problem with EPROMs concerns the physical problems with getting the erasing UV light to the chip. Most UV erasers have room for the chip, but not for your circuit board (PCB). This means you have to pull the chip out of its socket, erase it, put the chip in the programming fixture, program it, then plug the chip back into its socket.

It doesn't take too many cycles of this before the socket's pins begin to weaken, losing their ability to make a gas-tight connection with the EPROM's leads. Eventually, your circuit will begin to show weird failures that can be very difficult to track down.

Static RAM solves most of these problems, but introduces a big one in exchange. It can be programmed in-circuit at very high speeds. This eliminates the need for an external programmer and removes the problem of socket fatigue. Unfortunately, static RAM is volatile storage; its data disappears when power goes away, even briefly.

Solving this volatility issue yields a memory device that is cheap, easily reprogrammed, and non-volatile. Several methods exist to make static RAM non-volatile (or nearly so) and deserve some discussion. The simplest method involves never switching the RAM's power off. This can take several forms.

For example, the Rug Warrior circuitry, discussed in the book *Mobile Robots*, by Joe Jones and Anita Flynn, leaves the 62256 32K static RAM permanetly wired to the battery holder. Almost all other computer circuitry on the board routes through the power switch.

The only other component that receives battery power at all times is the 74hc10 IC, used to generate the chip-select signal for the static RAM. Keeping the 74hc10 alive is vital to protecting the RAM's data. If the chip-select signal going into the static RAM were not pulled high by the 74hc10's output, the RAM

would see electrical noise on its pins as commands to write data. This in turn would trash the data.

So what happens when the batteries run down? The Rug Warrior's circuit includes a 47 µf capacitor across the battery terminals. With the computer switched off, this capacitor provides enough voltage for you to change the batteries quickly without losing data.

Another technique for preserving the contents of a static RAM uses a small on-board battery of at least 3 volts. The battery, usually lithium, connects to the Vdd pin of the static RAM through a diode. See the accompanying battery backup schematic (Figure 7).

Here, the 1n34 diodes isolate both the +5 and +3 VDC power sources. The static RAM will see whichever voltage is higher. When the system's power switch is turned off, the battery will supply +3 VDC, less the forward voltage of the diode. This works out to at least 2.5 volts. Note that most static RAMs require a minimum of 2.0 volts to retain data.

If you cannot locate 1n34 germanium diodes, you can usually use 1n4148 silicon devices. These have a forward drop of about 0.7 VDC, leaving you with about 2.3 volts into the RAM. If necessary, test a few diodes to find one with a low forward voltage for use in your circuit.

The circuit distinguishes between the Vdd supplied to the static RAM and the +5 VDC supplied to all other ICs in the circuit. When the power goes off, you want only the static RAM to draw current from the battery. Similarly, the 10K pullup resistor must connect between the RAM's chip-select pin and the RAM's Vdd. This ensures that the chip is deselected whenever the +5 VDC is switched off, preventing accidental data corruption. Finally, C1 smooths out any power fluctuations that might occur as the +5 VDC is switched off.

You can avoid using this circuit by installing a static RAM chip with built-in battery backup. SGS-Thompson's ZeroPower RAM devices contain the battery and all switching circuitry to maintain the contents of the on-chip RAM. The MK48z08b-15 device is an 8K static RAM in a JEDEC standard package.

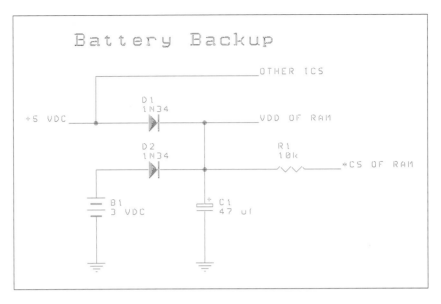

Figure 7. Battery backup schematic.

Using battery-backed static RAM speeds up development cycles something wonderful. It only takes about 1 minute to move a newly-compiled piece of code into my 68hc11 target and start testing.

One problem with testing code, however, occurs when your new software "runs away." The errant code can easily trash the RAM, leaving you with no clue as to where the problem occurred. This quickly leads to a frustrating series of download-test-crash cycles. The addition of a write-protect switch on your battery-backed RAM turns it into a pseudo-EPROM. You simply throw the switch, and the RAM becomes write-protected. Refer to the accompanying write-protect schematic (Figure 8). Now, runaway code can't trash the RAM. When your code takes off, just reset the 68hc11, install a breakpoint in the RAM, write-protect the RAM, and try again.

I usually use PCBUG11 to move my test code into the 68hc11's static RAM. The technique is simple, but involves a small trick worth discussing. I start by setting jumpers or throwing a switch to put

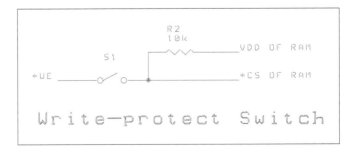

Figure 8. A simple write-protect switch.

the 68hc11 into special bootstrap mode. Electrically, this requires that both MODA and MODB pins on the 68hc11 are grounded.

Then, I reset the 68hc11 and fire up PCBUG11 on my PC. PCBUG11 downloads its communication code (called the talker) into the 68hc11, and I am ready to move my code over. But the 68hc11 is still in bootstrap mode. This means its external address and data busses are disconnected. Any attempt to read or write to external addresses will fail.

Fortunately, you can activate the external busses by simply changing one bit in the proper 68hc11 I/O register. This is MDA, bit 5 of the HPRIO register at $103c. With the 68hc11a1 chip, this usually translates into the following PCBUG11 commands:

control base hex
ms 103c f5

Now the external busses are active. All that remains is to **LOADS** my code into the 68hc11 and start running.

EEPROM

The 28c64 electrically-erasable PROM offers a different form of non-volatile memory. This chip contains 8K bytes of fast EEPROM in a JEDEC standard pinout. Prices range from $3 to $9 each, depending on where you shop.

This chip simply plugs into a 28-pin DIP socket wired per the table in Figure 2. Reads of this device act just like reads of EPROM or static RAM. Writing to this device, under certain conditions, changes the contents of the selected address. The chip retains its data indefinitely, even if the power is off.

The certain conditions mentioned above concern the wiring of pin 22, *OE. You can usually connect *OE to ground on almost any JEDEC device and it will function properly. I routinely wire static RAMs (such as the Sony CXK5864) this way.

But the 28c64 uses the state of *OE to control its writing of data. For details, consult the data sheets for the Microchip 28c64a device. Basically,

you can wire the 28c64's *OE to the *RE signal described in these circuits to reliably control the write operation. You can also use the RE signal on the CGN 1101 board.

When wired this way, writes to the 28c64 take about one msec to complete. This means that your software can store a byte in the 28c64, but must wait at least one msec before a read operation returns that value.

Oddly enough, the 28c64 works perfectly with PCBUG11. You simply use PCBUG11's EEPROM command to flag the 28c64's address range as EEPROM. Then you use LOADS to transfer your code into the 28c64. PCBUG11 will try to write the data using the 68hc11's internal charge-pump resource. This doesn't matter, since the 28c64 is external and therefore unaffected.

But PCBUG11 will also wait at least 12 msec after writing a byte to verify that address. This delay is more than enough for the 28c64 to complete its byte-write operation. If you want to protect the 28c64 from accidental writes caused by runaway code, you can install the write-protect switch described above.

Philips DS-750 Kit

Electronics trade journals are goldmines for robotics hobbyists. Besides the high-quality technical articles, you sometimes connect with opportunities too good to pass up. One such chance appeared in the April 14 1994 issue of *Engineering Design* News (EDN). Philips Semiconductors makes a line of 80c51-style microcontrollers with on-chip EPROM and RAM. Dubbed the 87c75x line, these devices are single-chip solutions to several robotics problems.

To introduce their chips, Philips is running a design contest. You simply submit a design idea, based on their chips, to Philips; rules are available from the company. First prize is a new Chevrolet Camaro coupe. Second and third prizes are color and monochrome laptops.

To help people get experience with these new chips, Philips is offering a low-cost development kit, the DS-750. It comes with a programmer, power supply, cables, software, documentation, and three samples of the MCUs. Total cost: $47.50 plus shipping.

I have already received my design kit, and am thoroughly impressed. This is an excellent piece of work, rivaling anything Motorola has done with their 68hc11 chips. I am especially impressed with the technical books supplied with the kit. The *Application Notes for 80c51-based 8-bit Microcontrollers*, in particular, is fabulous.

It contains many finished, tested 80c51 applications, complete with assembly language source listings. Projects cover on-chip features such as the I2C bus. There is even a complete micromouse robot design. It uses small stepper (not servo) motors, and the software fits into a single 87c751 MCU.

Part VIII
Way Cool
Robots

I've seen a million of 'em. Well, at least a hundred or so. And this section describes some of the more interesting of the many robots I've checked out while doing these columns. Probably my favorite column in the section describes the week I spent at MIT, catching the action at the 6.270 Robot Design class. The class pitted 50 teams of students in a design competition, which ended in a series of one-on-one duels between the students' robots. It was great fun to watch, and I fed off the noise and the energy for weeks afterwards.

Another column in this section describes a group project by members of the SRS to design and build an interactive robotic display for Seattle's Pacific Science Center (PSC). Other columns in this book mention facets of this exercise, but this column covers the initial work. It was a learning experience, and the finished robots were a hit with the PSC visitors.

As difficult as it is to build a robot for your living room, imagine the design problems in making a deep-sea submersible. I got to spend a day with the design team at the Monterrey Bay Aquarium Research Institiute (MBARI), as they finished up the construction of *Tiburon*, a tethered robot rated for 4000 meters. Talk about design challenges! This article covers some of their ingenious solutions for dealing with the intense pressures at such depths.

One group of SRS members has taken two first place prizes in robotic contests of international importance, and I describe some of the problems they had to overcome in creating their award-winning designs. One article covers a robot that picks up tennis balls in an arena, and another discusses the first robotic soccer tournament (MIROSOT). I think you'll find their designs educational and inspiring.

You can build interesting robots that don't even have microcontrollers, as the last article in this section details. The photovore robots use solar cells and ingenious electronics to show life-like insect behavior. Their lack of batteries places a premium on good design, and the robot I describe here, available as a kit, makes an excellent first robot. The second half of this article describes the monsters of the Robot Wars, radio-controlled behemoths that wage a winner-take-all battle. These are heavyweight machines, where gas engines and hydraulic rams are standard equipment.

A Visit to the MIT Campus

This is one of my favorite columns, both for the material it covers and for the memories it recalls. Printed in March of 1994, it describes that year's 6.270 class at Massachusetts Institute of Technology (MIT). This class, which culminated in a two-day contest among robots built by 50 teams of students, featured large, complex robots that duked it out in a series of one-on-one contests. The students who designed the robots were under intense time pressure, thrown into a working environment none of them had ever seen before.

I spent a week at MIT watching the students work, talking to the advisors, taking pictures, and soaking up the atmosphere, which often was frigid. Boston was getting hammered by wave after wave of snow storms that January, and I wasn't prepared for the bitter cold. But the event was great fun, and the MIT campus, with its strong foundation in robotics of all types, was awe-inspiring. This was a truly enjoyable week, and I will remember it always.

The two robots sat at either end of the contest platform, the TV lights gleaming off their red, yellow, and blue LEGO bodies. The handlers stepped back, the starter threw the switch, and both machines swung into action. Sixty seconds later, it was all over. Amid the din of the more than 500 cheering spectators, the judges proclaimed Wastoid the winner of the 1994 MIT LEGO Design contest.

The contest, which finished about 9:00 PM on 27 January 1994, marked the end of a month-long robotics design class held each year at the Massachusetts Institute of Technology in Cambridge, Massachusetts.

MIT 6.270

This year's 6.270 class was limited to 50 teams of students, with two or three students on a team. Each team began the course with an identical box of robot parts, a manual describing the course, and a set of rules for competition. The students on each team pooled their talents to design and build a robot capable of solving the course problem.

This year's problem involved a one-on-one contest between two robots. Ultimately, all the robots competed in a double-elimination tournament to determine the most successful design. The tournament's final rounds always draw the most attention, including local video coverage, but working on a team to solve a tough design problem may prove the most valuable aspect of the event. This trial-by-fire teaches students real-world skills for working with others in a creative environment.

Students may choose to take the course for six units of credit, though many enter just for the fun and experience. Those taking the course for credit must keep journals and submit weekly video-taped reports showing their project's status. Additionally, such students must attend 6.270 classes on related robotics subjects, submit a program listing of the finished code, and show a robot at a preliminary contest.

The Robot Parts

Each team got a box of robot parts, including bags of LEGO blocks, gears, and wheels, circuit boards and components for building a 68hc11 computer, and assorted other parts. Each kit was identical, and a team could use only the supplied parts to build its machine, supplemented by no more than $10 worth of extra parts.

The bags of LEGO parts would have filled a small toy box. This year, the assortment included plenty of the standard blocks, axles, and linkages. It also contained some less common parts, such as rubber wheels, pulleys, hinged blocks, turntables, and a wide variety of gears. It even boasted two complete differential gearing kits, for building sophisticated drive trains.

The 6.270 robots were all based on the 68hc11a1 microcontroller (MCU), built on five printed-circuit boards (PCBs) supplied with the parts kit. The teams built all the boards, using the supplied electronic parts, then tested the finished computer.

The main computer board sports 32K bytes of battery-backed static RAM, a two-line liquid crystal display (LCD), a beeper, several LEDs, ICs for driving four small motors, and assorted IDC connectors for hooking up the other circuit boards.

Figure 1. The electronics parts and blank PCBs in the 1994 6.270 robot kit. (Photos by Ken Birdwell and Karl Lunt.)

Figure 2. The bags of LEGO parts in each robot kit.

The expansion board contains additional LEDs and connectors, 16 channels of A/D converters, and a L293D motor-driver IC, capable of controlling two small motors.

The infrared (IR) transmitter board carries eight IR LEDs and eight red LEDs. It serves as a beacon, broadcasting a signal that the opposing robot can use for location.

Two of the PCBs are not actually part of a working robot. The battery charger board contains LEDs, a high-current diode bridge, and dropping resistors. This board recharges the supplied Gates lead-acid D-cells used for the robot's power supply. Finally, the motor switchboard PCB serves as a quick test jig for manually controlling motors. It has proven to be an invaluable tool for testing early design concepts. The switchboard PCB contains four DP3T switches, some diodes, and some connectors.

The 6.270 Manual

The inch-thick *6.270 Course Notes* manual serves two purposes. First, it gives step-by-step instructions for assembling all five circuit boards. More importantly, this manual provides terrific background information on robotics topics such as batteries, motors, a variety of sensors, LEGO construction, and using MIT's 6.270 interactive C (IC) compiler, provided with the course kit.

This manual, like much of the 6.270 course, has evolved over the last few years. It has been rewritten and refined by this year's organizers to the point that it is itself a first-class text on building small robots. Undoubtedly, next year's organizers will improve the manual even more.

I cannot over-emphasis the value of this manual. If you want to build small robots, even if you don't use LEGOs, you need this manual. You can get the postscript file for the manual (including schematics and diagrams) via ftp access from a machine on the MIT campus.

Interactive C

The students wrote all software for the 6.270 robots in a custom C system developed by several MIT programmers over the last few years. The system, known as interactive C or IC, allows you to write C programs on a host machine, then download the tokenized object into the 68hc11 computer board over the serial port.

The system contains a large library of tested, working source code for using the 6.270 68hc11 computer boards. This really eases the development task, as you can build upon work done by previous students. The 6.270 students compiled their robot code on the large Unix system available in the MIT laboratory. Plans are currently underway to create a robust PC version of the interactive C compiler. This would add yet another programming option for 68hc11 robot builders, expanding on the assembly language, tiny4th, and icc11 options already available.

Robo-Raiders!

This year's contest had a pirate theme, played out on a large, rectangular table with a white painted surface. The two competing robots became pirates, each starting out on its own shore or island. Refer to the accompanying diagram (Figure 3).

The table measures about nine feet by four feet, standing about three feet high. Each robot's end of the table contains a shelf holding eight empty plastic bottles. These bottles represent "pirate punch." Underneath each shelf are four foam blocks, serving as "treasure chests." Across the center of the table is a small trough, about two inches deep and six inches wide, painted blue. This represents the ocean. The ocean is marked on either side by a broad, black band that the robots can distinguish from the rest of the table's surface.

Resting near the ocean on each half of the table is a red block of wood, known as a dinghy. This block can be pushed into the ocean to provide a smooth bridge across the central gap, allowing a robot to roll easily across the ocean to the opponent's side of the table.

Standing vertically at each end of the ocean is a large piece of sheet metal with a wooden block fastened to one side. This sheet metal can be tipped over towards the table's center, where it will form a bridge across the ocean.

The ends of the table are marked by large, round pillars containing bright 120V lamps. These pillars stand on the floor beyond the table's edge; the lights face inwards toward the table's surface. The robots can use these light pillars to determine the direction toward the ends of the table. Additionally, one of the light pillars is covered with a polarizing filter. The robots can use this polarized light to determine which end of the table they are facing.

The starting point for each robot is marked with a two-inch round cutout in the tabletop, covered with clear plastic and holding three small incandescent lamps that face upwards.

The handlers place their machines directly over the round starting points, then press their machine's reset buttons. When the starter turns on the incandescent lights at the starting point, each robot's down-looking photoresistors sense the increased light and the robots begin the run.

Each contest run lasts exactly sixty seconds. The robots have on-board timers that begin counting from the moment the starting lights come on. A robot must stop immediately when the sixty seconds elapses; running beyond the sixty-second limit draws immediate disqualification.

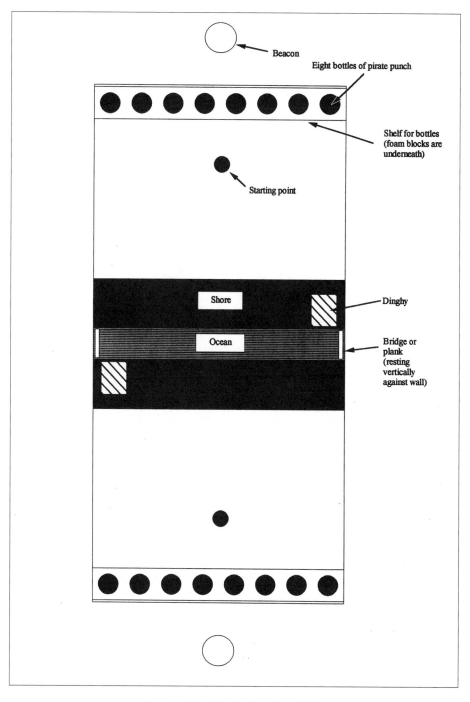

Figure 3. The Robo-Raiders playing field.

Figure 4. The practice table, going full blast.

After the run finishes, the judges add the points accumulated by each robot. The robot with the higher point total is declared the winner. Any robot that loses two runs is removed from competition.

Points are awarded using a complicated system that tends to favor aggressive software and design. Those teams that chose to build attacking robots came up with some ingenious methods for removing bottles and blocks from their opponent's side of the ocean.

Several machines would push their dinghy into the ocean, then roll across the platform and begin knocking the bottles and blocks from the opponent's area. Other machines would roll (swim?) across the ocean, bypassing the dinghy entirely. Once on the other side, they would begin pillaging their opponent's hoard.

A few machines shot ping-pong balls, representing cannonballs, into their opponent's area, attempting to knock bottles or blocks off the table. Since the targets were quite light, this strategy usually grabbed a few points. It always got the crowd excited.

The ultimate winner, Wastoid, took the attacking concept to its extreme. Wastoid would roll across the

ocean, move to the opponent's array of foam blocks, then grab one of the blocks in a large set of pincers. Next, Wastoid would lift the block onto a second set of arms, which would then fling the foam block backwards, hopefully landing on Wastoid's side of the ocean. Capturing an opponent's treasure chest and moving it to your side of the table netted five points, making it the most valuable option of the contest.

This maneuver, which took several seconds to complete, always brought the crowd to its feet. Wastoid performed this sequence in each of its runs, and was always successful. After throwing the block onto its side of the ocean, Wastoid would then proceed to clean the opponent's shelf of bottles.

No matter how often I watched Wastoid perform this block-flinging sequence, I was always impressed with how well it worked. The three students who designed Wastoid, Benjamin Calderon, Yishai Lerner, and Mihir Shah, raised LEGO robotics to a new level. They should be proud of the ingenuity their machine demonstrated.

Other teams opted for a defensive strategy. This typically involved collecting their own bottles and blocks, stashing them in a holding area on the robot itself. Robots that successfully collected bottles or blocks gained bonus points for their efforts.

The easiest objects to collect were the bottles. Usually, a robot would move to one end of its shelf, then use a long arm or stick made of LEGOs to try and sweep the bottles onto a flat, fenced-off area on the top of the robot.

One machine had a horizontal arm that it could swing over a pair of bottles, sweeping two bottles at a time onto a holding platform. Another machine would not only pick up the bottles, but would pull the foam blocks into a lower storage area. I enjoyed watching this machine, as the motion sequences involved were complicated.

But defensive techniques proved to be very high risk. Often, the robot would fail to line up properly with the shelf. In going through its sweeping or collecting motions, the robot would actually knock bottles or blocks off of its side of the table, incurring some heavy penalties.

This resulted in a lower score, leading to some unfortunate losses. Particularly embarrassing were those times when robots lost to a cardboard box. Since some of the contest runs involved an odd number of contestants, one robot would have to compete against a stationary cardboard box, known as the placebo.

The placebo was fitted with an IR emitter so it looked to its opponent like a working robot. Obviously, however, the placebo never moved. Quite often, a defensive robot matched against the placebo would accidently knock some of its own bottles off the table. Since the placebo would get 16 points (all of its bottles and blocks were still in place), the moving robot would end up losing.

Microsoft is a large corporate sponsor of the 6.270 class, and sends a team each year to the event. The same team has gone to at least two previous 6.270 events, so the team members have a lot of experience with building LEGO 'bots. Their robot could not compete in the actual 6.270 contest, but the students loved to watch it run.

Figure 5. Closeup of a LEGO robot.

Figure 6. Pi Tau Zeta leaving the starting point.

The Microsoft machine started by firing six ping-pong balls across the table into the opponent's area. The balls really flew, with a very flat trajectory, and the robot usually succeeded in knocking a couple of bottles onto the floor.

Next, the Microsoft machine would scan for the opponent's IR beacon. Tracking on this beacon, the machine would always move to block its opponent, protecting its own blocks and bottles. Since the Microsoft machine only made one demonstration run during the actual contest, the audience never got to see how successful the strategy could have been.

I think the 6.270 students were most impressed, however, by how quickly the three Microsoft programmers built their machine. Total time for design and construction: two days!

The Motorola team also built a LEGO machine, but this was their first appearance and their lack of 6.270 experience showed. Their machine turned in a credible performance, but it wasn't up to the level of the Microsoft robot. I am sure, however, that the Motorola team will be back next year.

The Students

I expected a very high level of ability from the 144 MIT students taking part in this event. After all, MIT does have a certain reputation for engineering excellence. Even so, the quality of designs was outstanding. Almost every robot showed some special feature that was compelling or interesting on its own. Whether it was an imaginative tread design, a clever arm or claw mechanism, or some well-designed software, the students really showed they had the right stuff.

Students enrolled in this course put in a month of very intense work. Many of them had little experience in vital areas of robot design and construction, so teams had to draw on the members' strengths. Those who knew how to solder did the board construction, while others with programming skills usually ran the compiler.

As time ran out and the contest day loomed closer, teamwork became more critical. Often, a team would have to abandon a nearly-complete robot design, having reached an insurmountable problem.

Even so, there were no apparent breakups or team-threatening arguments.

In the last couple of days, teams were putting in all-nighters, working 24 hours a day to complete their machines. I usually left the lab about 2:00 AM during those nights. Even at that hour, the Unix terminals were always manned, the practice table still had a crowd around it, and I could plainly hear motors and gears running in the other rooms.

About the time everyone needed a break, the Microsoft team sprang for a pizza feed. Around 11:00 PM one night, some of the organizers brought in a carload of pizzas from a local restaurant, and everyone grabbed a soda and a slice or two. After an hour or two of casual conversation and good food, it was time to get back to the robot building. The teams had to prepare their machines for the preliminary contest, held in a small auditorium in the engineering building.

Some of the teams were weeded out in the preliminary round of competition, early casualties in the contest. But many teams no longer in contention showed up at all the ensuing runs anyway, to cheer on those still competing.

Some teams were disqualified for rules or design infractions. These disqualifications were painful; being beaten by a better design was one thing, but no one wanted to see a team tossed out if it could be avoided. Still, the judges were more than fair in their decisions, and though disappointed, I never saw a disqualified team display poor sportsmanship or react angrily.

One team, that had a very good design and had been quite successful, had to be disqualified when its robot's IR beacon failed to perform properly. Even so, this team demonstrated its machine in several trial runs, and the audience got to see a very well-designed robot strut its stuff.

The Final Contest

The final runs took place in a 450-seat auditorium at 6:00 PM on 27 January. It was more than standing-room only; people crowded the aisles, stairways, and doors, trying to watch the competition. The organizers had set up two identical tables, so the contest took on the flavor of a two-ring circus. As quickly as

Figure 7. Final inspection before a contest run.

one run finished, attention shifted to the other table and the next run started. Scores were displayed on roll-up chalk boards and on a large overhead screen. An MIT crew televised the event, so people throughout the campus could watch the matches on the school's cable system.

As each run finished, the judges tallied the points, then announced the scores. Winners were noted on the overhead display; two-time losers were eliminated. After about two hours of competition, so few teams remained that the runs all took place on a single table. Finally, only three teams remained. The judges ordered a round-robin contest to decide the winner.

In the first run, Wastoid took on a machine called Slapper, a defensive-style robot. By now, only the best robots remained, and this run was very close. So close, in fact, that the run ended in a draw at 16 points each, even though Wastoid was the more aggressive robot in the contest. Since this was a round-robin event, each robot was awarded one-half point.

The next run pitted Slapper against rm *'s. This cryptic name refers to a Unix command used to remove files from a directory. rm *'s took the run, winning by a single point, 19 to 18. This set the stage for the showdown between rm *'s and Wastoid. The bedlam that broke out when Wastoid lofted an opponent's block across the ocean, ensuring victory, was deafening.

Following the contest, the three students who had designed Wastoid were mobbed by a local television crew, the MIT television crew, several photographers, and a reporter from the East coast bureau of the Los Angeles Times. When asked what they planned to do next, one student replied, "Sleep."

Summary

This was my first opportunity to see the MIT 6.270 in person. I was there only for the final six days, but the energy throughout that time was intense. The practice sessions, which took place on a table in the lab area, were a constant source of excitement and suspense. As team members tweaked a robot's design, they would bring the machine to the practice table and run against whatever robot was already there.

These informal matches served as a testing ground for design ideas. Often, a proposed idea would fail on the test table, never to be seen again. These tests always drew a crowd of interested students from other teams, anxious to scope out the competition. If these students saw a design idea that worked, they would scurry back to their own lab areas to work on counter-measures.

Eventually, they would test these counter-measures in trial matches of their own. If successful, the original team would have to return to the drawing board to deal with the newly discovered defense or attack. This work and rework simply escalated the quality level of all robots, until the machines that ran in the final contest were far more sophisticated than those I saw my first day.

Oddly enough, those students who tried to work in secret, without revealing their plans to any other teams, usually suffered the most. Their designs never saw high-quality competition until the final runs, so these students never had a chance to react to their opponents' designs.

For many students, this was the first exposure to actual design problems. Many had already taken various freshman and sophomore theory classes, but here they saw firsthand the difficulty in dealing with the real world.

It was also the first time many of them had faced tight schedules and intractable design problems simultaneously. In the first few days, many students seemed to regard the robot design as a "no-brainer;" after all, they had plenty of time.

As the deadline approached and other students began showing off some very clever designs, everyone realized that time was running out. A day or so after I arrived, the tension in the lab increased noticeably. Teams began putting in longer hours, paying more attention to others' designs, and spending more time on the test table.

Several teams learned perhaps the hardest lesson of all. Their design would do well against other robots in the test matches, but as the preliminary contest round drew near, they couldn't resist the urge to add one last change to the software. This often created some totally new bug elsewhere in the code. And since they had by now run out of time for code changes, the net result was an early loss and elimination.

Old-timers in the software game learned early that "perfect" doesn't exist; that "good enough" code beats over-improved code every time. Unfortunately, beginners never seem to understand or accept this idea. Some of the 6.270 teams paid a heavy price to learn this concept.

Recognition

This event was wonderful fun. I enjoyed meeting the students and organizers, watching the designs evolve, and taking part in the competition. The 6.270 class is really a labor of love, organized and conducted by a team of post-grads and teaching assistants (TAs). Their hard work and dedication helped ensure the class' success.

Although I had no direct involvement with the 6.270 class, except as a member of the press, I strongly feel that those TAs and organizers need some special recognition for work above and beyond the call of duty. TAs Carlos Aneses, Bill Baker, Anil Gehi, Darren Hsiung, Sam Lui, Mike Prior, Mike Seidel, Phillip Tiongson, Mike Wesler, and Scott Willcox rode herd on the students, answering questions, and providing help, tools and guidance. They also served as judges during the various runs.

6.270 organizers Matt Domsch, Pankaj Oberoi, Karsten Ulland, Sanjay Vakil, and Anne Wright provided the high-tech tools, documentation, software, and resources needed to pull off this event. Students taking the 6.270 class for credit had to turn in reports and video tapes of their work; these organizers verified that the work got done. The organizers also served as judges.

An event such as this always contains moments that stay with you. I remember PK leaving about 1:00 AM one morning. As he was passing a group of organizers and TAs, one of them asked PK how much sleep he was getting. "I've had four hours in the last 62," he said, as he headed out the door.

One night, as the preliminary contest approached, about fifteen of us decided we needed some pizza. We formed a group, then headed for the subway station in nearby Kendall Square. We "tube-surfed" to Boston, then hiked about half a mile through the North End to a small pizzaria whose name I never got. The two guys at the counter lit up when we trooped in and started ordering. We filled their little shop and turned a slow night into a major event.

During the ensuing feed, Phillip Tiongson fell asleep from exhaustion, leaning his head against the restaurant wall. We hatched a plan to sneak out and leave him sleeping in his chair, but the sudden silence must have clued him in and he woke up.

A local television crew covered the final contest runs, and after filming several runs, the cameraman felt obligated to get a sound bite. He stepped up to a victorious team, aimed his camera at them, and asked, "What did you guys do to win?" One of the students gave him a big grin and replied, "Data abstraction!" I had to laugh. How was data abstraction going to play on the channel 4 newscast?

Now, where did I put that tub of LEGOs?

Designing an Interactive Robot Display

This column, which appeared in April of 1994, covers the design of the Seattle Robotics Society's PSC-bots. These small, upright robots were part of an exhibit developed jointly by the SRS and the Pacific Science Center in downtown Seattle. The SRS effort occupied about a dozen people and untold man-hours, but we had a great time and learned a lot about putting together museum-quality exhibits. And the final robots were pretty cool, too.

A project this ambitious covers a lot of ground, and you can find information on everything from modulated light beacons to power supplies. All told, this turned out to be a very useful article, and could serve as the jumping-off point for many different robot designs

A few months ago, the Seattle Robotics Society (SRS) agreed to build a set of robots for use in an upcoming Pacific Science Center exhibit. The Pacific Science Center (PSC), located in downtown Seattle, offers many large, hands-on and interactive science exhibits for kids and adults alike.

As I write this, several members of the SRS are gathered around a work table in my living room, finishing up six sets of electronics for powering our robots. Dubbed the PSC-bot, each machine uses four simple circuit boards to provide an interesting and educational display of autonomous robotics.

I described the PSC exhibit in some detail in my column two months ago. Interested readers should leaf through that issue for a full explanation on what we have planned. The following paragraphs will discuss the robot, but will likely not provide a full description of the exhibit's workings.

Each robot consists of a clear plastic frame standing about 14 inches tall overall, topped with a six-inch-diameter clear plastic dome. This dome sits on a clear, hollow column about five inches tall. The bottom of this column is fastened to a ten-inch diameter base containing the robot's battery, circuitry, and bumper switches.

The robot is driven by two modified Futaba S148 servo motors, and contains a set of sensors such as IR object detectors, a Hall effect sensor, and a sensor for detecting an external IR beacon signal. The robot also contains a large collection of LEDs. Some are just for show, others serve as IR emitters for the object detection circuitry or as power-on or behavior indicators.

The robots motor around on a 5-foot by 10-foot table, which is enclosed in Lexan and framed in hardwood. The table was designed and drafted by Andy Olney and Martin Sloan, two of the many SRS members involved in this project. The PSC wood shop actually built this beautiful exhibit; I hope some of you get a chance to visit Seattle soon and see it.

The PSC-bot Electronics

We used Marvin Green's 68hc11 BOTBoard for the brain of our robot. I have already described Marvin's board in previous columns (see "Introducing the BOTBoard.") The low price and large list of features make this board an excellent foundation for very small 'bots. Its small size was also a compelling reason for choosing the BOTBoard. At just 2 inches by 3 inches, it helps minimize the amount of frame space devoted to holding electronics.

Each BOTBoard holds a 68hc811e2, providing a total of 2K of EEPROM for code space. Adding in the 256 bytes of on-chip RAM gives enough room for writing all our control software. One circuit board contains circuitry for detecting the associated IR beacon signal, emitted by a special IR LED mounted at each end of the exhibit table. This board, known as the tone detector board, measures about 1.5 inches by 2.5 inches and was designed by Frank Haymes. Refer to the accompanying schematic (Figure 1).

Frank's circuit uses an IR phototransistor to drive a LM386 amplifier IC. The LM386 provides a gain of about 20, yielding enough signal to properly drive a following NE567 tone detector. The NE567 samples the provided signal and compares it to an internal reference oscillator. Whenever the NE567 sees a signal of the same frequency as its reference signal, it pulls its output line low, indicating it is locked to the incoming signal.

This signal is fed back via J1, a 4-pin KK-style male connector, to the BOTBoard, where the controlling software can sense when the PSC-bot is facing the selected beacon.

We can select either the red or green beacon by changing the NE567's reference oscillator. The red beacon emits an IR signal of 3281 Hz, while the green beacon emits a 7109 Hz frequency signal. We can align one of these tone detector boards using either of two techniques. One method involves hooking up an oscilloscope to the reference oscillator of the NE567 (pin 6) and adjusting the timing potentiometer until the 'scope shows a waveform of the proper frequency.

Another method requires a pulsed IR source of the proper frequency. In this case, we would set a robot in sight of the proper beacon, then adjust the timing potientiometer until we get a lock indication. The NE567 will show a lock condition through a considerable range of the timing pot, so we would simply set the pot to the middle of this lock range.

This circuit has proven quite sensitive. Preliminary tests showed that the NE567 can reliably lock onto the beacon from over 12 feet away; not bad for a simple, two-IC design.

(Author's note: We ended up changing this circuit somewhat. See "A Look at the SPI" for details.)

Another circuit board houses all of the display electronics, as well as the 40 kHz IR object detector circuitry. This 2-inch by 3-inch board, known as the control board, represents a joint effort between Tom Dickens and me. Refer to the accompanying schematic (Figure 2).

U3, a 74hc138, uses three signals from the 68hc11 to select a single red LED (D1 through D8). Thus, the 68hc11 can use the three lines ETL0, ETL1, and ETL2 to activate any one of the eight LEDs, known as thinking lights.

The 68hc11 software cycles through these lines in a binary fashion, so each LED is on momentarily. This results in a group of eight flashing LEDs, in-

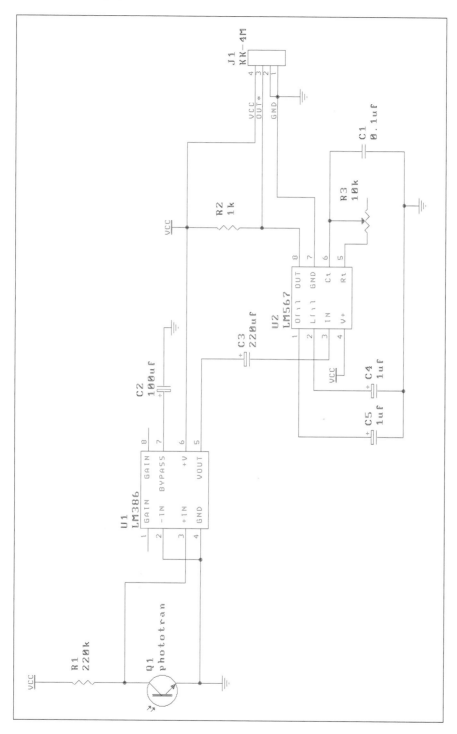

Figure 1. Schematic of the tone detector board.

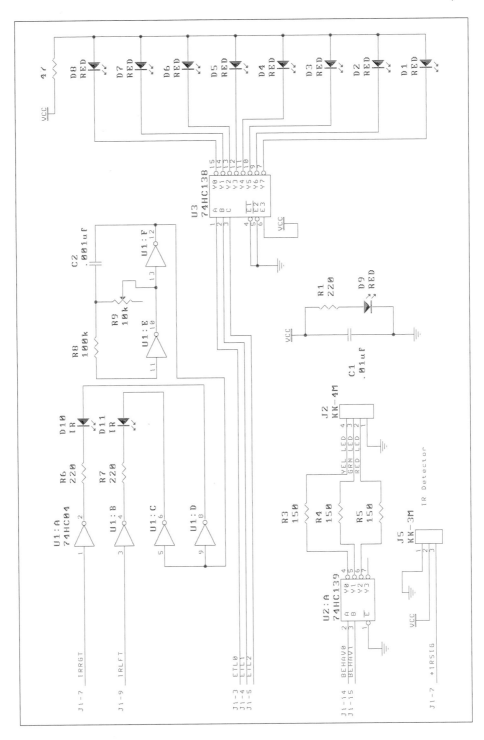

Figure 2. Schematic of control board.

tended to make the robot appear to be thinking. U1, a 74hc04, forms the now-familiar 40 kHz IR object detector circuit. This circuit originally appeared in the book, *Mobile Robots*, by Joe Jones and Anita Flynn.

The 68hc11 uses the two lines labeled IRRGT and IRLFT to activate the right and left IR emitters, respectively. Used in conjunction with the robot's IR detector (mounted facing forward), the software can detect objects to the right, left, or straight ahead. It does this by turning on the right IR emitter, then looking for a return with the IR detector. If it sees a return, the software notes that an object exists on the right side. Performing the same test with only the left IR emitter on checks for an obstacle on the left side. If an obstacle appears on both the right and left sides, the software assumes there is something straight ahead.

The 74hc139 at U2 activates one of three LEDs, depending on the state of the two input lines from the 68hc11. These three LEDs are known as behavior LEDs, and are used by the 68hc11 to provide a visual clue as to the robot's current behavior.

For example, the 68hc11 will light the green LED when the robot is scanning around, looking for its beacon. It will light the yellow LED whenever the robot needs to avoid an oncoming obstacle. Finally, the 68hc11 turns on the red LED to signal that it is backing away from an object that it contacted with its bumper switches.

Although not shown on the schematic, these LEDs are wired to J2, a 4-pin KK-style male connector. This cable terminates with a cluster of three LEDs, one of each color, that is inserted inside a ping-pong ball and mounted inside the robot's dome. This ping-pong ball changes color, based on which LED is on at any time. The rest of the circuitry is wiring for the IR detector, filtering and a power-on indicator, and protective wiring for the unused half of the 74hc139.

The fourth board in the set is the switch board, which also measures 2 inches by 3 inches. I haven't included the schematic here, but will discuss the board's features in general. I designed this board as a common point for all wiring and cables. Power

comes into this board through J3, a 2-pin Molex KK-style male connector that runs to a 6 VDC 4 Ahr gel-cell battery and switch. A 26-pin ribbon cable connects this board's J1 to the BOTBoard's large I/O connector. Similarly, a 16-pin ribbon cable runs from this board's J2 to the control board.

J9, a 4-pin Molex KK-style male connector, takes a cable from the tone detector board. The switch board also carries a few 10 Kohm pullup resistors for the switches, and a filtering capacitor. This board also contains four 2-pin KK-style male connectors (J4 through J7) that accept wires from the four bumper switches. Finally, J8, a 3-pin KK-style male connector, accepts wiring for the Hall effect sensor, used to detect the inch-wide goal line buried at either end of the exhibit table. Time for a slight digression ...

Hall-Effect Sensor

The Hall-effect sensor provides a signal whenever the robot passes over either goal line, a 1-inch wide strip of soft steel embedded under the table top. Dick Martin originally intended to use a magnetic strip for the goal line, but his research turned up some interesting features of these strips.

Available at most hardware stores for just a few bucks, these strips are normally used for hanging signs or holding doors closed. They are not, however, a single long magnet, though they may seem that way. Internally, these strips are a long chain of small magnets, each about 1/8-inch wide, stacked with opposite poles aligned, like this:

N S N S N S N S
S N S N S N S N

where the magnets run vertically. This arrangement works fine for holding signs up, but it really confuses Hall-effect switches. Since the Hall-effect switch is only about 1/16-inch wide, it can return nearly any signal level when it contacts the magnetic strip, depending on exactly where it hits one of these internal magnets.

In fact, the signals returned in Dick's tests were so unpredictable that he had to abandon using the magnetic strip as a goal line. Instead, he stuck a small magnet directly to the back side of a Hall-effect sensor, then passed this magnet-sensor pair over a piece of soft steel. Now he could get a consistent 350 mV signal at each pass. Granted, this is only about 16 counts when read by a 68hc11's analog port, but the sensor signal is very quiet (no electrical noise), so he doesn't expect any problems using the sensors this way. OK, back to the electronics...

More Electronics

Besides the 4-board set of electronics for each robot, this project uses a single box with circuitry for driving the IR beacons and related lights. This box, designed and built by Keith Payea, fits inside the exhibit table and accepts wiring for pushbutton switches, 120 VAC power, and other wiring.

The two IR beacon signals are generated with two 4060 binary counter/dividers, each fed by a separate ceramic resonator circuit. The board also carries two 4538 monostable multivibrators, for stretching the pushbutton closures to about 30 seconds each.

It also contains a set of 75452 buffers for driving on-board relays and off-board lamps; the latter serve as visible markers for the two different beacons. This lets the crowd see a red or green beacon, even though the robots are keying off the two IR signals.

Circuit Boards

We couldn't have done this project without some means of quickly creating printed circuit boards (PCBs). Dan Mauch and his homebrew CNC drill press, plus his experience in etching photoresist PCBs, proved invaluable. You may have already seen the article, in a previous issue of *Nuts & Volts*, describing Dan's homebrew drill press. Dan normally uses the Easytrax and DANCAM public domain CAD software to create his PCBs.

In this case, however, Frank and I did our designs with the shareware version of PADS. Dan simply downloaded our .job files via modem, created a drill file for the circuit boards, then drilled out the photosensitive PCB stock as needed. Next, he plotted the board's artwork out on his laser printer, using a transparency film. Finally, he exposed the film onto the drilled PCB stock, then developed and etched it.

Dan's help really made this project go. During the development phase of some of these circuits, Dan was providing one- and two-hour turnaround for test boards. It was so cool having that type of support in developing robotics projects.

Despite the success of the project, Dan wants to improve the quality of boards he makes. He has already paid to get a photo transparency of one of his own boards. Circuit boards made from this film are, according to Dan, "commercial quality." Dan feels we can get all of our robot artwork onto a single 8.5-inch x 11-inch film, which means we can do high-quality robot boards for about $10 for a set of master artwork. Not bad!

The PSC-bot Software

We wanted our robots to exhibit several behaviors, based on their environment. Behaviors include scanning for the beacon, moving towards the beacon, avoiding an object sensed with the IR detectors, fleeing an object that hits the bumper skirt, and resting at the goal area.

Each behavior carries an associated LED color or pattern, so that the crowd can pick up on the robot's current state. Additionally, the eight thinking LEDs have to cycle through their pattern, lighting in an interesting sequence. The PSC-bot must also control the IR object detector LEDs, and monitor the IR detector, the Hall effect sensor, four bumper switches, and the beacon tone detector signal.

All of these functions have to fit into the 2K of EEPROM available in a single-chip 68hc811e2. Although the code could have been written in assembly language, Kevin Ross decided to try his hand at writing the software in tiny4th. I was especially impressed by this, as Kevin is a C/C++ guru working for Microsoft. I have found most C programmers either don't know about Forth, or know about it and hate it.

Even though Kevin had no prior experience with Forth, he grabbed a copy of Leo Brodie's *Starting Forth*, my tiny4th package, and a BOTBoard, and went at it. We started the code on the flight home from Boston at the end of January, where we both attended the MIT LEGO robot design contest (see last month's column). Kevin continued the work, and finished the code in just a few days.

Besides all of the above functions, Kevin's code also includes quite a bit of serial I/O for debug purposes. Even so, his final program only uses about $7c0, or 1984 bytes. Removing the serial I/O code would probably free up an additional 256 bytes.

We use pcbug11 and a special serial cable to move the robot code from a PC into each BOTBoard. Refer to earlier columns for details on using pcbug11, building the cable, and working with the BOTBoard. All that remains is to reconfigure the BOTBoard to run the new program, then turn on power.

The PSC-bot Frame

These robots are real show-stoppers, due largely to the excellent plastics work done by Bill Harrison, working from a design developed with David Adams. Bill worked with clear acrylic plastic and two different adhesives to fashion the robot bases.

He suggests getting two manuals from Plexiglas, *Forming Manual* and *Fabrication Manual*. The latter will likely prove more useful, as it deals with machining and gluing acrylic. Note that these are geared for industrial users, but much of the info can be applied by the home builder. Check with your local plastics shop.

The thicker of the two glues is called WELD-ON #16, available in a tube from most large plastics houses. It works well for fitting together roughly cut joints. The solvent-type adhesive, which Bill prefers using, is WELD-ON #4. This solvent is very thin, evaporates quickly, and holds by wicking into a tight joint between two pieces of cleanly cut plastic.

This means that the #4 solvent needs plastic machined with very accurate cuts; it will not work if your joints have large gaps or cracks in them. Bill recommends using a table saw with a triple-chip blade for plastics, available from specialty saw shops. You need to use the proper drill bits as well. Bill suggests you get a special drill bit intended for use with plastic. Check the plastics shop for recommendations.

Carefully read and follow the instructions on both of these adhesives, as well as the instructions in the two recommended manuals. Plan on a long learning curve, but the final results are well worth the extra time and effort.

Bill notes that you can usually get surplus plastic from scrap bins at the local plastic shop. Normally sold by the pound, this will save you lots of bucks as you build up your plastic stockpile.

And most important... wear eye AND face protection while working with plastic! Flying chips and splashes of solvent can cause serious injury. Be extra careful working with this stuff!

Bill was also responsible for creating the only truly dependable bumper skirt I have ever seen, anywhere. He formed a complete ring of plastic, wide enough to go completely around the base of the PSC-bot. He added four thick plastic tabs, fastened to the inside of the ring at 90-degree intervals. These tabs in turn fit inside slots cut into the robot's frame. The back end of each tab (towards the robot's interior) rides against the leaf of a microswitch.

Inward pressure on the outside wall of this skirt, at any point on the ring, will trip at least one switch. Yet the skirt is so solidly mounted that you can lift the entire robot by the skirt.

Putting It All Together

All of these pieces do not a robot make. You have to get all the boards wired, mount them onto the frame, and add on the switches, motors, and other goodies. This takes time and careful construction. Ken Birdwell, Zack Moore, Kevin Ross, and I spent several hours at impromptu soldering parties, wiring up and testing the electronics.

Bill Harrison then drilled the boards and mounted them to the frames, bolted on the remaining hardware, and created a robot.

A typical SRS meeting runs about 60 people, while this PSC robot effort only involved about a dozen members. But everyone in the club followed the monthly progress reports with interest, and really got a kick out of watching the robot design evolve.

This has been an excellent learning experience for all of us. Each contributed to the project his special skills and talents, and we all got a chance to work on a large team project.

You will hear more of the PSC-bot in future columns, as I will refer to some of these elements again. For now, however, I want to close with special thanks to Ken Williams, Jennifer Lewis, and David Taylor of the Pacific Science Center staff. They have been very supportive and very patient in working with us. Hopefully, this is the beginning of a long relationship between the PSC and the SRS.

Special Mentions

As usual, Digi-Key proved to be a great source of quality components, delivered quickly to our doorsteps. Keith and I both turned to Digi-Key for some of our PSC-bot electronics. They can be a bit pricey, but if time is more of an issue than money, I recommend looking through a Digi-Key catalog. Give them a call and ask for the latest catalog.

This project has required us to locate large quantities of hard-to-find parts quickly. Often, we didn't have a detailed specification for exactly what we wanted. We just knew we needed "LEDs about this bright" or "bumper switches shaped kinda like this." In my case, Electronic Goldmine has proved to be exactly that; a goldmine.

Their catalog #221, for example, offers Stanley "high brightness" LEDs at 10 for $1 in red and 8 for $1 in green or yellow. I had a heckuva time locating matched brightness LEDs in three colors, so I ordered some of these devices from EG. They arrived on schedule, were exactly what I wanted, and the price was unbeatable. Call them and get a catalog.

Deep-Sea Submersible Robots

In August of 1995, I wrote a column that completed the hack of the Practical Peripheral's PP144 Pro-Class modem. This little jewel, available in most computer stores and mail-order outlets of the time, contains a 68302 microcontroller and plenty of I/O and memory. With just a little bit of work, you can turn it into a battery-powered robot brain. In this column, I show you how to rewire the on-board DB-25 connector (formerly used for RS-232 I/O) into an I/O port for your own experiments.

Next, I review a few surplus stores that I visited during a vacation trip to San Diego. At least one of these stores advertised in *Nuts & Volts* at the time, and I enjoyed an "up-close and personal" look at a store whose advertisements I have so often scanned.

But the high point of my trip to southern California was a tour of MBARI, the Monterey Bay Aquarium Research Institute. Thanks to two very kind engineers on the MBARI staff, I got to spend an entire day checking out two of the most sophisticated deep-sea research vessels in the country. *Ventana*, designed for depths of 1000 meters, was soon to be replaced by *Tiburon*, capable of working at the astounding depth of 4000 meters. The enormous pressures faced by any robot working at such depths pushes the engineering teams to their limits, and the MBARI staff showed a lot of ingenuity in overcoming the problems.

This article concludes with a short review of *Star Trek Creator*, the authorized biography of Gene Roddenberry, the driving force behind Star Trek. I thoroughly enjoyed this book and, given how popular the various Star Trek shows are with engineers in all fields, considered its review entirely appropriate for my column.

Last month, I went well down the path to creating a robot-ready computer from a Practical Peripherals 14.4 ProClass modem (PP144Pro). Refer to that article for details on the hacks necessary to convert this small circuit board into a powerful 68302 embedded controller. (See "Hacking a 68302 Modem, Part 2.")

As you'll recall from that article, I had removed three small surface-mount technology (SMT) chips just to the right of the DB-25 connector. Removing these chips (U1, U2, and U3) completely disabled the normal RS-232 communications, allowing me to add in my usual 4-pin KK serial connector.

Removing these chips had a secondary purpose, however. These chips sensed or provided several RS-232 signals to the DB-25 connector, signals used in normal modem communications. With the chips removed, the connections are now exposed at the empty pads, ready for rewiring and modification.

For example, removing U2 isolated DB-25 pin 2, which is wired to U2-1 and supplied RS-232 data out to the host PC. We have already rerouted the serial data to the KK connecter, so we are free to wire another signal to the pad at U2-1, making that signal available at pin 2 of the DB-25.

By adding suitable cuts and jumpers, we can turn the now-defunct DB-25 serial connector into an I/O expansion connector. We'll just wire selected 68302 signals to pins on the DB-25 connector. Later, in your robot, you can use these signals as needed for I/O.

We will leave two pins unchanged. DB-25 pins 1 and 7 already connect to the PP144Pro's ground plane, and changing them would be too much hassle. They will provide the ground signal for our expansion port.

Next up is the +5 VDC line. Cut two 4-inch lengths of 30 AWG wirewrap wire, and remove 1/8-inch of insulation from the ends of both. Remove the solder from pin 11 of the DB-25 connector, using a good solder sucker or copper braid. Connect one wire to this pin, then solder in place. Repeat the above to connect the second wire to pin 9 of the DB-25. Solder both wires to the upper lug on the power switch, SW1. Pins 9 and 11 of the DB-25 connector now supply +5 VDC, controlled by the on-board power switch.

On the underside of the board, cut the trace between pins 12 and 23 of the DB-25 connector. This isolates pin 12 completely, and leaves pin 23 connected to pad U1-3. The following mods each require a tiny jumper, made of bare 30 AWG wirewrap wire. Make each jumper by clipping a 1/4-inch length of bare wirewrap wire. You will use these jumpers to connect between two pads on selected SMT layouts. When

soldering these jumpers in place, be very careful that you don't allow them to contact any nearby traces or pads. You can dress these jumpers away from the board, so that they clear any problem areas.

Connect a jumper between pads U3-10 and U3-8. This hooks the 68302 signal *IACK6 (PB1) to pin 6 of the DB-25 connector. Connect a jumper between pads U3-13 and U3-11. This wires the *IACK1 signal (PB2) to pin 8 of the DB-25 connector. Take care, when soldering this jumper to U3-11, that you don't accidentally bridge the trace-cut you made earlier when wiring up the Molex KK-style connector.

Connect a jumper between pads U1-5 and U1-6. This hooks the 68302 signal TCLK2 (PA3) to pin 15 of the DB-25 connector. Connect a jumper between pads U1-8 and U1-10. This wires the RCLK2 signal (PA2) to pin 17 of the DB-25 connector. Connect a jumper between pads U1-11 and U1-13. This hooks the 68302 signal *TOUT1 (PB4) to pin 22 of the DB-25 connector. Connect a jumper between pads U1-2 and U1-3. This wires the TIN1 signal (PB3) to pin 23 of the DB-25 connector.

This completes my initial mods to the DB-25 expansion connector. Refer to the accompanying expansion port diagram for details on the new pin assignments (Figure 1). Note that six of the DB-25 pins connect to pads that are presently empty. You can bring your own signals to these pads as needed, provided you do so carefully. It helps to take advantage of existing wiring on the PP144Pro board.

For example, "Hacking a 86302 Modem, Part 2" lists signals that drive some of the panel LEDs. You might choose one of these signals, say PB0 which drives the leftmost green LED, and wire it to pin 24 of the DB-25 connector. To do so, cut a piece of 30 AWG of the proper length and trim 1/16-inch insulation from each end. Solder one end of this wire to pad U2-13, which is associated with pin 24 of the DB-25 connector. You have to be a little careful selecting the connection point for the other end of this wire. You have to find a trace, via, or pad connected directly to the 68302 pin of interest. In this example, the proper connection point is the pad labeled T114 on the underside of the board.

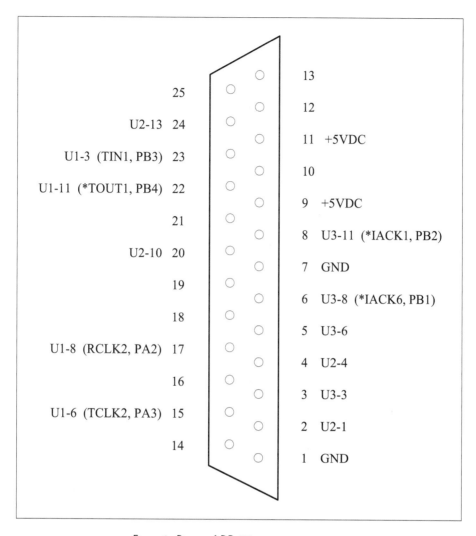

Figure 1. Pinout of DB-25 expansion connector.

Using an ohmeter, you can verify that the resistance between pad T114 and pin 112 of U13 reads nearly 0 ohms. Solder the free end of your 30 AWG wire to the T114 pad; this completes the connection to the I/O expansion port.

Tracking down traces or pads to pins of interest on the 68302 can prove entertaining and boost the options available to you in your robot design. Be sure to double- and triple-check your signal traces,

to make sure you get a trace that goes directly to the desired pin.

Also, watch out for side-effects. For example, 68302 signal RXD2 hooks to U21-2. You could probably run a wire from this pin to a likely I/O port connection, such as U2-10. But doing so means that U21, a 74HCT251, will feel the effects of whatever you do to RXD2. You have to allow for these side-effects whenever you make connections to the board.

If your chosen signal is already hooked to an output pin of some device, you will have to lift that pin from the circuit board, then make your connection to the freed pad. If, however, your signal is hooked to the input side of a chip, you may not have a problem. You can simply drive the selected 68302 pin, and the input to the affected chip will just go high or low, as necessary. You will, of course, need to allow for any resulting changes on the chip's output pins.

Simply lifting input pins of CMOS devices can cause internal oscillations in the modified device. This will result in increased power drain and, in extreme cases, latch-up and destruction of the device. To protect lifted input pins, you can wire them to ground or tie them to +5 VDC through a 10K resistor.

An easier fix to this floating input problem involves simply removing the chip. This, of course, will affect all lines connected to output pins on the removed chip, so you will have to check for ripple effects of this fix.

I'll leave you with the PP144Pro board at this stage. I've had a blast hacking this board, and will eventually work it into a robot. For now, look on it as the starting point of a larger, more powerful robot brain. Have fun!

Shopping Around

During a recent trip to San Diego, I spent time roaming some of the electronic stores, in search of robot parts. If my results are any indication, this part of the West Coast offers the robot builder lots of opportunities.

I first stopped at California Electronics and Industrial Surplus, in a San Diego suburb called El Cajon. CEIS has been around for several years, and carries a large selection of older electronics, as well as heavy industrial surplus and bulk materials. The store reminds me of the surplus outfits I used to frequent while I lived in Phoenix.

The aisles are kinda cramped, and the labels on many of the parts bins miss being current by at least one shipment. Some of the stock is down-right un-

recognizable, at least by this software guy. But the bargains are there if you know what you need and are patient.

I found a real-live traffic light for $100, but passed it up as my traffic light needs right now are not pressing. I noticed a vintage Motorola Digital Frequency Meter for $95. This is one of those tank-like metal boxes with a very simple user interface: a few switches, a couple of dials, and a 7-digit Nixie readout. Those readers too young to remember Nixies missed an exciting era of electronics.

I found plenty of relays, wire, project boxes, gearhead motors, lasers, and smaller components needed in a well-stocked junk box. I loaded up on trimpots, and even found a power cord for my PC monitor that I just had to have. I counted 225 plastic bins loaded with pulleys, belts, grommets, and nylon components. You could probably design a pretty good robot drive system while digging through these bins.

About two blocks down the street from CEIS, you can find Murphy's Electronic and Industrial Surplus Warehouse. Murphy's apparently opened within the last year or so, as the aisles are relatively open and uncluttered. The shelves, however, stand floor to ceiling and are filled to bending with power supplies, computer gear, test equipment, and miscellaneous small components.

I spotted several large micro-fiche viewers at $35 per, and a table-top fiber-optic splicing station for $450. I passed up the $7.50 weather balloons, though, to get a better look at a Data-I/O Model 22B EPROM programmer. The 22B was formerly owned by the US Navy, and contains an on-board UV eraser. As near as I could tell, it was complete and ready to go, and at $150, I worked hard to talk myself out of buying it.

Next up was Gateway Electronics, a regular advertiser in *Nuts & Volts*, and a required stop on any robo-visit to San Diego. The Gateway store is well-lit and spaciously laid out, but filled with vitally important robot parts. When I walked through the door, I was greeted by an electronic voice welcoming me to the store. Someone had hooked up a proximity detector and voice recorder board to an amplifier and speaker. When the detector saw me walk by, it started up the 12-second message. Needless to say, you can buy all the components for this system at Gateway.

I just had to pick up a couple of 12 VDC gel-cell batteries. Rated at 1.3 AHr, the batteries measure just 3.75 inches by 1.75 inches by 2 inches, and are surprisingly light. They will fit well in my next robot, and the $6.95 price was reasonable.

I also found several of the Ramsey and Electronic Rainbow kits for sale. I bought the Ramsey TV transmitter kit ($27.95), as I see a looming need for it, and one of the regulars in comp.robotics had recommended it. I'll build it up soon and let you know what I think of it.

I picked up a few more trimpots (you can't have too many trimpots), and some fuse clips for an upcoming project. Then I spent a lot of time wandering around the store, checking out the displays. I couldn't overlook the glass cabinet at the front of the store, complete with two working video displays. One monitor showed the typical black-and-white (B/W) image from a CCD pinhole camera. The camera that supplied the signal was so TINY; it looked to be only 1 inch by 1 inch.

But the monitor on the top shelf really caught my eye. It showed a color video picture from a small CCD camera sitting on the same shelf. I had not seen the output from one of these cameras before, and was impressed. The color camera is larger than the B/W pinhole camera, but not excessively so; it looked to be 2 inches by 2.5 inches. The price was a bit high for me, at $280, but that should drop as these cameras become more widely available.

I enjoyed visiting all three of these stores, and definitely rate Gateway as a "must-see."

Visiting MBARI

The Monterey Bay Aquarium Research Institute's Marine Lab sits at the end of Monterey Bay, in the little town of Moss Landing. Stretching out from this point into the Pacific Ocean runs the Monterey Canyon, an immense underwater chasm that drops rapidly to a depth in excess of 6000 meters, or over 19,500 feet.

The proximity of such deep water makes Monterey Bay unique in the United States, and possibly in the world. Monterey Bay is an ideal focal point for deep-water research, which explains the presence of MBARI.

Several years ago, MBARI developed the remotely-operated vehicle (ROV) *Ventana*, a deep-water submersible used to explore the Monterey Canyon to depths of 1000 meters. *Ventana* (which means "window" in Spanish), uses a three-chip Sony color camera and pure-white HMI lights to return real-time video of the ocean depths. The ROV also carries specimen collecting gear and grippers for performing deep-water experiments.

In 1991, MBARI engineers began the design of *Ventana's* successor, *Tiburon*. This vehicle, whose name means "shark," measures 9.5 feet long by 6 feet wide and will descend to 4000 meters (about 13,000 feet). It is due for "wet test," or actual ocean submersion, in September of 1995 and real launch in mid-1996.

Powered by three large electric motors and armed with sophisticated computers, cameras, and collection equipment of all types, *Tiburon* will open up vast new areas of exploration to deep-water researchers. I spent a day at MBARI, discussing *Tiburon's* design with two of the Institute's engineers. Andy Pearce walked me through many of the software and electronics issues involved in controlling such a complex piece of machinery. Bill Kirkwood pointed out many of the mechanical innovations needed to cope with the extreme pressures of deep-water work, and even took me on a tour of the Institute's research vessel.

And I always thought a living room was a hostile robot environment! At 4000 meters, the pressure on *Tiburon's* frame will hit 7200 pounds per square inch (PSI). When I asked Bill for some type of comparison, he said, "Imagine standing two Cadillac Coupe de Villes on your big toe."

Then there is the corrosion factor. The salt water, and resultant galvanic reaction, limits the effective life of even such well maintained craft as *Ventana* and *Tiburon*. Despite meticulous care, large sections of *Ventana* have already been replaced, and *Tiburon* will probably not last, in its original form, much beyond five years.

Both machines rely on a tether to deliver power for motor control, and to exchange video and data with the support ship. For *Tiburon*, this tether consists of a 5000 meter cable 0.68 inch in diameter. The cable contains an inner core of three large copper wires and three tiny fiber-optics cables. This inner core is protected by two thick steel jackets. Each jacket consists of many steel wires; the wires of the inner jacket are wrapped opposite to those of the outer jacket. This reversed winding minimizes any kinking effects as the cable rolls and unrolls from the spool.

I called them "wires," but the three large power conduits inside the tether cable more closely resemble solid copper rod. The three wires need to be large; they must carry 25 kWatts of power to the submersible, passed as 1800 volts at 400 Hz.

The bulk of MBARI's design efforts focus on canceling the effects of the tremendous pressures involved. The engineering labs in MBARI's downtown office, in nearby Pacific Grove, are a testament to the ingenuity brought to bear on the problem.

I saw several polished titanium containers, beautifully machined and welded, designed to protect fragile power systems and electronics. Each cylinder measured about eight inches in diameter by about 20 inches long; just the machining budget for these items would have built thousands of Huey robots.

A foam shell, artistically sculpted into smooth lines, covers *Tiburon's* upper body. To deal with the extreme pressures, the MBARI engineers built the shell, called the foam pack, from a special epoxy resin. The resin is mixed with millions of tiny glass spheres, then molded to the desired shape. Even at 4000 meters, the foam pack will only compress a tiny fraction of an inch.

Tiburon's frame construction uses a different method for handling the high pressures of deep submergence. The frame, made of hollow 2-inch aluminum tubes only 1/8-inch thick, is riddled with holes. The holes permit the sea water to flood all parts of the tubular frame, equalizing pressures on the tube walls as the craft changes depth.

Even the cabling between subsystems on the ROV gets special treatment. In some cases, engineers employ specially-designed high-pressure cables to carry power and data wires. Sometimes, however, *Tiburon's* designers use a novel oil-filled jacket to protect the wiring. In this case, the wires run through a flexible, clear plastic outer shell. The shell is filled with a light vegetable oil, similar to cooking oil. The oil barely compresses at depth, protecting the enclosed wires. Should the jacket leak or burst, the non-toxic oil floats to the surface, then evaporates.

It takes a lot of computing power to run the *Tiburon* ROV. The vessel carries a VME cage with three 68040 processors and a digital signal processor (DSP). These computers exchange data with MBARI's onshore computers over a TCP/IP network.

Having network access to all of the computers on board the *Tiburon* creates some powerful debug and development capabilities. MBARI engineers can do a full system reload over the network, even if the vessel is at depth. The ability to monitor or change system variables, while the control software is actually running on the ocean floor, really improves debug efficiency.

The 68040 computers handle the high-level tasks for controlling *Tiburon*. The actual motor control and sensor data collection fall to 15 80c196 processors scattered throughout the vessel and hooked to the VME cage with a custom RS-485 network.

MBARI engineers developed nearly all of the electronics used in *Tiburon*, including the controllers for the three drive motors. Here, the circuit board design posed a special problem, again caused by the intense pressure.

The controller circuitry is housed in an oil-filled container that is kept at pressure. Initial designs used commercially available potted power modules, but the intense pressures would crush the module's epoxy housing wherever an internal void existed. Ultimately, MBARI had to design a custom power module, built with discrete devices whose housings could withstand the pressures.

Ventana, the current ROV, runs nearly every day. In fact, images from *Ventana's* sled-mounted camera are fed live to monitors in an auditorium of the Monterey Bay Aquarium, located a few blocks from MBARI's headquarters.

The MBARI offices and labs also contain numerous TV monitors, each showing the current view from *Ventana's* camera. The video images, transmitted by microwave from the support ship, apparently run continuously, so long as the camera has power. Thus, you can tell at a glance what *Ventana* is doing, from nearly anywhere in the MBARI facilities.

While I was visiting, Bill Kirkwood noticed that the monitor showed a dock-side image; *Ventana's* support craft was tied to the pier across the street. He asked me if I would like a tour of the ship; naturally, I agreed.

The support ship, *Point Lobos*, contains a small deck that is nearly filled by *Ventana* and its associated machinery. Normally at sea during work hours, *Point Lobos* returned to Moss Landing because of a ruptured O-ring somewhere inside *Ventana*. When I walked on board, several engineers and technicians were poking and prying about the ROV.

Ventana's sled holds many different mechanisms designed to capture deep-sea data. These gadgets are bolted, clamped, or welded to all sections of the sled, and range from cutting-edge to surprisingly low-tech.

I noticed a small, cigar-shaped appliance bolted to the frame's upper rail; a cable ran from the device back into *Ventana's* innards. Bill mentioned that it was a high-quality color video camera. I found that impressive; the camera's outside diameter was little more than an inch. I asked about the camera's alignment, as it was aimed at one of *Ventana's* side rails, rather than forward. Bill pointed to a vertical row of three gauges, all bolted to the side rail. He said that the pilot onboard *Point Lobos* uses the camera to visually inspect the old-fashioned gauges, as a backup to the digital data sent up from the onboard computers.

Slung below the lower rail up front rides a beautiful, studio-quality Sony color camera in a high-pres-

sure container. This camera feeds all the monitors scattered throughout MBARI, and the picture quality is excellent.

On the very top rail, also up front, sat a white plastic barrel about five inches long and four inches high; its body contained several baffles and surfaces. It seemed oddly out of place, looking like something you would pick up at a garage sale. I reached up and gave it a spin, and it rotated freely on its horizontal axis. While it was spinning, I noticed a set of small magnets glued to one of the side rims, evenly spaced along the rim's circumference. When I asked Bill about the device, he said it was the current sensor.

Apparently, prior to *Ventana's* work, scientists had no reliable method of determining current flow at depths of 1000 meters. This little plastic device, with its reed-switch and magnets, gave researchers their first glimpse of deep-sea current rates.

Next, I went inside the ship to the control room. Here, the pilot, the chief scientist, and the assistant scientist sit in padded chairs before a group of eight relay racks crammed with gear. Several color monitors provide pictures from various cameras, both on board *Ventana* and from other sites on the ship or in the research lab.

The racks also contain several PCs and Unix boxes, linked via network to other machines throughout MBARI. While I was admiring the view, a technician walked in and rebooted one of the PCs. The usual boot messages scrolled up, superimposed on a high-quality color picture of the bow of *Point Lobos* as seen through *Ventana's* camera.

The pilot controls *Ventana* using a seat-mounted joystick that looks like it was pulled right out of a jet fighter's cockpit. Beside the pilot's chair sits a large, color touch screen display running a control program; the program allows the pilot to switch video feeds to any monitor in the ship.

A large readout at the top of one rack constantly displays the current longitude and latitude of *Point Lobos*, based on GPS technology. Behind the racks, a large steel trough carries a cable bundle, as thick as my thigh, back to the bowels of the ship.

Bill took me back to the Marine labs and, as a final treat, showed me a shelf of tiny styrofoam cups that one engineer has been collecting. This guy writes the day and depth of the dive on a foam coffee cup, then stashes the cup inside *Ventana* prior to a dive. When the vessel returns, he retrieves the cup and adds it to his mounting collection. Following the 1000-meter descent, each 8-ounce cup is squashed to the size of a sewing thimble; the 32-ounce soft drink cups wind up about 1 inch tall. In both cases, the writing is perfectly legible, though about 1/8 normal size.

I saw far more than this article can cover, and am indebted to Andy and Bill for spending so much time showing me around and answering my questions. The MBARI engineers are designing a beautiful machine for one of the most hostile robotic environments imaginable, and I was totally impressed with the ingenuity and quality of their work.

So Long, Bob

About the time you read this, the Seattle Robotics Society will say "Goodbye" to one of the club's first members and long a driving force. Bob Nansel and his wife, Karen, are packing up and heading to Japan, where they will teach English classes.

I first heard about Bob and the SRS through articles Bob wrote for the (sadly) defunct magazine, *Micro Cornucopia*. Bob's writings encouraged me to consider Seattle as a future home; after all, any city with a robot club has to be a great place, right?

Since Linda and I moved here, back in 1989, Bob did a tour of duty as the club's newsletter editor, and made the *Encoder* arguably the finest robotics newsletter in the country. He and Karen were also instrumental in organizing and running at least two major hobby robotics events in those years.

As the only club member I know of who is truly a robotics engineer (his degree is in robotics engineering), Bob was always explaining arcane electronic or mechanical problems to those of us more robotically challenged. And he used his skills in woodworking and machining to craft quality robot frames and bodies.

Best of luck to both of you, Bob and Karen, in your new Japanese lives. Hopefully, you'll keep the rest of us posted on robotics in Japan, and maybe come back to the States now and then for a visit.

Another "Goodbye"

The book *Star Trek Creator* describes the fascinating life of Gene Roddenberry. Shortly before his death in 1991, Gene asked writer David Alexander to prepare Gene's biography. David was given full access to the family archives, and drew upon literally thousands of pages of memos, letters, contracts, and reports, and countless interviews, in his efforts.

The resulting book, complete with an introduction by Majel Barret Roddenberry, covers the 70-year life of a man known to millions as the creator of the Star Trek shows. David pulled no punches in this book, as attested to by Majel's comments. Gene's life, the good parts and those less so, get full exposure.

David describes the background struggles and torturous efforts needed to develop the different Star Treks. Gene's allies, including noted authors Isaac Asimov and Ray Bradbury, helped him through the difficult years, as he fought battles that probably should have broken him (and might well have).

But throughout this book runs a constant thread: Gene's vision of a better tomorrow. Gene brought an eager enthusiasm to his view of humanity's future, a 23rd century in which our technology and our human-ness carries us to peace, and to the stars.

More than simply dream his dreams, Gene worked harder than most would have dared, to bring his visions into our living rooms and onto our movie screens. This book is filled with his struggles and triumphs, and with personal glimpses of those people intimately involved with the birth and growth of the various Star Treks.

Gene has left us, a fact I regret beyond words. But the 23rd century isn't that far away, and we have a lot more yet to do. Time to get on with it.

Cleaning up the Tennis Court

Over the years, members of the Seattle Robotics Society have earned the club a reputation for high-quality robots that compete well in major events. This is especially true of the four SRS members featured in this *Nuts & Volts* column published in October 1996. Anne Wright, Randy Sargent, Carl Witty, and Bill Bailey modified Bill's stepper-based robot, M1, for competition in a top-notch national event, and the little machine brought home first prize with a perfect run.

This column provides a behind-the-scenes look at some of the technical thrashing that went into building a first-rate competitor. The designers fought problems in all areas: hardware, software, and mechanics. Since M1 relied largely on a color vision system for its world view, the builders also had to solve some nasty optics problems. The difficulties they faced and the solutions they used should prove educational, even for designers of smaller, simpler machines.

Happily, M1's triumph was captured for posterity, as Alan Alda and the crew from *Scientific American Frontiers* were on hand to film the event. The show aired at the end of the season following this event, so many of us in the SRS got to see the local heroes in the spotlight for real. It's one thing to hear a description of the event at an SRS meeting; it's a whole different ballgame when you can watch fellow club-members in action.

The little robot rolled smoothly around the arena's interior, pausing occasionally to scan the floor for tennis balls. The machine had already collected nearly all of the original objects, including two motorized toys called squiggle balls that had been cruising around. But one tennis ball remained, set in the exact center of the 9 by 6 meter enclosure. By chance, it lay just beyond the small robot's sensor range, and the 'bot repeatedly missed locating it. With each sweep, the audience ringing the arena tensed in anticipation, then sighed in disappointment.

Then, with about 30 seconds remaining in the robot's allotted time, it made one last scan, found the tennis ball, and deposited it in the holding area. The crowd erupted in applause and cheers; the robot's designers, four members of the Seattle Robotics Society, had netted a perfect score in the 5th Annual American Association of Artificial Intelligence (AAAI) robotics contest.

In preparing for this contest, Anne Wright, Randy Sargent, Carl Witty, and Bill Bailey had put six long weeks into the design and modification of M1, Bill's original stepper-based robot. Long-time readers of this column will probably recognize Anne and Randy's names from my discussions of the MIT 6.270 contest two years ago. I've also mentioned Bill several times in these pages; I used his stepper motor controller board on a robot a couple of months

Figure 1. The proud contestants. Clockwise from upper left, Randy Sargent, Bill Bailey, Carl Witty, M1, and Anne Wright, the team that took 1st place in the 1996 AAAI Clean Up the Tennis Court contest.

ago. Carl is a principal in Newton Research Labs with Anne and Randy; NRL focuses on robot vision research, a key element in the group's success at AAAI. The collaboration of these four people on what would become a major win in a contest with worldwide attention began, appropriately enough, at one of the SRS' occasional tournaments.

Anne, Randy, and Carl showed up at the May 1996 Robothon with a modified R/C car. They had removed the R/C receiver from the 2-foot long car, replacing it with an NRL color vision system. The vision system basically took over the motor control for the R/C racer, steering it towards whatever object

of interest the vision system saw. Unfortunately, the car's motor controller did not lend itself to speed control, so the final design only had two speeds: stopped and flat out.

For demonstration purposes, Anne tuned the car's vision system to seek bright orange objects. She switched on the car and it sat on the floor, "running" at its stopped speed. Jet black with an open-body frame, it looked ominous, like a robo-raptor. Then Anne tossed a bright orange ball into the car's field of view and all hell broke loose.

The car went from stopped to full speed in a flash, hit the ball dead-center, then whipped around to track and realign on the target as it bounced away. Again and again the raptor tore after the orange ball, each time hitting it square and sending it off in another direction. Only solid furniture and Anne's IR remote control could stop the car once it had sighted the ball.

As a further demonstration, Randy and Carl set up a second, unmodified R/C car and handed the transmitter to one of the younger SRS members. Anne clipped a bright orange baseball cap to the back of this car, then gave the kid a head start before letting the raptor loose. The ensuing chase was short, violent, and not a contest. No one was able to out-maneuver the raptor, and usually the vision-equipped car hit the controlled car from behind in a matter of seconds after starting.

I found it easy to be impressed by the speed and the power of the robot, and lose sight of the real point. This car had only one sensor on it, a custom color video board. And although the car moved at an insane speed and had only one sensor, it **always** tracked and hit the target. A truly impressive display of a robust vision system.

At this same event, Bill demonstrated the ability of his M1 robot. Bill's stepper design, complete with a belt-drive speed reducer, meant that M1 had plenty of torque and was very precise. M1 moved very smoothly in all of its events, the stepper drive singing as the robot ran through its paces. M1's specialty event is the dead-reckoning contest, which requires a robot to move and steer a precisely defined (though unmarked) course. Points are awarded for how close a robot stops to the designated target, exactly 4 meters and 45 degrees away from the starting point. M1 usually hits within 1/2 inch of the target, and sometimes much closer than that.

So here were four robot builders, hanging out at Robothon. Anne, Randy, and Carl had a precise, sophisticated sensor system bolted onto a danger-

Figure 2. M1 shows its winning form. The robot holds a squiggle ball in its gripper. The gripper would next lift the ball off the floor for transport to the collecting pen.

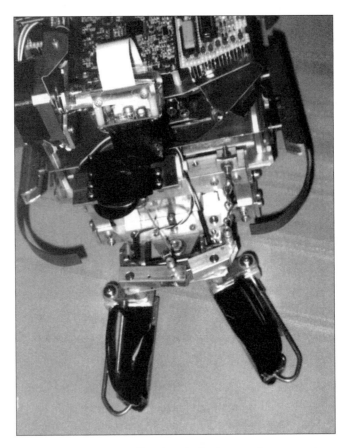

Figure 3. Another shot of the gripper. This photo looks down on the gripper, showing the hinge mechanism. The black box at the upper center of the photo is the light shield built over the color CCD camera; the white ribbon carries the CCD signals back to the larger PCB for conversion to NTSC video.

ously uncontrollable machine. Bill had a smoothly running, well-built machine carrying a traditional array of sensors, such as IR object detectors. It only seemed natural for the four of them to get together and discuss adding the vision system to M1.

In fact, the whole operation only took an afternoon. By the next meeting, one month later, the new M1 machine could follow a bright green line of plastic tape laid out in a maze pattern on the floor. The down-looking color camera kept the line in view, while the 68332 MCU fed stepper control signals to Bill's motor driver to keep M1 rolling forward and to keep

the line in sight. The result was a very smooth, fast robot that followed the line precisely.

But this was just a warm-up. The event that really interested the guys from Newton Research Labs was coming up August 3 through 8, 1996. The annual AAAI convention swaps ends of the country each year, and this summer found it in Portland, just down I-5 from Seattle. The AAAI event draws attendees and entrants from all over the world, and provides a showcase for real-world artificial intelligence applications. Besides the papers, discussions, and workshops, each convention includes at least one robotics contest. The NRL hackers had their sights set on the "Clean Up the Tennis Court" contest. Bill and M1 joined the team with little persuasion, and the four humans began designing the mods needed to turn M1 into a tennis ball collector.

The original rules for this AAAI contest called for a walled arena of undisclosed size, complete with a pen about one by two meters. The pen was to have a squiggle ball already rolling around in it, and the arena proper would contain several tennis balls and another roving squiggle ball. Within the 15 minutes allotted for each contestant, a robot would try to collect tennis balls and the roving squiggle ball, and deposit them in the pen. At the end of the run, the judges would award points based on the number of tennis balls and squiggle balls in the pen.

A robot could earn bonus points based on its behavior with the roving squiggle ball. For example, a robot would receive 30 bonus points the first time it demonstrated the ability to track a squiggle ball. This demonstration could appear as a voice or sound announcement. A robot earned an additional 30 points the first time it showed "purposeful contact" with the squiggle ball. That is, the robot announces it is moving toward the squiggle ball, then does so and contacts the ball. The first time a robot successfully captured a roving squiggle ball, the robot garnered an extra 50 point bonus.

Note that a robot could get the capturing bonus without getting either of the other bonuses. If a robot accidently scooped up a squiggle ball without announcing the ball's presence or moving purposefully toward it, that robot would earn only the capturing bonus.

From the beginning, the SRS team intended to take every bonus point and collect every tennis ball, a decision that ultimately earned them first place. For example, the rules permitted a team to mark or color the tennis balls, at a small penalty for each doctored ball, but this option was dismissed as it would cost points. The team wanted all possible points, and all of their work was based on this goal.

Figure 4. A side view of M1. This shot shows the belt-drive stepper motor system, the gripper, and some of the heavy metal construction Bill used in making M1.

Figure 5. M1's electronics. Here you see the stack of PCBs needed to make M1 a winner. The top board handles the IR object detection system, while the second layer includes the 68332 MCU board. The black cube left of center is a hobby servo motor that controls the gripper. The color CCD camera looks down from just above the servo motor.

Now began the tedious phase of determining the various behaviors the robot had to exhibit, and designing the system elements needed to support those behaviors. Some of these behaviors were readily apparent. M1 needed some method of transporting a tennis ball into the pen. After much discussion, the team settled on a front-mounted gripper that could clutch a tennis ball, then lift it off the floor. They didn't need a lot of lift, since their final pen design had a barrier across the front entrance only an inch tall, to keep the squiggle balls in the pen.

The final gripper design used a single hobby servo motor to close the gripper and to lift the gripper off the floor. Bill built a working model from aluminum, based on some proposed designs Randy did with Legos. The finished gripper contained molded plastic pieces to hold the ball in place, plus bumper switches to detect possible contact with the walls.

The team decided to use a wall-following behavior as part of M1's searching algorithm. This in turn meant that M1 needed electronics to find a wall, orient itself properly, and maintain a fixed distance from the wall as it moved. To support this behavior, Randy and Bill developed an eight-channel IR object detecting system and used it to drive a ring of sensors around M1's frame. This subsystem provided a single byte of data to the MCU, with each bit assigned to a specific sensor. Additionally, eight red LEDs mounted on M1's frame showed which sensors could see an object. If you passed your hand around M1, the red LEDs would light in a rippling pattern as each sensor in turn saw your hand.

Some behaviors required less obvious mechanical or electrical support. The vision system's color camera rode on the top of the frame, at the front of M1. After a few trials, the team realized that they needed a way to tilt the camera up and down, to aid in realignment. So Bill added a stepper motor with driver electronics. Since the MCU needed to realign the camera precisely, this addition included an optical interrupter that signaled when the camera had reached the desired position. Now the camera could tilt all the way down to track a tennis ball fully into the gripper, or tilt up to locate the pen or other targets across the floor.

By far the greatest number of behaviors centered on the vision system. The team ultimately cataloged over 20 different behaviors for this contest, many based on responses to seeing a yellow tennis ball, an orange squiggle ball, or the large blue square used to mark the pen's doorway. For example, M1 would move quickly towards the pen while carrying a tennis ball, but would ignore the pen when the gripper was empty. Even these behaviors required sub-behaviors. Thus, M1 only moved quickly towards the pen when it was a safe distance away. Once the robot got close to the pen, however, it automatically slowed down, so as to approach the pen at a safe speed and not dislodge the precious tennis ball with an inadvertent collision.

The four hackers relied on a suite of software tools developed by NRL over the years. The software running on M1's new 68332 brain supports some serious multi-tasking, and Carl and Randy were able to add and debug the various behaviors quickly while the robot ran. This kind of closely coupled software support for complex robotic systems can prove to be the crucial edge, whether you are working on a contest robot or an industrial application.

Carl Witty was instrumental in designing the software for controlling M1's motions. Carl's software was able to translate the vision system's output into the proper motion paths for intercepting a tennis ball or squiggle ball, then locating and moving toward the pen. The software design had to correct for the vision system's fish-eye lens, which distorted the robot's world view and complicated the mapping of objects onto an XY coordinate system for path planning.

The group worked feverishly in the weeks leading up to the Portland event. They tested and retested the vision system, the gripper, and other elements, and planned as best they could how all the behaviors would work together. But soon, too soon, the day arrived for the trip down to Portland and a look at the contest venue.

One of the first surprises the four encountered involved the lighting over the arena. The convention center used a high-intensity sodium lighting, which caused an immediate problem with the vision system. The light was so bright that it washed out the yellow tennis balls, and M1 was unable to see any yellow objects. Despite intense effort, the team could not solve this problem in time for the first preliminary run, and had to make this first run using two live squiggle balls and four dead ones, which served as tennis balls. The "modified" tennis balls cost the group several penalty points, but the results of this run did not count against the final score, so the penalties were of no consequence.

Shortly after this first run, Jeremy Brown, a grad student from MIT who had arrived to help out, hit upon the solution for the bright lights. He built a neutral filter out of several layers of plastic cut from High-8 video cassettes. When placed in front of the camera's lens, this simple filter reduced the glare enough that M1 could finally see the tennis balls.

Most robot sensor systems react oddly to seemingly innocent environmental changes, and M1's vision system obliged to add some harrowing problems for the group to solve. In one case, the robot would pick up a tennis ball, then head away from the pen, toward another corner of the arena, where it would wander back and forth. After a few trials, the team discovered that some of the audience standing along the three-foot wall were wearing T-shirts of nearly the same shade of blue used to mark the pen's door. The fish-eye lens distorted the camera's view to the point that it could see these shirts, and reacted as if they marked the pen. A slight change to the camera's alignment and some tweaked software fixed this problem.

Another problem surfaced with the IR wall detectors. The two front-looking IR detectors had been mounted next to some of the IR emitters, and one detector persisted in giving false positive signals. After considerable work, the team realized that the seams in the enclosure for this detector did not meet perfectly. The tiny gap let just enough IR leak in that the detector reported an object when there wasn't one. Some tape and remounting solved this problem.

The gripper did its part in complicating matters. Sometimes the gripper would carry a tennis ball too high. When the robot tried to drop the ball into the pen, the gripper would ride over the pen's front rail, then get hung up, trapping the robot. A slight redesign fixed this.

A second preliminary run, this time with tennis balls, turned up a few more problems that needed attention. Since the final run, the only run that counted, was scheduled for 2:30 the following afternoon, the group had no choice but to pull an all-nighter to complete the design.

The final contest would be held in an arena measuring six by nine meters, so the team set up a test arena of exactly this size. This uncovered a few minor problems with the vision system, since the walls were now farther away than they had been in the previous tests. I say minor, but had the SRS crew not run their tests in a full-sized arena, any of these "minor" problems could well have cost them dearly.

They worked through the night, tweaking and testing, always pushing the system a little harder, trying to ensure that they could get all the tennis balls and both squiggle balls. By now, the judges had decided that each run would have two squiggle balls roaming the floor and no squiggle ball in the pen at startup. This added a huge premium for collecting both squiggle balls, a feat the team was determined to pull off.

Finally, by 12:45 that morning, M1 was able to collect all the tennis balls and squiggle balls in three successive runs. At this point the team did something that marks an experienced group, and may well have sealed their victory; they froze the whole design. No software changes, no mechanical tweaks, no retuning. The discipline this requires seems beyond many robotic contestants, to their downfall. If

your robot works, and you change it, it no longer works. The years that Anne and Randy had spent in the MIT 6.270 robot contests had proven this repeatedly. So with the robot working perfectly, they shut everything down and caught some badly needed rest before the final competition.

The team assembled at the arena that afternoon, to the glare of TV lights, the crowd of spectators, and the sight of the other teams and their robots. Alan Alda and the crew for *Scientific American Frontiers* were on hand to film the event for later broadcast on the PBS show. After the warm-ups and introductions, the team watched the first competitor make its final run. Then center stage belonged to the four SRS teammates, Jeremy Brown from MIT, and M1.

Within seconds of starting, M1 had spotted its first tennis ball, moved directly to it, picked it up in its gripper, and hustled over to the pen. The smooth technique, with no wasted motion, signaled to the crowd that this robot knew what it was about. After picking up one or two more tennis balls, M1 had its eye on another tennis ball when a squiggle ball rolled into its field of vision. With a series of high-pitched chirps signaling M1's sighting of the ball, it turned and gave chase. The ball rolled this way and that, as if trying to avoid the machine, but M1 never lost sight and closed quickly. With the ball trapped in the corner, M1 deftly picked it up in its gripper and carted it away to the pen, to loud and prolonged applause from the crowd (Figure 6).

The first half of the collection went quickly, since many of the balls were visible to M1 from the pen's door. Thus, the robot had only to deposit one tennis ball and turn around to find another within view. After rounding up the easier tennis balls and the second squiggle ball, M1 had to rely more and more on its searching algorithm. This involved moving to the wall immediately to the right of the pen, then motoring around the arena's interior. Every four seconds or so, M1 would pause and turn 180 degrees looking for nearby tennis balls, then resume following the wall if it found no targets.

Using this technique, M1 located and retrieved all but one of the remaining tennis balls. The last ball

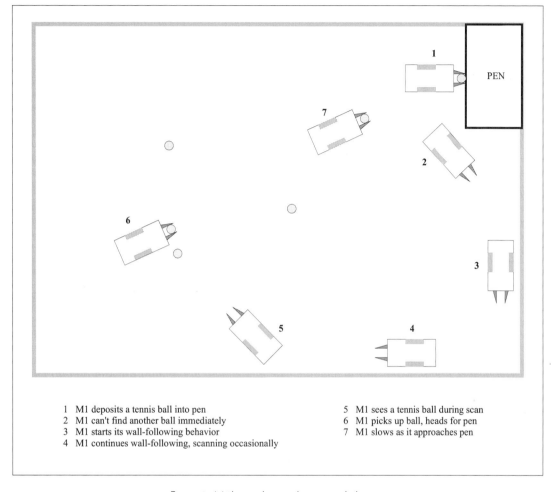

1 M1 deposits a tennis ball into pen
2 M1 can't find another ball immediately
3 M1 starts its wall-following behavior
4 M1 continues wall-following, scanning occasionally
5 M1 sees a tennis ball during scan
6 M1 picks up ball, heads for pen
7 M1 slows as it approaches pen

Figure 6. M1's searching and retrieving behaviors.

rested in the center of the arena, and M1 had to make several journeys around the interior before finally, with the clock winding down, the combination of timing and distance from the wall left the robot exactly where it needed to be to find the last ball. A quick excursion after the ball, another trip to the pen, and the rest, as they say, is history.

When asked what incident struck them most forcefully about this final competition, Anne and Carl both mentioned the last tennis ball in the center. The fish-eye lens, coupled with the placement of this tiny object, nearly proved too much for M1. It added drama to the event, no doubt, but it also aged the SRS team unnecessarily. Bill remembers watching M1 at the outset, moving quickly to the first tennis ball and picking it up within seconds; Alan Alda responded with a surprised, "Look at that!" Randy was struck by the crowd's repeated applause and cheers each time M1 snared a tennis ball or squiggle ball; this was, after all, what the robot was **supposed** to do.

Some final comments from this team seem appropriate. Anne Wright, who has been designing robotic vision systems for about five years now, noted that M1 was not the only robot using vision in this contest. However, M1 was the only robot sampling the video at 60 times per second. It was this high frame rate that made the difference for M1's winning performance. Slower rates, particularly 10 frames per second or less, prevent the robot from tracking a moving target, which immediately cuts that machine out of some big bonus points.

Bill and Randy both commented on the team's mindset. From the very beginning, the SRS group determined to solve every problem and to garner every bonus point. Other teams didn't take on some of the contest's aspects, perhaps deciding that something was just too difficult for anyone to do, so why go for it? Of course, given the NRL vision system and Bill's M1 base, the SRS team had the luxury of beginning from a proven and sophisticated technological starting point. Thus, the long hours they put into their designs went to defining behaviors and tuning M1, not, for instance, in getting a vision system to work at all.

But their victory was an impressive feat, and sets a very high standard for next year's competitors. It sets an equally high standard for the SRS team; after all, they got a perfect score this time out. How do you top that?

I have watched Anne's video tape of the SRS effort repeatedly, and each time find it thrilling and educational. The range of behaviors shown by M1, some quite subtle, make a great case study of subsumption architecture in action. I hope you have an opportunity to catch this event when it airs next season. Check your local TV listings for *Scientific American Frontiers*; the new season starts in October of 1996. You can find previews and comments about upcoming shows on the PBS web site.

Please note that Newton Research Labs' color video vision system, dubbed Cognachrome, is available commercially, and would make a wonderful addition to any university or industrial research facility. You can find detailed information on Cognachrome through links at the Newton Research Labs web site. Links on this page will also take you to information about ARC, the Labs' proprietary suite of software tools that support Cognachrome development.

Robot Soccer

The gang from Newton Research Labs (NRL) have figured in more than one high-profile championship robotics event; an earlier column discussed their first-place finish in the AAAI "Cleanup the Tennis Court" contest. This column, which appeared in the February 1997 issue of *Nuts & Volts*, describes a truly world-class win, their triumph at the First Annual MIROSOT robot soccer competition.

The team bested some very strong competitors but, as the article clearly shows, it was a grueling six weeks of hard work to get ready. They had to overcome many technical problems in all phases of robotics design before they were ready, and they never really knew that everything worked until they day they arrived at the venue in Korea.

This was a huge undertaking, but it's great fun to read about in the comfort of your easy chair. Next time you're agonizing about that Mini Sumo deadline four or five months out, reflect for a moment on the SRS team that had to design and build a small fleet of vision-controlled soccer-playing robots in just six weeks. Then, if you're like me, you'll go have a beer...

The goalie moved sharply to the left, staying between the ball and goal, while the right forward raced ahead, lining up a pass. A quick nudge sent the ball rolling across the field to the left forward, which redirected the ball into the now-open goal, well before the opposing goalie could react. This quick give-and-go play added yet another point, as the robots from the Newton Research Labs team powered their way to a 20-0 win in the finals of the first Micro-Robot World Cup Soccer Tournament.

The tournament, held November 9-12, 1996, took place in Taejon, Korea, and included 23 teams from all over the world. It was organized by Korea Advanced Institute of Science and Technology (KAIST), sponsored by the IEEE Robotics and Automation Society, and supported by LG Semicon Co., Ltd. Known by its acronym, MIROSOT, this event featured teams of three tiny robots in a single-elimination contest based on soccer.

The MIROSOT format plays very much like the regular game of soccer, but reduced in size and with added refinements designed to help its robot competitors. The pitch, or playing surface, is a smooth, dark-green wooden surface 90 cm wide and 130 cm long, covered with a grid of white lines 10 cm apart. The boundaries are marked with wooden vertical walls 5 cm high. The goals, set at opposite ends of the field's length, are 30 cm wide and 12 cm high, with a 5 cm wall behind each goal. An orange golf ball serves as a soccer ball.

Each team consists of two forwards and a goalie. Each robot must fit within a cube 7.5 cm (about 3 inches) on a side, excluding any antenna. If a robot uses a gripper or other appendage, it must still fit within the assigned size, even when fully extended. All robots must be fully self-powered, using self-contained motor systems. Only wireless communications between robot and host system are permitted.

All control signals must be generated solely by a host computer system. No human member of a robotics team may use joystick or keyboard input to control the motion of a robot. Should a team desire to use a vision system for scanning the playing area, the camera must be mounted directly over the pitch, no closer than 2 meters from the surface.

Additional rules cover topics such as fouls and penalties, goal-scoring, substitutions, time-outs, free kicks, and functions of the (human) referees. As of this writing (Thanksgiving, 1996), you could find all of the above information, and much more, from the MIROSOT web site. Oddly enough, two weeks after the event was held, I was not able to find any scoring information or even the name of the winning team.

Designing the Robots

Speaking of the winning team, the four humans involved had never even heard of MIROSOT until six weeks before the event was scheduled. Randy Sargent, Anne Wright, Carl Witty, and Bill Bailey, all members of the Seattle Robotics Society (SRS), were deep in the design of their winning AAAI entry (see "Cleaning up the Tennis Court") when Randy got word of the looming soccer contest. When presented with the idea of entering this contest as well, the other three team members responded with a loud, "No way!" The AAAI design effort was already back-breaking; no one wanted to pile on yet another project.

But Randy wouldn't let the idea die, and mentally tinkered with what it would take to build so much capability into such a small package. Finally, he decided that he would spend one day, no more,

trying to build a robot prototype to do the job. If, at the end of that day, he had a machine that was promising enough, he would once more ask the team about entering the MIROSOT event.

That evening, Randy had fashioned a small robot out of a cut-down 68hc11 Miniboard, some surplus gearhead motors, and two 9-volt batteries. The whole package fit within the required space, with 1/2 inch of headroom at the top, allocated for the not-yet-designed communications electronics. It was crude and didn't run straight, but it worked enough to convince the others that the design was indeed possible.

Following their triumph at the AAAI event and after moving tons of computer and robot gear into a new house in Bellevue, the four teammates settled down to tackle the difficult design elements of the soccer robots. Tops on the list was some method of sending information between a host PC and the three small robots. Since the host PC would use the Newton Labs vision system to monitor the playing field, the team only needed a one-way link. But that link had to run at least 9600 baud, or it couldn't keep up with the vision system or the action.

The group quickly ran through existing possibilities, discarding each in turn. IR was out, as it was too susceptible to ambient light; the sodium vapor lights at AAAI proved that. The team couldn't find a small enough RF modem in time, and was running out of options when Randy stumbled onto the answer. He returned home one day with a couple of wireless stereo headphones from the local Radio Shack. These are basically short-range one-way FM stereo radio systems, designed to pipe audio from some source, such as your stereo TV, to one or more headsets for private listening.

Randy and Bill were intrigued about using the headset as the RF link on a robot. The transmitting side seemed simple enough: just add a resistor divider to the serial output from a PC's COM port and run the reduced signal into the audio input of the wireless headset transmitter. Whatever serial data you sent to the COM port would be transmitted as audio tones over the 900 MHz RF link and picked up on the headset's receiver.

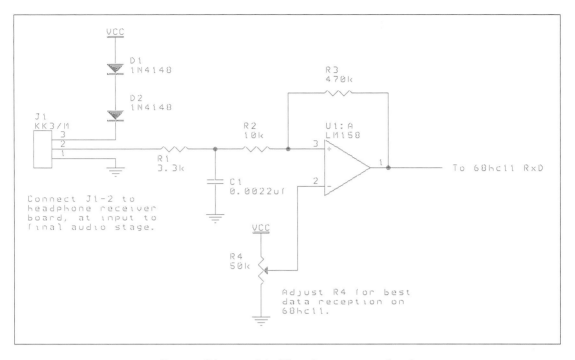

Figure 1. Schematic of the RF serial communications board.

Bill then designed a one-IC level detector and squaring circuit to convert the received signal into digital levels, for wiring directly into the Serial Communication Interface (SCI) of the Miniboard (Figure 1). Initial tests looked very good; the 9600 baud data stream appeared at the SCI port and was mostly accurate.

But not all of the characters came through cleanly. Tests showed that the headset's amplifier, which naturally treated the signal as audio, was trying to keep the output centered at 0 volts. Bill's circuit would then drop out some of the bits, since the signal had settled outside the detector's threshold.

Rather than give up on the whole design, the two determined by experimentation that if a data byte contained three, four, or five 1-bits, the byte came through intact. Apparently, these patterns kept the receiver's output signal in a suitable range for Bill's detector circuit. Armed with this data, they developed a table of the 182 different bit patterns that could be transferred successfully. They then built a conversion scheme that let them send the commands and data they needed, using only the permitted characters.

Further tests showed that the RF link worked perfectly at 9600 baud. Randy bought a few more headsets, Bill worked up a mechanical mounting scheme for the liberated receiver boards, and the RF link was in place.

Next up, the team needed some type of pulse-width modulation (PWM) control for the DC gearhead motors they had selected. These were prime units, but they were never meant to run on 18 VDC. And since the event called for both speed and control, the motor controller really needed some type of feedback for sensing actual speed. But there was no physical room left for adding wheel encoders and sensors. The designers needed another way.

Bill had been working for some time on just such a speed controller, based on an idea he had picked up from Lance Keizer, one of the SRS regulars. Bill's design uses a 68hc11 A/D port to monitor the volt-

Figure 2. A frontal view of a soccer-bot, showing the two-gear
Lego drive. Two 9-volt batteries, front and back, act as skids.

age on a motor's power leads, sensing the back EMF as a measure of the motor's running load. The circuit is quite simple, consisting of a few passives, including some diodes to protect the A/D port from out-of-bounds voltages. Coupled with the appropriate software, the little gearhead motors provided excellent performance. Bill's back-EMF motor sensor really deserves more attention than what little I'm providing here. Look for more info in an upcoming article.

Now Bill turned his attention to the mechanics needed to build the little robots. Three inches on a side doesn't leave a lot of room for hardware, and with Randy's help, Bill put together a robust platform with a minimalist design. Two small pieces of flat 1/8th-inch aluminum stock form the base. The motors, nestled side by side with the output shafts facing opposite directions, are held to the base with a thin, sheet steel clamp and metal screws. Four posts, made of threaded aluminum spacers, hold the stack of PCBs tight against the top of the insulated motor clamp.

The PCB stack consists of three boards. The cut-down Miniboard sits at the bottom, and holds the 68hc811e2 MCU, the motor power driver chips, and sundry LEDs for status. The middle board contains all of Bill's custom circuitry, including wiring for checking the remaining battery voltage. At the top sits the cannabalized RF board from the FM headset, held into place by short pieces of copper wire, acting as guys to nearby 4-40 hardware at the top of the spacer stacks.

Randy's original prototype had the drive wheels attached directly to the motors, resulting in off-center wheels. The unsymmetrical layout caused the robot to lurch sideways when it started up, aggravating the control and steering problems. While Bill worked on the mechanics, Randy looked into getting the wheels back on the frame's centerline where they belonged.

He ended up building a two-gear Lego drive system, using two Lego wheel-and-gear assemblies at the center axle. Fastening the Lego drive gear to

Figure 3. Looking down on a 3-inch soccer robot. The top board came from a Radio Shack wireless headset, the second board holds Bill Bailey's custom electronics, and the third board is a cut-down Miniboard.

Figure 4. The underside of this little robot shows the two 9-volt batteries, with the Lego axle and drive train running between them. Thin wire clamps hold the batteries to the custom metal frame.

the motor shaft proved difficult, however. Since the motor already had a substantial gear reduction built in, the resulting torque would pop the wheel off the shaft after just a few starts. Various glues couldn't hold the steel shaft to the wheel, so more drastic measures were needed.

Bill machined a narrow slot down the center of a motor shaft, into which he wedged a short wire. They fitted the wheel onto the shaft and tried again, but the motor's torque snapped the wire. A square metal wirewrap pin fared better; it bent but did not break. Finally, Bill used a small piece of hacksaw blade, cut to the proper shape. Randy fitted a Lego wheel onto this shaft and key arrangement, potting the wheel in place with hot glue. This mounting system has proved remarkably robust, and the team gets a charge out of seeing this item in the official bill of materials; "one hacksaw blade, heavily modified."

The robot power system proved another obstacle. The team had considered using small NiCd batteries for power, but abandoned that idea based on concerns over charger availability and charging time once the contest began. Instead, they opted to use a pair of 9-volt alkalines on each machine. Unfortunately, the power drain was so great that each robot could get only eight minutes of running time out of a pair of batteries. This didn't leave much spare time, since each game consisted of two five-minute halves and the team wouldn't be able to change batteries during either half. They solved the problem by sheer numbers, and boarded the flight to Korea with 100 alkaline batteries.

As the deadline raced inward, the group poured more and more energy into the project. The programming team, headed by Carl Witty, needed to build three suites of software. The 68hc11 code didn't need to do much by most robot standards; just basic command parsing and motor control. But they only had 2K of code space to work with, and translating the incoming characters into the proper numbers and letters used up a lot of space. Finally, with only bytes to spare, the 68hc11 code was ready to lock down.

The host system, however, was a different story. It consisted of two machines, the 68332 board that ran the vision system and a Pentium PC running Linux, used to calculate robot motions and direct traffic on the playing surface. The 68332 ran a suite of code for translating the color video signal into a data stream for use by the Linux box. Most of the 68332 code carried over from previous projects, but fitting it all together into a cohesive unit still took valuable time.

The Linux code to run the whole soccer team required a massive amount of design effort and computational capacity. The vision system, providing updates at 60 frames per second, pushed the original Pentium system to the limits as it tried to keep up with a rolling ball, the coordinate system, and all the robots running around on the field. Shortly before leaving for Korea, the group decided that they simply needed more horsepower, so they picked up a P6-180 machine.

And I do mean "shortly." In fact, the box arrived the morning before the group got on the plane for Korea. The Linux installation obviously worked, given the contest's outcome, but that was cutting it way too close.

After all this effort under vicious deadline pressure, the group was beginning to worry that maybe they weren't going to be able to pull this one off. Whereas a few weeks prior they had naively determined to grab first prize, the size of the competition and the problems they had struggled to overcome made them rethink their objectives. They freely admit that by the time they boarded the plane, their goal was simply "not to embarass ourselves."

The Big Event

After arriving in Taejon, Korea, the group went through registration and started to meet some of the competition. The MIROSOT contest drew several teams from Korea, as well as Singapore, Australia, France, Canada, Japan, Spain, Switzerland, Taiwan, Brazil, Italy, and Romania. Besides the Newton Labs team, the United States was represented by teams from University of California (Riverside), University of Southern California, and Carnegie Mellon University.

Repeated tests in the setup area showed the strength of the group's design, and they were quickly marked as the team to beat. Since they officially represented Newton Research Labs, they became known throughout the venue as "Newton Team."

The initial draw for pairings had them set to compete on the second day, but officials wanted to change that to the first day, to provide a showcase round for various government officials. Several of the competitors involved protested, and rightly so, about the last-minute proposal. The Newton Team resolved the matter by offering to hold a demonstration match on the first day. This was accepted, and the venue prepared for the first look at this American entry. Refer to the accompanying diagram for an overhead view of a typical MIROSOT contest (Figure 5).

It was (you've heard this before) not a contest. Newton Team took the demo event 12-3, and at least one of the goals scored against them was an own-goal, when the NT goalie booted the ball into its own goal. Still, the crowd applauded wildly at each score, and Newton Team got a chance to work out a couple of minor bugs before the real event began.

Looking back, one of the bugs was actually pretty funny. Both teams were set up for the kickoff and the referee blew her whistle to start the match. The NT goalie immediately raced the length of the pitch, elbowing its way through the crowd of robots, to take position at the opposite goal. Somewhere in the setup, someone had not initialized a variable properly, and the goalie was determined to guard the other goal. At the first stoppage of play, the problem was corrected and the goalie took up a more orthodox style.

The demo match was as close as it ever got. Newton Team played a total of four matches in the single-elimination event, and in those four matches gave up just one goal, while scoring anywhere from 15 to 21 points per game.

The video tape of that event highlights a number of differences between the Newton Team robots and those of the other competitors. Tops on the list

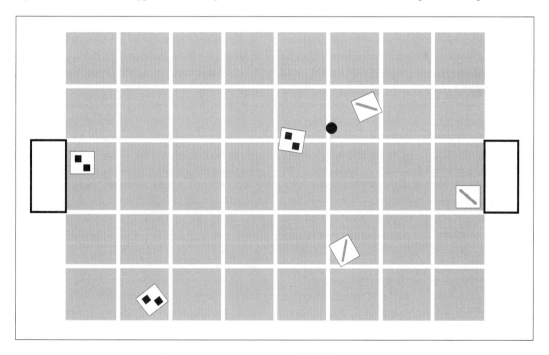

Figure 5. Overhead view of a typical MIROSOT robot soccer layout.

Figure 6. The winning soccer-bot designers - clockwise from lower left: Randy Sargent, Carl Witty (holding the trophy), Bill Bailey, and Anne Wright.

was the vision system frame rate. Every team used vision to guide their machines, but only the Newton Team was using a 60 Hz update rate; most settled for 10 to 15 Hz. This higher rate gave the NT machines a decided advantage in speed and reaction. Time after time, the opposing robots would either freeze in place at a crucial moment, or move in the wrong direction, reacting to a situation of an instant before.

In contrast, the NT robots were never still. Jittering around like windup toys on Jolt, they appeared to hover in place waiting for some cue. When the ball rolled free or the alignment was right, one of the robots would suddenly shoot off after the ball, smacking it with the front end of its frame and sending the ball towards the goal. Often, this move happened too fast for the opposing goalie, which got to the blocking position too late to stop the score.

Many times, the two NT forwards set up a passing attack or a simple give-and-go, using either a direct pass from player to player or, in one instance, a bounce shot of the wall. These would often materialize seemingly out of nowhere. At first the ball

was enmeshed in a knot of robots, then it was rolling down the field with an NT machine in pursuit. When a defender moved to intercept, the NT robot passed across to its partner, which either scored or passed back to end the play.

And these robots moved faster than I would have believed possible. After Newton Team's machines scored a goal, the opposing robots were set up to deliver a free kick from midfield. This meant that the ball would be placed in the center of the field, with an opposing robot immediately behind the ball and touching it. The NT machines, however, had to be set a minimum of six inches away. With monotonous regularity, an NT machine would streak to the ball and push it away from the opponent before that robot had even moved.

This doesn't mean the Newton Team design worked perfectly; only that the problems remaining were far less serious than those facing their competitors. For example, the power drain on the NT robots was so great that they slowed noticeably at the end of each half. In one case, the robots had actually dropped to about half speed, and it seemed doubtful that they would make it to the half before stopping altogether. They did, but it was a near thing.

In another contest, one of the robots developed a serious guidance problem that could not be fixed before halftime, so Bill Bailey, the head robot wrangler, simply turned off the power and left the robot sitting on the pitch. Even though it did figure in a later defensive play, the Newton Team played two on one for the final 45 seconds of that half; the opposing team, however, was unable to score.

This brings up the goalie. Carl's strategy for the goalie couldn't have been simpler. Just stay between the ball, wherever it goes, and the goal. That's all the goalie had to do, and that's all it took to keep the opponents from scoring. Doubtless, other teams had the same design, but the fast motors and high frame rate of the Newton Team's system meant that the NT goalie actually did its job. Other goaltenders would not track the ball quickly enough, or would freeze momentarily as if trying to understand their instructions. The NT attackers were so fast and accurate that the slightest hesitation on the part of the defending goalie usually resulted in an NT score.

And the Newton Team robots did a lot of scoring. By the final round, played for first place, a large trophy, and the generous prize money of US $2000, the audience was becoming pretty jaded. Hopes for the competitors were dashed almost immediately as

NT grabbed a couple of quick scores on the way to a 20-0 rout. The audience, overwhelmingly Korean, still applauded every score, but by the second half the applause was noticeably restrained. A human team would have thrown in the towel early on, but these were robots, and the contest ground along to its foregone conclusion.

Part of the MIROSOT event included a published paper from each team, describing some aspect of their design they felt most important. Accordingly, the Newton Team presented a paper entitled, *Use of Fast Vision Tracking for Cooperating Robots in the MIROSOT*. As I mentioned in my previous column on the AAAI event, the Newton Research Labs' vision system, dubbed Cognachrome, is available commercially. Stop by the Newton Labs web site to leaf through the technical info.

When last I talked to the Newton Team, they were planning to kick back and relax, to wind down from the grueling (and self-inflicted, I might add) pace of the past several months. Anne and Randy have taken up the Magic game, Bill wants to get back to writing software, and Carl has wads of Linux systems scattered all through the new house on which to work his own magic. But it won't take much to get them cranked up again. Randy, in particular, is usually just one email away from charging down yet another avenue, and he always brings his friends along. So stay tuned; you haven't heard the last from these SRS members.

The Extremes of Hobby Robotics

From the lightest, smallest robots to the heaviest, largest machines in the hobby, this column from the July 1997 issue of *Nuts & Volts* has it all. I start with a description of the BEAM robots, championed over the years by Mark Tilden, a pioneer in amateur robotics and a mondo 'bot builder. Mark's BEAM creations are simple, elegant, and often do not require an on-board microcontroller, making them ideal projects for those starting out.

In fact, the Photopopper, a small BEAM device sold by Solarbotics of Canada, qualifies as an excellent first robot. I had great fun putting together this robot kit, and go into some depth on its design and construction. The concepts behind this little solar-powered robot make it an excellent teaching tool, since you don't have to worry about writing software, charging batteries, or doing lots of wiring.

At the opposite end of the robo-spectrum lie the machines of Robot Wars, an annual event held in San Francisco. The sheer size and power of the beasts that compete in this free-for-all stagger the imagination. With a heavyweight class that tops out at 165 pounds, these metal monsters are built to take the punishment and to dish it out. I'm not sure how much technology will transfer directly from a Robot Wars design to your average robopet; most hobby machines don't use or need gasoline en-

gines for motive power. But watching these hulks duke it out is a kick, and just might inspire you to build a larger robot.

Finally, I describe an experimental drive platform that I lashed together from a couple of gutted cassette tape drives. The concept is pretty simple, but you can build on it to create a custom robot frame from surplus toys or other devices.

My recent articles have described small, lightweight robots suitable for building by beginners. But those robopets are hardly the final answer in low-power machines. To get a real taste for a simple, low-calorie robot you can build yourself, check out the BEAM Photopopper 4.2, available for $60 from Solarbotics in Canada.

I recently got to play with one of these little beasties. The Photopopper has no microcontroller, no software, no battery, and virtually no weight, yet its clever analog circuitry lets it show a limited set of behaviors. But before I get caught up in describing this robot, I need to provide a little history.

Mark Tilden first developed the idea of these small robots around 1990. He coined the term BEAM from the four driving elements of his philosophy regarding robotics: Biology, Electronics, Aesthetics, and Mechanics. His approach to making small, cool robots calls for reusing technoscrap to build efficient machines that mimic elements of Nature's designs. Mark's machines even turn to Nature for power, relying on a photocell to supply the miniscule amount of current needed to make a BEAM 'bot go.

Mark's web site carries lots of information about the various BEAM robots and competitions, plus a history of the annual BEAM Olympics that Mark started some years ago in Ontario, Canada. Solarbotics, which maintains a cool web page devoted to BEAM-robot products, sells the Photopopper kit based on one of Mark's designs. Drop by their web site for a look at pictures of several BEAM robots.

The Photopopper

The first thing I noticed about the BEAM robots in general and the Photopopper in particular was the lack of a microcontroller. Instead, these machines use clever but simple electronic circuits to control power and motion. After all, an MCU and its attendant electronics would add too much weight and burn too much power. Granted, we're only talking milliamps here, but for a BEAM machine, even that is too much drain.

The accompanying Photopopper schematic is typical of the many BEAM circuits I've seen, though this is the first one I can recall that controls two motors and allows for light sensing and tracking (Figure 2). In general, BEAM circuits use a high-efficiency photocell to soak up as much energy as possible from the ambient light, storing the accumulating charge in a large electrolytic capacitor.

When the stored voltage reaches a threshold, usually about 2.5 volts, a trigger circuit fires and dumps the entire accumulation into one of the two motor drivers; which driver receives the current depends on which of the two photodiodes "sees" more light at the moment. The corresponding motor spins a few revolutions before running out of gas and stopping. The capacitor, drained down to a few millivolts, resets the trigger and the soaking action starts all over.

Figure 1. Photopopper 4.2.

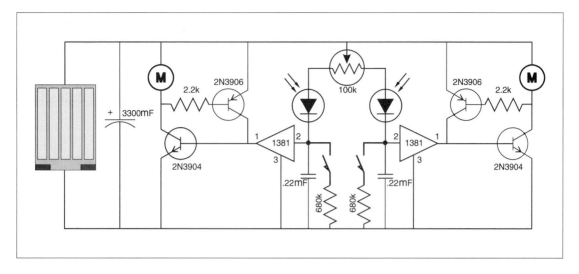

Figure 2. Schematic of Photopopper 4.2 robot (Copyright Solarbotics Ltd. 1996).

What you see as all this takes place is a series of arc-shaped lurches as the Photopopper pivots first one way, then the other, towards the brightest light nearby. If you set a bright light directly in front of and above the Photopopper, it will advance toward the light like a small robotic crab.

The Photopopper's motion profile depends on a great many factors, all carefully balanced in this design to use the tiny photoelectric voltage as efficiently as possible. The motors are miniature, high-efficiency devices that appear to have been pulled from pagers. Though the bodies of these motors measure just 5/8-inch long by 1/4-inch wide, they provide ample torque for moving the little 'bot. Rather than use wheels on the motor shafts, which add extra mass, the Photopopper rests directly on the downward-pointing motor shafts, and rolls forward as the motors spin. The robot itself uses a very thin, flexible circuit board as both robot base and electronics carrier. This dual use results in a much lighter 'bot, letting the Photopopper get as much mileage as possible from its photocell.

The robot's electronic design means that the Photopopper moves in a series of pulsed motions, as the capacitor alternately stores the photovoltaic charge, then dumps it into one of the motor drivers. Each time a motor driver fires, the robot pivots forward about 5/8-inch, and how frequently the robot moves depends largely on the ambient light. If I set the Photopopper down on my deck outside, in the (rare) bright Seattle sunlight, I get about four "pops" a second. This results in motion not unlike a giant green beetle spastically trying to lug a solar cell into the light.

But taking the Photopopper indoors doesn't mean it stops. Diffuse overcast sunlight through a living room window will still get a pop every second or so, while reflected light from a halogen lamp yields a pop about every ten seconds. In fact, the only way to stop the motion altogether is to put the little 'bot in a box and close the lid.

The Photopopper's designers added a very clever element in the form of a feeler or bump switch (Figure 3). This touch sensor consists of a single wire-wrap pin, coated with a piece of heat-shrink and inserted into the center of a tiny brass spring. The body of the spring is just long enough to extend the length of the pin, and the feeler end of the spring is nothing more than a five-inch length of the uncoiled brass wire. An electrical connection forms between

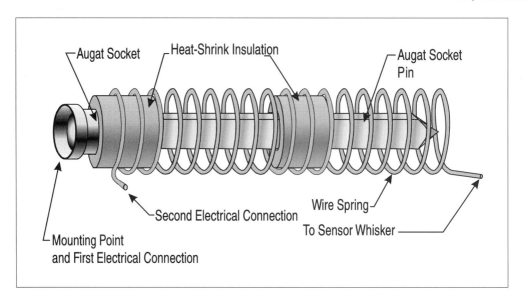

Figure 3. Bumper switch design.

the pin and the spring if the feeler wire is pushed hard enough to bend the end of the spring into the uninsulated tip of the wire-wrap pin. Since the inside of the spring is a circle, the feeler is sensitive through 360 degrees. This touch sensor, designed by Mark Tilden, stands as the most clever solution I've yet seen to the difficult problem of a cheap, elegant bumper switch with 360-degree coverage.

The two bumper switches, one for each side of the 'bot, appear in the schematic as the switches at one end of the 680K resistors. If the robot hits an obstacle and a bumper switch closes, the motor driver for the opposite side is shut down, regardless of the light's location. This effectively shuts down the light-tracking behavior, giving precedence to the obstacle-avoidance behavior. Mark was heavily influenced by designs of Rodney Brooks at MIT; this concept of overriding behaviors is known as subsumption architecture. I usually see subsumption architecture expressed in software designs. The Photopopper's creators have carried subsumption into the electronics realm in a very clean and elegant way.

I will admit to tinkering with the original design somewhat. You can't really blame me; since I couldn't

tweak the software (there isn't any), I had to find something to change. The robot originally used a smallish photocell. Though it appears to be one of the high-efficiency Sunceram™ types by Panasonic, I wanted to try a larger cell to see if I could get a power boost. So I wired in my own Sunceram cell that measures 2 1/4 inches by 3 1/8 inches, nearly the size of the entire Photopopper circuit board. The Sunceram cells are fabbed on unprotected glass, so I added a daub of hot-glue to each of the corners, to protect them against chipping. The new solar cell seems to add a little pep, but the Photopopper works just fine with the original cell.

I'll pass along a few tips learned from my construction of this kit. First of all, the kit's manual is wonderful. It is filled with accurate artwork and lots of well-done drawings, and the instructions are well-written and clear. The manual emulates the old Heathkit docs, complete with check-boxes and resistor color bands. The authors even sprinkled in a little humor here and there; did you ever see an intentional joke in a Heathkit manual?

With such a quality manual, this kit is within the range of an experienced builder, **IF** you follow the kit's instructions exactly! The authors state several times, read the manual completely before attempting to build this kit. I'll echo that, and stress its importance. All elements of the Photopopper's design must work together or the kit won't work at all. Get something in backwards or misaligned, and you'll have a devil of a time getting it fixed. (Go ahead, ask me how I know!)

All right, I'll confess. I got in a hurry to see my Photopopper running, and rushed through one step. Next thing I knew, I'd managed to install one of the fuse clips backwards. The fuse clips hold the two tiny motors in place, and have to be installed correctly or you can't mount the motors. I had to work very carefully to desolder the offending clip and remount it as required. From then on, I paid strict attention to the artwork and avoided any other problems.

I'll pass along one other comment for future Photopopper builders. The printed circuit board (PCB) that forms the Photopopper's body arrives already bent into nearly the proper shape for the finished robot. You, however, have to finish bending the board, and this takes a leap of faith. The PCB is very thin, and already looks like it's been badly warped. But to make the motors touch the floor properly, you have to bend the board some more and hold it in place with a strong copper wire stretched from one side of the robot body to the other.

Rather than simply flex the board and try to hold it in place while I soldered both ends of a wire, I did the process in stages. As recommended, I first soldered one end of the wire to one side of the PCB. Next, I stripped slightly more insulation from the opposite end of the wire than I knew I'd need, and ran the wire through the hole in the other side of the PCB. Then, I carefully flexed the board, crimped the wire in place, and checked the motors' spacing. If they didn't hit the floor properly, I released the crimp, reflexed the board, crimped the wire again, and checked the motors.

When I was finally satisfied with the motors' alignment, I soldered the second end of the wire in place. If a final check at this point shows that the wire has somehow slipped, you can quickly melt the solder on the second end of the wire, then use needle-nose pliers to reposition the wire and let the solder cool back down. Verify that the motors are properly aligned, with the wire soldered in place, BEFORE you cut the second end of the wire! Otherwise, you'll have a real problem getting the motors properly set up.

I finished this kit the night before Marvin Green and I headed for a Seattle Robotics Society meeting. Marvin had driven up from Portland that day to take in a meeting, and had brought the Photopopper kit with him. We had great fun building the kit, and learned a lot about this low-power side of hobby robotics. But I was a little worried about how such a simple device would play at the SRS. After all, this crew has built some pretty sophisticated machines, and I felt that a robot without an MCU might not hold their interest.

I shouldn't have worried; the Photopopper was a huge hit. We lucked out and the meeting occurred on one of the twenty days each year that Seattle actually gets sunlight. When I set the Photopopper down on the sidewalk outside the meeting room, the little bug really started hopping. The gang crowded around to get a better view, asked lots of questions, and left with plenty of ideas for their own solar-powered robots. I think the idea of lighter-than-battery robots struck a nerve, and the club may see other low-power robots in the near future.

Photopoppers and their ilk, a class of robot known as a photovore, offer hobbyists a chance to explore simple behavior with no software struggles. A photovore "race" towards a bright light might take a while to complete, but first place would mean your 'bot was lighter and/or more efficient than its competitors. You can build your own photovore from parts gleaned from gutted cassette tape drives or pagers, and your efforts to keep the weight down and the efficiency up will be visibly rewarded.

I can happily recommend the Photopopper kit to anyone looking to build a unique and interesting machine. I would suggest you have some experience wiring up electronics before tackling this kit, as some of the work is delicate and it is so important to get everything properly aligned. But if you work carefully, you should end up with a cute little robot that will run for years and never whine that its battery is low.

Robot Wars

At the extreme opposite end of the hobby robotics spectrum...

The May 1997 Robothon was the SRS' randomly arranged chance to get together, run some contests, and shoot some robo-breeze. This proved to be a typical event, with about 100 humans and 40 robots in attendance, and included contests such as Mini-Sumo, Line Following, Micromouse, and the Grand Maze. Besides plenty of finished robots, many SRS members brought works in progress, so everyone could check out the various ideas being used.

Kevin Ross, president of the SRS, did yeoman's service in pulling the event together. He has also been putting a lot of energy into the SRS' on-line newsletter, the *Encoder*. You can check out the newsletter, and get a more complete update on Robothon '97, by hitting the SRS web site and following the links.

As if doing major organization for such a celebration in entropy wasn't enough, Kevin brought along one of the day's top entertainments: a video tape of the 1996 San Francisco Robot Wars competition. Talk about your extreme robotics; this 20-minute tape is chockful of fast, heavy, dangerous robots waging mayhem on each other. The last robot left moving took first place for its creators, not to mention a ton of robotic fame, and the audience had a great time watching the event.

I'm talking about a whole different scale here. One team described their robot as being in the heavyweight class. They were quite proud that their machine had been trimmed to hit the exact upper limit of the permitted weight range... 165 pounds!

These machines are not meant for the average living room. Many were armed with high-speed diamond-tipped circular saw blades, several hydraulic rams showed up, and gas engines were the favored motors. All machines were controlled remotely using R/C radios, but most only showed two speeds: full blast and crashed. One of the favored heavy-

weights rolled into the arena with its creator standing on top, controlling the machine with the R/C unit and keeping the throttle 'way down.

High speed proved very important, what with Newtonian physics and all. When two of these juggernauts collided, which happened often, it was a miracle that either rolled away under its own power. But they did, then squared off and took a run at something else. The crashes were horrific, and the winning robot looked like something the Buick had dragged home. The losers were much thinner than when they started, and made mostly of dents. And then there was the walker that caught fire...

I can enthusiastically recommend this video, available from Mondotronics for about $20. Think of it as an alternate-reality check. The next time you think you're hot doo-doo with your kick-butt Mini-Sumo robot, pull this tape out and watch a horde of high-tech robo-monsters do the Crash-and-Burn on each other. You'll see your lawn mower in a whole new light.

Tidbits

Whenever Marvin shows up, we spend as much time as we can eating Chinese food and doing robots. After the SRS meeting, we headed back to my place to find something robotic to do. I dragged out a couple of surplus 12 VDC cassette drives that I had picked up God alone knows where, and we spent some time tearing these down and bagging the parts. These cassette drives are crammed with beautifully machined bits ideal for small robots and smaller mechanisms on larger machines.

In particular, each drive contained a small, high-energy solenoid, suitable for use as part of an actuator. Though I'm pretty sure the drive was supposed to run on 12 VDC, the solenoid gave plenty of snap at 6 VDC, and should work well as part of a release mechanism or a pen carrier on an artist robot.

The drives also yielded a good assortment of small springs. Anyone building tiny robots will end up needing a tiny spring now and then, and I've never been able to find a reliable, walk-in-the-door store for buying them. Now I've got a couple dozen of various strengths, waiting for that next design.

All told, the cassette drives gave me two gutted chassis and a collection of useful parts, for little more than an enjoyable afternoon of male bonding. You can usually find similar drives for sale in the mail-order surplus companies such as Marlin P. Jones and American Science & Surplus. Keep these drives in mind the next time you need that special, small mechanical part.

Blue LEDs have been out for some time now, but the last few I've seen left me very underwhelmed. I mean, how can I get excited about an expensive, one or two milli-Candela blue unit when all of the reds and greens I buy for pennies hit a Candela or more? An LED vendor recently showed up at the company where I work, lugging a suitcase full of LEDs. His sample set included some painfully bright blue LEDs, with a color so pure and intense I couldn't resist buying a couple, even at $3 each. The devices are rated at 1.5 Candelas each, and will throw a large spot of blue light on a wall a foot away in strong ambient light. I can't wait to add them to a show-bot.

These are Parralight NSPB500 LEDs. For those of a more technical bent, the specs for this single quantum well device are: dominant wavelength of 470 nm, typical DC forward voltage of 3.6 VDC at a forward current of 20 mA, 5 degree directivity, and typical luminous intensity of 1.5 Candela.

Marvin and I spent some time hacking a friction drive robot mechanism together. I started with a short length of 5/32-inch brass rod, cut to the proper length (about four inches). I slipped two model airplane wheels onto the rod; you can usually find a good selection of small wheels with the necessary 5/32-inch bore at most hobby stores. I then used a hand-held Dremel tool to hog out two cutouts in the underside of a plastic project box, so each wheel fit through a cutout and rested on the tabletop. I then used hot glue and foamcore to hold the axle in place across the inside of the box. See the accompanying drawings of the friction-drive robot chassis (Figure 4 at the end of the article).

The drawings don't show the motors mounted in place. Your design will vary, based on the dimensions and motor mounting scheme you use. Basically, you want to mount a motor next to each wheel, the shaft parallel to the axle, so that the motor's shaft presses firmly against the wheel. As the motor shaft turns, the friction between shaft and wheel acts as both a power transfer and as a gearing system. I suggest adding a sleeve of soft heatshrink or surgical tubing over each motor shaft, to improve the grip between the two surfaces.

I used 1 1/2-inch wheels in my design, and the typical small motors found in surplus cassette drives work fine for the drive motors. Given the difference in diameters between the drive motor shaft and the driven wheel, the friction-drive chassis can hit very useful speeds with a surprising amount of torque. My little test rig had no problem lugging a set of four AA-cell batteries over carpeted floors, with plenty of torque left over for the MCU and other stuff. Note that the final diameter of the motor shaft, including any sleeving you add, directly affects both the final speed and torque. The larger the final diameter, the faster the wheel speed but the lower the torque. If necessary, dabble with various sleevings and/or wheel diameters to get the right balance.

You don't need to limit yourself to a project box for a chassis, though they work very well. Try your hand at building a friction-drive system out of brass rod, with your motors and wheels hanging out there in mid-air. The robot drive will take on a kinetic sculpture appearance, and should net your machine even more crowd appeal.

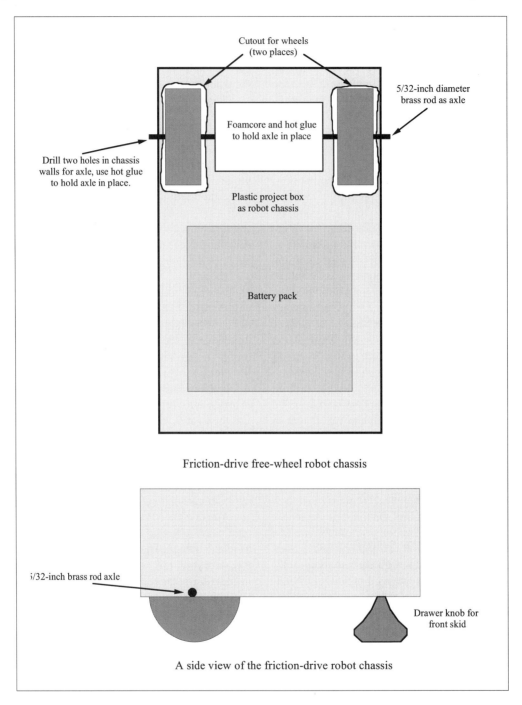

Figure 4. Design of a friction-drive robot chassis.

A Whole Lot of Robots

This article from the May 1998 issue of *Nuts & Volts* described the first Northwest Regional Fire-Fighting Robot contest, held by the Seattle Robotics Society as an adjunct to the national event held each year at Trinity College. We had several people who almost completed their robots in time, and two entrants with finished machines. The contest runs were exciting, the crowd was revved up, and both SRS robots turned in excellent performances.

I took some time in this column to describe a typical SRS meeting. The noise, the robots, the technical discussions, and the noise make it a high-energy morning, and the 50 or 60 people who show up each meeting always have a blast. If you're ever in the Seattle area on the third Saturday of a month, consider yourself invited; I'm sure you'll have a good time.

This column also includes a review of the Technological Arts' ADAPT-11C75 68hc11 MCU board. This board would make an excellent core for that next robot, given its on-board 8K of EEPROM, extra I/O ports, and small footprint. Check out my comments, then hit the TA web site for more information.

Finally, and I do mean finally, this column marked the end of my involvement with *Nuts & Volt*'s Amateur Robotics column. I covered a lot of robots during the years of my column, and I took some time to recap, and to thank the many readers for their interest and patience. Of course I also included a nod to the SRS membership, without whose involvement and contributions my column would not have been nearly as fun to write.

Ted Griebling won the first annual Northwest Regional of the Trinity College Fire-Fighting Home Robot Contest, held 28 March 1998 at Renton Technical College in Renton, Washington. Because of the short notice, only Ted and Gary Teachout had running robots, though several other teams demonstrated machines that were almost ready for prime time. Still, it was a hard-fought contest, and everyone involved deserves credit for their hard work and ingenuity. Congratulations to all!

Writing for a magazine always has a time-warp effect to it. I have to turn in copy so far in advance that the national event, to take place 19 April 1998 at Trinity College in Hartford, Connecticut, will be over by the time you read this. I, of course, don't know what happened, but if Ted and Gary enter the national contest, I trust they will do well.

But getting to the Northwest Regional, run by the Seattle Robotics Society (SRS), required some serious robo-skill on the part of all the entrants. Though the contest has been held at Trinity College for the previous four years, only one person I know of in the Puget Sound area had ever built a robot that actually competed. So nearly everyone in the SRS was starting cold on a very challenging event.

I began my own design a few months ago but, given my limited attention span and busy schedule, did not complete my machine. Recent columns have described some of the problems involved in making the small progress I did make, but finishing a real contender for this event meant overcoming many more difficulties than I faced. I got my first glimpse at the different robot designs when the SRS held a practice session during the March meeting.

Practice Day

I've long wished I could adequately convey what it's like to attend an SRS meeting. Phrases like "a robotics meeting" or "lots of interesting people" don't cut it. I just can't describe a classroom at Renton Technical College filled with over 60 robot builders and a dozen or so machines, plus the assorted test gear, laptops, cabling, frames, data books, and other hoohaw needed to support the gathering. Kevin Ross, our able president, keeps a lid on the entropy for an hour or two, and we spend the time exchanging questions, answers, tips, and information on robot building. But like a wound-up spring, this somewhat formal side of the meeting contains lots of potential energy waiting for a release, and when Kevin closes the meeting and what we call the "tech session" begins, the chaos is amazing.

The first thing that goes is your hearing; the noise of 30 simultaneous discussions and a dozen robots revving up will deafen you. Forget about talking; shouting is the only way of being heard by anyone more than a foot away. The white boards at the front of the classroom immediately draw several crowds as people grab markers and begin sketching motor control circuitry or sensor designs. Other crowds drift through the room in a tidal pattern, moving from one robot to another, checking out what cool additions the builder made since the previous meeting or examining the design of a new machine.

One very popular robot at our March meeting had nothing whatever to do with the Fire-Fighting contest. Rick DeWitt had built a marvelous motor-driven eclipse camera and had it all set up at the meeting; naturally, he took some time to run it through its paces for the crowd. Rick built his camera drive to satisfy a quest; he wanted to take many pictures, as in 100 or more, throughout a total solar eclipse. Since such an eclipse usually lasts about three and a half minutes, that averages a frame every two seconds. Obviously, we're talking a motor-driven film pack with about 250 frames of 35mm film in it, but that's the easy part.

The tricky bit involves tracking the sun as the earth rotates on its axis. If you don't compensate for the rotation, the very long exposures necessary during totality will be complete failures. But just rolling with the globe isn't enough. Rick wanted the camera's shutter under computer control, so he wouldn't even have to use a shutter cable or touch the camera during the eclipse. There were other design constraints, as well. Total solar eclipses, for some perverse reason, usually offer the best view far from your back porch. Rick not only had to make his rig battery powered, but it had to collapse down small enough that he could easily pack it onboard airplanes and into the hinterlands.

Rick got the drive system put together in time to head out for this year's hinterlands, the southern tip of the island of Aruba. He could have caught last year's eclipse, but ended up passing; possibly the mechanism wasn't ready for prime time, but more likely the notion of February in Siberia made the trip less appealing. Whatever... the day of the 1998 solar eclipse found Rick hunkered down in a World War II bomb crater on Aruba, waiting for the sun to hide. The bomb crater actually worked to Rick's advantage, since the prevailing trade winds would have kept the camera shaking had he not been sheltered from them.

I can't imagine what the tension must have felt like. It's one thing to build a robot, bring it to a Saturday meeting, and have a wheel fall off. OK, the bloody thing didn't work this time, but next month you'll come back with an extra $2 of hardware and everything will go smoothly. But filming a solar eclipse means you get exactly one chance every year or so, some place hell and gone from home, with who knows how much money tied up in the effort. If the robot doesn't work this time, it'll probably take a year to recover from the financial shock. Well, I've got good news and bad news. Rick's drive mechanism worked perfectly, but the film jammed. Rick plans to hit Europe next year for that solar eclipse, and we wish him the best of luck.

Although Rick didn't end up with any solar eclipse pictures to pass around, the SRS members enjoyed looking at the design of his camera drive mechanism. Rick built the drive system onto a 1/2-inch thick sheet of clear plastic, about one foot wide and 18 inches long. He started with a small stepper motor coupled to a 100:1 planetary gearhead purchased from Sterling Industrial. The gearhead's output shaft is connected to a 1/2-20 threaded rod, which is in turn threaded through a zero-backlash drive nut fastened to a homemade bracket. He fastened a wire to this bracket, then ran the wire through a complex pulley system that not only provided even more speed reduction, but damped out any vibration or backlash from the motor's stepping. Power for the motor and electronics comes from a pair of 12 VDC, 2.3 AHr gel-cell batteries.

To control this whole mechanism, Rick wired up a New Micros' 68hc11 single-board computer, complete with a two-line LCD display and a small keypad. He wrote his own software, using 68hc11 assembly language, to control the drive's tracking speed and direction and to work the camera's shutter. The computer provides some very sophisticated features, including displays of current position and step rate. Rick plans to add even more capability, such as sync'ing the shutter operation to the published time scale of the eclipse. He might even add a backlight to his LCD display; it gets dark during a total eclipse,

Figure 1. Rick DeWitt's solar eclipse tracking camera, ready for action. The chassis to the right of the camera platform holds a 68hc11 computer for controlling the camera's motion and shutter.

and right at the most critical part of this year's shoot, it got so dark he couldn't read the display.

Practice Robots

But the focus of the March meeting was trial runs of the Fire-Fighting robots, in preparation for the contest the following week. Several people demonstrated works in progress that had a lot of promise. Jeff Bronk made a wood-frame robot, sporting a pair of 12 VDC gearhead motors and two 12 VDC gel-cell batteries. The brains of this machine uses a V20 single-board computer, complete with 256K of EPROM and 128K of RAM. A serial port hooked to his computer provides on-the-fly debug capabilities during test runs. The robot carries a large complement of I/O, built around three 8255 parallel port chips. He also added eight channels of 8-bit A/D, and finishes with five sonar sensors, used for wall detection. For flame detection, Jeff's machine relies on an amplified photosensor with a custom baffle.

Another entrant, fielded by a three-man team of Martin Calsyn, David Buckley, and Mike Begley, relies on several processors for handling low-level tasks, all controlled by a larger, main computer. Their machine, built on a round, clear plastic frame of two levels, uses an MIT 6.270 board with a 68hc11 MCU as the main computer. A PIC 16C74 MCU handles all low-level sensor tasks and controls the two Hobbico CS-72 133 oz-in R/C servo motors that provide locomotion. A second 68hc11 board controls the robot's vision system, based on a hacked PC color QuickCam camera. This 68hc11 board emulates a PC's parallel port, controlling the QuickCam and collecting its data. To improve the camera's capabilities, the team removed the IR filter from the front of the camera; they warned that you need to be very careful doing this surgery, as the camera is very easy to destroy.

Jim Wright is the only SRS member with previous Fire-Fighting contest experience. He fielded an entry in last year's event; he figures he probably tied for 14th place "along with about 50 other ro-

Figure 2. A closeup of Jeff Bronk's wooden-frame robot. Here you can see two of the sonar ranging elements; the PCB below is a surplus I/O board for driving the sensors and motors.

Figure 3. Here's an entrant built by a three-man team: the robot sports sonar and a color QuickCam. The PCB on the table in front is an MIT 6.270 computer that controls two other, smaller computers.

bots." Jim built this year's entrant around one of Bill Harrison's Mini-Sumo robot designs. Jim's robot uses a pair of hobby servo motors powered by four AA batteries and controlled by a Basic Stamp. Once it locates the flame, the robot turns on a small fan, powered by a separate pair of 9 VDC alkaline batteries, to put out the fire.

Gary Teachout has earned a reputation for clever, robust designs built around simple technology, and his Fire-Fighting robot design keeps with that trend. He started with a 68hc11 BOTBoard-2, then added some external analog multiplexors to create 32 8-bit analog ports, all wired into the 68hc11's original complement of eight analog ports. He hooked IR phototransistors to every one of these analog inputs, and used some very clever optics and baffling to create the necessary sensors.

His wall detection scheme relies on the sharp delineation between the maze's white walls and black floors. Three vertical arrays of four phototransistors are mounted to look right, left, and forward. Each array is positioned in front of a small lens, so each

sensor in the array "sees" a different section of the floor ahead. For each of the three arrays, Gary's firmware turns on an IR LED, reads a phototransistor, turns off the LED, and takes a second reading. By analyzing the difference reported by each sensor in an array, his firmware can resolve the distance to the white wall to within 1/2 inch.

Gary's flame detection system also relies on IR and clever optics. He created two arrays of eight phototransistors each, and mounted each array in a separate, light-tight baffle. The baffles' openings, 1/4-inch wide slits, are aligned looking forward, aimed so that the two arrays converge on a point about ten inches in front of the robot. To detect and range on the candle's flame, Gary's firmware reads and records the light level detected by each phototransistor. The code then finds the phototransistor in each array that reports the brightest light and compares that level to the level reported by the adjacent sensors. If the delta between these two (or three) sensors is great enough, his firmware assumes that the bright spot is caused by the flame. When both arrays indicate seeing the candle at the same time, the robot is correctly positioned in front of the candle. To extinguish the flame, Gary's robot turns on a small fan to blow out the candle.

Ted Griebling based much of his design on elements from the modern classic, *Mobile Robots*, by Joe Jones and Anita Flynn. If you don't have a copy of this book in your library, by all means stop by the Mondotronics web site and check it out. Ted's robot design used a BOTBoard with a 68hc811e2, a pair of motors and gearheads from some HiTec hobby servos, and an L293 motor driver circuit from the Mobile Robots book. Ted did all of his firmware in assembly language, so he was able to cram a lot of functionality into his machine, but he still only had 2K of code space available. Getting a complete Fire-Fighting robot into just two thousand bytes is very impressive.

Ted's robot included a small electronic compass module, like the unit I described a couple of columns ago (see "Reworking the GameBoy®.") Ted was able to use the compass and the known configuration of the room layout to aid his robot's navigation. Whether intentional or not, each of the four rooms in the maze has a doorway on a different compass heading. Ted's firmware could detect when the robot crossed the white line marking a doorway, and use the compass' reading to verify which room it was about to enter, thus getting a new navigational fix at each doorway.

For flame detection, Ted used two sensors working in unison. The main sensor was a Hamamatsu UVtron (R2868) coupled with the matching electronics board (C3704). The small board generates the needed 350 VDC used to drive the sensor, and also conditions the signal for use by his 68hc11. The second sensor in the pair was a Radio Shack IR photodetector, mounted in a small brass tube to restrict the field of view. Working together, these sensor gave a reliable indication that the robot was aimed at the candle. To judge when the robot was within the required one foot distance, Ted's robot used down-looking photodetectors that picked up the one-inch wide white border around the candle. Finally, the robot's firmware activated a small fan to put out the fire.

The Big Day

The regional contest took place on the Saturday after the March meeting. It turned out that despite many long nights of work, most of the contestants did not have robots capable of running the entire contest. Only Ted Griebling and Gary Teachout had machines they felt could compete, and even Ted's robot was a little dicey. In fact, he was busily rewriting portions of the operating code during the test runs that morning. Normally, this is a sure invitation to disaster, but as events proved, Ted was able to pull off the changes cleanly.

The contest opened with Gary's robot, LC. The candle location for each run was determined by random draw from a program running on Kevin Ross' laptop, and Gary's first draw put the candle in the nearest room. Kevin said "Go!," the robot scurried down the hall, turned the corner, and put out the flame; the whole run was over in just 19.29 seconds. After snuffing the fire, the robot turned around and made its way back to the starting point, earning Gary a nice 10% bonus. Gary's second draw put the candle in a more difficult location, in a room farther away from the starting point. Even so, the robot performed perfectly. It moved down the hall, swept through the first room, rolled over to the second room, found the flame, put it out, and returned home; time for this round was 32.53 seconds.

Gary kindly allowed Ted some extra tweaking time on his code, and now it was Ted's turn. Like Gary, Ted's first run had the candle placed in the nearest room, and Ted's machine, Macbeth, made short work of the task, snuffing the candle in just 22.46 seconds. Macbeth was not programmed to return to the starting point, however, so Ted didn't get the 10% bonus. It wasn't until after Ted's first run that we remembered he had signed up for non-dead-reckoning mode, a format that rewards you with a huge 30% bonus if your robot can navigate without dead-reckoning navigation.

Next came Gary's third and final run. This time random chance put the candle in the fourth and last room of the robot's search pattern, ensuring that

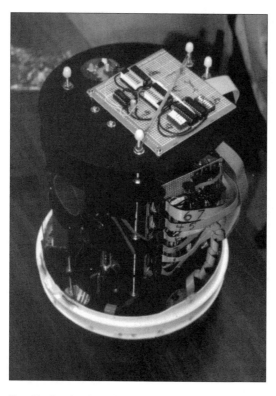

Figure 4. Second place went to Gary Teachout's robot LC, shown here with its white bumper skirt in place. The board on the right side of the frame handles some of the I/O; the round nozzle on the left side is the fan for blowing out the candle flame.

this run would be the most difficult. Still, the robot needed only 43.18 seconds to complete the task, and it returned to home base as usual: an impressive performance for the little machine. What's more, putting out the candle in all three runs netted an additional bonus for reliability.

Ted's second run again drew a very favorable placement, with the candle back in the first room. This run looked just like the first one, and the robot put out the candle in 22.42 seconds, a time nearly identical with its first run. But this time the non-dead-reckoning ramps were in place, and the performance meant a huge bonus for navigation. The third wasn't really necessary, as Ted's first two times, coupled with his navigation bonus, ensured victory, but we wanted to watch his 'bot strut its stuff one more time. This time, the draw put the candle in the fourth room, just like

Ted's final draw. Macbeth had a little more trouble, taking over 50 seconds to do the job. This included a timing error on my part while running the stopwatch, plus a five-point penalty assessed when Macbeth skimmed a wall. But, like Gary's LC, Macbeth put the candle out in all three runs.

This was an exciting contest, and the crowd of some sixty people really enjoyed watching these two machines duke it out for the title. Gary and Ted deserve high praise for their excellent designs and for their very low scores in a difficult contest. The other entrants who had not finished their robots in time vowed to get their 'bots done in time for a return match, to be held in August. Congratulations to all involved!

I'll include one other note here. Many members in the SRS worked behind the scenes to make this an

Figure 5. A mini-sumo fire-fighter? Jim Wright built this small, servo-powered robot from one of Bill Harrison's mini-Sumo frames. The board at the front of the robot holds the blower fan and a small sensor.

Figure 6. Macbeth in action! Ted Griebling's 68hc11-based robot goes to work, on its way to a first-place finish. You can see a small electronic compass module mounted on the upper-right section of the frame.

excellent first pass at the Fire-Fighting contest. Lance Keizer, Keith Payea, Frank Haymes, Anne Wright, and Kevin Ross, among others, contributed to building three different versions of the maze layout before we were done. Kevin also did his usual yeoman's duty in organizing the contest and making sure everything happened when it should. These people, and others who I'm sure I've left out, helped make this event a success, and their efforts deserve recognition. Thanks to everyone!

ADAPT-11

Technological Arts of Toronto, Ontario, Canada, makes a wide range of prototyping products of interest to the advanced hobbyists. A large part of their product line deals with sockets and connectors, with pins set on 0.3-inch rows specially designed to fit into those popular white plastic prototyping boards. For example, you can buy DB-9 and DB-25 connectors adapted to plug into your prototyping board. These cost between $5 and $7.50 each, a bargain considering the amount of wiring they save and the clean layout they provide. Other products include IDC headers (male) and matching dual-row receptacles (female). The PLCC socket adapters are of particular interest to robotics experimenters. These devices stand a PLCC socket of 28 to 84 pin-count vertically on the wide island of a prototyping board. Each pin on the PLCC socket is brought out on the base, giving you the greatest access to each pin with the minimum impact on board space. And I consider the prices very reasonable; a 52-pin PLCC adapter, suitable for most 68hc11 designs, costs just $12 each.

Even cooler than the line of adapters, however, is their Adapt-11C75 68hc11 single board computer. This tiny board, only slightly larger than a BOTBoard, contains a 68hc11e0 MCU, RS-232 level shifting circuitry, and a Xicor X68C75 8K EEPROM memory device. This may be the ideal low-end combination, since the 68C75 not only provides a bunch of nonvolatile memory, it restores the 16 I/O lines normally lost in memory expansion designs. What's more, the

Xicor device's two 8-bit ports can both be configured as input ports, an improvement over the 68hc11's original port B, which is an output-only port. Note, however, that the Xicor port arrangement does not replace the 68hc11's port B and port C exactly. The ports exist at different memory locations and the port setup differs from that provided by Motorola's 68hc24 Port Replacement Unit, the chip normally used for this purpose. Still, it's a big win over losing the ports entirely.

The board's design includes a dual-row 50-pin male header giving access to all 68hc11 and 68C75 I/O and control lines, making it easy to add this board into any project. Even cooler, the connector form-factor matches that of the 50-pin dual-row receptacle (surprise!), allowing you to plug the 'C75 board into your prototyping board. Imagine the projects you could prototype with a plug-in 68hc11 computer, complete with 8K of memory!

I've had problems in the past using the Xicor chip, so I was a little skeptical when I first powered up the Adapt-11C75. Long-time readers of this column will remember my last interaction with the smaller version of this chip, the X68C64. This device is similar to the 'C75, but lacks the I/O ports. The problem with both chips involves Xicor's vision of how an embedded firmware designer should use the memory in the device. Xicor embeds two chunks of 68hc11 code in memory space of each device; you use this code and a serial terminal to write your application firmware into the chip's EEPROM.

Obviously, this is a poor solution. The Xicor code must be left untouched or you lose the ability to download future firmware. So right off, you've lost 528 bytes of code space for your application. Worse, the breaks occur in two places, so you have to tailor your code layout to leave gaps in the right spots.

I didn't like this design the first time I saw it, and I still don't like it. Technological Arts provides a utility called xload, intended to sidestep this problem and allow you to download your code directly into the Xicor device. Since their solution does not require the Xicor loader code, you can safely overwrite the Xicor code and use all 8K memory.

Unfortunately, I was not able to get this program to work for me. But since I had already dealt with this problem, I had the solution at hand. I fired up my own Xicor loader program called, oddly enough, xload, and used it to download my test programs. Like the TA utility, my xload program takes over the whole download process, freeing up all of the Xicor EEPROM space. Both xload programs work by having you boot the 68hc11 in special boot-strap mode, then downloading a small talker program that knows how to write data to the Xicor EEPROM. You can get a copy of my xload utility by hitting my web page.

Once I cleared the hurdle of getting my code into the Xicor memory device, everything went as I expected. The X68C75 chip combined with a 68hc11e0 gives you a system with 8K of EEPROM and 512 bytes of static RAM. You can even substitute your own 68hc11e1 and use its 512 bytes of EEPROM, from $b600 to $b7ff, should you need a little more code space or data storage. Or, if you can get by with only 256 bytes of RAM, drop in your own 68hc811e2 to get a total of 10K of EEPROM.

TA has added a few nice touches to this board design. One element I like is their use of an on-board low-dropout +5 VDC regulator. Though the instructions recommend a supply of at least +8 VDC, I had no problems running on a set of 4 fresh C-cell alkaline batteries. If you intend to use your Adapt-11C75 in a prototyping breadboard, you can bypass the regulator and feed +5 VDC directly in from the prototyping leads. I also like the row of tiny switches across the top of the board; "top" here assumes you have the board standing vertically in your breadboard. Besides the reset pushbutton, TA has included a write-protect switch for the EEPROM.

Finally, they've left a PCB pattern so you can install your own voltage reference chip, should you want to supply an analog reference other than +5 VDC to the 68hc11's A/D inputs. Note, however, that your options here are somewhat limited. The 68hc11 requires a minimum of 3 VDC between Vrl and Vrh for reliable operation. The Adapt-11 documentation sug-gests a voltage reference of 2.5 VDC as an alternative to the normal +5 VDC, but you could have problems with such a small Vref.

Naturally, I have a few nits to pick. Although their user manual is filled with instructions and examples, it's a little dense and could use some simple graphics to break up the grey. The manual tries to cover all of the memory combinations possible with this board, often in the same paragraph, and it gets a little confusing trying to sort out what instructions apply to your board. I'd rather see one entire section devoted to downloading code into each of the major configurations. This might make the manual a little thicker, and some paragraphs might be exact duplicates of others, but it would make following the manual a lot easier. Lastly, TA was kind enough to include a copy of my SBasic system in their software disc, but their version is quite a bit outdated. As they pointed out in their manual, you can hit my web site for the latest and greatest.

The Adapt-11C75, used with a properly working xload utility, makes an unbeatable tool for the amateur robotics builder. For $87 each, the AD11C75-SP (starter package) gives you an assembled board with a 68hc11e0, Xicor X68C75, serial and power cables, a disc of free software, and manuals and data sheets. I consider this a good deal, and the board would make a great starting point for any robot. Recommended.

I'm Outta Here...

This is my 68th Amateur Robotics column for *Nuts & Volts* magazine, and it is my last. I can't believe that five-plus years can go by so quickly, or that the technology available to the robotics hobbyist could mutate so rapidly. Five years ago, building hobby robots to most people meant big iron: windshield-wiper motors, large batteries, and power-hungry computers, if any computers at all. There was no BOTBoard, and if you needed a software tool to help you, you wrote it yourself or did without. Hobbyists labored alone, usually with no one to share ideas with, save disinterested co-workers or a bewildered spouse.

The Internet changed all of that. With modems and computers, people can easily share technology with other builders, without meeting, talking, or even living in the same country. Software tools that speed robot development are as close as a web page, and plans, schematics, and building tips fill web sites around the globe. The list of people who've contributed major tools to the hobby robotics community spans the world, and just in my few columns I've drawn on ideas from people in Australia, Germany, Canada, and all across the United States.

Despite the hundreds of cool machines I've seen (I've even designed a couple of them myself), I enjoy the people in this hobby the most. I'm constantly amazed at the level of creativity and intelligence displayed by robot builders. Twelve or 75, or any age in between, all who have built 'bots have had to overcome some serious technical hurdles, and the ingenuity used always surprises me. Usually coupled with that ingenuity is a drive to pull the all-nighter, a nearly senseless desire to get the next part of the robot finished, just so you can show it off at the meeting that morning.

Speaking of meetings, I can't thank those of the Seattle Robotics Society enough for their contributions, direct and indirect, over the years. Many of the projects you have read about in these columns were conceived, developed, tested, redesigned, or in rare cases, broken, by some of the brightest people I've ever had the pleasure of knowing, the SRS membership. From the campfire talks that spawned tiny4th and SBasic, to the spontaneous tech sessions that yielded new motor mounts or frame designs, the SRS gang keeps the pot stirred and the juices flowing. One Saturday robotics meeting is worth a month of long labor alone at my workbench, and I count myself lucky to be part of the group.

As usual, I have a boatload of projects in front of me, all of which would make terrific articles for the column. I'm closing in on a GameBoy powered robot, though it might mutate into a home-control system before it's done. The solar-powered robot with on-board battery charger needs development time; any solar project in Seattle takes three times as long

to do as it would anywhere else, given our lack of sunshine. I still have to get my own beer from the refrigerator, as the beer-fetcher 'bot exists only as a vague set of plans in my mind.

I'd love to spend some time working on the smallest robot I can make. Given my lack of motor skills, "small" likely won't be very small, but I'm sure I can beat my current personal record of slightly larger than a set of four penlights. Another project before me is lighter-than-air robotics. Several people have built helium-lifted machines, and getting the most performance in the lightest weight sounds fascinating. A related project would be an intelligent rubber-band-powered glider, capable of responding to radio or IR commands. As with the blimp-bot, keeping the weight down is crucial, but seeing such a plane stay aloft and under control would be dynamite. Of course, there's the tree-climbing camera, an idea prompted by the squirrels playing in the fir trees in our backyard. I'd love to find out what they (the squirrels) look like in their own world.

Since I'm really a software person by trade, I've naturally got some software projects on my list as well. First in line is to clean up SBasic; it still has some irritating bugs in it. Once those are dealt with, I'd like to redo the code generator so it creates a second, intermediate file. This file would be machine independent, and would be processed in a separate operation to get the final executable file. This would make it very easy for me to port SBasic to virtually any machine. Even more entertaining, I could then publish the format of the intermediate file, so other software types could create their own file processor. That way, anyone could develop an SBasic cross-compiler for any machine, without my getting involved at all.

As you can see, I'm nowhere near running out of article ideas. But burnout is setting in, and I can tell it's time to head in another direction. I'll continue to write, however, and you may well see articles with my byline on them in future issues of N&V.

I began this column over five years ago, to help people get started with robotics and to describe a project or two that I had done. The months sped by almost unnoticed, my time measured instead in completed robots, SRS meetings, and article deadlines. Along the way, I've met some terrific people, seen an army of highly cool machines, and heard from hundreds of very nice readers. Thinking back on all the letters and email from people trying to build their first 'bot, I can't help but wonder how their projects turned out. I know that many started robots, but I'll likely never know if they finished them.

This hobby has come so far in the last few years, and the pace of change is accelerating. The big boys have sent machines to Mars and to the oceans' depths. But for every Sojourner or Tiburon, there are thousands of almost-working robopets or nearly-finished maze-runners. To all of you struggling to get that frame finished or software stabilized, hang in there. The technology and the tools are out there, and with patience and persistence, you can build machines today that few hobbyists would have tried just a few years ago.

Time to go. In the immortal words of Lance Keizer, SRS stalwart and past president, "Keep on keeping on..."

Part IX
Sidelights

The columns in this section cover topics a bit tangential to building robots. The first column contains a review of the Network Cybernetics Corporation's AI CD-ROM. I enjoyed looking through the mountains of files on this disc, which covered topics such as AI tools, data bases, and robotics.

The next article describes a typical SRS meeting. I have been going to SRS meetings for ten years now, and each has a character all its own, but this column will give you a feel for what they're like. I tried to tell this story through the eyes of two fictional SRS members; judge for yourself how well I did.

I close out this section with an article on building up a set of robotics software tools. I describe an excellent programming editor, PFE-32, which I use daily. I also discuss xgcchc11, a C-compiler released for DOS and Unix under the GNU general public license. These tools can help you in your attempts to build your own robots, and the article offers a starting point for finding other tools.

The NCC
AI-CDROM

This column, from August of 1994, features a lengthy review of the Network Cybernetics Corporation AI CD-ROM. I thoroughly enjoyed grazing through this collection of software, sampling a huge variety of programs, pictures, and data files on topics such as Artificial Intelligence, chaos theory, and robotics. Though today the World-Wide Web gives you access to far more information, the CD-ROM proved very convenient.

I also used this column to review a classic book in the embedded control field, Albert C. Leenhouts' *Step Motor System Design Handbook*. I first ran across this book years ago while working on a desktop appliance that used several stepper motors. This book contains valuable information for anyone designing stepper motor systems, and belongs in the labs of almost any robot builder.

I ended this column with a discussion of tiny4th, my ultra-small Forth system for the 68hc11. In particular, I described the notion of vectored execution, and how you can use it to support character I/O between a 68hc11 and many different types of I/O devices. The technique is widely used, and the presentation here can help guide you through similar projects, even on different MCUs or using different languages.

I recently got a chance to review Network Cybernetics Corporation's AI CD-ROM (revision 2). Although the title implies this is a disc filled with Artificial Intelligence software, many of the programs will prove useful (and fun) for robotics enthusiasts as well. This CD-ROM is an ISO-9660 format disc, compatible with CD-ROM drives on nearly any type of machine, including PCs, Macs, and Amigas.

The disc contains about 275 Mb of software, all of it compressed with PKZIP. The files are divided into directories based on subject matter. Some of the compressed files do not contain executable software. Instead, they hold .GIF images, text files, or databases useful to researchers or hobbyists. Occasionally, the compressed files contain not only PC-based executables, but source files suitable for compilation on other platforms, such as Macs and Unix systems.

The root directory on the CD-ROM contains a file named 00_index.txt. This is a directory listing of all files in all subdirectories on the disc, complete with a one-line comment describing each file. Printing this file generates about 40 pages of printout, but the listing comes in handy when you want to browse through the different directories.

Additionally, each subdirectory contains its own 00_index.txt file, listing the files, with comments, in that subdirectory. I browsed through some of the subdirectories, trying one or two programs that sounded interesting. What follows are some of my notes and observations.

\ROBOTICS

The file 6811srvo.zip contains a text file originally posted to the comp.robotics newsgroup on usenet. The comment in the 00_index.txt file implies that this file contains assembler source for controlling R/C servos from a 6811. Strictly speaking, this isn't true. The file contains text describing basically how the technique works, but the text certainly couldn't run through an assembler as is. Still, it is a starting point.

The file robofest.zip contains 28 snapshots from the May 1993 Robofest, held in Austin, TX, by the Austin Robot Group. The snapshots have been saved as .GIF images, suitable for displaying with many of the public-domain .GIF viewers. The pictures are fairly low-res, but quite presentable. I wish the Austin Robot Group had included some explanatory text for each snapshot. I would have enjoyed descriptions of the people and robots in the photos.

File ir_pd.zip contains a text file, posted to comp.robotics in February 1993, describing an IR proximity detector built with a 3900 op-amp and a 555 timer. The author, Bob Lee of Bowling Green State University, included a schematic of his design. The schematic was created with text characters, and clearly shows why text-based schematics are difficult to read, write and understand.

File sonar.zip contains a good technical article on sonar system design and theory, based on the Polaroid Ultrasonic transducer kit. The article was written by Jess Jackson and Jerry Burton of the Robot Society of Southern California. The article is well written and very informative, but it references schematics and figures that are not supplied on the CD-ROM.

File sonarart.zip contains the text from an article by Keith Payea, originally published in 1993 in the Encoder, the newsletter of the Seattle Robotics Society. The article, entitled "Sonar on the Cheap," describes a simple but effective sonar ranging circuit that can be driven by common MCUs such as the 68hc11. The text gives some Digi-Key part numbers for key components, but needed schematics and figures are not supplied.

This theme of missing schematics and other key documentation quickly became irritating. I understand why it happens. Often, an article's schematic exists only as line-art, or as a CAD drawing in some drafting package. It can prove very time-consuming to translate such a schematic into some common type of data file, suitable for printing by the end user. Sure, Postscript is a pretty universal file format, but getting from the author's original artwork to an accurate Postscript file can be difficult, if not impossible.

Still, the schematics are so vital that I wish the CD-ROM publishers would put more effort into providing both text and schematics. Keith's article, without the schematic, is little more than interesting reading. With the schematic, you could build a working sonar unit in a couple of evenings.

The file cmurobot.zip contains a text file listing major robotics papers published at Carnegie Mellon University (CMU) during 1991. This list was compiled by someone at CMU as a personal effort. The list contains citations, titles, and abstracts of each paper. The file also includes notes on ordering copies of any paper, though the method might have changed in the last three years.

The file joystick.zip contains C and ASM source files for reading a PC's joystick. The source code does not include an author's name, but it does carry the notation "Firmware Furnace #4." This implies the files are somehow related to Ed Nisely's column of that name; check out the early *Circuit Cellar INK* magazines. Note that this file contains the source code, makefiles, and executables for two joystick routines, one slow and one fast. It does not contain any text from the articles. It also does not provide information on which C compiler was used. I can tell you it wasn't Microsoft Quick C version 2.5. Although the file joyfast.c compiled, the makefile contained unrecognizable instructions for the link operation.

This is another area in which I wish the CD-ROM publishers would take the extra step. The original article must have described the compiler and linker used in creating the executables. Why can't the CD-ROM publisher pass that information along in the disc files? Better yet, I'd like to see the CD-ROM publishers work with either the original author or another highly technical person, to supply makefiles and related information for several popular compilers. Then the user could quickly customize the supplied software, starting from a program known to compile properly on her system.

The file robotfaq.zip contains the text, in two parts, of a sample Frequently Asked Questions file for the comp.robotics newsgroup. This file, commonly known as the FAQ file, is usually updated once a month or so and posted to comp.robotics. This copy of the comp.robotics FAQ file is dated 4 June 1993, and was prepared by Kevin Dowling of the CMU Robotics Institute. The FAQ file contains a wealth of information on robotics, including sources for parts and software, pointers to important papers, and FTP sites that carry robotics information.

Most widely-read newsgroups have a FAQ file. Its creation and maintenance is usually a labor-of-love, taken on by one or more contributors to help all readers of the newsgroup. I consider the FAQ files one of the most useful and powerful elements of the Internet culture. Thanks to Network Cybernetics for including a sample FAQ file on their CD-ROM.

\AI

The AI subdirectory contains "files on computer planning, reasoning, cognition, consciousness, and creativity." It also contains the source code listings for all programs listed in issues of *AI Expert* magazine, through the June 1992 issue. It also contains many additional files, all related to Artificial Intelligence.

The file aie9206.zip is one of about 50 files related to the *AI Expert* magazine. It contains the source code for the June 1992 issue of the magazine, as supplied on the CD-ROM.

I chose this file purely at random, to get a feel for the code provided. The code was written for the Zortech C++ compiler and, per the comments, "trains a CMAC to produce an odd parity bit when given an input vector with 3 components." Since my AI background is a little weak, I haven't a clue what this software does. The program accompanied an article written by George Burgin; the text of the article is not given, nor is a makefile supplied.

\AL (Artificial Life)

The file em22.zip contains a set of programs, developed in Denmark, that claim to be some type of evolution machine, version 2.2. Unfortunately, the compressed Turbo C source files were secured with a password. The accompanying readme file indicates that you should contact the authors, either by mail, email, or phone. Supposedly, you would then be given the password to unlock the files. I didn't bother.

I don't mind seeing demo versions of programs on a CD-ROM, provided they are high-quality and their number is kept to a minimum. But what possible good does it do to take up CD-ROM space with locked software such as this? This program's developers should either distribute a working demo version, distribute the unlocked source, or get out of the game. I also believe that Network Cybernetics should consider pulling this file, and any others like it, from its upcoming Revision 3 disc.

The file mcroants.zip contains a PC-based graphical simulation of an ant colony. Ants in the colony move around, locating "food pellets" and "eating" them. These ants can share food, mate with other ants, or, if they eat one of the poison pellets, die. Offspring share genetic information from both parents. The program records each ant's genetic coding and vital statistics. The file also includes an article on genetic algorithms. This is a well-done program, excellently designed and fun to watch.

\ANN (Artificial Neural Nets)

The files nets30a, nets30b, and nets30c.zip contain the C source code for the NASA NETS v3.0 neural net tool set. It compiled under Microsoft Quick C version 2.5 after minor changes to the makefile; I had to change all references in the makefile from "cl" to "qcl." The resulting executable loads and runs, but the menu options presented are beyond my understanding. The NETS package includes makefiles for Borland C and Microsoft C on the PC, Think C on the Mac, and other compilers/platforms.

\CA (Cellular Automata)

File dr_life.zip contains an excellent VGA Life game. This is the ever-popular computer game, first developed by John Horton Conway and introduced in *Scientific American* many years ago. dr_life is a superb rendering of this game, with a simple but powerful user interface. Setting the initial pattern is easy and painless, thanks to a nicely designed menu system and mouse interface. The program ran with lots of pep on my 40 MHz 386. You get the executable and a docs file, but no source.

\CHAOS

File fw108.zip contains a demo version of Fractal Witchcraft, a fast, robust Mandelbrot and Julia-set generator and viewer. The demo version includes a picture show made of several .pcx files, created with FW. The demo lets you use the keyboard to repeatedly zoom in on any selected region of either mathematical set. You can then specify the size (large, medium, or small) of the resulting display. The size basically controls the length of time the program will take to complete the requested display. Thus, you can use small or medium displays to get a quick view of your selected area, then request a large display when you are happy with the smaller ones.

This demo version does not allow you to save any displays you create as .pcx files; you'll need to register and buy the full-up version for that capability. Still, the slide show is excellent and the display speed was quite good. Very entertaining.

\EMBEDDED

This directory contains 20 MB of compressed files related to embedded control, a field similar to robotics.

File mcx11v15.zip contains the source code, documentation, and several example files for the MCX 68hc11 MicroControl Executive, version 1.5. This is a true multi-tasking executive suitable for use in any 68hc11-based robot or embedded system.

Note that the documentation file is dated April 1990. The manual states that you can download the MCX files from Motorola's FREEWARE BBS. Additionally, recent posts to the comp.robotics newsgroup indicate that at least one FTP site carries an echo of all FREEWARE BBS files. Thus, you could probably get a more recent version of these files, if you wanted to do some digging.

Assembling MCX generates two .s19 files, containing 68hc11 executable code. The file system.s19 contains the full MCX run-time executive. This file includes all support for the semaphores, queues, and services provided by MCX.

The file test.s19, as supplied in the .zip file, is a sample executable file that tests all of MCX's capabilities. You would normally replace test.s19 with your own executable file, containing code for the tasks you want your robot to perform. The .zip file also includes the source for the test program, so you can see how to write assembly language files for use with MCX.

Note that MCX is very small. The entire run-time executive fits in just 1.3K and requires less than 256 bytes of RAM. I haven't loaded it into a BOTBoard and tried it yet, but I can foresee lots of possibilities for such a small executive.

\DATA

By far the most bizarre directory in the entire CD-ROM, this directory contains 20 megs of compressed data files. These are usually ASCII text files, intended to serve as testing grounds for user-developed AI software. For example, suppose you are working on a stock market prediction program. Your program would use historical stock market behavior to somehow predict future trends. You could then invest, based on these predictions, retire early and play with robots all day long. But you will need test data, and lots of it, for your program to munch on. Which brings us back to this directory.

File nvest063.zip contains, according to its comment line, "Stock market data for years '89, '90, '91, & '92 in 2 formats." Other data files in this directory contain mushroom data from the Audubon Society Field Guide, a Landsat multi-spectral pixel value database, a solar flare database (with 1389 instances), and a wine recognition and chemical analysis database.

You also get an ASCII file containing the CIA World Factbook for 1991. I found this hugely entertaining. Using nothing more sophisticated than list.com, I could cruise through this file and read about countries I never knew existed. For each country, the file covers facts such as currency, political system, population groupings by religious, economic, and political factors, flag design, and more.

This directory also includes the complete text for several books, including *The Time Machine* and *War of the Worlds*, by Jules Verne, and *Moby Dick*. It also includes two versions of the Bible and a translation of Sophocles' *Oedipus* trilogy.

In all, this directory includes about 80 files of data for all kinds of subjects. If you are developing AI software for handling data, this could be a quick source of extensive data for you.

\PROG (programming languages)

The \PROG directory contains 44 megs of compilers, assemblers, and tools for all kinds of programming languages and platforms. You name it, it's probably in here somewhere. ABC, Algol-60, AWK, BOB, COBOL, dBASE, Fortran-to-C translators, IFP, J, MAX, Micro-COBOL, Oberon, OPS5, Estelle, Q-Pro, ADA, VisiCLANG, LOGO, Forth, Modula-2, Pascal, SNOBOL, ICON; the list of languages represented on this disc is amazing.

Two large groups of files contain the GNU C/C++ source files and tools for DOS and for OS/2. These were the most recent ports as of the CD-ROM's creation. For DOS, this means version 2.1.1 of the GNU assembler and version 2.3.8 of the FLEX sys-

tem. I would imagine more recent ports exist, which you can track down on the Internet.

Another large group of files contains several PC-based Forth compilers. Available systems include 4th-83, eForth, F83 version 2.1.0, Fifth version 1.8, Lake Ontario Forth-83 version 1.28, MiniFORTH version 2.5, Pygmy Forth version 1.3, and Zen Forth version 1.6. If you have been looking to try a PC-based Forth, you'll find what you want in these files.

Overall Impressions

The Network Cybernetics AI CD-ROM will keep you fascinated and occupied for weeks, if not months. You can check out comp.robotics, try out different computer languages, learn 80x86 assembly language, view graphics images, make your PC talk, or play with AI concepts and tools. The price for the Revision 2 CD-ROM was quoted as $129 plus $5 per disc for domestic shipping, $10 per disc for foreign shipping. You can order by contacting Network Cybernetics for the dealer nearest you. Note that shipments to foreign countries may be subject to export restrictions.

Stepper Motors

While I'm in review-mode, I'll mention the *Step Motor System Design Handbook*, by Albert Leenhouts. I don't do much stepper motor design, though several people in the Seattle Robotics Society (SRS) are heavily into it. *Nuts & Volts* has already run one article by SRS member Dan Mauch, who used stepper motors to create a three-axis, PC-driven CNC drill press from a stock Black and Decker-type unit.

Additionally, I did a contract job for a company whose product used stepper motors to drive a three-axis amorphous carver. I wrote the driving software, but wasn't involved in designing the mechanical or electrical systems behind the steppers.

Still, I've been involved enough with stepper motors over the years to know how difficult it can be to design a good stepper system. Issuing properly timed step pulses via software is the barest beginning. Designing a stepper system properly requires knowledge of the motor's internal design, its electrical characteristics, the mechanics of the motor, coupling, and load, and the design of the driving electronics.

True, you can usually slap an over-sized stepper, beefy electronics, and a honking power source together and get something to work. But understanding all the system elements can lead you to a more robust, efficient design: one you can be proud to show others and one you know will work reliably.

To learn more about all phases of stepper system design, pick up a copy of *Step Motor System Design Handbook*. This book contains 12 chapters, starting with an introduction to step motor control and ending with four real-world design examples. The book includes listings of several PC-based stepper system design tools, written in a generic version of BASIC. You can type the programs in from the book or, for an additional cost, purchase a PC-format disc with all the programs in source form. I don't know if the programs will run on a Mac or Amiga, but the BASIC seems generic enough that they should.

The book is well-written, the example designs are typical, real-world problems, and the systems elements covered are reasonably up-to-date; the book was rewritten in 1991. Engineers whose opinions I value consider this book to be excellent. I like the systems-level approach it takes. The math is fairly intense, but the supplied programs let your PC deal with that. You just need to understand the mechanics and electronics involved, and the book takes you step by step (no pun intended) through the details. Recommended.

The book alone costs $95 in singles; the source disc is an additional $200. Note that the SRS was able to negotiate a considerable discount for a group purchase of 11 books with discs, so it pays to get your club involved.

Exploring tiny4th

I've written a lot of 68hc11 assembly language code, mostly out of necessity. This is because I usually write for single-chip or small-memory systems, and

the extra code created by most compilers pushes the program length beyond the limits.

This need to write small but powerful programs led to my creation of tiny4th. tiny4th is a PC-based 68hc11 cross-compiler, designed around a subset of the Forth programming language.

tiny4th's strong suit is compact executable code. A tiny4th executable consists of two parts. The first part is the kernel, or run-time executive, written in 68hc11 assembly language. The tiny4th distribution package contains the source code for a standard tiny4th run-time executive. The second part of the executable is a series of eight-bit tokens, corresponding to the source code compiled by tiny4th.

Once you have created an executable file, which is simply a Motorola .s19 file, you download the entire executable into the code space of your target 68hc11 system. I usually use PCBUG11, a Motorola public-domain tool, for this operation. All that remains is to reset the 68hc11. It automatically initializes and runs the executive, which in turn begins running your tiny4th program.

The tiny4th compiler accepts as input an ASCII text file containing the source for your program. The compiler outputs an ASCII text file containing assembly language source. This output file is then passed to the Motorola asmhc11 public-domain assembler; the assembler in turn creates the final executable file.

The entire tiny4th package, including the compiler, assembler, and sample source files, is available from my web site. The latest released version of tiny4th is version 2.1.

tiny4th and the SCI

Forth's initial dictionary contains very few words for handling character I/O. At the most primitive level, you get KEY, ?TERM, and EMIT. These words get a character, test for the presence of an incoming character, and output a character.

Note that Forth does not specify any underlying mechanism for moving characters to and from the user.

For example, most PC-based Forths use the DOS BIOS routines for character I/O. The exact technique, however, is up to the implementer, and the BIOS concept certainly won't work on a small 68hc11 system.

tiny4th's run-time executive does not even support these three character I/O words. I didn't want to take up the code space if the application didn't need serial I/O. But tiny4th is like most Forth systems; if it doesn't have what you want, just add it.

The usual method of moving characters between the user and a 68hc11 is the Serial Communications Interface, or SCI. The SCI lets your code exchange 8- or 9-bit characters at high baud rates, and supports features such as error-checking and interrupt generation.

Motorola implemented the SCI as a group of registers in the I/O area, usually located from $1000 to $103f. The SCI registers are:

BAUD ($102b)	SCI baud rate register
SCCR1 ($102c)	SCI control register 1
SCCR2 ($102d)	SCI control register 2
SCSR ($102e)	SCI status register
SCDR ($102f)	SCI data register

Despite the number of registers involved, the initialization for most applications is simple and only takes a few instructions.

In its simplest form, initializing the SCI requires two steps. First, you must select the desired baud rate by writing the proper value to BAUD. Then, you must enable the SCI receiver and transmitter. Refer to the accompanying listing on SCI initialization in assembly language (Figure 1).

The same function, written in tiny4th, generates about the same amount of code. Refer to the accompanying listing on SCI initialization in tiny4th (Figure 1). All code examples in this article assume the target 68hc11 uses an 8 MHz crystal. Different crystal frequencies will generate different baud rates for the same BAUD register values. Consult the Motorola *M68HC11 Reference Manual* for details.

The three low-level I/O words, written in assembly language, show how easy it is to perform serial communications using the SCI. Refer to the accompanying listing on low-level SCI routines in assembly language (Figure 2, at end of article).

The routine QTERM checks for the presence of a character in the SCI receive buffer. Note that this routine does not fetch that character, it only indicates its presence. Upon exit from QTERM, the A register will be zero if no character is available, or non-zero if a character is available.

The routine KEY returns a character from the SCI. The incoming character is returned in the A register. Note that KEY sits in a tight loop, calling QTERM, until the SCI holds an incoming character. If, for some reason, QTERM never returns a non-zero value in this loop, your program will hang in KEY.

The routine EMIT outputs the character in the A register. Note that EMIT sits in a tight loop, polling the status register, until the SCI is ready to accept the outgoing character. If, for some reason, the status register never returns a ready condition in this loop, your program will hang in EMIT.

The corresponding code in tiny4th is also quite simple. Refer to the accompanying listing of low-level SCI routines in tiny4th (Figure 2).

tiny4th uses its data stack to pass data between words. Thus, a word that calls ?TERMINAL can expect a flag on the top of the stack (TOS) that is TRUE if a character is available, or FALSE if no character is available.

KEY invokes ?TERMINAL repeatedly, until ?TERMINAL returns a TRUE value. KEY then reads the SCI data register and returns the incoming character on the TOS. As with its assembly-language counterpart, KEY will lock up if ?TERMINAL never returns a TRUE.

EMIT reads the SCI status register repeatedly, waiting for a one in bit 7. This indicates the output buffer is empty. When this occurs, EMIT writes the value on the TOS to the SCI data register. EMIT locks up if the output buffer never reports as empty.

These three words provide enough low-level power to exchange characters serially between the 68hc11 and a host terminal. Other, more powerful

```
iobase          equ     $1000           start of I/O regs
baud            equ     $2b                 offset to BAUD
sccr2 equ       $2d             offset to SCCR2

                ldx     #iobase         point x at I/O regs
                ldaa    #$30            9600 baud with 8 MHz xtal
                staa    baud,x          write to BAUD
                ldaa    #$0c            enable rcvr and xmtr
                staa    sccr2,x         write to SCCR2


        SCI initialization in assembly language



            ------------------------------


hex

102b    constant    baud            ( addr of BAUD register )
102d    constant    sccr2           ( addr of SCCR2 register )

    :   init-sci
        30   baud   c!                   ( 9600 baud with 8 MHz xtal )
        0c   sccr2  c!                   ( enable rcvr and xmtr )
    ;


        SCI initialization in tiny4th
```

Figure 1. SCI initialization in assembly language and tiny4th.

Forth I/O words are built by invoking combinations of these three words.

For example, the word . (pronounced "dot") outputs the TOS as a string of ASCII digits, followed by a space. The output string will be in the current radix. That is, if the current radix is 16, the number will be output in hexadecimal. The current radix is traditionally controlled by the number stored in the Forth system variable BASE.

EMIT, KEY, and ?TERMINAL, as defined here, all use the SCI to exchange serial data. But this can prove limiting. Suppose, for example, you needed to read characters from the SCI and also from another device, perhaps an eight-bit parallel port. You could certainly define low-level words for exchanging data with the parallel port. These words would mimic the functions of KEY, ?TERMINAL, and EMIT, but they would have to be called by other names.

As a result, you would end up with duplicate high-level words, half of which work for the SCI and the other half for the parallel port, even though both sets of words basically perform the same functions. One way to avoid this unnecessary duplication involves vectored execution. This means that a routine, such as EMIT, gets its behavior from a previously modified variable, rather than from a hard-coded routine. All you have to do to change the behavior of EMIT is to change the value in the variable.

The accompanying listing of vectored SCI routines shows how easy this technique is to implement in tiny4th (Figure 3, at end of article). Note the changes to EMIT, KEY, and ?TERMINAL. Each routine now fetches a value from the appropriate variable and executes the function contained at that address. Change the address stored in a variable, and you change the behavior of the corresponding word.

Now you can create low-level I/O routines for as many devices as you want. To "activate" a device, simply use ['] and copy the addresses of the three low-level routines into the proper variables. All subsequent invocations of EMIT, KEY, and ?TERMINAL will now talk to the "active" device, ignoring all others.

This technique is hardly new, and is not limited to Forth. You can easily implement this same concept in assembly language or C. It results in flexible, robust code that is quite easy to work with. You pay a small penalty in speed, but that is usually not a problem in I/O routines.

The released tiny4th system includes a number of library source files, some of which contain the above code. These files serve as tested, working starting points, letting you quickly develop your own custom applications without first reinventing the wheel.

Again, this is hardly a new concept, but its power can be overlooked sometimes. One of the strongest features of C is the ANSI standard library set of functions, included in some form with nearly all compiler systems. Being able to draw on large libraries of functions with universally accepted behaviors really speeds up code development. I have tried to achieve some of this standardization with my tiny4th library files. By including tiny4th code that provides some of the more standard Forth high-level words, I hope to ease the creation of 68hc11 applications.

```
iobase          equ     $1000           start of I/O regs
scsr            equ     $2e                     offset to status reg
scdr            equ     $2f                     offset to data reg

qterm
        ldx             #iobase                 point x at I/O regs
        ldaa            scsr,x                  get the status
        anda            #$20                    leave only RDRF bit
        rts

key
        bsr             qterm           is char available?
        beq             key                     loop until yes
        ldaa            scdr,x                  get char
        rts

emit
        ldx             #iobase                 point x at I/O regs
emit1
        ldab            scsr,x                  check output buffer
        bpl             emit1           loop until empty
        staa            scdr,x                  send character
        rts

                Low-level SCI routines, assembly language

                ------------------------------

hex
```

Figure 2. Low-level SCI routines in assembly language and tiny4th.

```
102e    constant    scsr            ( addr of status reg )
102f    constant    scdr            ( addr of data reg )

:   ?terminal            ( - f )
    scsr  c@                        ( get the status )
    20   and                        ( leave only RDRF bit )
;


:   key             ( - c )
    begin
        ?terminal                   ( check the status )
    until
    scdr  c@                        ( get the char )
;


:   emit            ( c - )
    begin
        scdr  c@                            ( check output buffer )
        80   and                            ( loop until empty )
    until
    scdr  c!                        ( send character )
;

                    Low-level SCI routines, tiny4th
```

Figure 2 continued.

```
(
( The following three variables serve as vectors
( for all character I/O routines. The core words
( EMIT, KEY, and ?TERMINAL actually fetch the
( contents of the appropriate variable and execute the
( word found there to perform their function.
(
( Thus, by simply writing the address of a different
( word into one of these vectors, you can alter the I/O
( device used by EMIT, KEY, or ?TERMINAL.
(

variable   OutVector                        ( holds address of word to
                                    ( send character to active
                                ( output device


variable   InVector                         ( holds address of word to get
```

Figure 3. Vectored execution in tiny4th.

```
                                        ( a character from active
                            ( input device

variable   CheckVector          ( holds address of word to
                        ( test for character on active
                    ( input device


:   emit               ( c — )
    OutVector   @  execute
;


:   ?terminal              ( — f )
    CheckVector   @   execute
;


:   key             ( — c )
    begin
        ?terminal
    until
    InVector   @   execute
;


:   emit-sci      ( c — )          ( send char at tos to SCI port
    begin
        scsr   c@                       ( get the status
        80   and                    ( check transmit-empty bit
    until                           ( loop until SCI is ready
    scdr   c!                   ( send char out the SCI port
;


:   ?terminal-sci    ( — f )       ( test SCI for available char
    scsr   c@                   ( check SCI status
    20   and                    ( return 0 if nothing there
;


:   key-sci       ( — c )          ( get char, put on stack
    begin
        ?terminal               ( see if a char is available
    until                           ( loop until char is there
    scdr   c@                   ( get char and leave on stack
;
```

Figure 3 continued.

```
:   Init-SCI          ( — )          ( initialize SCI
    30   baud   c!                   ( set baud = 9600
    0c   sccr2   c!                  ( turn on xmtr and rcvr

    [']   emit-sci                   ( get address of sci output
    OutVector   !                    ( and initialize outvector

    [']   key-sci                    ( get address of sci input
    InVector    !                        ( and initialize invector

    [']   ?terminal-sci              ( get address of sci check
    CheckVector   !                  ( and initialize checkvector

;
```

Figure 3 continued.

A Typical(?) SRS Meeting

This column, from the October 1995 issue of *Nuts & Volts*, marked my first attempt at a fictional column. The two characters, Ted and the column's narrator, who is still unnamed, take the reader through a typical Seattle Robotics Society club event. I enjoyed writing the column and playing with the concept of a couple of robot-building friends, but the column received mixed reviews. I haven't given up on the concept, though, and you may yet see other adventures of Ted and his friend.

You have to see an SRS club event to believe the high-tech chaos we can cram into a single room. Our typical event draws fifty or more people, with about a dozen robots. The noise, the enthusiasm, the ideas, the noise, and all the robots make for an exciting day of hacking, and keeps us all motivated to build even more and better robots.

If you are interested in putting on a local event of your own, you will find several good event ideas in this column. The robotic art contest, in particular, is a great crowd pleaser. It is easy to stage and lets the crowd get involved in the judging, so you get a lot of return for your investment in space and materials. And the anything-goes nature of robot artist designs spawns some very creative entrants.

[Author's note: I thought I'd turn this month's column over to a couple of guys I know in the Seattle Robotics Society. I hope you'll find their story interesting.]

"So, do you have a machine ready for the big contest?" Ted asked by way of greeting, as he dropped his backpack onto the floor next to my workbench.

I was slumped in front of my '486, victim of a recent contract software Job from Hell. The pay's OK, but sometimes the hours cut into my hobby. "Sorry, Ted," I replied, "but I just ran out of time. The meeting's tomorrow, and Robothon '95 is the next day. I'm not going to have anything ready by Sunday."

His reply was typical Ted. "Good, that means you'll have time to do some software for my robot. Check this out." Ted placed a wire-frame robot on my computer table. The hobby servos for motors, the 68hc11 computer board, and a small caster up front clued me that he had modified a Dancer robot from a previous *Nuts & Volts* issue, "Build an Open-frame Robot Body."

But he had made a design addition. I pointed to the third hobby servo, nestled inside the robot's midsection. "Cute hack, Ted. What's it for?"

"That's where I'm going to fasten the pen holder. This is our entry for the Robot Art contest." Ted was referring to one of the six robotic events in

Sunday's Robothon '95. The SRS had set up this day-long competition so that everyone in the club could show off their machines, and spend some quality time with their hobby.

R-95 would take place at the Renton Technical College, a new site for the SRS. The club had approached one of the instructors with the idea of holding the event at RTC; the response was immediate and enthusiastic. In fact, it looks like RTC will be the new home of future SRS meetings.

For Sunday, Tom Dickens and his organizers had set up a club-wide competition with contests such as Robot Art, Sumo, the Rapid-Deployment Maze, the Grand Maze, Line-Following, and Dead-Reckoning Navigation.

Ted had already finished most of his Robot Art competitor, lacking only the pen holder and some software. The pen holder problem, being mechanical, didn't worry Ted. But as usual, he came to me for the software. "Here, I've sketched out how I'm going to add the pen holder. It's really pretty simple." Ted showed me his engineering notebook, opened to this new robot, dubbed Artist (Figure 1).

"The third servo is a Futaba S148, just like the two drive motors, only it hasn't been modified; it only moves back and forth. I fastened it down to that piece of copperclad on the underside of the frame, using some double-sided foam tape.

"Next, I'll cut a 4-inch length of thin brass rod, which I can bend into a pen holder. You can get an idea of the shape from the sketch. Finally, I'll mount the pen holder to this round servo control horn, using some #2 hardware and hot glue. Add some masking tape to attach the pen to the holder, and 'Bingo', it's ready for your software." Ted looked at me expectantly.

I tried hard to muster a frown, ready to claim overwork, but no dice. Our friendship goes back too far. Ever since our early pinewood derby days in Cub Scouts, we had pooled our talents on projects, each knowing the other would help out on the rough spots.

In this case, as in most robot projects, I contributed the software while Ted did the mechanics, and we worked together on the electronics. Which meant it was time for me to dust off the keyboard and start coding. "OK, Ted, what do you have in mind for our Artist robot?"

"I figured we could make the robot lower its pen, then draw a very simple design, maybe just three or four lines. The end of the design would leave the

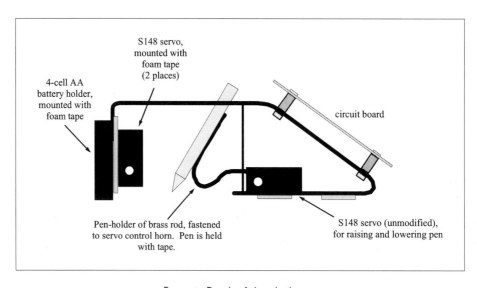

Figure 1. Details of Artist's design.

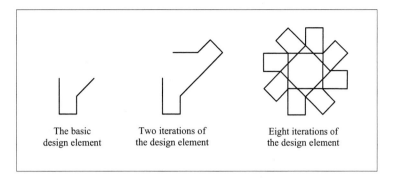

The basic
design element

Two iterations of
the design element

Eight iterations of
the design element

Figure 2. Technique used by Artist robot to create a drawing.

robot pointed at a slightly different angle than the starting alignment. Then, it would simply repeat the same design, over and over. The finished pattern should be a repetition of the basic figure, folded in upon itself many times. Kind of like this." Ted grabbed his pen and notebook, and started sketching (Figure 2).

I glanced at his notes, already juggling code in my mind as I talked. "Looks pretty simple. I'll need routines to move the drive motors, position the pen, and map out the lines. That means I'll also need some simple timer software, so I can track elapsed time."

"Well, I'll leave that part to you," Ted said. "I'm going to start on the pen holder. Where do you keep your brass rod?" I left Ted rooting through the closet in search of some thin brass stock. He already had the hot-glue gun plugged in and ready to go. I turned my attention to the computer.

Ted's Artist robot used a custom 68hc11 computer board, complete with 8K of external EEPROM. This was way more memory than the program would need, but Ted probably just grabbed whatever computer board was within reach.

Well before I was done with the first draft of my code, Ted had finished the pen holder. He taped a red washable marker in place, then manually moved the holder up and down, to gauge the range of motion. Satisfied, he returned to look over my shoulder.

"Good, you're using SBasic," he commented. I smiled; he had been half afraid I would write the code using tiny4th or assembly language. Ted could puzzle through most of my Forth programs, but was completely lost when it came to 68hc11 assembly.

"I figured SBasic would be simpler to follow, and easier to maintain. Besides, if I use SBasic, you can work on the program yourself, after the contest." I turned back to the keyboard.

Ted started downstairs, headed for the kitchen. "I'll find something for our lunch while you work on the code. Watching you type always makes me antsy." He returned after what seemed like only minutes, but was really an hour. "Your larder was pretty empty, so I biked down to the deli," he said, hoisting giant sandwiches out of his backpack. "You got salami, Muenster, and onion. Want half of my turkey, liver, and sauerkraut?"

Appalled as always by Ted's taste, I shook my head and reached for a Ballard Bitter. We both enjoy the ale, which started in a local microbrewery. Besides, you gotta love a brewery whose motto announces its Scandanavian heritage: "Ya, sure, ya betcha!"

After lunch, Ted zeroed in on the software. "Is it code yet?" he asked.

"Actually, I think I've got it. Here's a printout; you can see there isn't much to it." I started to walk Ted through the program, but he jumped in ahead of me (Figure 6, at end of article).

"I can see you used a bunch of **CONST**s to declare speeds for the right and left motors; that must be what all the rmtr_ and lmtr_ names are for. And I can see that you hooked the right drive motor to the 68hc11's TOC2, the left motor to TOC3, and the pen positioning motor to TOC4. But what about the deg90 constant?"

"Well," I explained, "the drive motors don't have encoders on them, so the only way the software can measure distance is to count the amount of time a motor is running. That delay contains the number of 4.2 msec ticks needed to turn the robot 90 degrees with both motors running full speed."

Ted picked up the thread again. "That means you must be using the 68hc11's Real Time Interrupt to count out the delays." He searched through the listing until he found the RTI interrupt service routine. "There it is, at the INTERRUPT $FFF0 line." He studied it for a moment. "Boy," he exclaimed, "there's nothing to it!"

"No, the code is amazingly simple when you see it written out in SBasic," I agreed. "All the ISR code has to do is look at some timer variable, WAIT in this case, and decrement it if it hasn't already hit zero.

"Next, the code rearms the RTI by setting the proper bit in the TFLG2 register. The END statement causes the 68hc11 to resume processing wherever it left off when the RTI interrupt occurred.

"This little ISR turns WAIT into a down-counting variable. Whenever WAIT is set to something other than zero, it will automatically decrement by one every 4.2 msecs, which is the RTI timer rate. When it hits zero, it stops decrementing. Really simple, and very effective."

Ted pointed to some code below the label MAIN. "Of course, you still have to set up the RTI initially. That must be what the POKEBs into TMSK2 and TFLG2 do. But that still doesn't look very complex."

"No, it isn't tough to figure out. The hardest part, actually, is wading through the Motorola technical literature, to figure out which bits in which registers need changing." I moved to the code inside the InitServos routine. "You recognize this code, don't you?" I asked.

"Well, I know what it needs to do." He pondered the code for a moment. "The software has to slave the three servo motors to the main timer, TOC1. After all, that's how all of the servos that we hook to 68hc11s work.

"And you've set it up so the three servo signals go low when TOC1 hits zero. Then, each servo signal goes high individually, when a TOC channel reaches its timeout count. But your other programs were never this easy to understand."

I had to agree. "This is the first real servo program you've seen in SBasic. I've written the same thing before, in assembler and in tiny4th, but SBasic really simplifies the code."

Ted was already looking at something else. "What's going on here inside the Delay routine?"

I noticed he was pointing to the POP() statement. "Well, SBasic doesn't support passing arguments to subroutines, so I used the data stack to handle arguments. Whenever code calls Delay, it first pushes an argument onto the data stack. The Delay routine then pops the value off the stack before setting up the delay. This lets me change the delay duration at each call."

Ted seemed to have run out of questions, so I hooked the 68hc11 board to my PC's serial port, fired up PCBUG11, and downloaded the compiled SBasic program into the 'hc11's memory. We set the robot down on a piece of paper, uncapped the pen, and pressed the reset switch. Our little robo-Picasso began drawing its masterpiece.

The SRS Meeting

Saturday morning, Ted picked me up in his old Toyota two-door, affectionately known as the Beater. His car looked as if it had been through a few serious wrecks, but Ted was actually a very careful driver. All of the Beater's "damage" resulted from large robotics experiments, many of which didn't pan out.

The parking lot already contained about 20 cars by the time we pulled in, so this looked to be a full house. Sure enough, by 10 o'clock the place was SRO, with about fifty people and at least a dozen robots attending.

Figure 3. Bill Bailey's M1 robot.

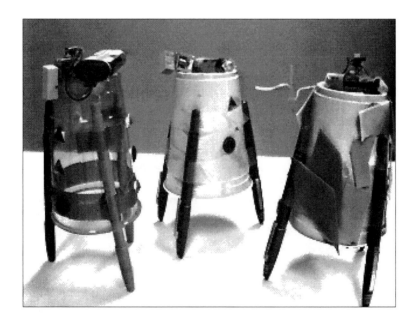

Figure 4. Marvin Green's simple art robots.

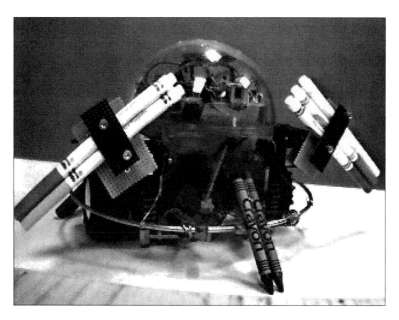

Figure 5. Tom Dickens' three-armed art robot.

The sheer number of robots at these meetings always surprises me. Ted and I go back to the late 1980's, when the SRS was lucky to muster ten people at a meeting, with just one or two robots under construction. More than half the discussion time at this day's meeting centered on the next day's contest. We hammered out final details on schedule, location, and event logistics.

The meeting also touched on the newly-installed SRS Web page. I hadn't had time to check this out, but Ted had browsed it, and was really excited. The club's Webmaster, Jared Reisinger, made sure everyone had the proper URL and gave us a quick tour of what the page offered. He also asked for a volunteer to bring a digital video camera to R-95, so he could get pictures of the event, and that was quickly arranged.

Several people showed off new robots, built specially for R-95. Each robot immediately triggered talks about motor controllers or sensors, so the meeting went into its "fits and starts" phase.

After a couple of hours, the meeting morphed into the usual robo-chaos, with about a dozen simul-taneous technical discussions taking place. At the lunch break, Ted and I headed off to the local surplus houses for some last-minute shopping.

Robo-thon '95

Bright and early Sunday morning, Ted aimed the Beater into a parking place at Renton Technical College. We collected our robots and other gear, then headed to the downstairs seminar rooms for R-95. The venue looked perfect. The room was well-lit, nearly 80 feet by 40 feet, with enough tables and chairs for at least 100 people. The floor sported a neutral color, short-nap carpet, ideal for running small robots.

Ted summed it up. "I like it; restrooms around the corner, vending machines up the hall, and it's got a video monitor and VCR hanging on the wall. Looks like a keeper."

As sponsor of the Line Following event, I had some early duties to handle. This mostly involved moving tables to clear an area for the event, then taping down an eight-foot length of blank 30 inch-wide drawing paper. Working carefully with a black marking pen, I drew a line the length of the paper, to mark out the contest's course. The 1/4-inch wide path contained some right-angle bends and some smooth arcs, with the starting point at one end and the finish line at the other.

Other sponsors were also setting up their contests, and before long we had a Sumo ring in one corner, and the Dead Reckoning course set up in another.

Ted wandered over about this time. "Somebody showed up with some robotic art he did. It doesn't move yet, but he wants the SRS to help him add motion to his work. He's a local sculptor, and he makes these neat pieces, about three feet high, welded out of springs and hardware he found at Boeing Surplus. The two he brought today are kind of like little metal people. One of them holds out a pan so you can throw in some pennies; the guy wants to add a tape deck so his robot can hit you up for loose change."

Ted helped me run the Line Following contest, the first event of the day. Four people entered machines, and Gary Teachout took first place with a speedy little machine that basically used two photodetectors straddling the line. Whenever a photodetector saw the line, it switched the drive motors to turn back towards the line. This simple bang-bang design usually results in a robot that crabs back and forth across the line, taking far too long to run the course.

Gary made up for this flaw by running his machine at an insanely high speed. As Ted said, "Heck, the machine probably only saw the line about ten times in the entire run!" The robot took just 6.5 seconds for the fastest of its three runs, down a twisty course that probably measured five feet long.

Contrast that time with Tom Dickens' second-place finish of 16.7 seconds. Karl Lunt came in third with his venerable CBE-1, clocked at 23.9 seconds; Bill Harrison's machine was not able to finish the course. The crowd definitely got into the LF contest, and each contestant got a good round of applause and lots of encouragement. It set the stage for a great robot day.

Ted glanced over at the far end of the room, then added, "Looks like the Dead Reckoning contest is getting started. Let's go take a look." The premise behind the DR contest seems simple. A robot must start at a marked location, then travel two meters in a straight line, turn 45 degrees to the right, and travel an additional two meters.

The target point, where the robot *should* stop, is marked with a tiny icon. A robot is judged by the absolute distance between its ending point and the target point. Obviously, the smaller the delta, the better the score.

All three entrants used stepper motors to drive their robot. After three heats, Bill Bailey was declared the winner; his machine's best run ended with a delta of 5/16 of an inch from the target point. Gary Teachout's design (different from his line-following 'bot) turned in a low of 11/16 of an inch, good for second place. Ken and Mike Birdwell's machine finished third with a two-inch difference.

After each contest, Tom Dickens had the entrants bring their robots over for a "family portrait." Tom had set up his Mac, complete with a QuickTime camera, and took a .GIF of the robots (see photos throughout this article). In fact, Tom took .GIFs of most of the robots at R-95, and Jared, the club's Webmaster, later put them on the SRS Web page.

By this time, Ted was feeling a mite peckish, and headed out to snare some lunch. I'll spare you the details, though I will say lunch was up to Ted's usual standards. I had almost recovered when Ted exclaimed, "OK, it's time for the Robot Art contest. Let's go!" He grabbed Artist and fast-walked to the far end of the room, where Tom Dickens was setting up the competition.

By the time I had arrived, a crowd of about 12 entrants and some 25 bystanders had gather around the three tables set up as drawing surfaces. Tom had taped a three-foot length of paper down to each table-top, to serve as a canvas.

The RA contest couldn't be simpler. Set your robot down, then have it draw, paint, chalk, spray, or whatever, some type of artistic creation. You get a maximum of five minutes for your run, then the paper is removed and hung up on the wall with the other drawings. After all robots have performed, the crowd votes by applause for the winner.

Since Ted had worked his way to the front of the crowd, Tom let him start the event. The little robot performed as expected, and got a nice hand for its effort. However, it quickly became apparent that Artist was too simplistic. Following robots used complex patterns and multiple colors to out-do our machine, and Ted's face fell a little more as each new picture went up on the wall.

Marvin Green, inventor of the 68hc11 BOTBoard that so many of us use, showed off a three-legged art machine so simple that several kids had entered similar devices. Marvin made his robot by fastening three felt pens, points down, to an upside-down plastic cup. The cup rested on the three pen tips, while a small battery-powered motor, mounted on the cup, quickly spun an eccentric weight. The offset weight made the cup vibrate across the table top, leaving three colored tracks as it went.

"Will you look at this?" demanded Ted, holding one of Marvin's machines. "How does he dream up this stuff?" Ted showed me the servo motor and BASIC Stamp that Marvin had added, to shift the position of the spinning motor and vary the robot's drawing. "This is a nice touch, and very simple. This robot would make a great school project."

The contest took about 30 minutes to run, and we ended up with about a dozen pictures on the wall. Kevin Ross, the SRS president, did the honors of polling the audience, and declared Bill Bailey the winner. Bill had taped a felt pen to the front of his Dead Reckoning entry, then done some hasty software writing during lunch. Now, his machine repeatedly drew a large, ornate pattern of lines and curves. Due to tiny differences in the drive system, the designs precessed as the robot drew.

It was basically the same concept Ted and I had envisioned for Artist, but on a much larger and more elegant scale. "Oh well," said Ted, "we'll just do a better job for the next contest."

So the day continued, with contests, lots of robots, and plenty of excitement. By 3:30, when we were supposed to clear out, the SRS members had enjoyed about seven solid hours of high-intensity robotics. We cleaned the place up, gave a big "Thank you!" to our host from Renton Technical College, and departed for home.

Ted summed up the day. "You know, we filled that room with robots and robot builders, we spent all day doing and talking robots, and it was only a club event! What's it going to look like in the spring, when the SRS holds a Robothon open to everyone?"

"No idea," I replied. "But I know we'll be there."

```
'
'    artist.bas        SBasic program to draw line art, using
'    a wire-frame robot similar to Dancer.
'

include   "regs11.lib"


'
'    Declare constants for various motor speeds, for both
'    the right and left motors.
'

const    rmtr_fwd  = $f800
const    rmtr_rev  = $f200
const    lmtr_fwd  = $f200
const    lmtr_rev  = $f800
const    rmtr_fwd75  = $f740
const    rmtr_rev75  = $f2c0
const    lmtr_fwd75  = $f2c0
const    lmtr_rev75  = $f740
const    rmtr_fwd50  = $f680
const    rmtr_rev50  = $f380
const    lmtr_fwd50  = $f380
const    lmtr_rev50  = $f680
const    rmtr_stop = $0
const    lmtr_stop = $0


'
'    Declare constants for pen positions.
'

const    pen_up  = $f500
const    pen_down = $f300


'
'    Declare constants for the TOCs attached to
'    the three servo motors.
'

const    rmtr_toc = toc2
```

Figure 6. Listing for the Artist robot.

```
const   lmtr_toc = toc3
const   pmtr_toc = toc4

const   seconds = 254
const   deg90 = 345

declare   wait                    ' universal timer
declare   n                       ' temp variable


'
'   InitServos        initialize the servo control system
'
'   This version uses three servos, hooked to toc2, toc3, and toc4.
All
'   servo control lines are pulled high when toc1 times out. Each
line
'   then drops low when the corresponding TOC channel times out.
'


InitServos:
poke   toc1, 0            ' clear toc1
pokeb  tctl1, $fc        ' set tocx when they time out
pokeb  oc1m, $70         ' connect toc2, toc3, and toc4 to toc1
pokeb  oc1d, $00         ' write 0s to tocx when toc1 = 0
gosub  StopMotors        ' shut off motors
return


StopMotors:
poke   rmtr_toc, rmtr_stop      ' stop the right motor
poke   lmtr_toc, lmtr_stop      ' stop the left motor
return


GoRight:
poke   rmtr_toc, rmtr_stop
poke   lmtr_toc, lmtr_fwd       ' turn left motor
return
```

Figure 6 continued.

```
GoLeft:
poke    rmtr_toc, rmtr_fwd              ' turn the right motor
poke    lmtr_toc, lmtr_stop
return




PlacePen:
poke    pmtr_toc, pop()                 ' make it so
return




Delay:
wait = pop()                            ' set the delay
do loop while wait <> 0                 ' wait it out
return




interrupt    $fff0                      ' RTI isr
if wait <> 0                            ' if need to drop it...
     wait = wait - 1                    ' drop it
endif
pokeb   tflg2, %01000000                ' rearm RTI
end




main:
pokeb   baud, $30                       ' 9600 baud
pokeb   sccr2, $0c                      ' turn on SCI

wait = 0                                ' clear the wait timer
pokeb   tmsk2, peekb(tmsk2) or %01000000
pokeb   tflg2, %01000000                ' start RTI system
interrupts   on                         ' allow interrupts

print   "Artist...";
gosub   InitServos                      ' start up servo system
push    pen_down                        ' flag for pen down
```

Figure 6 continued.

```
gosub   PlacePen                    ' place the pen
push    2*seconds                   ' get a delay
gosub   Delay                       ' and wait

for   n = 1 to 12
        gosub   GoRight
        push    deg90
        gosub   Delay

        gosub   GoLeft
        push    deg90 * 3
        gosub   Delay
next

gosub   StopMotors                  ' all done
push    pen_up                      ' raise the pen
gosub   PlacePen

print   "done"

end
```

Figure 6 continued.

Some Powerful Software Tools

I used this October 1997 column to describe some of the more powerful tools available to robot builders everywhere. The most powerful tool, by far, is the Internet, since it in turn gives you access to other high-quality tools from around the world. For example, PFE-32 is a top-notch GUI-based editor, available free from Alan Phillips of the Lancaster University Computer Centre in England. I use PFE-32 daily, and give it high marks in ease of use and versatility.

I also described some 68hc11 C compilers available to you over the web. In some cases, these are commercial products that you can order from companies given high marks by firmware developers worldwide. I also spent some time discussing the GNU 68hc11 C compiler, available free from the Internet. Since this is a GNU-licensed product, you not only get the working executable for free, you can even get the source code for free. With a little effort, you can roll your own compiler.

Success in this hobby depends on your tool set. I have little enough time to spend on robot building as it is; I can't afford to waste any of it making yet another version of the same computer board, mounting bracket, or motor driver software.

Longtime readers of this column are by now familiar with most of my tools. Nearly all of my small robots use Marvin Green's original BOTBoard, a gem of a 68hc11 single-board computer. Fitted with a 68hc811e2 microcontroller (MCU) and precious little else, this single-chip tool lets me get 2K bytes of robot code up and running in almost no time. This tool is so important to my career and my hobby that I spent a couple of afternoons building up three or four spares. Now, whenever I need to test a robot concept or try out some SBasic code, I grab a BOTBoard and some batteries; the test is done in minutes.

Speaking of SBasic, my homebrew compiler remains the most potent tool in my collection, to the point that I seldom write 68hc11 assembly language for robotics any more. Though I pay the mortgage with C, I can code working programs more quickly in SBasic, and they take up far less code space than comparable programs written in just about any other language. Sure, I can beat the code size and speed of SBasic if I write everything in assembly language, but I have a ton of working SBasic code to draw from, and the compiler handles all of the dirty work of assigning addresses and matching up labels. In all, SBasic has saved me days of painful programming effort and helped me keep my sunny disposition.

My tool set includes mechanical items as well. The all-time simplest way to get a robot running is to build it with hobby servo motors. The Futaba S148, among others, gives you a powerful motor in a small, easy-to-mount package, complete with a power cable and a standard control format, for only $15: truly a bargain. Two motors, some double-sided foam tape (another awesome tool), and a couple of wheels make most of a robot frame. You have to modify the servo motors to make them go 'round and 'round rather than back and forth, but that only takes a few minutes. You end up with a complete robot motor system for little time or cash.

Certainly the most flexible tool available to any robot builder is the Internet. An afternoon spent with a decent browser will turn up gobs of web pages filled with free software, application notes, technical information, links to other web sites, and descriptions of other, working robots. If you have any questions on how to do something in robotics, start by hitting the web.

Your tool set should include everything from real hardware tools, such as the Dremel hand-held moto-tool, to whatever programs ease the pain of writing software. If you find something that saves you time, simplifies your efforts, makes a more professional job, or saves you frustration, add it to your tool set and refine its use. You'll be much happier in this hobby.

PFE-32

Back in the days of DOS, I wrote all of my software using micro-Emacs, a PC-based subset of the Unix emacs editing system. This choice in editors was a carryover from an early contract job that required

use of the Unix operating system; I liked the flexibility of emacs so much that I moved the micro version onto my PC and never looked back.

As PCs became more Windows-based, sticking with emacs became more and more difficult. I spend so much time in the Windows desktop that popping back and forth between emacs and my other programs slowed down my productivity. And the emacs cut and paste buffers are not compatible with the other Windows programs, so I can't move blocks of text around as easily as I would like, if at all. So I began an idle search for an alternative to my well-worn tool. My search eventually led me to PFE-32.

PFE-32, which stands for Programmer's File Editor (32-bit version), has been around in some form or another for years. The version I pulled off the web (0.06.002) carries a copyright that covers 1992 through 1995. Written by Alan Phillips of the Lancaster University Computer Centre in England, PFE is a well-designed and seamlessly integrated Windows-based editor. Alan built this to be a programmer's editor, so it supports features not often found in word processors.

For example, PFE remembers the differences between editing an assembly language file and a C source file. I usually use tab stops of eight columns for my assembly language files and six columns for C. As I hop back and forth between the two different file types, PFE automatically tracks the tab stops so the formatting is correct. You set up PFE to track this and other editing elements of a file type by using pull-down menus of options and by defining groups of files by their extension. I told PFE that all files with an extension of .c are to be treated as C source files, and all C source files use 6-column tab stops. Now, anytime I create or edit a file with an extension of .c, PFE automatically uses the correct tab stop setup.

PFE uses an extensive Windows-style tool bar, with graphic buttons for activating any of its main functions. Besides the common functions such as New File, Cut, and Paste, PFE also has buttons that support launching other Windows or DOS applications. I can setup the Launch button (which has a picture of a rocket on it) to fire off a DOS application such as SBasic, complete with the necessary command line and working directory. When the application finishes, PFE retains the DOS screen for you so that you can see the results of the execution. You can customize the application command and the working directory for each launch, and even choose these items from drop-down lists in the launch setup window.

Sometimes you need to launch a DOS application, then edit the output screen for that application. PFE has a separate tool button for this function, which writes the entire output stream from the launched application and saves it to a command window for your use. This comes in very handy for fixing errors in compilations or assemblies.

One element of PFE that I really like is the ability to change the command keys to suit your own requirements. From my emacs days, I'm accustomed to hitting F1 whenever I want to start a forward search for a text string. In most editors, including the default keyset for PFE, this just gets you a help splat. Using PFE's Options menu, I can change the command key used for nearly any function to be anything else I want. I can even save my command key definitions into a keyfile, and select whichever keyfile I want to use during an editing session. If you long for the days of the Sidekick editor, just redo your PFE command keys and bingo, you're back to the past.

And for those really bizarre editing operations, PFE sports a macro capability. I've often pulled up emacs to edit a text file simply because its macro feature saved me minutes or hours of work. PFE supports a macro feature as good as any I've seen on a Windows editor, though it still falls short of what emacs can do. In PFE, I can start a macro recording by clicking a tool button with a picture of a cassette tape on it. PFE remembers all of my subsequent keystrokes until I click the button again, shutting off the record feature. To replay the saved macro, I just press F7. The macro can record Find and Repeat Find commands, making it easy to step through a table or list. You can even embed control characters such as tabs and line-feeds in your find string, using the C con-

vention of \t and \f. Unlike emacs, however, I can't set off N automatic repetitions of the macro. Still, PFE's macro capability is easy to use and will prove a real time-saver.

I've only scratched the surface of what PFE offers. The level of customization is amazing, and the editor performs smoothly and reliably, no matter how many times I bounce between various DOS and Windows applications. I still use emacs now and then, but PFE can do almost everything emacs does, and its multi-window mode is easier to use. Even if I don't abandon emacs entirely, PFE has become a powerful addition to my tool set.

You can find PFE-32 at Alan Phillips' web site. As I write this, Alan has released a newer version, which I will probably get around to testing pretty soon. Alan also maintains a superb FAQ list, describing in clear terms answers to some of the most common PFE questions. (Makes me think I probably ought to start an SBasic FAQ.)

Note that PFE is freeware, not shareware or a crippled version of a commercial program. You can use PFE at home, at work, or on as many machines as you like. Alan asks that if you intend to embed PFE in a commercial release, you contact him for a licensing agreement. I consider PFE a superb piece of work, easily of commercial quality, and I applaud Alan's attitude of releasing his excellent editor for free. Speaking as one who has authored a few freeware releases in the past, I encourage you to contact Alan and give him feedback on what you like and don't like about his program, and to thank him for his generosity. Highly recommended.

xgcchc11

If you want to write software for a 68hc11-based robot, you have a wide variety of assemblers and compilers from which to choose. Few of them, however, will generate code small enough to fit into the 2K bytes of room in a 68hc811e2. You can put a lot of functionality into 2K using assembler, and slightly less using SBasic or tiny4th. Beyond that, however, your options begin to drop off. I earn my living using C and I'm always on the lookout for C compilers that make the grade.

Two compilers in particular deserve mention as possible robot tools. Though they don't pass my cost test (cheap, if not free), people whose opinion I value like and use them both. Dave Dunfield has been selling compilers for a variety of microcontrollers for years now. His web site has an excellent catalog, at $100 the compilers are reasonably priced, and Dave has an excellent reputation on the Intenet for supplying good technical support and quick responses to questions. If I were to drop $100 of my hard-earned robo-bucks on a C compiler, I'd probably give the nod to Dave's Micro-C just on the strength of comments on the 'net.

A newer compiler, ImageCraft's ICC11, also gets frequent mention on the Internet, and is worth considering if you have the $130 or so for a Windows version of their product. The compiler is billed as "near-ANSI" with updates planned by year-end. Check the ImageCraft web site for details.

But when I look for a compiler, I want totally cheap, as in free. So I turned to the GNU family of software products. Several flavors of the widely used GNU cross-compilers exist on the web; besides the 68hc11, you can find compilers for the high-end Motorola 68000 processors and most processors that support a Unix platform, including Sparcs.

The GNU software suite started years ago, partly as a natural outgrowth of the "software should be free" attitude endemic to the Unix practitioners. GNU programs, such as the GNU compiler gcc, are available free of charge to any who want them, with a few minor restrictions. These restrictions appear in the GNU General Public License (GPL); I've copied the salient paragraph to show you the gist of the GNU mindset:

When we speak of free software, we are referring to freedom, not price. Our General Public Licenses are designed to make sure that you have the freedom to distribute copies of free software (and charge for this service if you wish), that you receive source code or can get it if you want it, that you can change the

software or use pieces of it in new free programs; and that you know you can do these things.

I took the above paragraph from version 2 of the GNU GPL included with the source files I downloaded for the gcc cross-compiler.

For those robot builders interested in hacking serious software, the GNU projects are a godsend. Where else can you not only find a high-quality, well-designed suite of compilers and tools, but get your hands on the source code and build instructions as well?

For some years, Coactive Aesthetics maintained a gcc compiler for the 68hc11. True to the GNU public license, you could easily get full source files for this product from CA. I have not been able to find the gcc source files at the CA web site recently, and it looks as if they have stopped supporting the compiler. But you can still find the full set of source files at other Internet sites. If you would like to try your hand at building your own top-notch C compiler, check the Free BSD web site.

I'll give you fair warning, however. Building such a compiler is doable but not easy. For example, you will have to un-tar the original collection of files, and if that doesn't mean anything to you, you have a steep learning curve in front of you.

If you would like to dabble with the gcc compiler, but don't want the hassle of rebuilding the executable, snap up a copy of xgcchc11. This 68hc11 cross-compiler is based on the Coactive port, but is readily useable from DOS or Windows. I picked up a copy from Oliver Kraus' 68hc11 web page. This well-maintained site contains plenty of good 68hc11 information, plus links to other useful 68hc11 web sites.

Downloading from Oli's site will get you a large file called xgcchc11.zip. Stick this in a new directory named something appropriate, such as gcc, and unzip it using the **-d** option, which preserves the original subdirectory structure. Next, read the file Xgcchc11.txt in the /doc directory. It contains some very terse instructions for setting up the needed environment variables and paths, and for modifying the \bin\igcc.bat file, which initializes xgcchc11 before you can run it.

After you finish the setup, reboot your system so the new environment variables take effect, then execute the \bin\igcc.bat file. Now you are ready to begin skimming through the various sample files in the \src and \include directories.

The \src directory contains a few sample programs, including the most important of all, crt0boot.s. This file holds the 68hc11 startup and run-time code that is executed whenever your target system comes out of reset and runs a gcc C program. The crt0boot.s file supplied with the xgcchc11 release was written for the Coactive Aesthetics GCB11 computer board, so you will likely need to rework the crt0boot.s file somewhat.

To help you in getting started, I've included below a crt0boot.s file suitable for use on a BOTBoard with a 68hc811e2 (Figure 1, at end of article). Yes, this means you can actually write (small) C programs and get them to run in 2K bytes of code space. Note that the code is larger than a comparable SBasic program would be, so you won't get a lot of functionality in that small space. However, if you are running on an expanded 68hc11 board, you won't need to worry about the larger code size.

A quick glance at the my crt0boot.s file should make one very nice feature of this system clear immediately: this is a linker-based compiler. Thus, you can build object libraries of relocatable code, then use the linker to assign addresses for any memory map you want to use. This feature alone justifies the time I've spent in reworking the various system files.

For example, refer to the line in crt0boot.s:

.area   _BSS

This statement creates a relocatable area of memory named _BSS, which will hold the program's stack, among other items.. You can use commands to the gcc linker to assign the _BSS area wherever you need it. If your program will run on a 68hc811e2 with its meager 256 bytes of RAM, then you can tell the linker to put _BSS at address $0000. If, however, you are using a board with expanded memory, you can move _BSS to $2000 to take advantage of the larger space. A typical linker command for building a program for the 68hc811e2 might be:

xgcc -c -mshort_branch blinky.c
ld -b_CODE=0x0f800 -b_BSS=0x0000 -m
blinky.o crt0.o -o blinky

Here, the **ld** command assigns the _CODE area to $f800, which is the start of the 68hc811e2 EPROM memory, and _BSS to $0000, which is the start of the 256 bytes of on-chip static RAM.

Another element worth noting in the crt0boot.s file concerns the large number of psuedo-registers. The gcc compiler model originally assumed a processor with a large number of registers, such as the 68000. When the compiler was ported to the much smaller 68hc11, the designers decided to preserve this model by assigning addresses in memory to act as the missing registers.

Using this design meant that gcc could be ported fairly quickly to the 68hc11. One unfortunate side effect, however, is a fair amount of code bloat. Operations that would be carried out between registers on a 68000 now require a lot of shuffling between the 68hc11 registers and these pseudo-registers, slowing down the code and increasing its size.

Notice how interrupts are handled in the crt0boot.s file. Whenever an interrupt occurs, control goes through the normal $ffxx vector area to a dispatch table starting at label **isr_table**. Since this table is located in area _BSS, it resides in RAM. The startup code fills this table with the address of a dummy interrupt handler at label **dummy_rts**. Thus, the default action of any interrupt handler is simply to return immediately. Should your C program need to do anything special in servicing an interrupt, your code should overwrite the proper address in the dispatch table with a pointer to the appropriate C function. Refer to the comments above label **isr_table** for an example.

Finally, see how the section at the end of crt0boot.s assigns the vectors in the $ffxx area. The associated addresses all appear in an area named BOOTLIST, which is declared to the linker as being ABS, or absolute. This means that the linker cannot move this area away from its assigned location at $ffd0. Thus, all the 68hc11 interrupt vectors appear where they belong in the memory map. Since the linker cannot move this area, there is no need to define its location in the linker command example given above.

The I/O modules provided with the gcc release work with hardware on the GCB11 board, so they aren't very useful to me. I'm currently developing some I/O routines for use with the BOTBoard and the 68hc11 SCI. I already have several test programs up and running, and it's quite a kick to write code on a free compiler, run it through a free linker, and get out a useable object file. Sure, the code isn't nearly as fast or as small as SBasic would make, but if you want C without the dollar expense, this method is worth a look.

That's a Wrap

This article covered some of the most powerful tools you can bring to bear on your robot projects. Although I emphasized software this time, the concept applies across the whole spectrum. Search out those mechanical, electronic, and software tools that give you the most bang for your buck or hour, get good with them, and use them to improve your robot building.

Above all, use the Internet search engines, such as AltaVista to locate the tools you need and the people who have built them. If you are just starting out in this hobby, resist the urge to start building something RIGHT NOW. Instead, take some afternoons and surf the web, looking at what others have done and what tools they used. The time spent in your surfing and collecting will be repaid tenfold or more in increased productivity and enjoyment. After all, we **are** in this hobby to have fun.

```
;
;       crt0boot.s
;
;       The C run-time startup code for the 68hc11.
;       Modified by Karl Lunt from the original code
;       written by Oliver Kraus.
;


        .module    crt0boot.s


;
;       A short note about labeling convention:
;
;       Labels inside an assembly language file such as this
;       one are declared with either one or two trailing
;       colons, as in:
;
;       foo::                    ; a global label
;       bar:                     ; a local label
;
```

Figure 1. crt0boot.s for the 68hc811e2 BOTBoard.

```
;    If the label ends in two colons, it is marked as
;    a global label and can be referenced by any external
;    routine, be it assembly language or source. If the
;    label ends in a single colon, the label is local
;    and is not visible to any external routines.
;
;    Additionally, all assembly language labels referenced
;    from a C routine must have an underscore ('_') prepended
;    to it. So if you want to refer to an assembly language
;    label or variable named  'hoohaw', use the following
;    convention:
;
;    _hoohaw::                    ; global assembly label
;    j = hoohaw * 2;             ; referenced from C
;    jmp    _hoohaw         ; referenced from assembly
;


;    Declare the direct-page area of memory, which for
;    the 68hc11 is page 0 ($00 - $ff). The linker must
;    treat this area as absolute so it always appears
;    at the proper address.
;

     .area    DIRECT (ABS,PAG)

;
;    Leave address 0 alone, so a null-pointer doesn't
;    trash the C psuedo-registers.
;

     .org   2
;
;    Declare the C psuedo-registers. These registers
;    are referenced by various C internal routines; also,
;    the xgcc code generator assumes these registers exist.
;    Thus, every C program compiled by xgcc must make these
;    registers available, so they might as well appear
;    in this file.
;
```

Figure 1 continued.

```
ZD0::   .blkb  2
ZD1::   .blkb  2
ZD2::   .blkb  2
ZD3::   .blkb  2
ZD4::   .blkb  2
ZD5::   .blkb  2
ZD6::   .blkb  2
ZD7::   .blkb  2
ZD8::   .blkb  2
ZB0::   .blkb  1
ZB1::   .blkb  1
ZB2::   .blkb  1
ZB3::   .blkb  1
ZB4::   .blkb  1

;
;    A scratch area for other functions.
;

ZXT::  .blkb  2




;
;    If you want to assign a fixed address for the stack
;    at edit-time, you can use the following technique:
;
;            .area STACK   (ABS)
;            .org  0x0100
;    __stack_end::
;            .blkb 96    ; or whatever
;    __stack_begin::
;
;
;    But assigning the stack space at link-time makes the
;    software more flexible. The following technique lets
;    you assign the stack location in the loader command
;    line, based on where you assign segment _BSS:
;
```

Figure 1 continued.

```
        .area  _BSS
__stack_end::
        .blkb 96              ; stack size in bytes, change if you want
__stack_begin::


;
;   Declare the area reserved for the RAM-based interrupt
;   handler table. This area will be filled by a call to
;   function __build_isr_table() below, before control
;   jumps to main().
;
;   This table provides your program with hooks into the
;   68hc11 interrupt system. At run-time, your program can
;   write the address of a C routine in one of these addresses.
;   When the corresponding interrupt occurs, your C routine
;   will be invoked to handle the interrupt.
;
;   Note that your C routine does not need to be declared as
;   an interrupt handler with the typical #pragma-interrupt
;   syntax. You are free to use a normal, non-interrupt
;   C routine as an interrupt handler. The RTI needed for
;   returning from an interrupt routine is buried inside
;   the low-level code below for each interrupt vector.
;
;   For example, to use your C routine service_rti() as
;   the interrupt handler for the RTI interrupt, put the
;   following lines of code in your main() function:
;
;   extern   unsigned int       isr_rti;
;   void                service_rti();
;
;   isr_rti = (unsigned int)service_rti;
;

        .area  _BSS

_isr_table:
_isr_none1::        .blkb 2
_isr_none2::        .blkb 2
```

Figure 1 continued.

```
_isr_none3::        .blkb 2
_isr_sci::  .blkb 2
_isr_spi::  .blkb 2
_isr_paie:: .blkb 2
_isr_pao::  .blkb 2
_isr_to::   .blkb 2
_isr_toc5:: .blkb 2
_isr_toc4:: .blkb 2
_isr_toc3:: .blkb 2
_isr_toc2:: .blkb 2
_isr_toc1:: .blkb 2
_isr_tic3:: .blkb 2
_isr_tic2:: .blkb 2
_isr_tic1:: .blkb 2
_isr_rti::  .blkb 2
_isr_irq::  .blkb 2
_isr_xirq:: .blkb 2
_isr_swi::  .blkb 2
_isr_illegal::      .blkb 2
_isr_cop::  .blkb 2
_isr_clock::        .blkb 2




;
;   These are the default interrupt handlers, used any time
;   a 68hc11 interrupt occurs; they are initialized by
;   a call to the function __build_isr_table() below.
;


        .area _CODE

v_none1:
        ldx   _isr_table
        jsr   0,x
        rti


v_none2:
        ldx   _isr_table+2
        jsr   0,x
```

Figure 1 continued.

```
        rti

v_none3:
        ldx     _isr_table+4
        jsr     0,x
        rti

v_sci:
        ldx     _isr_table+6
        jsr     0,x
        rti

v_spi:
        ldx     _isr_table+8
        jsr     0,x
        rti

v_paie:
        ldx     _isr_table+10
        jsr     0,x
        rti

v_pao:
        ldx     _isr_table+12
        jsr     0,x
        rti

v_to:
        ldx     _isr_table+14
        jsr     0,x
        rti

v_toc5:
        ldx     _isr_table+16
        jsr     0,x
        rti

v_toc4:
        ldx     _isr_table+18
        jsr     0,x
```

Figure 1 continued.

```
        rti

v_toc3:
        ldx     _isr_table+20
        jsr     0,x
        rti

v_toc2:
        ldx     _isr_table+22
        jsr     0,x
        rti

v_toc1:
        ldx     _isr_table+24
        jsr     0,x
        rti

v_tic3:
        ldx     _isr_table+26
        jsr     0,x
        rti

v_tic2:
        ldx     _isr_table+28
        jsr     0,x
        rti

v_tic1:
        ldx     _isr_table+30
        jsr     0,x
        rti

v_rti:
        ldx     _isr_table+32
        jsr     0,x
        rti

v_irq:
        ldx     _isr_table+34
        jsr     0,x
```

Figure 1 continued.

```
              rti

v_xirq:
        ldx     _isr_table+36
        jsr     0,x
        rti

v_swi:
        ldx     _isr_table+38
        jsr     0,x
        rti

v_illegal:
        ldx     _isr_table+40
        jsr     0,x
        rti

v_cop:
        ldx     _isr_table+42
        jsr     0,x
        rti

v_clock:
        ldx     _isr_table+44
        jsr     0,x
        rti

;
;    __build_isr_table
;
;    Load the interrupt handler table with the default values
;    for all of the interrupt handlers. By default, each
;    handler will contain the address of __dummy_rts below. Thus,
;    the default action of each handler is to do nothing.
;
;    This means that by default, no interrupt, including
;    the RTI, touches any register flags. If you need
;    interrupt support, write your own interrupt service
;    routine and stash its address in the appropriate table
;    entry during run-time.
```

Figure 1 continued.

```
;

__build_isr_table:
      ldy    #__dummy_rts
      ldx    #_isr_table
      ldab   #24
1$:
      sty    0,x
      inx
      inx
      decb
      bne    1$
__dummy_rts::
      rts

;
;                      Start of program execution
;
;   Control reaches this point immediately following reset.
;
v_reset:
      lds          #__stack_begin-1     ; initialize stack pointer

;
;   If you need to change any time-critical registers, put
;   your code for those functions here.
;


;
;   The following code sets up the interrupt handler table.
;   This is a dedicated area of RAM reserved for addresses of
;   ISRs.
;
      bsr          __build_isr_table    ; build the interrupt table

;
;   Setup is complete; execute the function main().
;
      jsr          _main                ; main()
```

Figure 1 continued.

```
;
;    The following code acts as a safety net, in case your code
;    accidently leaves main(). NEVER allow control to exit
;    main()!
;

_exit::
       bra         _exit


;
;    The label ___main serves as a prologue point for the
;    code that calls main(). If you need to do any setup
;    after the stack frame for main() has been created
;    but before the code in main() actually starts, put
;    that setup code here. Normally, a simple RTS is
;    sufficient, and that is the default here.
;

___main::
       rts                   ; return from function



;
;    Here are two small functions for enabling and disabling
;    interrupts on the 68hc11.
;

_enable_interrupts::
       cli                   ; allow interrupts
       rts

_disable_interrupts::
       sei                   ; shut off interrupts
       rts




;
;    The following code declares the interrupt vectors and
```

Figure 1 continued.

```
;     fills them with the addresses of the appropriate interrupt
;     handlers. The interrupt vectors must reside in a fixed
;     location of memory.
;

        .area     BOOTLIST        (ABS)
        .org      0xffd0              ; vectors must start here!

        .word     v_none1   ; 0xffd0
        .word     v_none2   ; 0xffd2
        .word     v_none3   ; 0xffd4
        .word     v_sci           ; 0xffd6   SCI
        .word     v_spi           ; 0xffd8   SPI
        .word     v_paie          ; 0xffda   Pulse Accum Input Edge
        .word     v_pao           ; 0xffdc   Pulse Accum Overflow
        .word     v_to            ; 0xffde   Timer Overflow
        .word     v_toc5          ; 0xffe0
        .word     v_toc4          ; 0xffe2
        .word     v_toc3          ; 0xffe4
        .word     v_toc2          ; 0xffe6
        .word     v_toc1          ; 0xffe8
        .word     v_tic3          ; 0xffea
        .word     v_tic2          ; 0xffec
        .word     v_tic1          ; 0xffee
        .word     v_rti           ; 0xfff0   Real Time Interrupt
        .word     v_irq           ; 0xfff2   Interrupt Request (IRQ)
        .word     v_xirq          ; 0xfff4   Nonmaskable Interrupt
Request (XIRQ)
        .word     v_swi           ; 0xfff6   Software Interrupt (SWI)
        .word     v_illegal ; 0xfff8   Illegal Opcode
        .word     v_cop           ; 0xfffa   COP Watchdog Time-Out
        .word     v_clock   ; 0xfffc   Clock Monitor Fail
        .word     v_reset   ; 0xfffe   Reset vector

;
;   This completes the boot portion of the C run-time module. The
above code
;   serves as the bare minimum necessary to build a working C pro-
gram for the
;   68hc11.
;
```

Figure 1 continued.

```
;    If you need added low-level functionality for your C program,
build a
;    second module and add your extra code there. Keeping project-
specific
;    low-level code in separate modules lets you use the same
crt0boot.s file
;    and graft on different files for different projects.
```

Figure 1 continued.

Appendix A
Contacts

Active Electronics (Future-Active, Inc., or FAI)
Products: Electronic components, test gear, supplies
Phone: 800-677-8899 (Main number for US and Canada)
Check your telephone book for local outlets.

A K Peters, Ltd.
Products: A K Peters specializes in books on technical and scientific topics, robot kits, and the occasional robot book; they have a great catalog, too. And I don't say that just because they published this book...
Phone: (508) 655-9933
Fax: (508) 655-5847
Mail: A K Peters, Ltd.
63 South Ave.
Natick, MA 01760
email: service@akpeters.com
Web: http://www.akpeters.com

Alberta Printed Circuits, Ltd.
Products: Fast-turn prototype blank PCBs from your Gerber files
Phone: (403) 250-3406
Web: http://www.apcircuits.com/

American Science and Surplus (formerly Jerryco)
Products: Weird assortment of nearly everything; a must-have catalog!
Phone: 847-982-0874
Fax: 800-934-0722
Mail: American Science and Surplus
3605 W. Howard St.
Skokie, IL 60076
Web: http://www.sciplus.com/

B. G. Micros, Inc.
Products: ICs, sockets, EPROMs, tools, surplus
Phone: 800-276-2206 or 972-271-5546
FAX: 972-271-2462
Mail: B.G. Micro
PO Box 280298
Dallas, TX 75228
Web: http://bgm.bgmicro.com/

CGN Technology Innovators
Products: Small, self-contained modules with on-board MCUs
Phone: 408-720-1814
Mail: CGN
1000 Chula Vista Terrace
Sunnyvale, CA 94086
email: custserv@cgntech.com
Web: www.cgntech.com

California Electronics and Industrial Supply
Products: Surplus electronics
Address: 221 N. Johnson Ave.
           El Cajon, CA

DigiKey
Products: Huge range of current electronic compo-
           nents, very fast delivery
Phone: 800-344-4539 or 218-681-6674
Fax: 218-681-3380
Mail: DigiKey
           701 Brooks Avenue South
           Thief River Falls, MN 56701-2757
Web: www.digikey.com

Edmond Scientific
Products: Scientific apparatus of all types, motors,
           optics, raw materials
Phone: 609-573-6250
Fax: 609-573-6295
Mail: Edmond Scientific
           101 East Gloucester Pike
           Barrington, NJ 08007-1380
Web: http://www.edsci.com/

Electronic Engineering Times
Products: Magazine articles on current electronics
           technology
Phone: 847-291-5215
email: For subscriptions, contact Michelle Gertz
           (mgertz@cmp.com)
Web: http://www.eetimes.com/

Electronic Goldmine
Products: Excellent source of surplus electronics
           parts, computer gear
Phone: 800-445-0697
FAX: 602-661-8259
Mail: Electronic Goldmine
           PO Box 5408
           Scottsdale, AZ 85261
email: goldmine-elec@goldmine-elec.com
Web: http://www.goldmine-elec.com/

Gateway Electronics
Products: Wide range of kits, electronics supplies,
           and surplus equipment
Phone: 619-279-6802 or 800-669-5810
Fax: 314-427-3147
Mail: 9222 Chesapeake Dr.
           San Diego, CA 92123
Web: www.gatewayelex.com

W. W. Grainger, Inc.
Products: A huge assortment of tools, equipment,
           and materials for the machine and electri-
           cal shop. Graingers is nationwide, with
           retail stores in most major cities.
Phone: Check your local phone directory.
web: www.grainger.com

**HSC Electronic Supply of Santa Clara
(Halted Specialties)**
Products: Electronic gadgets and high-tech surplus
Phone: (408) 732-1573
Fax: (408) 732-6428
Mail: HSC
           3500 Ryder Street
           Santa Clara, CA 95051
email: hscmail@halted.com

Herbach & Rademan (H&R Company)
Products: Motors, batteries, electrical components,
           mechanical items
Phone: 800-848-8001 or 609-802-0422
Fax: 609-802-0465
Mail: H&R Company
           16 Roland Ave.
           Mt. Laurel, NJ 08054-1012
email: sales@herbach.com
Web: http://www.herbach .com

ImageCraft

Products: C compilers for several MCUs, including
the 68hc11
Phone: 650-493-9326
Fax: 650-493-9329
Mail: ImageCraft
706 Colorado Ave., Ste. 10-88
Palo Alto, CA 94303
email: info@imagecraft.com
Web: www.imagecraft.com/software/index.html

Marlin P. Jones & Associates, Inc.

Products: Surplus electronics, power supplies,
batteries, motors
Phone: 800-652-6733 or 561-848-8236
Fax: 800-432-9937 or 561-844-8764
Mail: Marlin P. Jones & Assoc. Inc.
P. O. Box 12685
Lake Park, FL 33403-0685
email: mpja@mpja.com
Web: http://www.mpja.com/

Marvin Green

Products: BOTBoard, BOTBoard-2, BBOT frame
Mail: Marvin Green
190 SE Hacienda Ct.
Gresham, OR 97080
Web: http://www.rdrop.com/users/marvin/

Mondotronics

Products: Perhaps the definitive source of items for
the robotics hobbyist
Phone: 800-374-5764 or 415-491-4600
Web: www.robotstore.com

New Micros, Inc.

Products: Top-notch single-board computers for
many MCUs
Phone: 214-339-2204
Mail: New Micros, Inc.
1601 Chalk Hills Rd.
Dallas, TX 75212
Web: www.newmicros.com

Newton Research Labs

Products: High-end robotic color vision systems,
robotics-oriented software
Phone: (425) 251-9600
Fax: (425) 251-8900
Mail: Newton Research Labs
4140 Lind Ave SW
Renton, WA 98055
Web: www.newtonlabs.com

Powell's City of Books

Products: Enormous selection of books of all kinds,
including many out-of-print
Phone: 800-291-9676 or 503-228-4651 ext 482
Web: http://www.powells.portland.or.us/

Precision Navigation, Inc.

Products: Vector-2X electronic compass module
Phone: 707-566-2260
Fax: 707-566-2261
Mail: Precision Navigation
5550 Skylane Blvd., Ste. E
Santa Rosa, CA 95403
email: sales@PrecisionNav.com

Software Development Systems, Inc.

Products: Excellent 68000 compiler/assembler tool
set, with free demo available
Phone: 630-971-5900, 800-448-7733
Fax: 630-971-5901
Mail: SDS, Inc.
333 East Butterfield Road
Lombard, Illinois 60148
Web: www.sdsi.com

Solarbotics, Ltd.

Products: BEAM-based robots, including solar-
powered machines
Phone: (403) 818-3374
Fax: (403) 226-3741
Mail orders: Solarbotics Ltd.
179 Harvest Glen Way N.E.
Calgary, Alberta, Canada T3K 4J4

Technological Arts

Products:	68hc11 (and others) computer boards, prototyping tools
Phone:	(416) 963-8996
Fax:	(416) 963-9179
Mail:	Technological Arts
	26 Scollard St.
	Toronto, Ontario, Canada M5R 1E9
email:	sales@technologicalarts.com
Web:	www.technologicalarts.com

Tern, Inc.

Products:	Single-board computers for many MCUs
Phone:	530-758-0180
Fax:	530-758-0181
Mail:	Tern, Inc.
	1724 Picasso Ave., Ste. A
	Davis, CA 95616
Web:	www.tern.com

Tower Hobbies

Products:	Superb selection of raw materials for robot frames, servo motors, tools
Phone:	800-637-4989
Mail:	Tower Hobbies
	P.O. Box 9078
	Champaign, IL 61826-9078
Web:	www.towerhobbies.com

WinSystems, Inc.

Products:	Small PC-compatible single board computers
Phone:	817-274-7553
Mail:	WinSystems
	715 Stadium Dr.
	Arlington, TX 76011
Web:	www.winsystems.com

Appendix B
Hobby Servo Mods

Modifying a Futaba FP-S148 Hobby Servo Motor for Use as a Robot Motor

In the following instructions, "front" means the part of the motor case that encloses the motor's output shaft (and has the Futaba label on it);"back" means the opposite side of the motor case.

You will need:

Jeweler's screwdrivers (Phillips)
Small solder iron
Solder sucker
Needlenose pliers
Diagonal cutters
Two 2.7K ohm, 1/4-watt resistors
    (you could probably use 2.2K ohm resistors in a pinch; those are Radio Shack 271-1325)

1. If your motor already has some form of mechanical coupler device screwed onto the end of the output shaft, remove it.

2. Remove the four screws from the back of the case.

3. Remove the front and back covers.

4. Remove nylon center (top) gear and nylon gears on output shaft and motor shaft. Try not to disturb or wipe off any of the white grease on the gears.

5. Using diagonal cutters, carefully trim and remove the nylon spur on the surface of the large output gear. This spur normally limits the servo's movement to an arc of about 270 degrees. Make sure you remove the spur completely. You must not leave any chunks of nylon that might prevent the output gear from rotating freely.

6. Pry off the bronze sintered bushing from the plastic hub around the potentiometer (pot) shaft.

7. Remove the two small screws on either side of the motor shaft.

8. Firmly press on the pot's shaft to push it back through the servo's case. This should push the pot and the printed circuit board (PCB) out the back of the case.

WARNING: DO NOT pry on the PCB at all! DO NOT push on the motor's spindle!

9. Remove the pot from the PCB by carefully heating its connections, then removing the excess solder with a solder sucker. Work carefully and do not damage the PCB's traces.

10. Install two 2.7K resistors, wired in series, in place of the pot. The two resistors will appear to the servo's circuit as a 5K pot rotated to its center position. Refer to the following schematic:

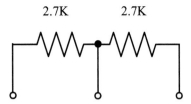

where the Os represent the solder pads that previously held the pot's leads. Make sure you install the junction of the two resistors in the center pad on the PCB. Trim the leads so that the resistors will fit inside the case when you later reassemble the motor. Make sure you don't accidently short any traces on the underside of the PCB when you solder the resistors in place.

11. Carefully reassemble the servo motor. Simply reverse the above steps for disassembling the motor.

Appendix C
Web Pages

Web pages of interest to robot builders

NOTE: As you are probably already aware, the Internet changes constantly. By the time you read this, I've no doubt that some of these links will have broken. Still, I've done my best to verify all of these links. If you have updated information on a link that turns up broken, drop me some email.

Karl Lunt's home page
Address: www.seanet.com/~karllunt
Comments: Where this whole thing started. You can get my SBasic and tiny4th compilers, lots of working robotic code, and descriptions of my current projects.

Seattle Robotics Society
Address: www.seattlerobotics.org
Comments: Well, I think it's the best robotic club web site in the Internet, but then I'm biased. Full of information for robot builders of all ages and levels of expertise. Be sure to scope out the club's on-line newsletter, *the Encoder*.

Portland Area Robotics Society (PARTS)
Address: http://www.rdrop.com/users/marvin
Comments: This is the main page for the PARTS robotics club, and for Marvin Green, developer of the popular BOTBoard series of circuit boards. Marvin's pages contain plenty of good robotics information, including a complete web-accessible copy of the SBasic user's manual. Check here also for information on ordering BOTBoards and other robotics goodies.

Dallas Personal Robotics Group (DPRG)
Address: http://199.1.173.112/index.html
Comments: The Dallas group has been around for a long time, and the membership has done some very good work in the hobby robotics area. I like the group's web site; the links page is very extensive and should keep you browsing for days.

Alta Vista
Address: http://www.altavista.com/
Comments: There are lots of web search engines,

but I use Alta Vista almost exclusively. If you haven't found a favorite yet, give this one a try.

Tern, Inc.
Address: http://www.tern.com/
Comments: Tern sells a good selection of single-board computers suitable for embedding into larger robots.

Mars Pathfinder Information (mirror)
Address: http://mars.catlin.edu/default.html
Comments: One of many mirror sites for the NASA web site providing coverage of the Mars Pathfinder mission. From this site, you can reach many other NASA-related sites, including the NASA home page at **www.nasa.gov**.

Intel
Address: http://support.intel.com/support/controllers/mcsx51/51tools/
Comments: In the old days, Intel was a good source for 80x51 tools. With today's focus on high-end PC chips, this source has pretty much dried up. You can use this address as a jumping-off point for a search for high-end tools.

PPPL (Plasma Physics Laboratory, Princeton)
Address: ftp://ftp.pppl.gov/pub/8051/
Comments: Like the Intel site, this one has pretty much dried up for embedded developers. Still worth looking through the /pub directory, though.

Bode
Address: ftp://bode.ee.ualberta.ca/pub/
Comments: An excellent source for all kinds of embedded control and robotics development tools.

MIT Wearable Computing Web Page
Address: http://wearables.www.media.mit.edu/projects/wearables/

Comments: The lead-in image is terrific, and the site contains pages and links of interest to anyone working on low-power or mobile computing. The Hardware and Software sections can help you find low-cost or rugged peripherals that have stood the test of Cyborg use, and other sections will guide you to worthwhile commercial and academic sites.

Motorola's FREEWARE web site
Address: http://www.mcu.motsps.com/download/
Comments: Filled with tons of applications and tools for all of the Motorola processors.

Motorola's 68hc11 list server
Address: Majordomo@oakhill-csic.sps.mot.com
Comments: To subscribe to this informative 68hc11 resource, send email to the above address. Include the word HELP as the body of the email's text. You will receive a reply with full details on subscribing. You can find more information about this listserver at: **http://www.mcu.motsps.com/major.html**.

Maxim's web site
Address: http://www.maxim-ic.com/
Comments: Maxim has a terrific samples program and a huge variety of special-purpose ICs. If your browser has Java enabled, you can search their database for different chips.

Atmel's web site
Address: http://www.atmel.com/
Comments: Atmel has long been famous for high-density flash memory devices. Lately, they have released the AVR series of microcontrollers, which promise to make a big impact in the hobby robotics field, among others.

National Semiconductor's web site
Address: http://www.national.com/index.html
Comments: Filled with data on National's chips, application notes, and product announcements.

Microchip's web site

Address: www.microchip.com

Comments: The definitive web site for anything based on the PIC microcontroller. Loaded with tools, application notes, and data sheets.

Eric Smith's web site

Address: www.brouhaha.com/~eric/pic/index.html

Comments: A Web page containing many finished PIC designs, some complete with code. Eric Smith outlines projects such as a clock with scrolling text for your television, and a PIC-based sonar ranging system. He adds notes discussing technical aspects of the projects, including why he selected a particular PIC for each design. He also provides links to other pages of PIC projects done by his friends. Good quality and very informative.

The ScanCam web site

Address: http://www.scancam.com/

Comments: This Web page describes a wonderful hack that turns a handheld scanner into a wide-field camera for use with optical telescopes. Peter Armstrong takes you step by step through the process of selecting a scanner, rearranging the electronics into a suitable enclosure, and wiring the (minimal) supporting electronics. He even includes high-quality photos of the PCBs used in his scanner, with edited arrows and text explaining what goes where. This is a detailed presentation of magazine quality, and his sample pictures of the moon are incredible. If you're looking for a wide-field CCD camera for your telescope, and want to see what you can get for about $200, check out this page.

Marc Tilden's web site

Address: http://nis-www.lanl.gov/robot

Comments: Marc spawned the BEAM concept of small, low-power, elegant robots. This site contains projects, philosophy, and concepts to help you get the most out of your robotic junkbox. A must-see site.

Alan Phillips, creator of the PFE-32 editor

Address: http://www.lancs.ac.uk/people/cpaap/pfe/default.htm

Comments: Alan has done the robotics community (and others) a huge service by releasing a very complete programmer's editor for free. This is a work of art, far more powerful than most IDE editors sold with commercial products. Well maintained and robust.

Dave Dunfield (Dunfield Development Systems)

Address: www.dunfield.com

Comments: Sells C compilers and other software tools for a variety of microcontrollers. Maintains a web site of useful tools and source files.

ImageCraft

Address: www.imagecraft.com/software/index.html

Comments: Sells C compilers for a variety of microcontrollers, including the 68hc11.

Mondotronics

Address: http://www.robotstore.com/

Comments: An excellent source of robotics parts, kits, books, and information.

Free BSD

Address: http://www.fi.freebsd.org/ports/master-index.html

Comments: Repository of numerous programs built under the GNU Public License.

WinSystems, Inc.

Address: www.winsystems.com

Comments: Sells a broad range of PC SBCs in PC-104 and STD form-factors. Their web site includes some good graphics images of their products.

Dan Mauch's CCNC web page

Address: www.seanet.com/~dmauch
Comments: Dan has developed some BIG stepper-based kits for retrofitting inexpensive machining tools for computer control. His driver electronics would make it easy to build large, PC-based robots and machines.

Jeff Frohwein's GameBoy® page

Address: http://fly.hiwaay.net/~jfrohwei/gameboy/home.html
Comments: Everything you could ever need to know about hacking a GameBoy is either on this site or at one of the links from this site. Guaranteed to keep the robot hackers out there busy for weeks.

Kevin Ross (member of the Seattle Robotics Society)

Address: http://www.nwlink.com/~kevinro
Comments: Contains some great robotics info for beginner and intermediate builders alike. Kevin sells 68hc12 project boards, including his BDM interface board, through this site as well.

Bill Harrison (member of the Seattle Robotics Society)

Address: www.halcyon.com/www3/sinerobt/
Comments: Bill champions the Sumo and Mini-Sumo robot competitions in the SRS.

Pete Dunster's web page

Address: http://mirriwinni.cse.rmit.edu.au/~fl/
Comments: Pete designed an excellent set of PCBs around the Motorola 68hc11f1 chip. You can also contact Pete by email at:
pdunster@connectivity.net.au.

Trinity College (Fire-Fighting Home Robot Contest)

Address: www.trincoll.edu/~robot
Comments: Information center for the annual fire-fighting contest. Contains latest contest rules, layouts, and other news items.

Oli Kraus' web page

Address: www.e-technik.uni-erlangen.de/~kraus/olihc11.html
Comments: Oli maintains a 68hc11 port of the GNU gcc C compiler, suitable for developing C code for your 68hc11 projects.

Massachusettes Institute of Technology (MIT) robotics pages

Address: ftp://cher.media.mit.edu/pub/projects/6270
Contents: cheupahka is a goldmine for the robotics experimenter. The FTP site above contains information on the MIT 6.270 class. Browse through the other directories on this machine for more tools and information on robotics.

Public Broadcasting System (PBS)

Address: www.pbs.org
Contents: Schedules of upcoming shows, including *Scientific American Frontiers*, and information on technological and scientific topics.

Newton Research Labs (NRL)

Address: www.newtonlabs.com
Contents: Technical information on the Cognachrome vision system, a key element in some impressive wins in international robotics events. NRL also sells robotic tools such as the Interactive-C compiler.

MIROSOT (robot soccer)

Address: www.mirosot.org
Contents: Home page for the Micro-Robot World Cup Soccer Tournament. Contains rules, results of recent events, and lists of organizers.

MIT Library

Address: http://libraries.mit.edu/docs/
Contents: This site provides a valuable service to the robotics community; the means to purchase copies of MIT master's theses, reports and articles at a reasonable cost.

Index

A

A K Peters, Ltd. 7, 32, 61, 354–355, 559
Active Electronics 8, 115, 142
Actuators
  gearhead motor viii, 5, 16, 31, 73, 75, 131–132,
    169, 183, 187, 218, 221, 224–225, 236,
    267, 296–298, 313, 470, 486–488, 506
  hobby servo motor vii, ix, xii, 5, 8, 16–17, 20–21,
    23–24, 28, 31, 33, 35–37, 39, 47–48, 73,
    112, 125, 131–132, 135, 161, 164, 168,
    171, 187, 218–219, 222, 236, 238–239,
    243–246, 249, 257, 262–264, 286, 292,
    304, 306, 348, 353–354, 376, 378, 417,
    432, 460, 480, 506–507, 532, 536–537,
    542, 564
Adams, David 465
AI Expert magazine 519
Alberta Printed Circuits 171, 176, 199
Alda, Alan xii, 300, 475, 482–483
Alexander, David xi, 474
Andersen, Dale 129
Aneses, Carlos 457
Armstrong, Peter 567
ARRL Handbook 188, 363
as11 assembler 43, 126, 165
ASIC Basic compiler 98
Asimov, Isaac 474
asm51 8051 assembler 371
asmhc11 assembler 4, 28, 43, 45, 52, 56, 58, 66,
    165, 178, 184, 273, 321, 522
Austin Robot Group 518

B

Bailey, Bill 91–92, 98, 155, 187, 190, 285–286,
    289, 331, 475–476, 486, 489, 492–493,
    535–536
Baker, Bill 457
Barch, Don 129
BAS051 Basic compiler 374
Basic Stamp 161, 507
BASIC52 Basic interpreter 374
Batteries
  NiCd 6, 8, 13, 35–36, 59–60, 139, 141,
    160, 257, 490
  Renewal alkaline
    vii, 36, 51, 59, 139, 141, 153, 158, 194,
    202, 287, 364
BBOT, BBOT jr. frames x, 34, 347, 348, 353
BBot frame 34

B

BEAM robotics 495, 496, 567
Begley, Mike 506
Berry, James 129
Birdwell, Ken 222, 448, 465
Birdwell, Mike 219, 535
BOTBios viii, 91–98
BOTBoard viii, xi, 33, 37, 39–41, 45–48, 121,
    125–126, 131, 135–137, 145, 160–161,
    163–164, 166, 169, 184, 198–199, 203,
    206, 209, 222, 224, 228, 269, 282,
    287–289, 293, 295–298, 303–308, 323,
    331, 348–351, 353, 365, 369, 399,
    401–404, 411–412, 460, 463, 465,
    507–508, 511–512, 520, 536, 541,
    545–546, 565
BOTBoard-2 163–167, 313, 315
Bradbury, Ray 474
Brandt, Wally 333
Brodie, Leo 58, 119, 465
Bronk, Jeff 506
Brooks, Rodney 3, 61, 498
Brown, Jeremy 481–482
Buckley, David 506
BUFFALO monitor 166–167
Burgin, George 519
Burton, Jerry 518
Butterfield, Jim 182
Byte magazine 366

C

Calderon, Benjamin 452
Calsyn, Martin 506
Carbon Copy Card 415–416
CGN 1001module ix, 111, 239, 258
CGN 1101module viii, 111–113, 115, 228, 434,
    437–439, 443
CGN microcontroller module ix, 239, 258
chopper stepper motor driver ix, 187, 190–192
Circuit Cellar INK magazine 357, 518
Coactive Aesthetics 544–545
Cognachrome vision system 484, 493, 568
Cogswell College 348
Conway, John Horton 520
Cyborgs x, 295, 301

D

Dain, Bob 348
DANCAM CAD software 464
de Goede, Hans 416